VOLUME ONE HUNDRED AND TWENTY ONE

CURRENT TOPICS IN DEVELOPMENTAL BIOLOGY

Fly Models of Human Diseases

CURRENT TOPICS IN DEVELOPMENTAL BIOLOGY

"A meeting-ground for critical review and discussion of developmental processes"
A.A. Moscona and Alberto Monroy (Volume 1, 1966)

VOLUME ONE HUNDRED AND TWENTY ONE

CURRENT TOPICS IN DEVELOPMENTAL BIOLOGY

Fly Models of Human Diseases

Edited by

LESLIE PICK
*University of Maryland,
College Park, MD, United States*

ACADEMIC PRESS
An imprint of Elsevier
elsevier.com

Academic Press is an imprint of Elsevier
50 Hampshire Street, 5th Floor, Cambridge, MA 02139, United States
525 B Street, Suite 1800, San Diego, CA 92101-4495, United States
The Boulevard, Langford Lane, Kidlington, Oxford OX5 1GB, United Kingdom
125 London Wall, London, EC2Y 5AS, United Kingdom

First edition 2017

ISBN: 978-0-12-802904-6
ISSN: 0070-2153

For information on all Academic Press publications visit our website at https://www.elsevier.com/

Publisher: Zoe Kruze
Acquisition Editor: Zoe Kruze
Editorial Project Manager: Shellie Bryant
Production Project Manager: Vignesh Tamil
Cover Designer: Greg Harris

Typeset by SPi Global, India

CONTENTS

CONTRIBUTORS

H.J. Bellen
Baylor College of Medicine; Jan and Dan Duncan Neurological Research Institute, Texas Children's Hospital; Program in Developmental Biology; Howard Hughes Medical Institute, Houston, TX, United States

P. Bergman
Clinical Microbiology F68, Karolinska Institutet, Karolinska University Hospital, Huddinge, Stockholm, Sweden

C. Burke
Neurology Research, Biogen, Cambridge, MA, United States

R.L. Cagan
Icahn School of Medicine at Mount Sinai, New York, NY, United States

L. Chakravarti
Perelman School of Medicine at the University of Pennsylvania, Philadelphia, PA, United States

R.T. Cox
Uniformed Services University, Bethesda, MD, United States

J.M.I. Daenzer
Emory University School of Medicine, Atlanta, GA, United States

Y. Engström
The Wenner-Gren Institute, Stockholm University, Stockholm, Sweden

J.L. Fridovich-Keil
Emory University School of Medicine, Atlanta, GA, United States

D.C.I. Goberdhan
University of Oxford, Oxford, United Kingdom

P. Graham
University of Maryland, College Park, MD, United States

F. Hamdy
University of Oxford, John Radcliffe Hospital, Oxford, United Kingdom

V.L. Hewitt
Medical Research Council Mitochondrial Biology Unit, Cambridge, United Kingdom

S.X. Hou
The Basic Research Laboratory, National Cancer Institute at Frederick, National Institutes of Health, Frederick, MD, United States

M.S. Kayser
Perelman School of Medicine at the University of Pennsylvania, Philadelphia, PA, United States

R.E. Kreipke
University of Washington; Institute for Stem Cell and Regenerative Medicine, School of Medicine, Seattle, WA, United States

M. Krench
The Picower Institute for Learning and Memory, Massachusetts Institute of Technology, Cambridge, MA, United States

Y.V. Kwon
University of Washington, School of Medicine, Seattle, WA, United States

A. Leiblich
University of Oxford; John Radcliffe Hospital, Oxford, United Kingdom

G. Lin
Baylor College of Medicine; Jan and Dan Duncan Neurological Research Institute, Texas Children's Hospital, Houston, TX, United States

J.T. Littleton
The Picower Institute for Learning and Memory, Massachusetts Institute of Technology, Cambridge, MA, United States

D. Mao
Jan and Dan Duncan Neurological Research Institute, Texas Children's Hospital; Program in Developmental Biology, Baylor College of Medicine, Houston, TX, United States

E.H. Moscato
Perelman School of Medicine at the University of Pennsylvania, Philadelphia, PA, United States

V. Nadar
Neurology Research, Biogen, Cambridge, MA, United States

L. Pick
University of Maryland, College Park, MD, United States

H. Ruohola-Baker
University of Washington; Institute for Stem Cell and Regenerative Medicine, School of Medicine, Seattle, WA, United States

S. Sanyal
Neurology Research, Biogen, Cambridge, MA, United States

A. Sen
Uniformed Services University, Bethesda, MD, United States

S. Seyedoleslami Esfahani
The Wenner-Gren Institute, Stockholm University, Stockholm, Sweden

H.R. Shcherbata
Max Planck Research Group of Gene Expression and Signaling, Max Planck Institute for Biophysical Chemistry, Göttingen, Germany

S.R. Singh
The Basic Research Laboratory, National Cancer Institute at Frederick, National Institutes of Health, Frederick, MD, United States

M. Sonoshita
Icahn School of Medicine at Mount Sinai, New York, NY, United States; Kyoto University Graduate School of Medicine, Kyoto, Japan

K. Trinh
Neurology Research, Biogen, Cambridge, MA, United States

A.J. Whitworth
Medical Research Council Mitochondrial Biology Unit, Cambridge, United Kingdom

C. Wilson
University of Oxford, Oxford, United Kingdom

PREFACE

The fruit fly, *Drosophila melanogaster*, is arguably the most sophisticated model organism developed to date. With over a hundred years of genetics and thousands of researchers worldwide, tools to study gene expression and function—from detailed molecular analysis of single genes to genome-wide analyses of regulatory phenomena—are available at an unprecedented scale. Results are shared freely among *Drosophila* researchers through informational databases and repositories such as FlyBase, collections such as the Bloomington Stock Center that distributes fly lines at a minimal cost, and the historic collegial nature of interaction among fly researchers. In addition to both past and ongoing contributions to basic research in almost every area of science, *Drosophila* has now been established as a model to study human disease. With orthologs of more than 60% of human disease-causing genes, often present in single copy in flies but duplicated in vertebrates, and the wealth of tools available to assess gene function, genetic interactions, and environmental influences, *Drosophila* is an ideal system for elucidating mechanisms underlying pathologies. This special volume provides examples of some of the best-developed fly models for a variety of human diseases. Many of these affect the nervous system, but the types of diseases that have been effectively modeled span a broad range, from metabolic disorders to cancer to bacterial infections. In each chapter, the basic pathologies in human patients and the parallels to *Drosophila* biology are discussed. The strengths of *Drosophila* as a model are presented, as well as weaknesses of its use as a model system, important to keep in mind for accurate translation to the human condition.

Mitochondrial diseases are a unique and complex set of diseases that are maternally inherited and difficult to diagnose and treat. Sen and Cox review the use of *Drosophila* to study *mitochondrial disease and mitochondrial inheritance.* They review existing fly models for several mitochondrial diseases and methodology available in the fly to generate targeted mutations in mitochondrial DNA. Finally, they discuss recent findings on the role of mitochondrial purification during oogenesis and experiments in *Drosophila* that are unraveling how this is achieved.

While insects do not have adaptive *immune* systems, they share with vertebrates an innate immune system. In fact, much of our understanding of innate immunity comes from *Drosophila* research. As reviewed by Bergman,

Esfahani, and Engström, flies and humans also share microbial flora that influence health. Studies of host–microbe interactions, which include shared intracellular signaling pathways and effector mechanisms, are revealing basic mechanisms utilized by animals to both protect against microbial invasion and to promote the growth of a healthy microbiome.

Ruohola-Baker and coworkers present a discussion of contributions made using *Drosophila* to study pathophysiological mechanisms underlying *muscle degenerative disorders*, including Duchenne muscular dystrophy and spinal muscular atrophy, as well as muscle wasting seen in cancer and other human disease. They point to genetic interaction screens in flies that have identified pathways which could be targeted for therapeutic intervention in these and other muscular diseases.

A number of chapters focus on the use of *Drosophila* to study *neuronal disease*. Lin, Mao, and Bellen focus on one devastating disease, amyotrophic lateral sclerosis (ALS). A number of genes have been implicated in familial ALS and, given their conservation, *Drosophila* provides a model to study their mechanisms of action. The authors propose that defects in RNA and protein homeostasis build upon each other to cause accumulating protein aggregation that is pathological in ALS. Results to date implicate abnormalities in pre-mRNA splicing, protein folding, ER stress, proteasomal degradation, and protein mislocalization as underlying bases for disease. This review highlights the use of *Drosophila* as an assay system to uncover the function of genes associated with disease in human studies, and the potential the fly system has to reveal mechanism of action of mutations found in human patients.

Hewitt and Whitworth similarly focus on a single neurodegenerative disease, Parkinson's disease (PD). Interactions between and mechanism of action of several genes associated with and causal for PD have been studied in *Drosophila*. Among others, the authors discuss the role of *parkin* in regulating mitochondrial dynamics, a mechanism first elucidated in studies in *Drosophila*. Krench and Littleton discuss using flies to study the mechanisms underlying neurodegenerative disorders associated with expanded polyglutamine tracts, such as Huntington's diseases. They highlight progress made using *Drosophila* models to explain the underlying basis of disease and suggest approaches to using these models to identify and assess therapies, including genetic suppressor screens that have the power to both dissect genetic interactions and identify potential drug targets.

Sanyal and coworkers take a slightly different approach to discussing neurodegenerative diseases, such as PD, by reviewing the importance of interactions between genes, age, and environment in disease onset and progression. They look toward future "intersectional experiments" that combine genetic approaches in the fly model with systems biology and "omics" to reveal complex interactions that lead to full-blown disease. While highlighting many strengths offered by fly genetics, they also remind the reader about the need for deeper understanding of the molecular basis of cellular defects before extrapolation to the human condition based upon grossly similar phenotypes.

One of the more unexpected areas of fly research that proved to effectively model human conditions is *sleep research*. This area exploded in recent years with the identification of genes impacting circadian rhythms and sleep activity in flies. Kayser and coworkers review studies of specific regions of the brain, neuronal circuitry, and signaling pathways that are highly conserved between flies and humans that play roles in sleep–wake cycles, response to deprivation, caffeine, and other stimulants, impacting sleep patterns as well as learning and memory. They point to early studies in this field that identified the *period* (*per*) mutation in flies—a gene later implicated in human sleep disorders, and describe additional fly mutants that effectively model human neurodevelopmental, psychiatric, and sleep disorders.

Drosophila has served as an excellent model for studies of *cancer*, as flies harbor similar oncogenes and tumor suppressor genes to humans, and mutations result in overgrowth and metastasis, similar to that seen in human disease. Sonoshita and Cagan review the identification of genes responsible for overgrowth in *Drosophila*, which contributed to the finding that loss of cell polarity is a key feature of malignancy. They give examples of genes identified in *Drosophila* through genetic modifier screens that can be placed in ordered pathways using fly genetics, providing insight into mechanisms of action. They next review models of different types of cancer that have been developed in flies and highlight interactions between diet and genotype that are likely to prove important for cancer treatment. Finally, these authors emphasize the potential of *Drosophila* for drug screening in whole animals, citing, among others, the contribution of a *Drosophila* model to the approval of vandetanib as a targeted therapy for human patients, and projects underway to use fly models expressing specific oncogenes to test new, rationally designed chemistries.

The role of stem cells in cancer is a major focus of recent research. *Drosophila* is a particularly good model here, not only because of the powerful genetics and numerous classes of identified stem cell lineages, but because there are no regulatory issues constraining research. Hou and Singh review a number of recent findings on the role of stem cells in tumor formation and progression in different stem cell populations in *Drosophila*. They discuss the roles of signaling molecules and pathways, providing examples of how *Drosophila* research has contributed to our understanding of how stem cell tumors initiate and propagate. They discuss molecular differences between normal and tumorigenic stem cells, and resistance of stem cells to standard therapies. Wilson and coworkers focus on a specific cancer type—prostate cancer—and ways in which the *Drosophila* accessory gland has been used to model this disease. They highlight the rather unexpected parallels between *Drosophila* and human organ systems here and highly conserved cellular mechanisms that control exosome formation and secretion, a class of secreted vesicles that have been implicated in tumor microenvironments as well as metastasis.

Lastly, *Drosophila* have also been used effectively to model *metabolic disorders*. Daenzer and Fridovich-Keil discuss diseases of galactose metabolism, primarily resulting from abnormalities in Leloir pathway enzymes which are shared between flies and mammals. Functional similarities have also been demonstrated: for example, similar to classic galactosemias in human, flies carrying mutations in the orthologous gene fail to thrive in the presence of galactose and this failure was rescued with a human transgene, demonstrating functional conservation. These studies and others, including identification of motor defects and impacts on life span that parallel phenomena seen in patients, make flies and excellent model to study genetic, environmental, and pharmaceutical interventions for this important class of disease. Graham and Pick discuss *Drosophila* models of diabetes and related disorders. They describe fly models that have been generated for both Type 1 and Type 2 diabetes, and approaches to study metabolic disorders that are influenced by diet and other environmental factors.

While this volume was being assembled, a similar collection of articles was published (Perrimon, Bonini, & Dhillon, 2016). In that special issue of DMM, Bellen and coworkers provide a "Poster" illustrating three ways in which *Drosophila* research contributes to the study of human disease: reverse genetics, forward genetics and a diagnostic strategy, or pipeline that utilizes the UAS/GAL4 system to assess the roles of candidate genes (Ugur, Chen, & Bellen, 2016). Most of the chapters in this volume of CTDB illustrate the use of these approaches to tackle the underlying bases of a range of human disease. I refer the reader to both sets of collections to learn more about general approaches to disease modeling and to progress made on specific disease types. Overall, enormous headway has been made in recent years modeling human diseases in flies. This work paves the way for the development and refinement of additional models and future progress in translating these findings to effective drug development and disease intervention.

Leslie Pick

REFERENCES

Perrimon, N., Bonini, N. M., & Dhillon, P. (2016). Fruit flies on the front line: The translational impact of Drosophila. *Disease Models & Mechanisms*, *9*(3), 229–231.

Ugur, B., Chen, K., & Bellen, H. J. (2016). *Drosophila* tools and assays for the study of human diseases. *Disease Models & Mechanisms*, *9*(3), 235–244.

CHAPTER ONE

Fly Models of Human Diseases: *Drosophila* as a Model for Understanding Human Mitochondrial Mutations and Disease

A. Sen, R.T. Cox[1]
Uniformed Services University, Bethesda, MD, United States
[1]Corresponding author: e-mail address: rachel.cox@usuhs.edu

Contents

Abstract

Mitochondrial diseases are a prevalent, heterogeneous class of diseases caused by defects in oxidative phosphorylation, whose severity depends upon particular genetic mutations. These diseases can be difficult to diagnose, and current therapeutics have limited efficacy, primarily treating only symptoms. Because mitochondria play a pivotal role in numerous cellular functions, especially ATP production, their diminished activity has dramatic physiological consequences. While this in and of itself makes treating mitochondrial disease complex, these organelles contain their own DNA, mtDNA, whose products are required for ATP production, in addition to the hundreds of nucleus-encoded proteins. *Drosophila* offers a tractable whole-animal model to understand the mechanisms underlying loss of mitochondrial function, the subsequent cellular and tissue damage that results, and how these organelles are inherited. Human and

Current Topics in Developmental Biology, Volume 121
ISSN 0070-2153
http://dx.doi.org/10.1016/bs.ctdb.2016.07.001

Drosophila mtDNAs encode the same set of products, and the homologous nucleus-encoded genes required for mitochondrial function are conserved. In addition, *Drosophila* contain sufficiently complex organ systems to effectively recapitulate many basic symptoms of mitochondrial diseases, yet are relatively easy and fast to genetically manipulate. There are several *Drosophila* models for specific mitochondrial diseases, which have been recently reviewed (Foriel, Willems, Smeitink, Schenck, & Beyrath, 2015). In this review, we highlight the conservation between human and *Drosophila* mtDNA, the present and future techniques for creating mtDNA mutations for further study, and how *Drosophila* has contributed to our current understanding of mitochondrial inheritance.

1. MITOCHONDRIA PLAY DIVERSE ROLES

Mitochondria are thought to have arisen through endosymbiosis (Margulis, 1970). As such, in metazoans mitochondria are the only organelle, other than the nucleus, which contains its own DNA, mtDNA. The mtDNA in different species encodes a variable number of products; however, animal mtDNA represents the most stripped-down version of the genome (Gray, 2012). The 16 kb human mtDNA codes for 13 proteins, 22 tRNAs, and 2 rRNAs. All 13 proteins are components of the electron chain complexes (ETC) I, III, IV, and V (the ATP synthase). Almost all the DNA is coding—for example, there are no introns in the resulting mRNAs and very few gaps in coding sequence. *Drosophila* mtDNA encodes the same transcripts as human mtDNA, albeit in a slightly different genomic order (Fig. 1). This fundamental similarity makes the fly an excellent model for studying mitochondrial function.

The evolution between the nucleus and mitochondria has culminated in mitochondria taking on highly specialized functions, offering an environment to support a variety of biochemical reactions required for the cell. Because mitochondria have such a small genome, they rely heavily on imported proteins encoded in the nucleus. The best known mitochondrial product is ATP, produced via the ETC and oxidative phosphorylation (OXPHOS). However, mitochondria are also required for fatty acid beta oxidation, heme, steroid, and nucleotide biosynthesis and are integral to apoptosis. In fact, under specialized conditions it is possible for yeast and cultured cells to survive without functional OXPHOS (i.e., loss of mtDNA); however, they cannot survive in the complete absence of the organelle (Chandel & Schumacker, 1999; Goldring, Grossman, Krupnick, Cryer, & Marmur, 1970; Morais, Gregoire, Jeannotte, & Gravel, 1980; Nagley & Linnane, 1970). All of these

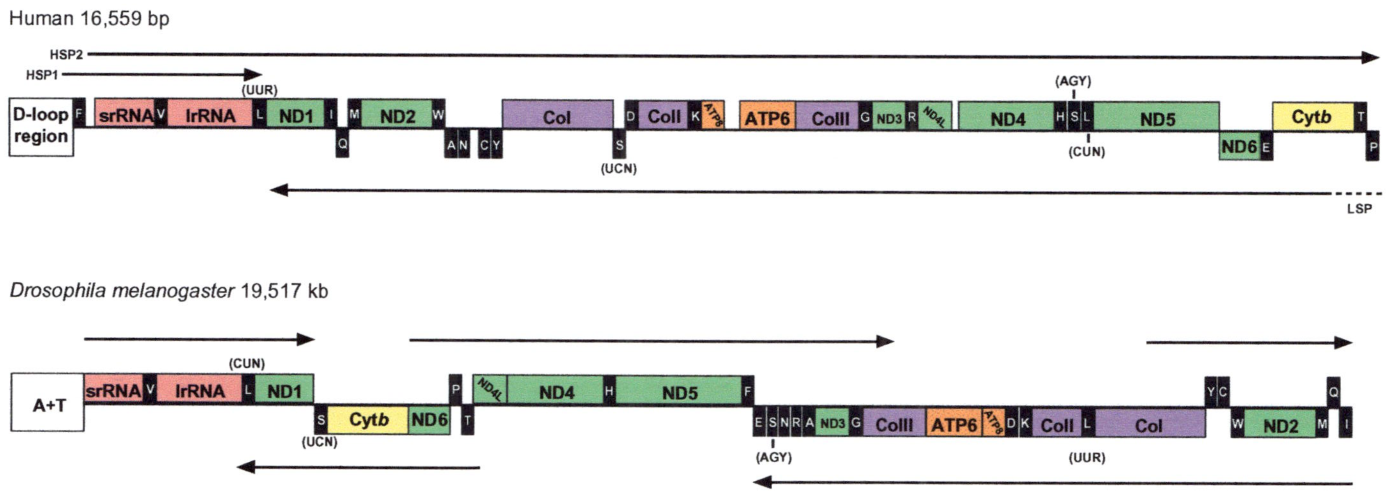

Fig. 1 Human and *Drosophila* mitochondrial DNA encode the same products. Human mtDNA (*top*) is approximately 3 kb shorter than *Drosophila* mtDNA (*bottom*). The size difference is predominantly due to the expanded "A+T-rich" region in *Drosophila*, which varies among different *Drosophila* species. Human mtDNA is transcribed as three polycistrons (*arrows*): two for the heavy strand (HS), which encodes most of the products, and one for the light strand (LS). The LS promoter (LSP) starts in the D-loop region (indicated by *dashed line*), where the origin of replication is found. HSP1, which includes the rRNAs, expresses at higher levels compared to HSP2. *Drosophila* mtDNA is thought to be transcribed as five polycistrons. Note that the products are relatively evenly encoded on both strands, in contrast to human mtDNA. Some segments maintain the same sequence (e.g., ATP8 → mt:tRNA^{Gly}). *ATP*, ATP synthase (*orange*, Complex V); *CO*, cytochrome *c* oxidase (*purple*, Complex IV); Cytb, cytochrome *b* (*yellow*, Complex III); *ND*, NADH dehydrogenase (*green*, Complex I).

basic biochemical reactions are required for both *Drosophila* and human cells, and because *Drosophila* mtDNA encodes the same products as human mtDNA, *Drosophila* mitochondria function requires essentially the same nuclear genes as human mitochondria.

2. MITOCHONDRIAL DISEASES—CAUSES AND EFFECTS

Human mitochondrial diseases mostly result from a loss of OXPHOS. The term "mitochondrial function" is used broadly so it is important to be as specific as possible when characterizing any particular loss in biochemical function. There are estimated to be between 1000 and 1500 proteins encoded in the nucleus that are imported or associated with mitochondria (Area-Gomez & Schon, 2014). Thus, mitochondrial disease arises through mutations in either nuclear DNA or mtDNA. As with any nuclear gene, mitochondrial disease due to mutations in nuclear genes can be inherited in a Mendelian fashion as either a dominant or recessive trait. In comparison to mutations in disease-causing nucleus-encoded mitochondrial genes, there are over 250 verified disease-causing point mutations in mtDNA. Because mitochondria cannot be made de novo, they are inherited through the mother, and thus mtDNA mutations are exclusively maternally inherited.

The 13 proteins encoded by mtDNA are translated in the mitochondrial matrix using mtDNA-encoded tRNAs (mt:tRNAs) and the mitochondrial ribosome, which consists of the mt:rRNAs and nucleus-encoded proteins. This suite of mt:tRNAs is all that is required for translation in human and *Drosophila* mitochondria. Human and *Drosophila* mtDNA is transcribed as a series of polycistrons (Fig. 1). The mt:tRNAs are thought to function as "punctuation," with most of the mRNAs separated by at least one mt:tRNA (Ojala, Montoya, & Attardi, 1981). Thus, it is critical that each mt:tRNA and every product of the genome be properly excised allowing each mRNA to be further processed and translated. Of the known disease-causing point mutations in mtDNA, over half are found in mt:tRNAs. This is somewhat surprising, given that the 22 mt:tRNAs encode only 9% of the genome. One explanation could be that mutations in the protein-coding regions are too deleterious and incompatible with life.

The known mutations that cause human mitochondrial disease have been extensively reviewed (Area-Gomez & Schon, 2014; Chinnery, 1993; Dimauro & Davidzon, 2005; Lightowlers, Taylor, & Turnbull, 2015). In general, they give rise to defects in the musculature and nervous system. However, while mutations in nuclear DNA and mtDNA would be expected to cause a decrease in OXPHOS, the different mutations have variable onset and features, from symptoms as mild as eye muscle weakness (external ophthalmoplegia) to infant mortality. There is even evidence that defects in OXPHOS may be a cause of miscarriage (Tay, Shanske, Kaplan, & DiMauro, 2004). Why deficits in the proteins involved in ATP production give rise to such different outcomes largely remains a mystery. In the case of disease caused by mutations in nuclear genes, all cells should have the same genotype. However, for mutations in mtDNA, one reason for differences in tissue deficits could be the threshold effect (Picard et al., 2014; Rossignol et al., 2003; Stewart & Chinnery, 2015). Each cell contains many mitochondria, and each mitochondrion usually contains multiple copies of mtDNAs. Mutations in mtDNAs are usually heteroplasmic (a mixture of wild type and mutant), thus tissues with a higher mutation load would be expected to be more severely affected. But a recent finding suggests that many associated symptoms, secondary to a prominent trademark phenotype that arose beyond the threshold point, can appear even when a particular mutation load is well below the threshold mark. Picard and colleagues found that a predominant pathogenic mutation in a mitochondrial tRNA gene (mtDNA 3243A > G; mutation in mt:tRNA$^{Leu(UUR)}$) has an effect on nuclear gene expression when present at well below the threshold point (Picard et al., 2014). Heteroplasmic cells harboring different mutational loads of this particular mutation have striking variations in their gene expression profile. This could explain why patients with the same mutation manifest different clinical symptoms. For example, patients with mtDNA 3243A > G that have high levels of mutated mtDNA exhibit mitochondrial encephalomyopathy, whereas patients with lower levels can suffer from type II diabetes, deafness, or even be on the autistic spectrum (Goto, Nonaka, & Horai, 1990; Pons et al., 2004; van den Ouweland et al., 1992).

3. HOW CAN STUDYING *DROSOPHILA* CONTRIBUTE TO OUR UNDERSTANDING OF HUMAN MITOCHONDRIAL DISEASES?

As a model organism, *Drosophila* can help our understanding of mitochondrial diseases in several specific ways. The first takes advantage of the rich history of cell biology in *Drosophila*. From dissected tissues taken from wild-type and mutant flies, mitochondria can be clearly visualized at the organelle level in fixed or live tissues (Cox & Spradling, 2006; Sen, Damm, & Cox, 2013). This allows for a detailed and accurate view of location, numbers, and morphology. In addition, with molecular and biochemical assays, researchers can assay which mitochondrial functions are compromised in vivo in *Drosophila* carrying various gene mutations.

The second advantage to using *Drosophila* to study deficits in mitochondrial function is that they have complex central and peripheral nervous systems, and contain the various organ systems that are frequently affected in humans suffering from mitochondrial disease, such as skeletal and heart muscles. These organ systems are far simpler than those in humans, allowing for detailed analysis of the progression of tissue degeneration either during development or aging. Seizures are one feature of mitochondrial diseases such as myoclonic epilepsy with ragged-red fibers (MERRF) and mitochondrial encephalomyopathy with lactic acidosis and stroke-like episodes (MELAS). This clinical feature is easily mimicked in *Drosophila* by assaying for bang-sensitivity, the paralysis and seizures that can occur after mechanical stimulation (Engel & Wu, 1994; Fergestad, Bostwick, & Ganetzky, 2006; Ganetzky & Wu, 1982). Bang-sensitivity occurs in mutants for several genes involved in metabolism in general, and mitochondrial function in particular (Burman et al., 2014; Celotto et al., 2006; Fergestad et al., 2006; Royden, Pirrotta, & Jan, 1987; Zhang et al., 1999). Another symptom of certain mitochondrial diseases is brain degeneration which can cause cerebellar ataxia, for example, found in MELAS, Leigh syndrome and myoclonic epilepsy, myopathy, and sensory ataxia. *Drosophila* with mtDNA mutations with bang-sensitivity also exhibit progressive brain degeneration (Burman et al., 2014; Celotto et al., 2006). Finally, cardiomyopathy occurs with mutations of multiple nucleus-encoded and mtDNA-encoded genes (Antonicka et al., 2003; Jaksch et al., 2000; Loeffen et al., 2001; Majamaa-Voltti, Peuhkurinen, Kortelainen, Hassinen, & Majamaa, 2002; Papadopoulou et al., 1999; Wahbi et al., 2010). In *Drosophila*, knockdown of several proteins found in the ETC Complex I (NDUFS2, NDUFS7, and NDUFC2) specifically in the heart caused significant, abnormal heart dilation (Tricoire, Palandri, Bourdais, Camadro, & Monnier, 2014).

An additional advantage is the ease of genetic manipulation in *Drosophila*. RNAi knockdown of single proteins in each ETC complex can be temporally and spatially controlled using the UAS/GAL4 system, which has been done for a subset of ETC and OXPHOS proteins (listed in table 1 of Foriel et al., 2015). In addition, mutant analysis has been very helpful in determining the molecular mechanisms underlying mitochondrial dysfunction, sometimes offering unexpected results. For instance, tko^{25t} (*technical knockout*) is a recessive mutation in the nucleus-encoded mitochondrial ribosomal protein S12. This bang-sensitive mutation confers a developmental delay and has long been considered a model for mitochondrial disease because the mutation targets the respiratory chain and causes seizures and deafness (Jacobs et al., 2004; Toivonen et al., 2001). The main pathological symptom identified in *tko*^{25t} was oxidative stress. Thus, it was assumed that decreasing oxidative stress would ameliorate the effect of the *tko* mutation. However, a recent finding shows that expressing neither the alternative oxidase (AOX, from *Ciona instestinalis*) nor NADH dehydrogenase (*ndi1*, from yeast) could rescue the phenotype of *tko*^{25t} mutant flies (Kemppainen, Kemppainen, & Jacobs, 2014). These results imply that the *tko*^{25t} mutation affects greater mitochondrial function such that merely targeting oxidative stress cannot afford a remedy.

A major challenge in mitochondrial disease therapy is establishing the root cause against which potential therapeutics can be generated. While the pathological manifestation of a mitochondrial disease looks simple, namely, loss of OXPHOS, it is difficult to dissect what component(s) of that complex structure is compromised. The basic functional moiety of a mitochondrion is its OXPHOS network. For example, if one of the 44 components of Complex I does not function properly, it may be reflected in abnormal Complex I activity. However, a traditional clinical approach may not be adequate to identify the primary cause of the nonfunctional Complex I, whereas a model system allows various forms of experimentation not possible in patients. Even if a particular subunit is mutated, it can be difficult to predict what specific aspect of the molecular complex is deficient. This creates difficulties for rational therapeutic design. Thus, using *Drosophila* offers tremendous potential for identifying the exact molecular mechanism behind any pathological symptoms and could aid in more targeted drug delivery and discovery.

4. MITOCHONDRIAL DISEASES CAUSED BY mtDNA MUTATIONS

Since mitochondria cannot be made de novo and contain their own DNA, they are maternally inherited through the egg's cytoplasm. Thus, mutations in mtDNA, and therefore mitochondrial diseases, are inherited through the mother. Common mitochondrial diseases occur from point mutations in all products encoded by mtDNA: protein-coding regions, tRNAs, and rRNAs. For example, maternally inherited forms of Leber hereditary optic neuropathy are due to point mutations in the subunits of the NADH dehydrogenase complex (Complex I) ND1, ND2, ND4, and ND4L. Point mutations in different mt:tRNAs cause different symptoms and disease (MELAS, MERRF, CPEO as a few of the examples). All of these mutations have effects on the stoichiometry of the protein complexes used for OXPHOS which is the underlying cause of a decrease in mitochondrial output.

What have we learned from flies harboring deleterious mutation in their mtDNA? Currently, there are three fly models to examine specific mutations; however, this is an area with great potential (discussed later). The first model is a serendipitously identified point mutation in mt:ATP6 (Celotto et al., 2006). The bang-sensitivity of *Drosophila* mutant for adenine nucleotide translocase type 1 (ANT1, called stress sensitive (SesB) in *Drosophila*) was separable from a second, cytoplasmically inherited bang-sensitivity found to be caused by a single point mutation in $mt{:}ATP6^{1}$. The $mt{:}ATP6^{1}$ mutation exists at near homoplasmy and gives rise to neurodegenerative phenotypes reminiscent of those found in mitochondrial diseases caused by mutations in *ATP6*. This model has been useful for studying the bioenergetic changes that occur before and after neurological symptoms occur (Celotto, Chiu, Van Voorhies, & Palladino, 2011).

The second model used a mitochondrially targeted restriction endonuclease to generate a single-site cleavage in the mtDNA. This idea was first tested in tissue culture cells by the Moraes laboratory, then subsequently successfully used in *Drosophila* to create an intact organism containing a mutation in cytochrome *c* oxidase subunit I (mt:CoI, Complex IV) as well as a small insertion and small deletion in mt:ND2 ($mt{:}ND2^{ins1}$ and $mt{:}ND2^{del1}$, respectively) (Bacman, Williams, Pinto, & Moraes, 2014; Srivastava & Moraes, 2001; Xu, DeLuca, & O'Farrell, 2008). In *Drosophila*, a single *Xho*I site is located in *CoI*. Xu et al. conditionally expressed a transgene encoding mitochondrially targeted *Xho*I exclusively in the germline. Upon germline expression, most mtDNA was irreparably cleaved, giving rise to sterile females; however, at a low level, mtDNA with mutations which rendered the DNA impervious to *Xho*I cleavage were selected for. These mutations were able to replicate and repopulate the germline. As would be expected, certain mutations were silent; however, others caused amino acid changes that affected CoI function (Xu et al., 2008). Adults homoplasmic for $mt{:}CoI^{R301S}$ mtDNA had 50% of the normal CoI levels, age-related reduction in ATP levels, and neurological and muscular defects, thus largely recapitulating many of the general symptoms exhibited in human mitochondrial diseases.

The $mt{:}ND2^{del1}$ mutation, which removes three highly conserved residues at positions 186–188, was recently further characterized (Burman et al., 2014). This work is a good example of how *Drosophila* exhibit tissue phenotypes associated with human mitochondrial disease, while also helping define the molecular function of a particular protein in one of the ETC Complexes. Complex I is the largest complex, with ~44 proteins. Burman et al. found that $mt{:}ND2^{del1}$ mutants were bang-sensitive and had reduced life spans, similar to $mt{:}ATP6^{1}$ mutants. The mutant adults had neurodegenerative vacuoles in aged fly brains, suggesting a progressive neurodegenerative phenotype, while the musculature remained intact. To pinpoint any defects in respiration, the authors determined that under maximally demanding conditions, the mutants showed a Complex-I-dependent respiratory defect, but no defect in Complex-II- and Complex-IV-dependent respiration. The total amount of assembled CoI was reduced, as was the mitochondrial membrane potential and amount of ATP. Together, these data support a role for mt:ND2 in the proton pumping mechanism of Complex I, the first time this has been shown in a eukaryote.

There are also disadvantages to using *Drosophila* to study mtDNA mutations. One disadvantage is that any organism with a high mtDNA mutational load will potentially eventually die. Therefore, while it is trivial to maintain a lethal mutation in a nuclear gene in *Drosophila*, maintaining mtDNA mutations can be challenging over many generations. In the case of the $mt{:}ATP6^{1}$ mutation, for reasons that are unknown, this mutation is sustained nearly homoplasmically in a background of the $SesB^{1}$ allele of ANT1. Thus, to examine and characterize the effect of a 100% $mt{:}ATP6^{1}$ mutation load, the *SesB* mutation need simply be crossed out of the strain. In the case of the CoI mutation located at the *Xho*I restriction site, while this stock will eventually die, it can be generated reproducibly over and over again by expressing the mitochondrially targeted *Xho*I in the germline.

5. DISEASE-CAUSING POINT MUTATIONS ARE MOST PREVALENT IN mt:tRNAs—CONSERVATION BETWEEN HUMAN AND *DROSOPHILA*

A thorough and continuously updated compilation of human mtDNA mutations and polymorphisms indicates that there are a total of 305 disease-causing modifications in mt:tRNAs and mt:rRNAs, with the numbers likely growing (Lehmann et al., 2015; Lott et al., 2013). This is an unusually large number given that only about 9% of mtDNA codes for tRNAs. There are various nucleus-encoded factors that are involved in processing mt:tRNA from their immature transcripts, as well as those that are responsible for post-translational modifications. Mutations in mt:tRNA residues that are sites for processing and modification can cause mitochondrial disease (Powell, Nicholls, & Minczuk, 2015). What are the possible effects of these mutations? mt:tRNA mutations may affect tRNA processing. Mature mt:tRNAs are embedded within the newly synthesized polycistronic transcript (Fig. 1). Due to this "punctuation" model, precursor mt:tRNA processing is critical not only to generate mature mt:tRNAs but also to cleave out the mt: mRNAs (Ojala et al., 1981). Recently, the mitochondrial cognate of RNase Z, the endoribonuclease responsible for 3′-end tRNA cleavage, was identified in *Drosophila*. The authors showed that specific loss of the mitochondrial form affected mt:RNA processing, causing larval lethality, cell-cycle defects, and an increase in reactive oxygen species (Xie & Dubrovsky, 2015). A mitochondrial protein-only RNase P, containing no enzymatic RNA, performs the 5′-end cleavage of mt:tRNAs (Holzmann et al., 2008). Recently, the three *Drosophila* orthologs comprising this complex have been identified and abolishing any of them causes lethality, loss of ATP, and aberrant mt:RNA processing (Sen et al., 2016). Potentially, mutations in the nucleotide residues that participate in the interaction between the mt:tRNA and either of these enzyme complexes responsible for cleavage could hinder processing, leading to not only a reduction in mt:tRNAs but also normal polycistron processing.

A second possible effect of mt:tRNA point mutations is on their unique stem–loop hairpin structure since the primary sequence is responsible for forming this secondary conformation. Mutations in residues contributing to this stable structure may lead to unstable mt:tRNA molecules that could be susceptible to enzymatic degradation. For example, a recent study by Duff et al. described how a single mutation in mt:tRNA^{Trp} caused a wide range of defects (Duff et al., 2015). A homoplasmic 5559A > G mutation in cells from a family affected with Leigh syndrome not only altered the processing and stability of mt:tRNA^{Trp} but also affected the stability of many other mitochondrial tRNAs, mRNAs, and rRNA.

Another tRNA-specific problem could be due to improper charging. The main function of tRNAs is to read the genetic code and transfer the respective amino acid residues onto the nascent polypeptide. But to do so, tRNAs must get charged with the appropriate amino acid by aminoacyl tRNA synthetase (ARS). In mitochondria, this charging process requires cross talk between nucleus-encoded mitochondrial tRNA synthetase (mtARS) and mt:tRNAs (Tyynismaa & Schon, 2014). As these two components rely on each other for proper function, mutations in either may lead to a cascade of functional abnormalities. For example, mutations in 9 of 19 total mtARs, encoded by the nuclear genome, are associated with mitochondrial disease in a tissue-specific manner (Konovalova & Tyynismaa, 2013). Despite having wide-spread consequences in causing mitochondrial disease, mutations in mt:tRNAs are not yet treatable. Repairing these mutations is not possible, due to the maternal inheritance of mtDNA. Recent studies have shown that their functional incompatibility can lead to developmental defects in *Drosophila*, and cell growth defects in yeast and mammalian cell culture (Meiklejohn et al., 2013; Perli et al., 2016; Wang et al., 2016). However, researchers have shown that engineered nucleus-encoded mtARSs can be used to suppress the effect of deleterious mt:tRNA mutations, which is easier to accomplish than altering mtDNA.

A comprehensive list of diseases related to mutations in mt:tRNAs is available at Mitomap (Brandon et al., 2005; Ruiz-Pesini et al., 2007). Due to their prevalence, there is a great deal of interest in mt:tRNAs point mutations and how they contribute to disease (Yarham, Elson, Blakely, McFarland, & Taylor, 2010). Using *Drosophila* to model human mitochondrial diseases caused by mt:tRNA mutations would be useful to determine the specific effect each mutation has on mt:tRNA processing, stability, and/or modification. To highlight the conservation between *Drosophila* and human mt:tRNAs, Fig. 2 shows a pair-wise structural alignment of all 22 tRNAs using the LocARNA alignment tool (Smith et al., 2010; Will, Joshi, Hofacker, Stadler, & Backofen, 2012). This program compares primary sequences as well as the structural compatibility of input RNA sequences, which is especially important as this indicates any potential disruptions caused by different mutations on secondary structure. Using LocARNA, all mt:tRNA pairs produce highly compatible canonical cloverleaf tRNA structures (examples shown in Fig. 2C and D), except for mt:tRNA$^{Ser(AGY)}$ and mt:tRNA^{Pro}. The confirmed disease-causing point mutations are marked with asterisks below each pair. We have also marked the residues where unique mutations have been reported (Genebank frequency 0, thus not due to polymorphism). Of a total of 145 mutant residues, 77 are conserved in *Drosophila*, and some of these conserved residues, which are structurally more compatible than others, would be excellent targets for mutagenesis. Using these alignments, along with weighted-based pathogenicity scoring models, allows prioritization of which residues would be best targeted for mutation studies (Blakely et al., 2013; Yarham et al., 2011).

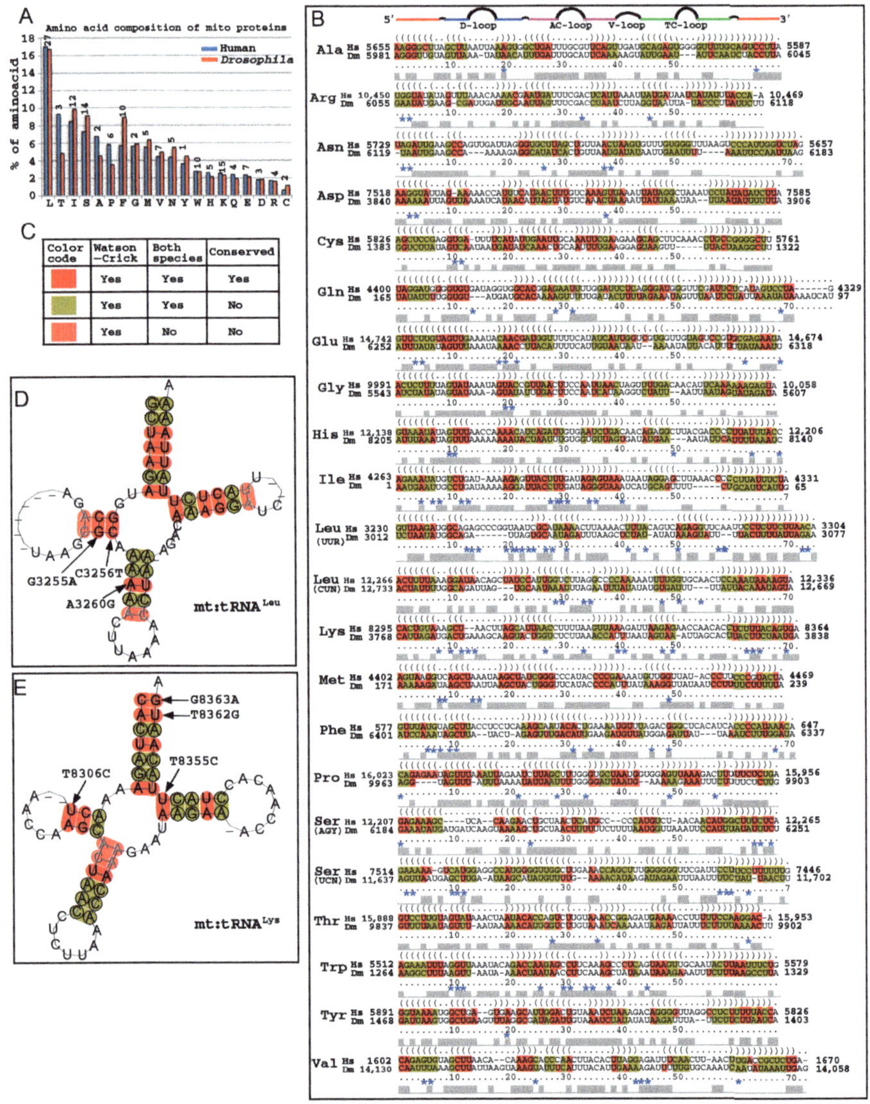

Fig. 2 mt:tRNA comparison between human and *Drosophila*. (A) A *graph* showing the amino acid composition (in percent) of mitochondria-encoded proteins in human and *Drosophila*. Amino acids are represented in single letter code. The number of disease-causing point mutations in each mt:tRNA are indicated at the *top* of each column. (B) Pair-wise alignment between human and *Drosophila* mt:tRNAs using the web-based LocARNA tool. The tRNA sequences were obtained from human (accession # NC_012920) and *Drosophila* (accession # U37541) mitochondrial genome sequences. The nucleotide start of each sequence is on the *left*. The gray boxes underneath each alignment indicate conserved nucleotide identity. To show the general location of the stems

Arrows point to conserved disease-causing residues on the consensus structures of mt:tRNA^{Lys} and mt:tRNA^{Leu} (Fig. 2D and E). These two tRNAs, along with mt:tRNA^{Ser}, are the most frequently mutated mt:tRNAs in mitochondrial disease (Lott et al., 2013; Ruiz-Pesini et al., 2007).

6. *DROSOPHILA* MODELS OF mtDNA-INDUCED DISEASE: UNTAPPED FUTURE POTENTIAL

There are multiple, devastating maternally inherited mitochondrial diseases. Developing additional fly models containing mtDNA mutations would be very useful for understanding the effect of each specific mutation on assembly and level of ETC complexes, on different tissues, and for determining how and what level of mutation load gives rise to deficits in organ and cell-type function. In addition, being able to generate specific mtDNA mutations at will would allow researchers to determine at the cellular and developmental level the molecular mechanisms governing inheritance. Given that there are only three models for mtDNA mutations in flies (*ATP6*^{1}, *mt:CoI*, and *mt:ND2*), what are the future prospects for generating more?

There are several potential ways to generate models of mtDNA-dependent mitochondrial disease in *Drosophila*. The first way is to use the restriction endonuclease method described earlier to generate mutations in mtDNA (Xu et al., 2008). This method has the advantage that it appears robust in manufacturing escaper flies harboring mtDNA mutations through germline selection. Furthermore, this method can be used repeatedly to regenerate the fly stock, since the genetics underlying the technique is relatively simple. However, the disadvantage is that these single-cut endonucleases recognize specific locations in the fly mtDNA genome, which limits the number of positions that would be affected (Table 1). In addition, any given location is not guaranteed to have a deleterious effect.

---

and loops, a schematic of a canonical tRNA cloverleaf structure in stretched-form is shown at the *top*, with the complementary stems the same color. (C) The *color* coding indicates whether the nucleotides are conserved and if they form a Watson–Crick base pair. The color-coding matrix for sequence compatibility was obtained from the LocARNA site (Smith, Heyne, Richter, Will, & Backofen, 2010). (D and E) mt:tRNA$^{Leu(UUR)}$ (D) and mt:tRNA^{Lys} (E) show generalized compatible secondary structures for each tRNA based on the sequence alignment. The *arrows* indicate some common point mutations found in human mitochondrial diseases that are conserved in *Drosophila*.

Table 1 Restriction Endonucleases with Only One Site in *Drosophila melanogaster* Mitochondrial DNA

Endonuclease	Site	Gene
*Bsm*FI	287	ND2
*Bsr*BI	558	ND2
*Bgl*II	800	ND2
*Eco*RV	1359	$tRNA^{Cys}$
*Nru*I	1473	COI?
*Bsg*I	1642	COI
*Bst*1107I	2005	COI
*Tsp*45I	2182	COI
*Ava*I	2368	COI
*Xho*I	2368	COI
*Nsi*I″	3158	$tRNA^{Leu}$
*Sap*I	3310	COII
*Nci*I	3646	COII
*Drd*I	4245	ATPase 6
*Bss*SI	4922	COIII
*Sty*I	4938	COIII
*Ple*I	5305	COIII
*Ahd*I	5462	COIII
*Hpa*I	6751	ND5
*Afl*II	7417	ND5
*Pst*I	7514	ND5
*Bsu*36I	9613	ND4L
*Bsa*BI	10,671	Cyt*B*
*Bsm*BI	10,711	Cyt*B*
*Nde*I	11,656	$tRNA^{Ser}$
*Bbs*I	13,160	lrRNA
*Bsr*GI	14,207	srRNA
*Ban*I	14,741	srRNA

8 out of the 13 protein-coding regions have sites and only 3 out of 22 mt:tRNAs have sites.

To get around this problem, a method that is beginning to be explored involves the evolving technology of genome editing combined with mitochondrially targeted nucleases. Transcription activator-like effector nucleases (TALEN) technology has recently been shown to abolish neurogenic weakness with ataxia and retinitis pigmentosa-associated mtDNA mutations in patient heteroplasmic cells by targeting and cleaving mutated mtDNAs (Reddy et al., 2015). TALEN technology targeting nuclear genes has been shown to be robust in *Drosophila*, thus adapting the modification of mitochondrially targeted TALEN developed for mammals could be used to cleave the *Drosophila* mtDNA genome at any site (Beumer et al., 2013; Katsuyama et al., 2013; Liu et al., 2012; Zhang, Ferreira, & Schnorrer, 2014). This method could potentially generate escaper flies repopulated with nuclease-resistant mtDNA, as is the case with mitochondrially targeted *Xho*I (Xu et al., 2008).

A method with the potential to create any mtDNA mutation on demand in *Drosophila* involves clustered regularly interspaced short palindromic repeats (CRISPR)/Cas9 technology (Sander & Joung, 2014). Recent work indicated this method can cleave mtDNA at the CoI locus in HEK cell mitochondria using a mitochondria-targeted Cas9 protein and introducing guide RNAs specific for mtDNA (Jo et al., 2015). While the authors show CoI is cut, it is not clear how the guide RNAs get into the mitochondria to mark the CoI locus, though various RNAs are known to be imported into eukaryotic mitochondria (Wang, Shimada, Koehler, & Teitell, 2012). CRISPR/Cas9 works very effectively in *Drosophila* on nuclear genes, and this technology appears to be more effective than gene targeting by homologous recombination in flies (Rong & Golic, 2000; Rong et al., 2002). In the nucleus, this genome editing involves homology-directed repair that uses an exogenously supplied oligo DNA encoding the desired change as a template for repair. Thus, for this technology to work on mtDNA for directed mutagenesis, there must be the appropriate repair mechanisms. Homologous recombination between mtDNA molecules has been clearly demonstrated for the first time in *Drosophila*, and the proteins required for double-strand break repair are present in mitochondria (Duxin et al., 2009; Ma & O'Farrell, 2015; Sage, Gildemeister, & Knight, 2010; Tann et al., 2011; Thyagarajan, Padua, & Campbell, 1996). Thus, it may be possible to use CRISPR/Cas9 to induce specific nucleotide changes in mtDNA to mimic human disease-causing mutations.

Point mutations in mtDNA lead to decreases in the proteins comprising the ETC complexes. While this likely occurs through a variety of mechanisms (e.g., too few tRNAs, unstable mRNAs, nonfunctional proteins), targeted knockdown of individual mt:mRNAs would be useful in order to understand and characterize the resulting developmental and tissue-specific effects. Nucleus-encoded noncoding RNAs are imported into mitochondria in all species (Sieber, Duchene, & Marechal-Drouard, 2011). Wang et al. successfully targeted wild-type mt:tRNAs to mitochondria using a 20-ribonucleotide stem–loop sequence from H1 RNA, the RNA component of RNase P (Wang, Shimada, Zhang, et al., 2012). They demonstrated this method could correct deficits in mt:tRNAs in cultured cells containing mtDNA mutations. Using a variation on this theme in *Drosophila*, Towheed et al. combined a similar approach with the idea of RNA silencing (Towheed, Markantone, Crain, Celotto, & Palladino, 2014). The 5S rRNA was originally thought to be a component of only cytoplasmic ribosomes; however, it is also imported into mitochondria where its function is not entirely clear (Magalhaes, Andreu, & Schon, 1998; Yoshionari et al., 1994). Towheed et al. identified the *Drosophila* ortholog of mitochondrial 5S rRNA and used the stem–loop leader sequence to target antisense RNA to mitochondria (Towheed et al., 2014). This technique resulted in translational inhibition of *mt:ATP6* and a 40–50% reduction in protein levels, which phenocopied $ATP6^1$ mutant flies. Called mitochondrial-targeted RNA expression system (mtTRES), the authors used the GAL4/UAS system to conditionally express the mt: ATP6 antisense mRNA, thus giving them spatial and temporal control.

The final method to create mutated mtDNA that has not been exploited in *Drosophila*, but has much potential, is creating a so-called mutator fly using a proof-reading-deficient mitochondrial polymerase gamma (PolG). PolG, the catalytic subunit of mtDNA polymerase, is a highly processive enzyme that contains three exonuclease domains responsible for excising and repairing mismatched nucleotides during replication (Kaguni & Olson, 1989; Wernette, Conway, & Kaguni, 1988). First described in yeast, PolG mutations were created by mutating conserved residues in the exonuclease domains, which led to an increase in mtDNA mutations as assayed by increased erythromycin resistance (Foury & Vanderstraeten, 1992). Erythromycin, an antibiotic that targets bacterial ribosomes, affects mitochondrial ribosomes due to the conserved mode of action between mitochondrial and bacterial ribosomes. mtDNA mutator mice have been successfully generated that lead to an increase in mtDNA mutations (Trifunovic et al., 2004). The *Drosophila* ortholog of PolG is called Tamas (Iyengar, Roote, & Campos, 1999). Mutations in *tamas* are lethal, and mutations in human PolG are known to cause mitochondrial disease (Wong et al., 2008). One problem with creating a mutator fly is that ectopically overexpressing *tamas* from a transgene leads to a decrease in mtDNA and lethality (Lefai et al., 2000). Using CRISPR/Cas9 would get around this problem by altering *tamas* at the endogenous locus, and thus the same exonuclease mutations used in yeast and mouse could be introduced under control of the endogenous promoter. A mutator fly would be useful for generating random mtDNA mutations and studying their effect on specific tissues, as well as inheritance. In addition, this would potentially be a good model for examining the effect of increased mtDNA mutation load on aging.

7. MITOCHONDRIAL INHERITANCE AND QUALITY CONTROL CHECKPOINTS

MtDNA has a higher mutation rate than nuclear DNA. In *Drosophila*, for example, it is 10 × higher than the nucleus (Haag-Liautard et al., 2008; Vermulst et al., 2007). Coupled with seemingly more rudimentary DNA repair mechanisms, an outstanding question is how oocytes generally maintain high levels of highly functional mitochondria. During inheritance, mitochondria can undergo a rapid change in genotype, giving rise to the hypothesis that there is a genetic bottleneck. Evidence supports that this bottleneck may take place during oogenesis; however, where and how this happens is not fully understood (Wallace & Chalkia, 2013). Studies in bovine indicated that a change in mtDNA genotype can be rapid, and data from mouse have tried to pinpoint the developmental timing of the bottleneck by estimating changes in mtDNA copy number at different times during fetal oogenesis (Cao et al., 2007; Hauswirth & Laipis, 1982; Jenuth, Peterson, Fu, & Shoubridge, 1996). This bottleneck was thought to be due to random genetic drift; however, there is increasing evidence that it may serve as a purifying mechanism to ensure only the most fit mitochondria populate the oocyte (Fan et al., 2008; Freyer et al., 2012; Stewart et al., 2008).

Women with disease-causing point mutations in mtDNA have a high probability of having children affected by the disease (Taylor & Turnbull, 2005). They may not manifest any disease symptoms until later in life, or at all, and thus may already have children. Prognoses in these cases are hard to make. A woman whose germline is heteroplasmic can have viable oocytes with different levels of mutated mtDNA, thus siblings can inherit different disease severity. There are only limited tools to determine which oocytes have a lower mutation load. One possible remedy recently advanced is nuclear transfer, using the affected oocyte's nucleus and an enucleated donor egg containing normal mitochondria (Mitalipov & Wolf, 2014). While this method is now legal in the United Kingdom, a better understanding of the molecular mechanisms underlying mitochondrial inheritance is required to provide their patients with an accurate prognosis.

Drosophila are an ideal model in which to study the mechanisms responsible for mitochondrial inheritance. *Drosophila* oogenesis is well characterized (Spradling, 1993). The stem cells can be unambiguously identified, each developmental stage is present for examination, and mitochondria can be visualized at the single-organelle level (Cox & Spradling, 2003). *Drosophila* oocyte formation shares a surprisingly large number of similarities with vertebrate oogenesis (Matova & Cooley, 2001). For example, both *Drosophila* and vertebrate germ cells spend part of their life as a cluster of interconnected cells called cysts (Pepling, de Cuevas, & Spradling, 1999). The presence of cysts allows the germ cells to share cytoplasmic components, such as microtubules, Golgi, centrosomes, and mitochondria (Cox & Spradling, 2003; Lei & Spradling, 2016; Pepling & Spradling, 1998). A prominent structure in oocytes is a mitochondrial cloud or Balbiani body (Kloc, Bilinski, & Etkin, 2004). In *Drosophila*, this highly conserved structure forms when a subset of mitochondria from connected sister germ cells moves into the oocyte using the microtubule cytoskeleton and molecular motors (Cox & Spradling, 2003, 2006). Since only a subset of mitochondria is transported into the oocyte to populate the oocyte for the first half of oogenesis, this raises the possibility that these mitochondria may be the most highly functional. Microtubule motor complexes appear to be important, suggesting that the ability of a particular organelle to bind to the motor and be transported may be part of the mechanism; however, this has not been directly tested.

Models of mitochondrial inheritance in *Drosophila* have given insight to the potential mechanisms underlying mitochondrial inheritance during oogenesis. The original studies examining the mitochondrial bottleneck in mouse and *Drosophila* took advantage of natural size differences and neutral polymorphisms between mtDNAs and did not look at competition between deleterious mutations and wild-type mtDNA (Jenuth et al., 1996; Kann, Rosenblum, & Rand, 1998; Solignac, Génermont, Monnerot, & Mounolou, 1984; Solignac, Genermont, Monnerot, & Mounolou, 1987). Recent work has used cytoplasmic injection to create heteroplasmic flies containing wild-type and mutated mtDNAs, followed by monitoring of mitochondrial purification over time (Hill, Chen, & Xu, 2014; Ma, Xu, & O'Farrell, 2014). The authors of these studies have exploited *Drosophila* to pinpoint the stages when mtDNA purification is taking place during oogenesis.

By carefully analyzing the percent heteroplasmy in flies over multiple generations, Ma et al. were able to demonstrate that there is a mtDNA purifying mechanism that takes place during oogenesis (Ma et al., 2014). They created heteroplasmic flies containing either two mutant mtDNAs (*mt:CoI* and *mt:ND2*) or wild type and mutant. This recreated a more physiological situation, where most mtDNA is wild type, and a small proportion contains a deleterious lesion in the mtDNA. In agreement with previous work, deleterious mtDNA mutations paired heteroplasmically with wild-type mtDNA were lost, giving clear-cut evidence that there is a purifying mechanism for mutated mtDNA. When two mtDNAs for lesions in two different genes were combined, they complemented each other and were maintained, resulting in viable flies. This observation is satisfying, since each cell contains many mtDNA molecules that should be able to complement function; however, this had not been demonstrated. As with mouse, the change in heteroplasmy took place quickly between mothers and their eggs, supporting that mtDNA genotype shifts happen during oogenesis. As one of the mutations was temperature sensitive, the authors were able to perform temperature shift experiments to test when during oogenesis any selection may be occurring. By doing this, they found that the selection occurs after germ cell mitotic expansion, and thus a large proportion of the selection occurs in the later germarium stages or later during oogenesis. This coincides with when the motor-driven Balbiani body formation occurs.

What mechanism could cause this mutant mtDNA selection during oogenesis? One possibility is that wild-type mtDNAs have a replicative advantage over mutated mtDNAs. Hill et al. developed a method using 5-ethynyl-2′-deoxyuridine (EdU), a thymidine analog, to examine mtDNA replication in dissected ovaries (Hill et al., 2014). This was the first time that mtDNA replication had been visualized during oogenesis. In wild-type flies, they found mtDNA replication was particularly high very early in germ cell development. This occurs right after the germ cells have completed their mitoses and have started their meiotic program at stage 2b. In addition, the mtDNA replication was dependent on mitochondrial function and membrane potential as germ cells containing the temperature sensitive, deleterious point mutation $mt{:}CoI^{T300I}$ had greatly reduced replication. The timing of increased mtDNA replication occurred around the same developmental time as Ma et al. (2014) postulated selection occurs. These two elegant studies together demonstrate the advantages of using *Drosophila* to study mtDNA inheritance: the combination of genetic manipulation (through injection to create heteroplasmic flies), the short generation time which allows mtDNA genotypes to be followed over many generations, immunofluorescence, and the ability to generate large sample sizes.

8. CONCLUDING REMARKS: LOSS OF MITOCHONDRIAL FUNCTION BROADLY IMPACTS HUMAN DISEASE

Mitochondrial disease is usually defined by loss of OXPHOS. Human mitochondrial disease is thought to affect as many as 1 in 5000 people, and there are no cures and few effective treatments (Schaefer, Taylor, Turnbull, & Chinnery, 2004). Because *Drosophila* mtDNA is so similar to human mtDNA, there is much potential to study the cell and developmental consequences of loss of nucleus-encoded mitochondrial proteins, and also mutations in mtDNA. Recent manipulation of deleterious mutant mtDNA has allowed *Drosophila* researchers to start to uncover the molecular mechanisms governing mtDNA inheritance and selection. Of course, mitochondria are responsible for generating many other important metabolites and are also pivotal in cell biological processes such as apoptosis and signaling. Due to a high demand for energy, muscle and neurons are cell types particularly sensitive to alterations in mitochondrial output. Decreases in mitochondrial function can lead to cardiomyopathy and heart problems, Parkinson's disease (reviewed in this issue), as well as other neurodegenerative diseases. Studying mitochondria in *Drosophila* will continue to inform and enlighten researchers about human mitochondrial diseases.

ACKNOWLEDGMENTS

We would like to thank Dr. Frank Shewmaker for critically reading the manuscript.

Funding Sources: This work was supported by the National Institutes of Health [1R21NS085730 to R.T.C.] and the National Institutes of Health/Department of Defense [CHIRP HU0001-14-2-0041 to R.T.C.].

REFERENCES

Antonicka, H., Mattman, A., Carlson, C. G., Glerum, D. M., Hoffbuhr, K. C., Leary, S. C., … Shoubridge, E. A. (2003). Mutations in COX15 produce a defect in the mitochondrial heme biosynthetic pathway, causing early-onset fatal hypertrophic cardiomyopathy. *The American Journal of Human Genetics*, *72*(1), 101–114. http://dx.doi.org/10.1086/345489.

Area-Gomez, E., & Schon, E. A. (2014). Mitochondrial genetics and disease. *Journal of Child Neurology*, *29*(9), 1208–1215. http://dx.doi.org/10.1177/0883073814539561.

Bacman, S. R., Williams, S. L., Pinto, M., & Moraes, C. T. (2014). The use of mitochondria-targeted endonucleases to manipulate mtDNA. *Methods in Enzymology*, *547*, 373–397. http://dx.doi.org/10.1016/B978-0-12-801415-8.00018-7.

Beumer, K. J., Trautman, J. K., Christian, M., Dahlem, T. J., Lake, C. M., Hawley, R. S., … Carroll, D. (2013). Comparing zinc finger nucleases and transcription activator-like effector nucleases for gene targeting in Drosophila. *G3 (Bethesda, Md)*, *3*(10), 1717–1725. http://dx.doi.org/10.1534/g3.113.007260.

Blakely, E. L., Yarham, J. W., Alston, C. L., Craig, K., Poulton, J., Brierley, C., … Taylor, R. W. (2013). Pathogenic mitochondrial tRNA point mutations: Nine novel mutations affirm their importance as a cause of mitochondrial disease. *Human Mutation*, *34*(9), 1260–1268. http://dx.doi.org/10.1002/humu.22358.

Brandon, M. C., Lott, M. T., Nguyen, K. C., Spolim, S., Navathe, S. B., Baldi, P., & Wallace, D. C. (2005). MITOMAP: A human mitochondrial genome database—2004 Update. *Nucleic Acids Research*, *33*(Database issue), D611–D613. http://dx.doi.org/10.1093/nar/gki079. 33/suppl_1/D611 [pii].

Burman, J. L., Itsara, L. S., Kayser, E. B., Suthammarak, W., Wang, A. M., Kaeberlein, M., … Pallanck, L. J. (2014). A Drosophila model of mitochondrial disease caused by a complex I mutation that uncouples proton pumping from electron transfer. *Disease Models & Mechanisms*, 7(10), 1165–1174. http://dx.doi.org/10.1242/dmm.015321.

Cao, L., Shitara, H., Horii, T., Nagao, Y., Imai, H., Abe, K., … Yonekawa, H. (2007). The mitochondrial bottleneck occurs without reduction of mtDNA content in female mouse germ cells. *Nature Genetics*, *39*(3), 386–390. http://dx.doi.org/10.1038/ng1970.

Celotto, A. M., Chiu, W. K., Van Voorhies, W., & Palladino, M. J. (2011). Modes of metabolic compensation during mitochondrial disease using the Drosophila model of ATP6 dysfunction. *PloS One*, *6*(10), e25823. http://dx.doi.org/10.1371/journal.pone.0025823.

Celotto, A. M., Frank, A. C., McGrath, S. W., Fergestad, T., Van Voorhies, W. A., Buttle, K. F., … Palladino, M. J. (2006). Mitochondrial encephalomyopathy in Drosophila. *The Journal of Neuroscience*, *26*(3), 810–820. http://dx.doi.org/10.1523/JNEUROSCI.4162-05.2006.

Chandel, N. S., & Schumacker, P. T. (1999). Cells depleted of mitochondrial DNA (rho0) yield insight into physiological mechanisms. *FEBS Letters*, *454*(3), 173–176.

Chinnery, P. F. (1993). Mitochondrial disorders overview. In R. A. Pagon, M. P. Adam, H. H. Ardinger, S. E. Wallace, A. Amemiya, L. J. H. Bean, T. D. Bird, C. T. Fong, H. C. Mefford, R. J. H. Smith, & K. Stephens (Eds.), *GeneReviews(R)*. Seattle (WA): University of Washington.

Cox, R. T., & Spradling, A. C. (2003). A Balbiani body and the fusome mediate mitochondrial inheritance during Drosophila oogenesis. *Development*, *130*(8), 1579–1590.

Cox, R. T., & Spradling, A. C. (2006). Milton controls the early acquisition of mitochondria by Drosophila oocytes. *Development*, *133*(17), 3371–3377. http://dx.doi.org/10.1242/dev.02514.

Dimauro, S., & Davidzon, G. (2005). Mitochondrial DNA and disease. *Annals of Medicine*, *37*(3), 222–232. http://dx.doi.org/10.1080/07853890510007368.

Duff, R. M., Shearwood, A. M., Ermer, J., Rossetti, G., Gooding, R., Richman, T. R., … Filipovska, A. (2015). A mutation in MT-TW causes a tRNA processing defect and reduced mitochondrial function in a family with Leigh syndrome. *Mitochondrion*, *25*, 113–119. http://dx.doi.org/10.1016/j.mito.2015.10.008.

Duxin, J. P., Dao, B., Martinsson, P., Rajala, N., Guittat, L., Campbell, J. L., … Stewart, S. A. (2009). Human Dna2 is a nuclear and mitochondrial DNA maintenance protein. *Molecular and Cellular Biology*, *29*(15), 4274–4282. http://dx.doi.org/10.1128/MCB.01834-08.

Engel, J. E., & Wu, C. F. (1994). Altered mechanoreceptor response in Drosophila bang-sensitive mutants. *Journal of Comparative Physiology A*, *175*(3), 267–278.
Fan, W., Waymire, K. G., Narula, N., Li, P., Rocher, C., Coskun, P. E., … Wallace, D. C. (2008). A mouse model of mitochondrial disease reveals germline selection against severe mtDNA mutations. *Science*, *319*(5865), 958–962. http://dx.doi.org/10.1126/science.1147786.
Fergestad, T., Bostwick, B., & Ganetzky, B. (2006). Metabolic disruption in Drosophila bang-sensitive seizure mutants. *Genetics*, *173*(3), 1357–1364. http://dx.doi.org/10.1534/genetics.106.057463.
Foriel, S., Willems, P., Smeitink, J., Schenck, A., & Beyrath, J. (2015). Mitochondrial diseases: Drosophila melanogaster as a model to evaluate potential therapeutics. *The International Journal of Biochemistry & Cell Biology*, *63*, 60–65. http://dx.doi.org/10.1016/j.biocel.2015.01.024.
Foury, F., & Vanderstraeten, S. (1992). Yeast mitochondrial DNA mutators with deficient proofreading exonucleolytic activity. *The EMBO Journal*, *11*(7), 2717–2726.
Freyer, C., Cree, L. M., Mourier, A., Stewart, J. B., Koolmeister, C., Milenkovic, D., … Chinnery, P. F. (2012). Variation in germline mtDNA heteroplasmy is determined prenatally but modified during subsequent transmission. *Nature Genetics*, *44*(11), 1282–1285. http://dx.doi.org/10.1038/ng.2427.
Ganetzky, B., & Wu, C. F. (1982). Indirect suppression involving behavioral mutants with altered nerve excitability in Drosophila melanogaster. *Genetics*, *100*(4), 597–614.
Goldring, E. S., Grossman, L. I., Krupnick, D., Cryer, D. R., & Marmur, J. (1970). The petite mutation in yeast. Loss of mitochondrial deoxyribonucleic acid during induction of petites with ethidium bromide. *Journal of Molecular Biology*, *52*(2), 323–335.
Goto, Y., Nonaka, I., & Horai, S. (1990). A mutation in the tRNA(Leu)(UUR) gene associated with the MELAS subgroup of mitochondrial encephalomyopathies. *Nature*, *348*(6302), 651–653. http://dx.doi.org/10.1038/348651a0.
Gray, M. W. (2012). Mitochondrial evolution. *Cold Spring Harbor Perspectives in Biology*, *4*(9), a011403. http://dx.doi.org/10.1101/cshperspect.a011403.
Haag-Liautard, C., Coffey, N., Houle, D., Lynch, M., Charlesworth, B., & Keightley, P. D. (2008). Direct estimation of the mitochondrial DNA mutation rate in Drosophila melanogaster. *PLoS Biology*, *6*(8), e204. http://dx.doi.org/10.1371/journal.pbio.0060204.
Hauswirth, W. W., & Laipis, P. J. (1982). Mitochondrial DNA polymorphism in a maternal lineage of Holstein cows. *Proceedings of the National Academy of Sciences of the United States of America*, *79*(15), 4686–4690.
Hill, J. H., Chen, Z., & Xu, H. (2014). Selective propagation of functional mitochondrial DNA during oogenesis restricts the transmission of a deleterious mitochondrial variant. *Nature Genetics*, *46*(4), 389–392. http://dx.doi.org/10.1038/ng.2920.
Holzmann, J., Frank, P., Löffler, E., Bennett, K. L., Gerner, C., & Rossmanith, W. (2008). RNase P without RNA: Identification and functional reconstitution of the human mitochondrial tRNA processing enzyme. *Cell*, *135*(3), 462–536.
Iyengar, B., Roote, J., & Campos, A. R. (1999). The tamas gene, identified as a mutation that disrupts larval behavior in Drosophila melanogaster, codes for the mitochondrial DNA polymerase catalytic subunit (DNApol-gamma125). *Genetics*, *153*(4), 1809–1824.
Jacobs, H. T., Fernandez-Ayala, D. J., Manjiry, S., Kemppainen, E., Toivonen, J. M., & O'Dell, K. M. (2004). Mitochondrial disease in flies. *Biochimica et Biophysica Acta*, *1659*(2–3), 190–196. http://dx.doi.org/10.1016/j.bbabio.2004.07.004.
Jaksch, M., Ogilvie, I., Yao, J., Kortenhaus, G., Bresser, H. G., Gerbitz, K. D., & Shoubridge, E. A. (2000). Mutations in SCO_2 are associated with a distinct form of hypertrophic cardiomyopathy and cytochrome c oxidase deficiency. *Human Molecular Genetics*, *9*(5), 795–801.

Jenuth, J. P., Peterson, A. C., Fu, K., & Shoubridge, E. A. (1996). Random genetic drift in the female germline explains the rapid segregation of mammalian mitochondrial DNA. *Nature Genetics*, *14*(2), 146–151. http://dx.doi.org/10.1038/ng1096-146.

Jo, A., Ham, S., Lee, G. H., Lee, Y. I., Kim, S., Lee, Y. S., ... Lee, Y. (2015). Efficient mitochondrial genome editing by CRISPR/Cas9. *BioMed Research International*, *2015*, 305716. http://dx.doi.org/10.1155/2015/305716.

Kaguni, L. S., & Olson, M. W. (1989). Mismatch-specific 3'——5' exonuclease associated with the mitochondrial DNA polymerase from Drosophila embryos. *Proceedings of the National Academy of Sciences of the United States of America*, *86*(17), 6469–6473.

Kann, L. M., Rosenblum, E. B., & Rand, D. M. (1998). Aging, mating, and the evolution of mtDNA heteroplasmy in Drosophila melanogaster. *Proceedings of the National Academy of Sciences of the United States of America*, *95*(5), 2372–2377.

Katsuyama, T., Akmammedov, A., Seimiya, M., Hess, S. C., Sievers, C., & Paro, R. (2013). An efficient strategy for TALEN-mediated genome engineering in Drosophila. *Nucleic Acids Research*, *41*(17), e163. http://dx.doi.org/10.1093/nar/gkt638.

Kemppainen, K. K., Kemppainen, E., & Jacobs, H. T. (2014). The alternative oxidase AOX does not rescue the phenotype of tko25t mutant flies. *G3 (Bethesda)*, *4*(10), 2013–2021. http://dx.doi.org/10.1534/g3.114.013946.

Kloc, M., Bilinski, S., & Etkin, L. D. (2004). The Balbiani body and germ cell determinants: 150 Years later. *Current Topics in Developmental Biology*, *59*, 1–36. http://dx.doi.org/10.1016/S0070-2153(04)59001-4.

Konovalova, S., & Tyynismaa, H. (2013). Mitochondrial aminoacyl-tRNA synthetases in human disease. *Molecular Genetics and Metabolism*, *108*(4), 206–211. http://dx.doi.org/10.1016/j.ymgme.2013.01.010.

Lefai, E., Calleja, M., Ruiz de Mena, I., Lagina, A. T., 3rd., Kaguni, L. S., & Garesse, R. (2000). Overexpression of the catalytic subunit of DNA polymerase gamma results in depletion of mitochondrial DNA in Drosophila melanogaster. *Molecular and General Genetics*, *264*(1–2), 37–46.

Lehmann, D., Schubert, K., Joshi, P. R., Baty, K., Blakely, E. L., Zierz, S., ... Deschauer, M. (2015). A novel m.7539C > T point mutation in the mt-tRNA(Asp) gene associated with multisystemic mitochondrial disease. *Neuromuscular Disorders*, *25*(1), 81–84. http://dx.doi.org/10.1016/j.nmd.2014.09.008.

Lei, L., & Spradling, A. C. (2016). Mouse oocytes differentiate through organelle enrichment from sister cyst germ cells. *Science*, *352*, 95–99. http://dx.doi.org/10.1126/science.aad2156.

Lightowlers, R. N., Taylor, R. W., & Turnbull, D. M. (2015). Mutations causing mitochondrial disease: What is new and what challenges remain? *Science*, *349*(6255), 1494–1499. http://dx.doi.org/10.1126/science.aac7516.

Liu, J., Li, C., Yu, Z., Huang, P., Wu, H., Wei, C., ... Jiao, R. (2012). Efficient and specific modifications of the Drosophila genome by means of an easy TALEN strategy. *Journal of Genetics and Genomics*, *39*(5), 209–215. http://dx.doi.org/10.1016/j.jgg.2012.04.003.

Loeffen, J., Elpeleg, O., Smeitink, J., Smeets, R., Stockler-Ipsiroglu, S., Mandel, H., ... van den Heuvel, L. (2001). Mutations in the complex I NDUFS2 gene of patients with cardiomyopathy and encephalomyopathy. *Annals of Neurology*, *49*(2), 195–201.

Lott, M. T., Leipzig, J. N., Derbeneva, O., Xie, H. M., Chalkia, D., Sarmady, M., & Wallace, D. C. (2013). mtDNA Variation and analysis using MITOMAP and MITOMASTER. *Current Protocols in Bioinformatics*, *1*, 1–26. http://dx.doi.org/10.1002/0471250953.bi0123s44.

Ma, H., & O'Farrell, P. H. (2015). Selections that isolate recombinant mitochondrial genomes in animals. *eLife*. *4*. , http://dx.doi.org/10.7554/eLife.07247.

Ma, H., Xu, H., & O'Farrell, P. H. (2014). Transmission of mitochondrial mutations and action of purifying selection in Drosophila melanogaster. *Nature Genetics*, *46*(4), 393–397. http://dx.doi.org/10.1038/ng.2919.

Magalhaes, P. J., Andreu, A. L., & Schon, E. A. (1998). Evidence for the presence of 5S rRNA in mammalian mitochondria. *Molecular Biology of the Cell*, *9*(9), 2375–2382.
Majamaa-Voltti, K., Peuhkurinen, K., Kortelainen, M. L., Hassinen, I. E., & Majamaa, K. (2002). Cardiac abnormalities in patients with mitochondrial DNA mutation 3243A>G. *BMC Cardiovascular Disorders*, *2*, 12.
Margulis, L. (1970). *Origin of eukaryotic cells*. New Haven, CT: Yale University Press.
Matova, N., & Cooley, L. (2001). Comparative aspects of animal oogenesis. *Developmental Biology*, *231*(2), 291–320. http://dx.doi.org/10.1006/dbio.2000.0120.
Meiklejohn, C. D., Holmbeck, M. A., Siddiq, M. A., Abt, D. N., Rand, D. M., & Montooth, K. L. (2013). An incompatibility between a mitochondrial tRNA and its nuclear-encoded tRNA synthetase compromises development and fitness in Drosophila. *PLoS Genetics*, *9*(1), e1003238. http://dx.doi.org/10.1371/journal.pgen.1003238.
Mitalipov, S., & Wolf, D. P. (2014). Clinical and ethical implications of mitochondrial gene transfer. *Trends in Endocrinology and Metabolism*, *25*(1), 5–7. http://dx.doi.org/10.1016/j.tem.2013.09.001.
Morais, R., Gregoire, M., Jeannotte, L., & Gravel, D. (1980). Chick embryo cells rendered respiration-deficient by chloramphenicol and ethidium bromide are auxotrophic for pyrimidines. *Biochemical and Biophysical Research Communications*, *94*(1), 71–77.
Nagley, P., & Linnane, A. W. (1970). Mitochondrial DNA deficient petite mutants of yeast. *Biochemical and Biophysical Research Communications*, *39*(5), 989–996.
Ojala, D., Montoya, J., & Attardi, G. (1981). tRNA punctuation model of RNA processing in human mitochondria. *Nature*, *290*(5806), 470–474.
Papadopoulou, L. C., Sue, C. M., Davidson, M. M., Tanji, K., Nishino, I., Sadlock, J. E., ... Schon, E. A. (1999). Fatal infantile cardioencephalomyopathy with COX deficiency and mutations in SCO_2, a COX assembly gene. *Nature Genetics*, *23*(3), 333–337. http://dx.doi.org/10.1038/15513.
Pepling, M. E., de Cuevas, M., & Spradling, A. C. (1999). Germline cysts: A conserved phase of germ cell development? *Trends in Cell Biology*, *9*(7), 257–262.
Pepling, M. E., & Spradling, A. C. (1998). Female mouse germ cells form synchronously dividing cysts. *Development*, *125*(17), 3323–3328.
Perli, E., Fiorillo, A., Giordano, C., Pisano, A., Montanari, A., Grazioli, P., ... d'Amati, G. (2016). Short peptides from leucyl-tRNA synthetase rescue disease-causing mitochondrial tRNA point mutations. *Human Molecular Genetics*, *25*(5), 903–915. http://dx.doi.org/10.1093/hmg/ddv619.
Picard, M., Zhang, J., Hancock, S., Derbeneva, O., Golhar, R., Golik, P., ... Wallace, D. C. (2014). Progressive increase in mtDNA 3243A>G heteroplasmy causes abrupt transcriptional reprogramming. *Proceedings of the National Academy of Sciences of the United States of America*, *111*(38), E4033–E4042. http://dx.doi.org/10.1073/pnas.1414028111.
Pons, R., Andreu, A. L., Checcarelli, N., Vila, M. R., Engelstad, K., Sue, C. M., ... DiMauro, S. (2004). Mitochondrial DNA abnormalities and autistic spectrum disorders. *The Journal of Pediatrics*, *144*(1), 81–85. http://dx.doi.org/10.1016/j.jpeds.2003.10.023.
Powell, C. A., Nicholls, T. J., & Minczuk, M. (2015). Nuclear-encoded factors involved in post-transcriptional processing and modification of mitochondrial tRNAs in human disease. *Frontiers in Genetics*, *6*, 79. http://dx.doi.org/10.3389/fgene.2015.00079.
Reddy, P., Ocampo, A., Suzuki, K., Luo, J., Bacman, S. R., Williams, S. L., ... Izpisua Belmonte, J. C. (2015). Selective elimination of mitochondrial mutations in the germline by genome editing. *Cell*, *161*(3), 459–469. http://dx.doi.org/10.1016/j.cell. 2015.03.051.
Rong, Y. S., & Golic, K. G. (2000). Gene targeting by homologous recombination in Drosophila. *Science*, *288*(5473), 2013–2018.
Rong, Y. S., Titen, S. W., Xie, H. B., Golic, M. M., Bastiani, M., Bandyopadhyay, P., ... Golic, K. G. (2002). Targeted mutagenesis by homologous recombination in D. melanogaster. *Genes and Development*, *16*(12), 1568–1581. http://dx.doi.org/10.1101/gad.986602.

Rossignol, R., Faustin, B., Rocher, C., Malgat, M., Mazat, J. P., & Letellier, T. (2003). Mitochondrial threshold effects. *The Biochemical Journal*, *370*(Pt. 3), 751–762. http://dx.doi.org/10.1042/BJ20021594.

Royden, C. S., Pirrotta, V., & Jan, L. Y. (1987). The tko locus, site of a behavioral mutation in D. melanogaster, codes for a protein homologous to prokaryotic ribosomal protein S12. *Cell*, *51*(2), 165–173.

Ruiz-Pesini, E., Lott, M. T., Procaccio, V., Poole, J. C., Brandon, M. C., Mishmar, D., … Wallace, D. C. (2007). An enhanced MITOMAP with a global mtDNA mutational phylogeny. *Nucleic Acids Research*, *35*(Database issue), D823–D828. http://dx.doi.org/10.1093/nar/gkl927. gkl927 [pii].

Sage, J. M., Gildemeister, O. S., & Knight, K. L. (2010). Discovery of a novel function for human Rad51: Maintenance of the mitochondrial genome. *The Journal of Biological Chemistry*, *285*(25), 18984–18990. http://dx.doi.org/10.1074/jbc.M109.099846.

Sander, J. D., & Joung, J. K. (2014). CRISPR-Cas systems for editing, regulating and targeting genomes. *Nature Biotechnology*, *32*(4), 347–355. http://dx.doi.org/10.1038/nbt.2842.

Schaefer, A. M., Taylor, R. W., Turnbull, D. M., & Chinnery, P. F. (2004). The epidemiology of mitochondrial disorders—Past, present and future. *Biochimica et Biophysica Acta*, *1659*(2-3), 115–120. http://dx.doi.org/10.1016/j.bbabio.2004.09.005.

Sen, A., Damm, V. T., & Cox, R. T. (2013). Drosophila clueless is highly expressed in larval neuroblasts, affects mitochondrial localization and suppresses mitochondrial oxidative damage. *PloS One*, *8*(1), e54283. http://dx.doi.org/10.1371/journal.pone.0054283.

Sen, A., Karasik, A., Shanmuganathan, A., Mirkovic, E., Koutmos, M., & Cox, R. T. (2016). Loss of the mitochondrial protein-only ribonuclease P complex causes aberrant tRNA processing and lethality in Drosophila. *Nucleic Acids Research*, *44*(13), 6409–6422. http://dx.doi.org/10.1093/nar/gkw338. gkw338 [pii].

Sieber, F., Duchene, A. M., & Marechal-Drouard, L. (2011). Mitochondrial RNA import: From diversity of natural mechanisms to potential applications. *International Review of Cell and Molecular Biology*, *287*, 145–190. http://dx.doi.org/10.1016/B978-0-12-386043-9.00004-9.

Smith, C., Heyne, S., Richter, A. S., Will, S., & Backofen, R. (2010). Freiburg RNA tools: A web server integrating INTARNA, EXPARNA and LOCARNA. *Nucleic Acids Research*, *38*(Web Server issue), W373–W377. http://dx.doi.org/10.1093/nar/gkq316.

Solignac, M., Génermont, J., Monnerot, M., & Mounolou, J.-C. (1984). Genetics of mitochondria in Drosophila: mtDNA inheritance in heteroplasmic strains of D. mauritiana. *Molecular and General Genetics MGG*, *197*(2), 183–188. http://dx.doi.org/10.1007/bf00330961.

Solignac, M., Genermont, J., Monnerot, M., & Mounolou, J. C. (1987). Drosophila mitochondrial genetics: Evolution of heteroplasmy through germ line cell divisions. *Genetics*, *117*(4), 687–696.

Spradling, A. C. (1993). *Developmental genetics of oogenesis: Vol. 1*. Cold Spring Harbor, NY: Cold Spring Harbor Press.

Srivastava, S., & Moraes, C. T. (2001). Manipulating mitochondrial DNA heteroplasmy by a mitochondrially targeted restriction endonuclease. *Human Molecular Genetics*, *10*(26), 3093–3099.

Stewart, J. B., & Chinnery, P. F. (2015). The dynamics of mitochondrial DNA heteroplasmy: Implications for human health and disease. *Nature Reviews. Genetics*, *16*(9), 530–542. http://dx.doi.org/10.1038/nrg3966.

Stewart, J. B., Freyer, C., Elson, J. L., Wredenberg, A., Cansu, Z., Trifunovic, A., & Larsson, N. G. (2008). Strong purifying selection in transmission of mammalian mitochondrial DNA. *PLoS Biology*, *6*(1), e10. http://dx.doi.org/10.1371/journal.pbio.0060010.

Tann, A. W., Boldogh, I., Meiss, G., Qian, W., Van Houten, B., Mitra, S., & Szczesny, B. (2011). Apoptosis induced by persistent single-strand breaks in mitochondrial genome: Critical role of EXOG (5'-EXO/endonuclease) in their repair. *The Journal of Biological Chemistry*, *286*(37), 31975–31983. http://dx.doi.org/10.1074/jbc.M110.215715.

Tay, S. K., Shanske, S., Kaplan, P., & DiMauro, S. (2004). Association of mutations in SCO_2, a cytochrome c oxidase assembly gene, with early fetal lethality. *Archives of Neurology*, *61*(6), 950–952. http://dx.doi.org/10.1001/archneur.61.6.950.

Taylor, R. W., & Turnbull, D. M. (2005). Mitochondrial DNA mutations in human disease. *Nature Reviews. Genetics*, *6*(5), 389–402. http://dx.doi.org/10.1038/nrg1606.

Thyagarajan, B., Padua, R. A., & Campbell, C. (1996). Mammalian mitochondria possess homologous DNA recombination activity. *The Journal of Biological Chemistry*, *271*(44), 27536–27543.

Toivonen, J. M., O'Dell, K. M., Petit, N., Irvine, S. C., Knight, G. K., Lehtonen, M., … Jacobs, H. T. (2001). Technical knockout, a Drosophila model of mitochondrial deafness. *Genetics*, *159*(1), 241–254.

Towheed, A., Markantone, D. M., Crain, A. T., Celotto, A. M., & Palladino, M. J. (2014). Small mitochondrial-targeted RNAs modulate endogenous mitochondrial protein expression in vivo. *Neurobiology of Disease*, *69*, 15–22. http://dx.doi.org/10.1016/j.nbd.2014.04.017.

Tricoire, H., Palandri, A., Bourdais, A., Camadro, J. M., & Monnier, V. (2014). Methylene blue rescues heart defects in a Drosophila model of Friedreich's ataxia. *Human Molecular Genetics*, *23*(4), 968–979. http://dx.doi.org/10.1093/hmg/ddt493.

Trifunovic, A., Wredenberg, A., Falkenberg, M., Spelbrink, J. N., Rovio, A. T., Bruder, C. E., … Larsson, N. G. (2004). Premature ageing in mice expressing defective mitochondrial DNA polymerase. *Nature*, *429*(6990), 417–423. http://dx.doi.org/10.1038/nature02517.

Tyynismaa, H., & Schon, E. A. (2014). Mixing and matching mitochondrial aminoacyl synthetases and their tRNAs: A new way to treat respiratory chain disorders? *EMBO Molecular Medicine*, *6*(2), 155–157. http://dx.doi.org/10.1002/emmm.201303586.

van den Ouweland, J. M., Lemkes, H. H., Ruitenbeek, W., Sandkuijl, L. A., de Vijlder, M. F., Struyvenberg, P. A., … Maassen, J. A. (1992). Mutation in mitochondrial tRNA(Leu)(UUR) gene in a large pedigree with maternally transmitted type II diabetes mellitus and deafness. *Nature Genetics*, *1*(5), 368–371. http://dx.doi.org/10.1038/ng0892-368.

Vermulst, M., Bielas, J. H., Kujoth, G. C., Ladiges, W. C., Rabinovitch, P. S., Prolla, T. A., & Loeb, L. A. (2007). Mitochondrial point mutations do not limit the natural lifespan of mice. *Nature Genetics*, *39*(4), 540–543. http://dx.doi.org/10.1038/ng1988.

Wahbi, K., Larue, S., Jardel, C., Meune, C., Stojkovic, T., Ziegler, F., … Laforet, P. (2010). Cardiac involvement is frequent in patients with the m.8344A>G mutation of mitochondrial DNA. *Neurology*, *74*(8), 674–677. http://dx.doi.org/10.1212/WNL.0b013e3181d0ccf4.

Wallace, D. C., & Chalkia, D. (2013). Mitochondrial DNA genetics and the heteroplasmy conundrum in evolution and disease. *Cold Spring Harbor Perspectives in Biology*, *5*(11), a021220. http://dx.doi.org/10.1101/cshperspect.a021220.

Wang, G., Shimada, E., Koehler, C. M., & Teitell, M. A. (2012a). PNPASE and RNA trafficking into mitochondria. *Biochimica et Biophysica Acta*, *1819*(9–10), 998–1007. http://dx.doi.org/10.1016/j.bbagrm.2011.10.001.

Wang, G., Shimada, E., Zhang, J., Hong, J. S., Smith, G. M., Teitell, M. A., & Koehler, C. M. (2012b). Correcting human mitochondrial mutations with targeted RNA import. *Proceedings of the National Academy of Sciences of the United States of America*, *109*(13), 4840–4845. http://dx.doi.org/10.1073/pnas.1116792109.

Wang, Y., Zhou, X. L., Ruan, Z. R., Liu, R. J., Eriani, G., & Wang, E. D. (2016). A human disease-causing point mutation in mitochondrial threonyl-tRNA synthetase induces both structural and functional defects. *The Journal of Biological Chemistry*, *291*, 6507–6520. http://dx.doi.org/10.1074/jbc.M115.700849.

Wernette, C. M., Conway, M. C., & Kaguni, L. S. (1988). Mitochondrial DNA polymerase from Drosophila melanogaster embryos: Kinetics, processivity, and fidelity of DNA polymerization. *Biochemistry*, *27*(16), 6046–6054.

Will, S., Joshi, T., Hofacker, I. L., Stadler, P. F., & Backofen, R. (2012). LocARNA-P: Accurate boundary prediction and improved detection of structural RNAs. *RNA*, *18*(5), 900–914. http://dx.doi.org/10.1261/rna.029041.111.

Wong, L. J., Naviaux, R. K., Brunetti-Pierri, N., Zhang, Q., Schmitt, E. S., Truong, C., ... Copeland, W. C. (2008). Molecular and clinical genetics of mitochondrial diseases due to POLG mutations. *Human Mutation*, *29*(9), E150–E172. http://dx.doi.org/10.1002/humu.20824.

Xie, X., & Dubrovsky, E. B. (2015). Knockout of Drosophila RNase ZL impairs mitochondrial transcript processing, respiration and cell cycle progression. *Nucleic Acids Research*, *43*(21), 10364–10375. http://dx.doi.org/10.1093/nar/gkv1149.

Xu, H., DeLuca, S. Z., & O'Farrell, P. H. (2008). Manipulating the metazoan mitochondrial genome with targeted restriction enzymes. *Science*, *321*(5888), 575–577. http://dx.doi.org/10.1126/science.1160226.

Yarham, J. W., Al-Dosary, M., Blakely, E. L., Alston, C. L., Taylor, R. W., Elson, J. L., & McFarland, R. (2011). A comparative analysis approach to determining the pathogenicity of mitochondrial tRNA mutations. *Human Mutation*, *32*(11), 1319–1325. http://dx.doi.org/10.1002/humu.21575.

Yarham, J. W., Elson, J. L., Blakely, E. L., McFarland, R., & Taylor, R. W. (2010). Mitochondrial tRNA mutations and disease. *Wiley Interdisciplinary Reviews. RNA*, *1*(2), 304–324. http://dx.doi.org/10.1002/wrna.27.

Yoshionari, S., Koike, T., Yokogawa, T., Nishikawa, K., Ueda, T., Miura, K., & Watanabe, K. (1994). Existence of nuclear-encoded 5S-rRNA in bovine mitochondria. *FEBS Letters*, *338*(2), 137–142.

Zhang, X., Ferreira, I. R., & Schnorrer, F. (2014). A simple TALEN-based protocol for efficient genome-editing in Drosophila. *Methods*, *69*(1), 32–37. http://dx.doi.org/10.1016/j.ymeth.2014.03.020.

Zhang, Y. Q., Roote, J., Brogna, S., Davis, A. W., Barbash, D. A., Nash, D., & Ashburner, M. (1999). Stress sensitive B encodes an adenine nucleotide translocase in Drosophila melanogaster. *Genetics*, *153*(2), 891–903.

CHAPTER TWO

Drosophila as a Model for Human Diseases—Focus on Innate Immunity in Barrier Epithelia

P. Bergman*, S. Seyedoleslami Esfahani†, Y. Engström†,[1]
*Clinical Microbiology F68, Karolinska Institutet, Karolinska University Hospital, Huddinge, Stockholm, Sweden
†The Wenner-Gren Institute, Stockholm University, Stockholm, Sweden
[1]Corresponding author: e-mail address: ylva.engstrom@su.se

Contents

Current Topics in Developmental Biology, Volume 121
ISSN 0070-2153
http://dx.doi.org/10.1016/bs.ctdb.2016.07.002

Abstract

Epithelial immunity protects the host from harmful microbial invaders but also controls the beneficial microbiota on epithelial surfaces. When this delicate balance between pathogen and symbiont is disturbed, clinical disease often occurs, such as in inflammatory bowel disease, cystic fibrosis, or atopic dermatitis, which all can be in part linked to impairment of barrier epithelia. Many innate immune receptors, signaling pathways, and effector molecules are evolutionarily conserved between human and *Drosophila*. This review describes the current knowledge on *Drosophila* as a model for human diseases, with a special focus on innate immune-related disorders of the gut, lung, and skin.

The discovery of antimicrobial peptides, the crucial role of Toll and Toll-like receptors, and the evolutionary conservation of signaling to the immune systems of both human and *Drosophila* are described in a historical perspective. Similarities and differences between human and *Drosophila* are discussed; current knowledge on receptors, signaling pathways, and effectors are reviewed, including antimicrobial peptides, reactive oxygen species, as well as autophagy. We also give examples of human diseases for which *Drosophila* appears to be a useful model. In addition, the limitations of the *Drosophila* model are mentioned. Finally, we propose areas for future research, which include using the *Drosophila* model for drug screening, as a validation tool for novel genetic mutations in humans and for exploratory research of microbiota–host interactions, with relevance for infection, wound healing, and cancer.

1. INTRODUCTION

The fruit fly *Drosophila melanogaster* has been used as a research model for over a century. *Drosophila* was one of the first multicellular organisms to have its genome sequenced and well annotated (Adams et al., 2000). Releasing the human genome sequence a few years later revealed that 75% of disease-related genes in human have functional orthologs in flies (Lloyd & Taylor, 2010; Reiter, Potocki, Chien, Gribskov, & Bier, 2001). This strengthened the role of *Drosophila* as a model to study biological processes with relation to human diseases.

Much of today's general knowledge about innate immunity has developed from research that was initially carried out in *Drosophila* and other insects. The basis for this successful approach of using an invertebrate model for studies of human immune responses is motivated by the similarity and evolutionary conservation of fundamental aspects of the underlying processes. This can be exemplified with the well-conserved signaling pathways that regulate innate immune responses, gut epithelium regeneration, and wound healing. Also, cellular immune responses, such as phagocytosis and autophagy, are evolutionarily conserved; bactericidal and fungal effector mechanisms, such as production of antimicrobial peptides (AMPs) and reactive oxygen species (ROS), are shared, as well as some antiviral responses (Lamiable & Imler, 2014). The fact that insects lack an adaptive immune system in the form it is present in vertebrates has simplified the dissection of innate immunity per se by genetic and molecular analyses. Furthermore, humans and flies both have a commensal microbial flora and can be infected partly by the same pathogenic bacteria, fungi, and viruses, and both are hosts for protozoa and nematode infections. This has not only enabled discoveries of many crucial components of innate immune responses against these pathogens but also disclosed *Drosophila* as a useful model for human diseases, where host–microbe interaction plays an important role, such as intestinal inflammation and tumorigenesis. In addition, the *Drosophila* model has been used for unraveling microbial pathogenesis and virulence mechanisms of various microbes in gut, lung, and skin. More recently, the fly is being used as a primary or complementary whole-animal target for chemical drug screening to discover novel antibiotics. For this, the large number of flies that easily can be tested in high-throughput screens make it a cost-effective and logistic choice, not the least for incorporating the replacement, reduction, and refinement (3R) principles of alternatives in vertebrate animal drug testing regimes.

The genetic and molecular tool box for *Drosophila* is excellent. In addition to large collections of well-characterized mutants, a plethora of genetically modified flies has been engineered, which enables detailed manipulation of gene activity both temporally and spatially. This provides great possibilities for functional analysis of genes and pathways that have been linked to a human disease but where the molecular mechanism is unknown. One of the major advantages with a genetic model such as *Drosophila* is that well-planned genetic screens almost always give unexpected and unbiased results, which in essence is a hallmark of new discoveries. While the human genome usually carries multiple gene copies for regulatory proteins, the fly genome typically contains single genes, meaning less redundancy and more straightforward genetic analyses of gene function in genetically modified flies. For studies of host–microbe interaction and immunity, whole animal responses can easily be followed in flies by a multitude of measures of tolerance, resistance, or fatal outcome of the interactions (Ayres & Schneider, 2012). In addition, it is simple to create germ-free flies, or flies with mono-association of specific commensal or pathogenic microbes.

For which biological questions or human immune-related diseases is the *Drosophila* model a less appropriate choice? First, for diseases where adaptive immunity is dominating in humans, the fly cannot give a complete picture and some questions may not even be realistic to address. This includes diseases with strong B-cell responses, such as rheumatoid arthritis and antibody-dependent diseases; T-cell responses, such as autoimmune diseases and multiple sclerosis, and immunological disorders of T- and NK-cells. That said, research in the fly has often provided completely unexpected insight into immune processes that initially have been considered to be vertebrate specific. Second, many viruses are highly host specific and cannot infect *Drosophila* without prior manipulation, while infection with insect viruses can be used for answering general questions on innate antiviral responses. Third, some human pathogens use the human body temperature of 37°C in combination with serum factors to trigger expression of virulence factors. Regular *Drosophila* husbandry uses temperatures of 18–29°C, while 37°C for longer periods is lethal. Thus, infection with such pathogens will only cause harmless interactions in the fly. A fourth point is that host–microbe interactions in the gastrointestinal tract of mammals involve obligate anaerobic bacteria, and those are not found in the fly gut.

In the first part of this review, we will describe the systemic innate immune response in *Drosophila* in a historical perspective, as the discoveries made in *Drosophila* were crucial for today's understanding of innate immunity in humans, and paved the way for the important characterization of Toll-like receptors (TLRs) in mammals. We will also highlight other conserved signaling pathways involved in different aspects of the immune response. The second part will focus on *Drosophila* as a disease model to study host–microbe interaction and innate immunity in epithelial tissues, with a focus on bacterial and fungal infections in gut, lung/trachea, and skin/epidermis. Finally, we will emphasize for which human diseases *Drosophila* already has been used or could be a good model to answer fundamental questions on disease mechanisms, with possible impact for prevention and treatment in humans in the future.

2. EVOLUTIONARY CONSERVATION OF INNATE IMMUNITY

2.1 The Innate Immune System of *Drosophila*

The immune system of *Drosophila* is multifaceted and involves many cellular and humoral processes, which show high similarity or direct evolutionary relationships with the ones observed in humans. While all cannot be covered here, we refer to broad general reviews (Buchon, Silverman, & Cherry, 2014; Lemaitre & Hoffmann, 2007) and to more specialized reviews describing responses to viral infections (Lamiable & Imler, 2014), cellular immunity including phagocytosis (Honti, Csordas, Kurucz, Markus, & Ando, 2014; Parsons & Foley, 2015); hemocyte development, with parallels to the two myeloid systems in vertebrates (Gold & Bruckner, 2014), and coagulation and clotting systems (Theopold, Krautz, & Dushay, 2014). In addition to these well-conserved immune system processes, *Drosophila* and other insects mount a strong melanization reaction upon wounding or infection that is not found in mammals (Tang, 2009); as well as encapsulation of large intruders, such as parasitic wasp eggs, which can be considered functional equivalents of vertebrate granulomas (Honti et al., 2014). Extracellular serine proteinase cascades are crucial in activating many of the immune processes in *Drosophila*, such as the Toll pathway, melanization reaction, and hemolymph clotting reaction, the latter with functional analogy to the activation of the human complement system (Loof, Schmidt, Herwald, & Theopold, 2011).

2.2 The Discovery of Antimicrobial Peptides

A milestone in the history of innate immunity was the pioneering discovery made by Boman, Nilsson, and Rasmuson (1972) of an inducible, humoral antibacterial defense system in *Drosophila* and other insects. Following purification, primary structure determination, and activity measurements, it became clear that insects synthesize several families of peptides and proteins, such as cecropins, attacins, and lysozyme, with bacteriostatic or lytic activities (Hultmark, Steiner, Rasmuson, & Boman, 1980; Steiner, Hultmark, Engstrom, Bennich, & Boman, 1981). The term "antibacterial peptides" was coined, referring to secreted peptides/proteins with direct effects on bacterial membranes, leading to lysis or growth inhibition. It was later changed to AMPs to include also peptides with antifungal activity. Within few years, seven different gene families encoding AMPs were isolated from *Drosophila*, and many also in other insects (reviewed in Hultmark, 1993; Imler, 2014).

Mammalian peptides with analogous functions called defensins and cathelicidins were subsequently isolated from rabbit macrophages and human neutrophils, bone marrow, and testis (Agerberth et al., 1995; Ganz et al., 1985; Selsted, Brown, DeLange, & Lehrer, 1983; Selsted, Harwig, Ganz, Schilling & Lehrer, 1985; reviewed in Ganz, 2003). The membrane-activity and killing mechanisms of some of the AMPs have been well characterized, while the function of others still is not completely understood (Shai, 1999). In addition, a large number of immunomodulatory peptides, called host defense peptides or innate defense regulators, which do not kill microbes directly but show different modulatory effects on both innate and adaptive immune responses have been identified in mammals (Scott et al., 2007). These peptides have attracted much attention as possible drugs for host-directed therapies, as reviewed in Mansour, Pena, and Hancock (2014).

2.3 The Role of *Drosophila* Toll and IMD Pathways in Innate Immunity

In the beginning of the 1990s, the genes encoding several insect AMPs were found to harbor κB-like DNA sequence elements in their upstream regions (Reichhart et al., 1992; Sun et al., 1991), and then shown to be required for AMP gene expression in vivo in response to microbial challenge (Engström et al., 1993; Kappler et al., 1993). The κB motif was a known target sequence for the mammalian nuclear factor kappaB (NF-κB) transcription factor in regulation of immunoglobulin gene expression in B-cells (Sen & Baltimore, 1986). Thus, this was one of the first indications of evolutionarily conserved mechanisms in regulation of innate immune responses between *Drosophila* and mammals. Also, parallels between the *Drosophila* Toll pathway and the mammalian IL-1 pathway were gradually becoming evident (Gay & Keith, 1991; Heguy et al., 1992; Schneider et al., 1991). At that time, only one *Drosophila* NF-κB-type transcription factor, called Dorsal, had been described in for its role in dorsoventral pattern formation in the *Drosophila* embryo (Anderson & Nusslein-Volhard, 1984; Steward, 1987). Despite Dorsal's capacity to activate AMP gene expression in reporter assays (Reichhart et al., 1993), *dorsal* mutants were still capable of producing AMPs in response to infection (Lemaitre, Meister, et al., 1995). Instead, another NF-κB-type transcription factor, named Dorsal-related immunity factor (Dif), was isolated and found to be a potent activator of many AMP genes

(Ip et al., 1993; Petersen et al., 1995). Later it was shown that Dif is the predominant transactivator upon antifungal infection and that *Dif* mutant flies are susceptible to fungal and Gram-positive bacterial infections (Rutschmann et al., 2000). Meanwhile, a third NF-κB-type transcription factor, called Relish (Dushay et al., 1996), was cloned and shown to be the predominant downstream activator of the IMD pathway, as described further later.

During embryo development, the Toll pathway regulates NF-κB/Dorsal activity. Thus, it was tested if Toll could be involved also in regulation of the NF-κB factors Dorsal and Dif during an immune response. Linking a constitutively active form of Toll with AMP gene expression in larvae (Ip et al., 1993) and in cell culture (Rosetto et al., 1995) in response to microbial elicitors attracted additional attention to Toll as a likely immunoregulatory factor. Final proof for the importance of Toll and the downstream pathway in immune response activation came when it was shown that flies with mutations in several Toll pathway components were killed by fungal infection (Lemaitre et al., 1996).

Consequently, a search for mammalian orthologs of Toll started, which led to the cloning of the first human TLR 1 year later (Medzhitov et al., 1997). Subsequently, five human TLRs were cloned (Rock et al., 1998), and the important immune function of the TLRs and the involvement in sensing microbial ligands were demonstrated in mice mutant for the *Tlr4* locus (Poltorak et al., 1998; Takeuchi et al., 1999). The number of known mammalian TLRs has now increased to 13; 10 of them (TLR1–10) are expressed in human and mice and the 3 remaining (TLR11–13) only in mice. The intracellular signaling cascade of the *Drosophila* Toll pathway and mammalian TLR pathways is evolutionarily conserved and was recently reviewed in Lindsay and Wasserman (2014).

Several lines of evidence were indicating that more than one signaling pathway is involved in regulation of AMP gene expression in *Drosophila*. The breakthrough came with the isolation of a mutant for the *immune deficiency* (*imd*) gene, which was affecting the expression of several AMP genes, but had little effect on expression of the antifungal peptide Drosomycin (Lemaitre, Kromer-Metzger, et al., 1995). In addition, it was shown that loss-of-function mutations in several Toll pathway genes still could mount expression of Diptericin in response to infection with Gram-negative bacteria (Lemaitre et al., 1996). Thus, it became clear that flies could discriminate between infection with different classes of microorganisms (Lemaitre et al., 1997). In addition, flies with mutations in both *Toll* and *imd* were sensitive to infections by most classes of microorganisms, and also became susceptible to nonpathogenic microbes (Gottar et al., 2002). Importantly, transgene expression of a single AMP was shown to be sufficient for rescue and survival of such *Toll/imd* pathway double mutants, confirming the crucial role of AMPs for *Drosophila* immunity (Tzou et al., 2002).

Following the identification of *imd* mutant flies, the efforts of many labs led to identification of numerous components and regulators of the so-called IMD signaling pathway. The IMD pathway in *Drosophila* is homologous to the mammalian tumor necrosis factor receptor (TNFR) pathway and many signaling components are conserved between *Drosophila* and human (recently reviewed by Kleino & Silverman, 2014; Myllymaki et al., 2014).

The main downstream activator of the IMD pathway, the NF-κB-type transcription factor Relish was isolated as an immune-inducible gene itself (Dushay et al., 1996) and shown to be crucial for humoral immunity, as *Relish* mutant flies were extremely sensitive to infection (Hedengren et al., 1999). Like the mammalian NF-κB transcription factors p100 and p105, Relish is localized in the cytoplasm in an inactive form with a C-terminal domain containing multiple copies of ankyrin repeats (Stöven et al., 2000). Activation of Relish involves phosphorylation and cleavage to release the N-terminal fragment (REL-68), which translocates to the nucleus for DNA binding and transcriptional activation (Erturk-Hasdemir et al., 2009; Stöven et al., 2000, 2003). Whole genome expression analysis later revealed that most genes regulated by the IMD pathway in *Drosophila* utilize Relish for activation (De Gregorio et al., 2002).

More recently, factors that act downstream of the IMD pathway by recruitment of chromatin remodeling complexes have been identified, such as Akirin (Goto et al., 2008). It acts as a Relish cofactor for a subset of its target genes by SWI/SNF-Brahma complex (Bonnay et al., 2014). The mouse homolog, Akirin2, has been shown to play a similar role in bridging NF-κB and SWI/SNF complexes during activation of both innate and adaptive immune responses, such as activation of proinflammatory gene expression in mouse macrophages (Tartey et al., 2015, 2014).

2.4 Pattern Recognition Receptors

The microbial elicitors and the pattern recognition receptors (PRRs) upstream of both the Toll and IMD pathway were unknown until the beginning of this millennium. Bacterial lipopolysaccharide (LPS) was considered as a potent activator of the *Drosophila* pathways in analogy with the mammalian TLR-4 signaling pathway. The isolation of peptidoglycan recognition proteins (PGRPs) (Kang et al., 1998; Yoshida et al., 1996) and Gram-negative binding proteins (GNBPs) (Lee et al., 1996) from different insects indicated, however, that peptidoglycan and β-glucans are the true microbial elicitors, which was subsequently experimentally confirmed (Kaneko et al., 2004; Mishima et al., 2009; Takahasi et al., 2009). The PGRP gene family in *Drosophila* consists of 13 genes encoding 19 PGRPs, including secreted, transmembrane, and intracellular variants (Werner et al., 2000). Many parallel studies demonstrated that different PGRPs and GNBPs act as specific PRRs upstream of either the Toll or IMD pathway (Choe et al., 2002; Gobert et al., 2003; Gottar et al., 2002; Michel et al., 2001; Ramet, Manfruelli, et al., 2002), thereby triggering responses to different classes of microbes, as reviewed in Aggrawal and Silverman (2007). Some PGRPs bind to peptidoglycan and act as true PRRs, while others have catalytical amidase activity and act as immune scavengers (Mellroth et al., 2003), as recently reviewed in Kurata (2014) and Royet (2011).

Importantly, the PGRP family is conserved from insects to mammals. In mammals, four PGRP genes have been characterized: PGLYRP1–4, and they have all been shown to be bactericidal. In general, mammalian PGLYRPs are expressed in barrier epithelia with direct contact with commensal or environmental bacteria and, therefore, seem to play a role in protecting the host from enhanced inflammation, tissue damage, and colitis (reviewed by Royet, 2011).

In spite of the similarities between the *Drosophila* Toll pathway and mammalian TLR signaling, important differences were noticed. All mammalian TLRs have been shown to act as direct PRRs for bacterial-, viral-, and parasitic-produced ligands; and also to some host cell products (Kawai & Akira, 2011). In contrast, *Drosophila* Toll is a cytokine receptor, which is activated by a cleaved form of the endogenous polypeptide Spätzle (Valanne et al., 2011; Veillard et al., 2016; Weber et al., 2003). Thus, the role of Spätzle as an immune-stimulating ligand seemed to be unique to insects and absent in vertebrates. However, a recent study indicated that nerve growth factor β (NGFβ), which is a cystine knot protein and a putative vertebrate ortholog of Spätzle, plays an important role in immunity to *Staphylococcus aureus* (Hepburn et al., 2014). NGFβ is released by macrophages in response to *S. aureus* via activation of NOD-like receptors (NLRs) and shown to stimulate a broad range of responses and activities in macrophages and neutrophils. In addition, mutations in human NGFβ or in its receptor TRKA, and knockdown of *trkA* in zebrafish, were associated with susceptibility to *S. aureus* infections. This suggests an evolutionarily conserved role of cystine knot proteins in innate immunity against pathogenic Gram-positive bacteria, such as *S. aureus*.

An intense research field involving mammalian PRRs focuses on the role of inflammasomes in a number of diseases related to inflammatory disorders (Guo et al., 2015). Inflammasomes consist of multimeric protein complexes, which serve as PRRs and include NLRs and absent in melanoma 2 (AIM)-like receptors (ALRs). Stimulation of these receptors leads to oligomerization of the relevant NLRs/ALRs, followed by activation of caspase-1, and subsequent generation of active forms of the proinflammatory cytokines IL-1β and IL-18 (Vanaja et al., 2015). Inflammasome activation will also trigger a specific form of cell death called pyroptosis. Thus, activation of inflammasomes by host-derived factors can both initiate and exaggerate inflammatory reactions. Therefore, it is not surprising that autoinflammatory and autoimmune diseases have been linked to activation of inflammasomes in humans, including neurodegenerative diseases and metabolic disorders. Although inflammatory-like reactions have been observed in *Drosophila* in response to sterile tissue damage and tumor growth (Krautz et al., 2014; Shaukat et al., 2015), direct fly homologs of the NLR/ALR components of the mammalian inflammasome have not been identified (Martinon et al., 2009). Thus, genetic and mechanistic studies of inflammasome activation and function are presently not feasible in the *Drosophila* model. However, continued studies of inflammatory reactions in the fly may lead to discovery of other shared components and pathways, which regulate responses to tissue damage and inflammation in both the fly and humans.

2.5 The Role of Other Evolutionarily Conserved Signaling Pathways in Immunity

In addition to Toll and IMD pathways, a number of other conserved signaling pathways play important roles in *Drosophila* immune response processes directly or indirectly. The *Drosophila* Jun N-terminal kinase (dJNK) pathway is homologous to mammalian tumor necrosis factor (TNF) pathway. The first indications of JNK playing a role in the immune defense in *Drosophila* came from activation of dJNK by microbial elicitors (Sluss et al., 1996). It was later reported that the *Drosophila* IMD pathway bifurcates into two branches, activating Rel and Jun target genes, respectively (Boutros et al., 2002), in a similar manner as the TNFR pathway in mammals (Dai et al., 2012). In relation to infection and immunity, the *Drosophila* JNK pathway thus plays a role in regulation of AMP gene expression (Delaney et al., 2006;

Kallio et al., 2005; Kim et al., 2005), melanization through crystal cell rupture (Bidla et al., 2007), and bacteria-induced stem cell activation in the gut epithelium (Buchon, Broderick, Chakrabarti, & Lemaitre, 2009).

The Janus kinase/signal transducer and activator of transcription (JAK/STAT) pathway is well conserved between human and *Drosophila* (Li & Watowich, 2014; Myllymaki & Ramet, 2014). The *Drosophila* pathway has only one JAK (*Hop*) and one STAT (*STAT92E*) and therefore confers less redundancy compared to mammals that have multiple JAKs and STATs. The *Drosophila* JAK/STAT pathway is activated by cytokine-like proteins Os/Upd, Upd2, and Upd3 that bind to the single receptor Domeless (*Dome*). The pathway is involved in many processes linked to immunity, especially cellular immunity such as hematopoiesis, encapsulation, and lymph gland responsiveness (Hanratty & Dearolf, 1993; Sorrentino et al., 2004). In response to bacterial injury, hemocytes secrete Upd3, which stimulates JAK/STAT signaling in fat body cells, leading to immune gene expression (Agaisse et al., 2003). In addition, the JAK/STAT pathway is involved in gut epithelial responses to infection and is required for bacteria-induced stem cell proliferation in the gut epithelium (Osman et al., 2012). Just as in mammals, the *Drosophila* JAK/STAT pathway also participates in the control of viral infection (Lamiable & Imler, 2014; Myllymaki et al., 2014) and in tumorigenesis (Amoyel et al., 2014). In a recent *Drosophila* study, methotrexate was discovered as a strong inhibitor of the JAK/STAT pathway and suggested as a novel treatment for myeloproliferative neoplasm in humans (Thomas et al., 2015).

In addition to the signaling systems mentioned earlier, a wealth of studies has shown involvement of other pathways in immunity both in humans and in *Drosophila*. In flies these include the Duox pathway (Bae et al., 2010), insulin pathway (Becker et al., 2010), the Wingless/Wnt pathway (Gordon et al., 2005), the Pvr pathway (Bond & Foley, 2009), the p38 pathway (Chen et al., 2010; Davis et al., 2008), and the Hippo pathway (Liu et al., 2016). Furthermore, there is growing evidence for positive and negative cross talk between these and the Toll and IMD pathways. Similarly, mammalian TLR and TNF pathways were shown to interact (Kawai & Akira, 2011).

3. INNATE IMMUNITY IN BARRIER EPITHELIA

3.1 Epithelia as Physical and Chemical Barriers

Surface epithelia constitute physical and chemical barriers that separate internal tissues and organs from the surrounding environment. The epithelial linings of the skin, lungs, gut, and genitalia are normally exposed to a very broad range of microorganisms, including commensals as well as potentially harmful microbes. Therefore, these barrier epithelia serve important functions in protecting the organism from invasion of other organisms and protection against toxic and harmful molecules.

Epithelial cells also create a chemical barrier by releasing AMPs, chemokines, and cytokines. Human α-defensins are expressed in polymorphonuclear leukocytes (PMNs), Paneth cells in the gut, and epithelial cells in the genital tract. Human β-defensins are widely expressed in epithelial cells in all mucosal tissues, including intestine, lung, and skin (Ganz, 2003). Theta-defensins are only expressed in nonhuman primates and the genes are truncated in humans (Cole et al., 2004). The expression of AMPs in humans is tightly regulated and can be constitutive and/or induced by cytokines or microbial compounds, and even downregulated by certain virulent bacteria, like *Shigella* spp. (Gudmundsson et al., 2010). The barrier epithelia of *Drosophila* larvae and flies maintain basic expression levels of AMPs. It was shown, using transgenic flies carrying fluorescent reporter genes, that each epithelial surface expresses several AMPs (Tzou et al., 2000). In addition, local infection triggers increased expression of these AMPs in barrier epithelia, as reviewed in Davis and Engstrom (2012).

Thus, it is likely that both human and fly epithelia produce cocktails of AMPs to protect against invasion by pathogenic microbes. An alternative function for AMPs in barrier epithelia would be that they shape the local microbial community and promote certain commensals to become predominant. Such selected microbial communities may then in fact serve as a first line of defense by competing with pathogenic microbes. A protective role of the microbiota resident in epithelial surfaces has in fact been demonstrated in both insects and humans. It was shown in *Drosophila* that germ-free larvae are more susceptible to infection by pathogenic fungi, such as *Candida albicans*, than in the presence of the normal microbial flora (Glittenberg et al., 2011). In humans, this phenomenon is best illustrated by *Clostridium difficile*-associated diarrhea, which often occurs after antibiotic treatment. Thus, it is clear that a healthy microbiota protects against pathogens by occupying a niche in the intestinal mucosa (Britton & Young, 2014).

The regulatory networks controlling tissue specificity in epithelia of both fly and human AMPs are relatively poorly described, in comparison to the well-studied pathways regulating responses to systemic infection. Improved knowledge of the cues that control endogenous AMP expression should enable the development of novel approaches to strengthen the epithelial barriers and to boost responses to infection. With the high degree of evolutionary conservation of transcription factors and signaling between *Drosophila* and mammals, it is likely that studies of effector mechanisms and their regulation in *Drosophila* barrier epithelia will continue to provide important knowledge to the benefit of understanding these regulatory networks in humans.

3.2 Impact and Relevance of Innate Epithelial Infections in Humans

3.2.1 Bacterial Infections and Immunity

Bacterial infections in humans are a common clinical problem in all disciplines, ranging from primary care to advanced surgery in university clinics. In particular, the emerging resistance against common antibiotics has become a real threat to many surgical procedures. Spread of *Staphylococci* resistant to methicillin (MRSA), Enterobacteriaceae with extended spectrum of β-lactamases (ESBL), and carbapenemase-producing Enterobacteriaceae (CPE) constitute real clinical challenges due to their resistance to first and second line treatments (Pitout & Laupland, 2008; Watson, 2011). Infections with these bacteria require treatment with expensive drugs, which are not accessible in all countries, and thus, the infections cannot be treated properly. In fact, bacterial strains being resistant against colistin, the last treatment resort, were recently discovered in China (Stoesser et al., 2016). Combined, this new situation requires novel approaches to prevent and treat infections with multidrug-resistant bacteria. One such approach would be to harness the power of the innate immune system and to use *Drosophila* to screen for novel compounds, or screen existing drugs for new purposes, which then rapidly can be incorporated in clinical treatment regimes. In addition, *Drosophila* could be utilized to study the virulence of multidrug-resistant bacteria, an area that just recently has been addressed.

3.2.2 Fungal Infections and Immunity

Fungal infections are of huge medical importance, but the knowledge about fungal immune responses is not as developed as the knowledge for bacterial and viral infections. Fungal infections in humans are becoming a significant problem due to rising numbers of immune-compromised individuals. Drug resistance is also increasing and there is a large need for novel treatments and prevention against severe fungal infections.

The skin and mucosal surfaces of humans are inhabited by commensal yeasts and fungi, such as *Candida* and *Malassezia* species, and the skin and mucosa is an entry point for invasive fungal diseases (Underhill & Pearlman, 2015). The lungs also constitute an important route of infection as they are exposed to airborne spores of common molds, such as *Aspergillus* and *Fusarium*. Immunosuppression and genetic immune-deficiencies severely increase the risk for developing chronic and/or invasive fungal infections. Although T-cell responses are necessary for full defense against fungi, the innate immune system plays an important role (Lionakis et al., 2011). Recognition of yeast and fungi in humans depends on lectins that recognize fungal β-glucans. A large number of receptors for lectins exist, such as the C-type lectin receptor (CLR) clusters Dectin-1 and Dectin-2. These promote phagocytosis and production of inflammatory cytokines. This and related topics on recognition and responses to fungal infections in humans have been well covered in recent reviews (Sancho & Reis e Sousa, 2012; Underhill & Pearlman, 2015).

Fungal immune responses in *Drosophila* are also based on the recognition of β-glucans as described earlier, and on sensing of fungal virulence factors (Gottar et al., 2006). Signaling via the Toll pathway to the Rel factor Dif promotes expression of antifungal peptides, such as drosomycin, metchnikowin, and cecropin (reviewed in Lindsay & Wasserman, 2014; Uvell & Engstrom, 2007) and the recently characterized bomanins (Clemmons et al., 2015; Uttenweiler-Joseph et al., 1998). Phagocytosis and encapsulation by hemocytes are also important for antifungal defense in *Drosophila* (Lemaitre & Hoffmann, 2007).

3.2.3 Human Microbial Pathogens and Virulence Mechanisms Studied in Drosophila

Wild-type strains of *Drosophila* are easy to grow in a time-efficient manner and have successfully been used for virulence tests of human pathogenic bacteria that cause systemic infections in flies, such as *Mycobacterium marinum* (Dionne et al., 2003), *Salmonella typhimurium* (Brandt et al., 2004), *Serratia marcescens* (Cronin et al., 2009), *Francisella tularensis* (Ahlund et al., 2010), and *S. aureus* (Wu et al., 2012). Similarly, a number of human fungal pathogens are lethal when injected into wild-type *Drosophila*, such as *C. albicans* (Davis et al., 2011; Glittenberg et al., 2011) and *Cryptococcus* (Thompson et al., 2014). However, several human fungal pathogens, such as *Candida glabrata* and *Aspergillus fumigatus*, do not cause lethal infections in fully immune-competent *Drosophila*. For these fungi, flies/larvae with mutations in the IMD or Toll pathways have been used as hosts, as reviewed in Panayidou et al. (2014).

So far, relatively small size drug screens for antibacterial or antifungal chemicals have been carried out in *Drosophila*, primarily identifying the response to a combination of a few known or new antibiotics (Ben-Ami et al., 2013; Lionakis & Kontoyiannis, 2005; Oh et al., 2013, 2014; Thompson et al., 2014). These studies serve, however, as good proof of principles, indicating the great potential in using *Drosophila* as a primary in vivo target for high-throughput screening efforts for novel pharmaceuticals, and for retesting drugs already approved for human use (Tzelepis et al., 2013).

4. EPITHELIAL IMMUNITY IN THE GASTROINTESTINAL TRACT OF HUMANS AND *DROSOPHILA*

The gastrointestinal systems of *Drosophila* and human share many similarities in structural and cellular architecture of the gut epithelium, and its barrier functions include gut epithelial immunity and host–microbe interactions. With the unique possibilities for genetic manipulation, *Drosophila* has become an important model for studies of the underlying mechanisms regulating gut development, epithelial regeneration and stem cell activity, metabolism, and immunity. This is likely to bring more light into many unsolved questions of human gastrointestinal diseases that are caused by disturbances in these processes, such as intestinal barrier function, Crohn's disease, and colon cancer (Frosali et al., 2015; Merga et al., 2014).

4.1 Similarities and Differences in Human and Fly Gut Structure and Immune Systems

Both the human and *Drosophila* digestive systems are highly compartmentalized tubular structures with different anatomical/morphological, transcriptomic, and functional immune specialization (Buchon, Osman, et al., 2013; Lemaitre & Miguel-Aliaga, 2013; Marianes & Spradling, 2013; Mowat & Agace, 2014). The *Drosophila* gut is structurally divided into the foregut, midgut, and hindgut (Fig. 1), and the midgut serves the same functions as the human stomach, small intestine, and colon in food digestion and nutrient absorption. The *Drosophila* midgut epithelium is a single cell layer with two differentiated cell types, absorptive enterocytes (ECs), and enteroendocrine (EE) cells, which are renewed from intestinal stem cells (ISCs) via a nondividing transient cell types called enteroblasts (EBs) and

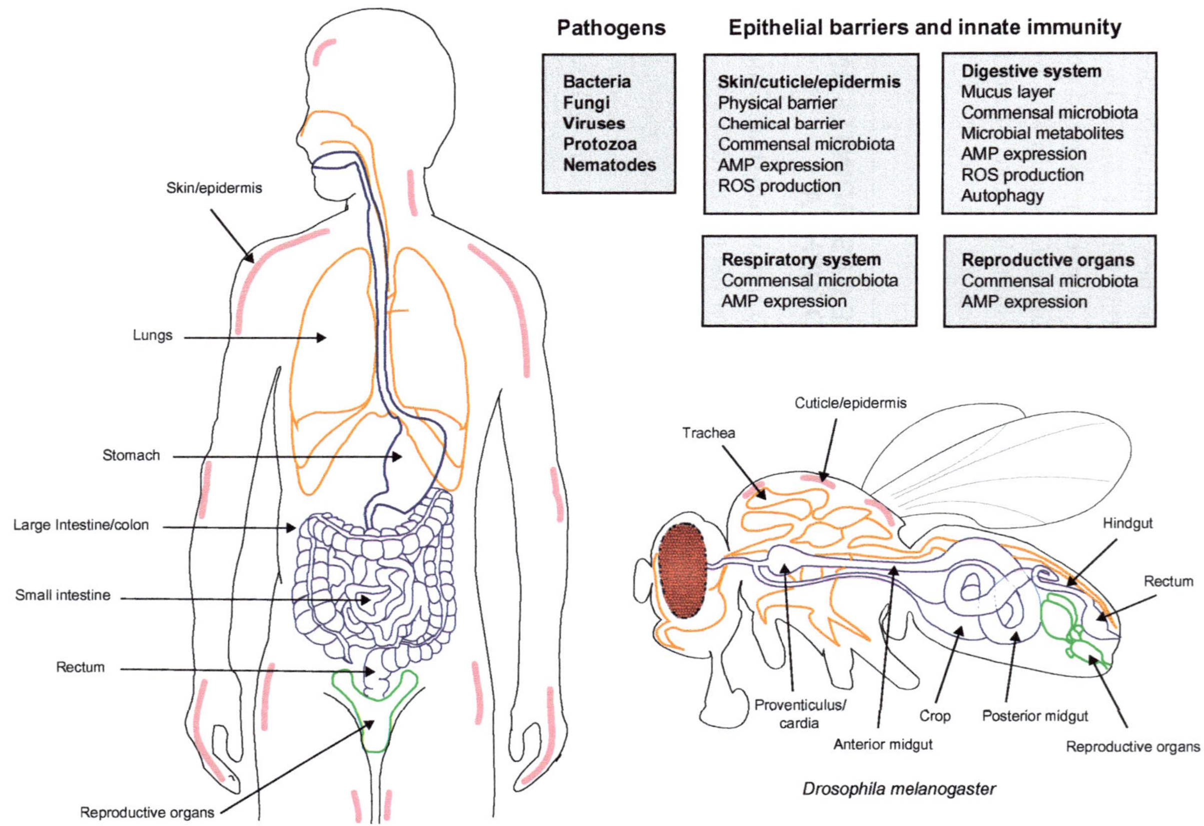

Fig. 1 See legend on opposite page.

pre-EE cells (Micchelli & Perrimon, 2006; Ohlstein & Spradling, 2006; recently reviewed in Li & Jasper, 2016). The epithelium of the human small intestine is more three-dimensional with finger-like villi and deep crypts, while the colon lacks the protruding villi and has a smooth epithelium (Mowat & Agace, 2014). The ISCs reside in the crypts and after cell division the transient cells move upward, further proliferate, and finally differentiate into absorptive ECs, goblet cells, or EE cells. The ISCs can also differentiate into Paneth cells, which reside at the bottom of the crypts and escape from

---

Fig. 1 Epithelial barriers and innate immunity. Overview of analogous organ systems in human and *Drosophila* that are exposed to common types of pathogens and share evolutionarily conserved defense reactions to prevent and fight such infections. In addition, all barrier epithelia harbor commensal bacteria that stimulate host immune competence and also protect the host by competing with more harmful microbes. The skin/epidermis of humans and cuticle/epidermis of insects serve as physical and chemical barriers that prevent infection. Insults that breach this barrier trigger AMP production, and in combination with other humoral and cellular reactions, promote local protection. The respiratory systems consist of tubular epithelial organs, which in flies directly transport oxygen throughout the body cavity, while the lungs in humans are connected to the vascular system. Nevertheless, the lungs and trachea share many immune defense reactions, such as constitutive and inducible expression of AMPs. The gastrointestinal system of humans and flies is functionally analogous in their digestive and excretory functions and shares a similar overall regionalized structure. While the human gut epithelium is covered by a thick protective mucus layer, the *Drosophila* foregut and hindgut are of ectodermal origin and their epithelia are covered by an impermeable cuticle. The fly midgut, which is analogous to the human stomach, small intestine, and colon, is covered by the peritrophic matrix and a thin mucus layer that together serve a similar function as the human mucus layer, to separate the cellular epithelium from bacteria and toxic compounds present in the gut lumen. The midgut of both human and fly is surrounded by visceral musculature, which is innervated and in the fly also supplied via fine tracheoles (Lemaitre & Miguel-Aliaga, 2013). The fly system also comprises the crop, which is a sack-like structure for food storage and detoxification. The intestinal epithelium of both human and flies consists of differentiated epithelial cells, the enterocytes (ECs), and enteroendocrine (EE) cells. The regeneration of the gut epithelium from intestinal stem cells that divide asymmetrically to form transient amplifying (TA) cells in human and analogous enteroblasts (EB) in flies, which then further differentiate into ECs and EEs, shows surprisingly high degree of evolutionary conservation, with homologous signaling pathways being involved. The excretory system of flies consists of the malphigian tubules, which are analogous to human renal organs/kidney, connected to the midgut/hindgut junction and of the ileum that regulates osmolarity by absorption of water and ions (Lemaitre & Miguel-Aliaga, 2013). The epithelia of the reproductive organs in both human and fly express AMPs but will not be further described in this review.

the upward migration (van der Flier & Clevers, 2009). Both the *Drosophila* and human intestine are surrounded by visceral musculature.

Human intestinal cells are covered by the glycocalyx (mucus), which is a thick and viscous fluid composed of negatively charged mucins. The mucus layer keeps the microbiota at a distance from the epithelial cells, and a deficient mucus layer leads to intestinal inflammation. The *Drosophila* gut epithelium is covered by a thin chitinous peritrophic matrix that serves the same function as the human glycocalyx, to separate the cellular epithelium from the contents of the gut lumen. The human gut differs from that of the fly in the presence of a lamina propria that contains cells from the adaptive immune system (Mowat & Agace, 2014). These immune cells play important roles in regulating intestinal immunity by producing cytokines and immunoglobulins of the IgA-type. Recently, a novel group of immune cells, the innate lymphoid cells (ILCs), have been intensively studied and found to coordinate many of the immune activities in the intestines of mice and humans (Mowat & Agace, 2014). Since *Drosophila* lacks adaptive immunity, these pathways are not possible to study in the fly system.

The enteric nervous system plays a key role for the physiologic response in the human intestine by releasing neurotransmitters in response to physiological stimuli, including bacterial metabolites (Kabouridis & Pachnis, 2015). Even though the fly enteric nervous system is different from the human counterpart, many aspects are actually conserved, which includes release of serotonin and neuropeptides (Kuraishi et al., 2015). It should therefore be possible to study nerve-immune cross talk in the *Drosophila* system with relevance for human physiology.

4.2 The Importance of the Gut Commensal Microbiota in Health and Disease

Humans contain rich and diverse microbial communities in their intestines, and our understanding of their importance in health and disease has increased vastly during the last decade. The commensal gut microbes are crucial partners in absorption of nutrients and as suppliers of essential nutrients, and their roles in shaping an organism's metabolic and immune status have become increasingly evident (Wu et al., 2015). In addition, we are just starting to understand their influence on development, and physiology, with direct effects on general health, aging, and lifetime expectancy (Sommer & Backhed, 2013). In fact, many of these processes can be mimicked in the *Drosophila* model (Buchon, Broderick, & Lemaitre, 2013). The human gut contains many orders of magnitude more bacteria than the fly gut, distributed over more than 500 taxa per individual. Interestingly, the bacterial composition in the *Drosophila* gut seems to be more diverse than the previously reported 5–30 taxa (Broderick & Lemaitre, 2012) and may in fact show more overlap with the human gut microbiome than anticipated (Dantoft et al., 2016). The human gut is dominated by Firmicutes and Bacteroidetes and also contains Proteobacteria, Actinobacteria, Fusobacteria, Verrucomicrobia, and Cyanobacteria (Lozupone et al., 2012), while the fly gut is dominated by Proteobacteria and Firmicutes and also contains Actinobacteria and Bacteroidetes (Broderick & Lemaitre, 2012). The ease with which *Drosophila* can be cultivated in germ-free conditions has stimulated a wealth of studies reporting on the impact of both pathogens as well as of commensals in regulating local and systemic immunity, gut epithelium regeneration, metabolism, physiology, and age-related tissue dysfunction. We will discuss some of these topics in the following sections, but refer to recent reviews for a comprehensive coverage of this intense research area in *Drosophila* (Buchon, Broderick, et al., 2013; Erkosar & Leulier, 2014; Lee & Lee, 2014; Li & Jasper, 2016).

4.2.1 Microbial Metabolites and Regulation of the Immune Responses

The role of microbial metabolites in regulation of human physiology is a large and very active area of research. For a detailed review about microbial metabolites and their role in human metabolism and immunity, see Donia and Fischbach (2015). Gut microbiota produce many important metabolites, including short-chain fatty acids (SCFAs) such as butyrate, acetate, and propionate. SCFAs are known to serve as nutrients for human colonic cells and also to suppress inflammation, proliferation, and the development of cancer in the human colon (Louis et al., 2014). In fact, reduction of butyrate-producing bacteria by antibiotics or in germ-free systems inevitably leads to more inflammation. The mechanisms for butyrate-mediated effects have partly been delineated and involve G-protein-coupled receptors on the surface of both epithelial and immune cells. Via binding to G-protein-coupled receptors, GPR43 and GPR109A, butyrate inhibits inflammation, reduces oxidative stress, and promotes mucosal defenses (Macia et al., 2015). In addition, butyrate induces AMP production in colonic epithelial cells (Schauber et al., 2003) and restores mucosal defenses during *Shigella* infections (Raqib et al., 2006).

The presence of *Lactobacillus plantarum* and *Acetobacter pomorum* in the *Drosophila* gut was shown to confer positive effects on metabolism and growth, by activating the insulin pathway (Shin et al., 2011; Storelli et al., 2011). Some of the effects could be restored by food supplementation with acetic acid, but additional, unknown metabolites are likely to also be important. When searching for virulence factors in *Vibrio cholera* in a *Drosophila* model, it was observed that the bacterial two-component system CrbRS played a major role on host organism symptoms and survival (Hang et al., 2014). The CrbRS regulates an acetate switch that activates acetate consumption, leading to downregulation of host insulin signaling and host lethality. Similar effects of SCFA on human metabolism and insulin sensitivity are likely to occur, but this is an area which not yet has been fully explored (Canfora et al., 2015).

4.3 Recognition of Microbes in Human and *Drosophila* Gut

As mentioned earlier, several of the key processes of microbial recognition, signaling, and responses are well conserved between human and fly. For example, both recognize microbes via PRRs and the different host receptor systems show conserved and nonconserved features. In human gut, such PRRs include TLRs at the cell membrane, the NLRs, CLRs, and RIG-I-like receptors inside the cell (Cao, 2015). The expression levels of these receptor systems vary along the gastrointestinal tract and are hardwired to specific response programs, including expression of cytokines, chemokines, and AMPs (Sperandio et al., 2015). Similarly, the large *Drosophila* family of different PGRPs, which include extracellular, membrane-bound, and intracellular members are expressed differently along the fly gut (Bosco-Drayon et al., 2012; Marianes & Spradling, 2013) and confer both activation and negative feedback regulation on expression of AMPs (Royet & Charroux, 2013; Royet et al., 2011).

An outstanding question in the field of intestinal immunity is how the host can differentiate between the innocuous normal flora and potentially pathogenic microbes. Several explanations have been proposed. First, in the human gut, an intact mucus layer keeps bacteria at a distance and pathogens may penetrate the mucus layer and cause inflammation and disease. Similarly, the *Drosophila* midgut is lined by a chitinous peritrophic matrix, which separates the bacteria within the gut lumen from the gut epithelium (Lemaitre & Miguel-Aliaga, 2013).

Second, TLRs have been exclusively found inside the epithelial cells of the mammalian gut (Hornef et al., 2003), which would create a tolerant extracellular environment for the microbiota and only respond to invading pathogenic bacteria. In the *Drosophila* gut, several layers of negative regulation of the IMD pathway fulfill a similar purpose of increasing the tolerance to the commensal flora (Buchon, Broderick, et al., 2013). For example, one of the predominant sensors of Gram-negative bacteria in the *Drosophila* midgut, PGRP-LE, is an intracellular PRR (just as mammalian gut TLRs) (Bosco-Drayon et al., 2012). Other PGRP family members, like PGRP-SCs and PGRP-LB, are amidases that degrade the immune elicitor peptidoglycan (Mellroth et al., 2003; Zaidman-Remy et al., 2006) so that under normal conditions the peptidoglycan concentration is kept low. The genes for negative regulators of the IMD pathway are activated upon infection and create a negative feedback loop (Buchon, Broderick, et al., 2013). The transcriptional repressors Caudal (homologous to human Cdx2/Cdx4) and Pdm1/Nub (homologous to human Oct1/Oct2) bind AMP gene promoters and prevent expression in different parts of the midgut in healthy conditions (Dantoft et al., 2013; Ryu et al., 2004). An important layer of recognition of pathogens vs commensals is prevalent in the *Drosophila* gut, where it was found that many pathogens release uracil (Lee et al., 2013). Uracil is a strong inducer of ROS secretion and other immune responses, as described further in Section 4.5.

Finally, it has been proposed that the normal situation in the human gut is dominated by immunoregulatory cells, sustaining an immunosuppressive environment and thus controlling inflammation. This hypothesis is supported by the fact that deletion of IL-10, an immunosuppressive cytokine, leads to spontaneous colitis. Also in humans with a mutation in IL-10, colitis is a common symptom (Glocker et al., 2009). Although experimental evidence for an immunosuppressive role of *Drosophila* hemocytes in gut immunity is missing so far, hemocytes are adhering to the gut epithelium and were found to stimulate phagocytosis, ISC activity, and to contribute to intestinal dysplasia in aging flies (Ayyaz et al., 2015; Zaidman-Remy et al., 2012). Thus, this indicates the possibility to use flies to study mechanisms of recruitment and adhesion of circulating immune cells to the gut epithelium, and of interactions that may influence gut pathologies in humans.

4.4 Innate Immune Responses in the Human and *Drosophila* Gut—Effector Molecules

In order to keep the microbiota in check, the intestinal epithelium is equipped with a plethora of responses downstream of microbial recognition. AMPs and proteins constitute families with similar bactericidal and bacteriostatic activities, as well as antifungal properties.

In humans, the Paneth cells of the small intestine produce the human AMPs defensin 5 and 6 (HD-5 and HD-6) with specific functions. HD-5 belongs to the α-defensin family and has broad antimicrobial activity against a range of bacteria. In particular, it is active against *Salmonella*. To test the functional importance of HD-5, a mouse model for overexpression of HD-5 was created. Notably, oral inoculation of these mice with *S. typhimurium*, which normally would have killed the mice, resulted in complete protection from disease. In contrast, a systemic challenge resulted in 100% mortality of both wild type and transgenic mice, clearly showing that HD5 protected the mucosa from *Salmonella* invasion. A follow-up study could also show that the normal flora of HD-5 expressing mice was fundamentally changed compared to wild-type mice. Thus, one single AMP has profound effects on mucosal immunity and determines the composition of the normal flora (Salzman et al., 2010). In contrast to HD-5, HD-6 has no antimicrobial activity, and its role in intestinal immunity has remained elusive. However, a recent study could show that HD-6 forms amyloid structures, which entangle bacteria and remove them from the mucosal wall, thereby preventing invasion without direct killing (Chu et al., 2012).

Human colonic epithelial cells produce the AMPs LL-37 and β-defensins. The role of LL-37 is illustrated by the fact that *Shigella* spp., a common human pathogen causing dysenteriae, downregulates LL-37 expression as a part of its invasion program (Islam et al., 2001; Sperandio et al., 2008). Notably, the SCFA butyrate can counteract this effect and upregulates LL-37 expression, which restores colonic immunity and improves symptoms in a rabbit model (Raqib et al., 2006), as well as in humans (Sayem et al., 2011). In addition to AMPs, colonic epithelial cells produce antimicrobial proteins, such as the lectin REGIIIgamma (Cash et al., 2006). Deletion of this protein from mouse intestine results in loss of bacterial-mucosal segregation and leads to mucosal inflammation caused by the microbiota (Loonen et al., 2014; Vaishnava et al., 2011).

In *Drosophila*, several AMP genes are constitutively expressed in different parts of the intestinal epithelium in a highly regionalized manner (Dutta et al., 2015; Marianes & Spradling, 2013; Tzou et al., 2000). Most AMPs are also strongly upregulated upon infection, primarily in an IMD pathway-dependent manner (Ryu et al., 2006). While the *Drosophila* Toll pathway is not active in the gut, the JAK/STAT pathway regulates some AMPs in response to epithelial damage (Buchon, Broderick, Poidevin, Pradervand, & Lemaitre, 2009). Flies mutant in the IMD pathway are more sensitive to oral infection by pathogenic bacteria, such as *Pseudomonas entomophila* (Liehl et al., 2006) and *S. marcescens* (Nehme et al., 2007), while wild-type flies are very resistant to the presence of bacteria in their food, except for dedicated insect pathogens that have evolved features to effectively evade the innate immune system.

4.5 Dual Roles for ROS in the Intestinal Epithelium

Another key response system in the gut is the production of ROS. Intestinal epithelial cells in both mice and flies produce ROS, and it has been suggested that during enteric infections, high levels of ROS act as bactericidal effectors, while during homeostatic conditions, the presence of commensals stimulates low levels of ROS that act as signaling molecules to promote ISC proliferation (Lambeth & Neish, 2014). ROS can be produced by two independent, but evolutionarily conserved pathways, one involving the NAPDH oxidase (NOX) and the other by dual oxidase (DUOX). The importance of ROS for an intact intestinal barrier in mammals is evident from studies in the mouse where NOX-deficient mice develop colitis in response to avirulent *Salmonella* (Felmy et al., 2013; Rodrigues-Sousa et al., 2014). Patients with chronic granulomatous disease (CGD) lack a functional NOX and suffer from frequent bacterial and fungal infections. In addition, CGD patients develop severe colitis, possibly due to lack of control of the normal microbiota, but the exact mechanism is still unknown (Broides et al., 2016; Leiding & Holland, 1993). Interestingly, it appears that impaired ROS production leads to deficient autophagy and an excess of IL1-beta, which presumably could drive inflammation and cause colitis (de Luca et al., 2014; van de Veerdonk & Dinarello, 2014). Most interestingly, and in correlation with the examples described earlier, Jones et al. (2015) reported that commensal *Lactobacillus* bacteria promoted NOX-dependent ROS production and subsequent ISC proliferation in both the *Drosophila* and murine gut, suggesting that NOX is important in regulating gut homeostasis in normal conditions, and that this requires the presence of commensal microbiota such as *Lactobacillus*. In contrast, activation of DUOX, which has been studied extensively in *Drosophila*, requires the presence of pathogens (Kim & Lee, 2014). In fact, DUOX acts in parallel with NF-κB-dependent AMP production in the *Drosophila* gut, and it was shown that either pathway protects against enteric infection and only when both pathways are impaired, the flies will succumb due to infection (Ryu et al., 2006). Both the ROS-producing enzyme activity of DUOX and its gene expression are activated by the presence of microorganisms (Ha, Lee, Park, et al., 2009; Ha, Lee, Seo, et al.,

2009). As mentioned earlier, this has been linked to pathogen-produced uracil, which was found to be a strong elicitor of immune responses in the *Drosophila* gut (Lee et al., 2013). The receptor for uracil is unknown, but is likely to be a G-protein-coupled receptor. The uracil-induced DUOX activation was recently shown to be modulated by Hedgehog signaling (Lee et al., 2015).

DUOX enzymes have so far not been studied extensively in mammals, but recent work suggests that DUOX2, which is expressed in gut epithelium and upregulated in patients with inflammatory bowel diseases, regulates interactions between the intestinal microbiota and the mucosa to maintain immune homeostasis in mice. Mucosal dysbiosis leads to increased expression of DUOX2, which might be a marker of perturbed mucosal homeostasis in patients with early-stage inflammatory bowel disease (Grasberger et al., 2015). Taken together, intact ROS production is crucial for an intact intestinal barrier, regulation of ISC proliferation and of pathogen-induced immune responses both in humans and in *Drosophila*.

4.6 Autophagy as an Effector Mechanism

Intracellular bacteria are killed and degraded by the autophagic system in both human and *Drosophila* intestinal cells. An intact autophagic system promotes *Drosophila* survival after *Listeria* infection (Yano et al., 2008) and is important for control of the *Drosophila* symbiont *Wolbachia* (Voronin et al., 2012). In humans, autophagy has been shown to be essential for mucosal protection against invasive *Salmonella* (Benjamin et al., 2013) and is also associated with Crohn's disease (Salem et al., 2015). Autophagy also acts as an antiviral response process, which is conserved between flies and mammals (Lamiable & Imler, 2014; Moy et al., 2014). The main autophagic pathways are well conserved between fly and human, and pharmacological modulation of autophagy has been analyzed in *Drosophila* models of neurodegenerative disease (Jaiswal et al., 2012).

The activation of autophagy can occur by starvation or via activation of the mTOR system in both the fly and human systems. In addition, activation of surface-associated PRRs as well as of intracellular recognition systems leads to increased autophagy in both species. In the fly, intracellular PGRP-LE activates autophagy and the mammalian counterparts NOD1/2 bind to ATG16L in human cells. Likewise, the transcription factors FoxO and TFEB (*Drosophila* homologue Mitf) increase transcription of several autophagy-related genes.

The *Drosophila* model has been very useful in defining key regulatory genes for autophagy activation, with importance for bacterial infection. One such study found the gene MORN2 to be essential for LC3-associated phagocytosis and to be conserved between fly and humans (Abnave et al., 2014). Another example is the gene CAP-D3, which was found to be important for the innate immune response in *Drosophila* by regulating the expression of AMP genes (Longworth et al., 2012). The same group set out to study the role of CAP-D3 in ulcerative colitis (UC) in humans. Interestingly, in colonic biopsies from UC patients, the level of this protein was significantly lower than in healthy controls. Moreover, CAP-D3 was found to regulate autophagy in human cells and decreased expression levels of CAP-D3 impaired clearance of *Salmonella*, suggesting a conserved role for this protein in intestinal immunity (Schuster et al., 2015). Finally, large genome-wide screens in Crohn's patients have revealed mutations in the pattern recognition receptor, NOD2 (Liu et al., 2015), and in the autophagy-related genes ATG16L1 and IRGM (Salem et al., 2015). Combined, these findings underscore the existence of defects in both recognition and autophagy effector mechanisms in Crohn's pathogenesis.

4.7 The Intestinal Barrier and Aging—Examples from Human and *Drosophila*

The intestinal barrier keeps microbial products away from the circulation. When this barrier fails in diseases such as gastrointestinal inflammation, HIV, or hepatitis, bacterial products leak from the intestine into the circulation, causing a "leaky gut syndrome." This process has been named "microbial translocation" and is considered to drive systemic inflammation and subsequent increase of cardiovascular disease and premature death.

There are several aspects of the leaky gut concept with relation to the *Drosophila* model system, and especially the links between aging and a leaky gut phenotype have been addressed in the fly. It has been shown that the intestinal barrier function correlates well with the expected life span of the fly. Markers of the aging fly include increased expression of AMPs in the intestine and impaired insulin-signaling pathways (Rera et al., 2012). Recently, these effects in the aging fly could be coupled to changes in the microbiota. In fact, the changes in the bacterial flora preceded and could predict the subsequent impairment of the intestinal barrier function (Clark et al., 2015). Aging has also been shown to lead to chronic activation of FoxO in the *Drosophila* intestine, which caused a reduced expression of PGRP-SC2, a negative regulator of IMD/Relish innate immune signaling.

This caused commensal dysbiosis, stem cell hyperproliferation, epithelial dysplasia, and reduced life span of flies (Guo et al., 2014). Also in humans, the intestinal barrier appears to be reduced during aging (Mabbott, 2015), but the link to the microbiota is less established than in the *Drosophila* system. Based on the many overlapping aspects of intestinal barrier function and aging between humans and *Drosophila*, the *Drosophila* model has been proposed to serve as a platform for further studies in this field (Jasper, 2015).

4.8 Gut Regeneration and Microbiota Interactions in Inflammation and Cancer

The cells of the gut epithelium of both human and flies are short lived and replaced constantly. The rate of shedding of old/dead epithelial cells has to be kept in balance with the renewal from a pool of long-lived ISCs. The balance is especially critical during damage or infection, when the acute need for cell replenishment leads to an increase in ISC proliferation and differentiation, with the risk of overproliferation unless it is well controlled. This regenerative homeostasis is controlled by a large number of signaling pathways that are evolutionarily conserved between human and *Drosophila*, such as JAK/STAT, RTK/Ras/MAPK, Hippo, JNK, Notch/Delta, wnt/wg, BMP, and insulin-signaling pathways (for a comprehensive review, see Jiang & Edgar, 2012). Mutations in components of some of these pathways lead to hyperproliferation both in flies and in mice, while others cause premature differentiation and loss of the ISC pool. Many also act as tumor suppressor pathways in humans and have been linked to the development of colorectal cancer (CRC). This process involves transformation of healthy epithelial cells into premalignant adenomas and sometimes malignant cancer. It is well accepted that recurrent damage caused by chronic inflammation, microbial dysbiosis, or presence of certain bacteria, such as *Helicobacter pylori* and—more recently—*Fusobacterium nucleatum*, is a risk factor for development of premalignant conditions in the stomach and colon, respectively (Gur et al., 2015; Lee et al., 2016).

When the *Drosophila* gut epithelium is damaged by wounding or pathogenic infection, the JNK pathway is activated in the stressed, dying ECs, which then secrete IL6-like cytokines, Upds, which subsequently activate JAK/STAT signaling in neighboring ISCs. Together with activation of receptor tyrosine kinase (RTK) signaling by ligands secreted from the visceral musculature, this leads to proliferation of ISCs. The Hippo signaling pathway, a central regulator of organ size in flies and man, is also activated in ECs and ISCs, leading to autocrine cytokine signaling, which further stimulates proliferation. Later in the process, the same Hippo pathway acts negatively to turn off cytokine production and to block the activity of the Yorkie transcription factor so that the system can return to homeostasis (Jiang & Edgar, 2012). Aberrant regulation of the Hippo pathway, as well as of the pathways feeding into it, is prone to uncontrolled hyper-proliferation and tumor-like phenotypes, particularly when combined with microbial dysbiosis, such as in the aged *Drosophila* gut. In fact, deregulation of Hippo signaling has been linked to many human tumors including CRC development, underscoring the importance of the discoveries made in *Drosophila*.

The microbiota is also an important factor in CRC and increasing evidence point out a protective role of bacteria producing SCFAs, including butyrate, propionate, and acetate. The mechanism whereby SCFAs protect against CRC is not fully elucidated but probably involves inhibition of inflammatory cells, reducing oxidative stress and promoting a healthy microbiota. Reciprocally, it is clear that some bacterial species produce metabolites with direct or indirect toxic effects on the epithelial cell, possibly also involving DNA damage, with direct consequences for malignant transformation. Such compounds include secondary bile acids, ROS, and *N*-nitrosamines (for a detailed review on the topic, see Louis et al., 2014).

Given the far-reaching similarities between human and fly with regard to microbiota-mediated effects on epithelial cell proliferation and differentiation, with links to adenoma and CRC development in humans, we suggest that this is an area where *Drosophila* should continue to provide fundamental insight into common processes with implications for human disease.

5. *DROSOPHILA* AS A MODEL FOR HUMAN RESPIRATORY ORGAN DISEASES LINKED TO INFECTION AND INFLAMMATION

Respiratory tract infections (RTIs) are very common in clinical practice and lead to significant morbidity and mortality worldwide. RTIs can further be classified according to viral and bacterial causes. Viral RTIs include the most common RTI viruses, i.e., influenza (adults and children) and RSV (children) but will not be further discussed here. Bacterial RTIs comprise pneumococcal pneumonia, which is the most common single etiology to community-acquired pneumonia. In contrast, hospital-associated pneumonia is mostly caused by other bacterial species, including the opportunistic pathogens *S. aureus*, *Klebsiella pneumoniae*, and *P. aeruginosa*, which often are multidrug resistant and difficult to treat. The number of patients with immunosuppression is increasing in modern medicine, due to more advanced surgery, organ transplantation, and use of antiinflammatory drugs, blocking important immune pathways. The result is that the bacteriological cause to severe RTI in patients is increasingly diverse. Moreover, the emerging epidemic of multidrug-resistant bacteria requires a deeper understanding of host–microbial interactions, processes that could be further studied in *Drosophila* as a model.

Rodents are the most commonly used animal models for lung diseases such as pneumonia, asthma, and lung cancer. Although the rodent models mimic human lungs well in terms of chemical and physical conditions and also provide the possibility to carry out in vivo lung infections and studies of other lung pathologies, the results cannot be extrapolated directly from these models to humans. In addition, experimental lung infections in mammals are controversial from an ethical standpoint, and development of complementary models is desirable. The *Drosophila* model is one of the most interesting ones, as it has an airway system that can be regarded as a lung equivalent. As described earlier, *Drosophila* is a cost-effective model that can be infected by human pathogens and screened in different genetic background to pinpoint important host factors both for immunity and for pathogen virulence, and used in drug screens for chemical compounds that can inhibit disease progression.

On a superficial level the respiratory organs of flies and humans may seem very different. However, there are in fact far-reaching similarities in the development, physiology, function, and in responses to microbes between insect trachea and the lungs of mammals. Although these are not homologous organs, both airway systems consist of epithelial tubular organs that supply the whole organism with oxygen. The insect trachea is, just as the human lungs, a gas-filled branched tubular organ consisting of primary, secondary, and terminal branches. The exchange surface area increases with branching; hence, most gas exchange occurs in the distal parts. The organogenesis of these branched tubular networks has been found to share many fundamental principles between the fly and mammalian airway systems, especially in genetic components and the signaling pathways that control their branching (Horowitz & Simons, 2008; Samakovlis et al., 1996). In this aspect, the *Drosophila* embryo has been an excellent model for manipulating gene activities and subsequent analyses of the consequences for development and maturation of the trachea. However, it has been much less utilized for functional assays of airway performance in the larval stages or in response to environmental factors. In contrast, the prenatal and early postnatal development of mammalian lungs, as well as exposure to environmental risk factors, have got much attention in recent years in being predictive for the development of chronic lung diseases later in life (Krauss-Etschmann et al., 2013). Further development of the fly model, with more physiological readouts of airway function, such as oxygen consumption, would improve its usefulness.

5.1 Human Lung Responses to Infection

The human lung depends on epithelial cells for keeping the barrier intact against the external environment. Just as other epithelia, lung epithelial cells provide both physical and chemical protection against microbes. Airway epithelia express most of the known PRRs including TLRs, NLRs, and CLRs, thus enabling recognition of bacteria, virus, and fungi. In analogy with the gut, the human lung contains a microbiome, which differs significantly between healthy individuals and those with an inflammatory lung disorder, such as asthma and cystic fibrosis (CF). It is clear that this component of the human lung has to be taken into account for a full understanding of any disease process affecting the lung. Most of the effector systems present in other epithelia are also present in the lung. For example, mucins, which constitute an important part of mucociliary clearance, are important for containing and removing pathogenic bacteria. Defect mucin production has been shown in CF, for example, and a lack of mucus transport, like in the cilia-deficient Kartagener's syndrome, is associated with chronic bacterial RTIs in these patients, thus lending support for a key role of mucins in lung immunity. In addition, airway epithelia produce AMPs of both the cathelicidin and defensin families during infections. The importance of LL-37 in lung immunity has been shown by using a mouse knockout model (Kovach et al., 2012). It is, however, important to remember that an important source of LL-37 in the lung is from incoming neutrophils. Neutrophils also deliver human α-defensins to the site of infection. In contrast, human β-defensins are exclusively produced by epithelial cells. Finally, the ROS-based effector system is necessary for an intact lung defense against microbes, as shown in the CF-lung where ROS levels are decreased (Hiemstra et al., 2015).

5.2 *Drosophila* Tracheal Responses to Infection

The lumen of *Drosophila* trachea is covered by a cuticular lining that serves as a physical barrier against dehydration, and also against microbes that may enter through the tracheal openings. When *Drosophila* larvae are forced to crawl in food infected with pathogenic bacteria or fungus, the immune-competent tracheal epithelial cells respond by expressing AMPs and other immune and stress response genes (Ferrandon et al., 1998; Tzou et al., 2000; Wagner et al., 2008). The IMD pathway is the major immuno-responsive pathway activated in trachea upon bacterial infection. It requires PGRP-LC or PGRP-LE, while PGRP-LF seems to act as a negative regulator (Maillet et al., 2008; Persson et al., 2007; Takehana et al., 2004). The subsequent transcriptional activation by NF-κB/Relish in *Drosophila* trachea elicits expression of a smaller set of immune-induced genes compared to other immunoresponsive tissues (Gendrin et al., 2013). This may be due to the presence of negative regulation, as suggested for different regions of the gut epithelium, or due to the requirement of trachea-specific positive regulators, or a combination of these.

The activation of the IMD pathway was found to not be strictly cell autonomous in tracheal epithelial cells as it could spread to neighboring cells (Akhouayri et al., 2011; Takehana et al., 2004). This nonautonomous spreading was enhanced in mutants of Toll-8/Tollo, its putative ligand Spätzle-2, and intracellular mediator ECT-4 (a TIR domain protein homologous to mammalian SARM), indicating a role of this Toll-8/Tollo pathway in negative regulation. In a study by Wagner et al. (2009), it was reported that prolonged infection of the *Drosophila* tracheal system initiates remodeling processes of epithelial structure, primarily as thickening of the epithelial cell layer. Microarray analysis indicated changes in expression of genes involved in tracheal development and cell cycle progression, and of genes known to modulate Hedgehog-, JNK-, JAK/STAT-, MAP/ERK kinase-, and Ecdysone-dependent signaling (Wagner et al., 2009). This is highly interesting in the light of the chronic inflammatory diseases of the human lung. However, more in-depth mechanistic studies will be required to understand the nature of the observed remodeling of the airway epithelium, and its usefulness as a model for human diseases that lead to lung epithelium remodeling and metaplasia.

5.3 *Drosophila* as a Model of Specific Lung Infections and Diseases

5.3.1 Asthma and Chronic Obstructive Pulmonary Diseases (COPD)

Asthma and chronic obstructive pulmonary disease (COPD) are the most prevalent chronic inflammatory diseases of human lungs. They share that structural alterations in the lung tissue lead to variable impairment of airflow.

In addition, a varying degree of inflammation is present, leading to a vicious circle of viral and bacterial infections followed by more inflammation and airway remodeling. The first line therapy is inhaled corticosteroids, which significantly improve the clinical condition, but also further suppress local immune responses, and thus may contribute to prolonged infectious susceptibility.

The cause of asthma is not known but the prevailing idea is that a genetic susceptibility interacts with environmental factors. A number of genes have been shown to be associated with asthma, including IL-33, PCHD1, and orosomucoid 1-like 3 (ORMDL3). Interestingly, these genes are expressed in lung epithelial cells, which suggest that the innate part of the immune system is more important than previously thought. The immunological profile in asthma is dominated by a strong Th2 dominance and release of IL-4 and IL-13. These cytokines impair the epithelial barrier, downregulate AMP expression, and provide a niche for respiratory viruses.

Although asthma previously has been regarded as a disease with strong links to adaptive immune responses, recent findings suggest that innate immune signaling within airway epithelial cells plays a primary role. In both asthma and COPD, NF-κB signaling is a central player in inflammatory gene expression, regulating cytokine activity, and airway pathology (recently reviewed in Schuliga, 2015). Asthma susceptibility genes have been identified by genome-wide association studies and good models are needed to clarify the roles of these genes in normal and diseased airway tissues. *Drosophila* has been suggested as a favorable model for elucidation of the physiological and pathophysiological significance of asthma susceptibility genes (Roeder et al., 2012). A recent report addressed the role of one of these predicted human asthma susceptibility genes in *Drosophila* (Kallsen et al., 2015). Polymorphisms in the human gene for an endoplasmic reticulum transmembrane protein, *ORMDL3*, have been highly associated with childhood asthma (Moffatt et al., 2007). *ORMDL3* was first studied in a mouse model, and its overexpression led to increased airway remodeling and airway responses typical of asthma (Miller et al., 2015). The *Drosophila* study corroborates these results and serves as an example to how the functional role of human asthma-linked candidate genes can be tackled in the fly model (Kallsen et al., 2015).

5.3.2 Hypercapnia

Hypercapnia is a condition of elevated blood and tissue concentrations of CO_2, which is common in patients with severe COPD. It has also been linked to exacerbations of bacterial and viral infections in patients with other lung diseases, such as pneumonia, adenoviral lung infections, and CF. It is known that hypercapnia blocks NF-κB activation and normal expression of a number of immunoregulatory factors, and that it suppresses phagocytosis, ROS activation, and autophagy. The mechanisms underlying the immunosuppressive conditions of hypercapnia have been highly obscure. Recent work in *Drosophila*, in which many of the immune suppressive characteristics of hypercapnia are conserved, led to the identification of the zink finger homeodomain 2 (Zfh2) as a mediator of the hypercapnic immune suppression (Helenius, Haake, et al., 2016). The mammalian orthologs of Zfh2 are ZFHX3/ATBF1 and ZFHX4. By using a genome-wide RNAi screen in *Drosophila* S2 cells, followed by functional assays in vivo, it was shown that mutation in *zfh2* enable flies to mount a stronger immune response and survive infection better after exposure to hypercapnia. Thus, Zfh2 suppresses immune responses after CO_2 exposure, but not in normal air conditions. In a follow-up chemical drug screen with a CO_2-responsive luciferase reporter in *Drosophila* S2 cells, the same group identified a plant alkaloid, evoxine, as an inhibitor of some of the hypercapnia-induced immune defects (Helenius, Nair, et al., 2016). Most importantly, evoxine did rescue immune response capacity not only in *Drosophila* cells but also in human THP-1 macrophages. This indicates a strong evolutionary conservation of the pathway(s) regulating hypercapnia-induced immune suppression and also demonstrates that pharmacological drug screening for CO_2 effects can be addressed in *Drosophila* cells.

5.3.3 Cystic Fibrosis

Cystic fibrosis (CF) patients are frequently infected with *P. aeruginosa*. Chronic infections are linked to biofilm formation, and there is a need for simple infection models in which biofilm formation and its consequences can be followed. *P. aeruginosa* was shown already in 1972 to be a virulent pathogen of *Drosophila* (Boman et al., 1972), while less virulent *P. aeruginosa* mutants have been characterized subsequently using this host (D'Argenio et al., 2001). Mulcahy et al. (2011) developed a *Drosophila* in vivo model of *P. aeruginosa* biofilm formation and could show that biofilm infections were less virulent than nonbiofilm infections. The *Burkholderia cepacia* complex is a group of related bacterial species that are especially problematic for CF patients. *Drosophila* has been used in several studies as a host to study the virulence of different *B. cepacia* strains and mutants, and to isolate specific virulence factors (Castonguay-Vanier et al., 2010; Schwager et al., 2013).

5.3.4 Tuberculosis

Tuberculosis is a serious human lung infectious disease, caused by *M. tuberculosis*. The mouse has been the dominating infection model used; however, mice are not a natural host for *M. tuberculosis* and disease progression including latency and reactivation has not been possible to study in this model. Instead, lethal infections of zebrafish and *Drosophila* with *M. marinum* have emerged as powerful models not only for acute infection stages but also for disease progression and immunopathology (Dionne et al., 2003, 2006). *Drosophila* has also been developed as a suitable host for testing new drugs against serious *M. abscessus* infections (Oh et al., 2013).

5.3.5 Fungal Lung Infections

The lungs are prone to fungal infections as they are exposed to airborne spores of common molds such as *Aspergillus* and *Fusarium*, which then can disseminate and lead to invasive aspergillosis in immune-compromised humans and also in flies (Lemaitre et al., 1996). As described earlier, *Drosophila* has been used to study virulence of a number of human fungal pathogens. The fly model has further been used for combinatorial drug tests and revealed synergistic effects of, for example, voriconazole and terbinafine against *Aspergillus* infection (Lionakis & Kontoyiannis, 2005).

5.4 The Role of Intestinal Microbiota in Lung Diseases

An increasing literature describes how the commensal gut microbiota affects lung immune responses in mammals. It has been reported that the microbiota composition of the gastrointestinal tract can affect allergy and asthma development, immune responses to lung infectious diseases, and trigger systemic inflammatory responses (reviewed in Samuelson et al., 2015).

When *Drosophila* larvae creep in the food and contaminate it with its excrements, gut microbiota will come in direct contact with the tracheal openings, the spiracles. It is likely that the composition of the gut microbiota will affect tracheal immune responses to pathogens. Although this seems as an interesting model to study the direct effects of gut microbiota on tracheal immunity and airway functions, as well as the response to other environmental factors, the possibility of using *Drosophila* has not yet been evaluated.

6. *DROSOPHILA* AS A MODEL OF HUMAN SKIN INFECTIONS AND WOUND HEALING

Although there are important structural differences between human and *Drosophila* skin (Harden, 2005), *Drosophila* can serve as a good model for skin development, barrier immunity, and wound healing, and as a screening tool for novel therapeutic targets and drug discovery (Munoz-Soriano et al., 2014).

The physical structure and relatively impermeable nature of the outer layers of skin and epidermis, like hair and nails in humans and the cuticle of insects, serve as efficient protection against many types of physical and chemical types of stress. However, the underlying skin and epidermis must be flexible and allow for exchange of gases, fluids, and molecules, which makes them vulnerable to insult. In addition, some microorganisms have evolved ways to breach the physical barrier. To further protect the underlying tissues, barrier epithelia are also equipped with chemical and immunological barriers, which creates unfavorable conditions for microorganisms, such as high salt concentration, low pH, production of lipid-rich sebum, ROS, and AMPs.

6.1 Expression and Regulation of AMPs in Skin/Epidermis

The most prominent innate immune effector molecules in the skin/epidermis of both humans and flies are AMPs. There is constitutive expression of some AMPs in the absence of microbial stimuli, while expression of other AMPs requires the presence of microbial products.

The dominating families of AMPs in human skin are the cathelicidins and β-defensins. Keratinocytes are the primary cells in the skin to produce AMPs under normal conditions, but resident mast cells also contribute, and upon infection AMP-producing neutrophils are recruited (Gallo & Hooper, 2012). In *Drosophila*, relatively few studies have addressed expression of AMPs in the epidermis. It was shown using reporter assays that the gene for *CecA1* is activated in infected wounds in larvae (Önfelt Tingvall et al., 2001), and that expression of *CecA1* and Diptericin (*Dipt*) can be induced by bacteria-derived molecules in the epidermis of embryos (Esfahani & Engstrom, 2010; Tingvall et al., 2001). This epidermal expression was dependent on the IMD pathway and on the downstream NF-κB transcription factor Relish.

The human cathelicidin gene is upregulated in skin in response to injury and infection, by the vitamin D receptor (VDR) and its ligand 1,25 dihydroxyvitamin D3 (Gombart et al., 2005; Liu et al., 2006; Wang et al., 2004). Interestingly, *Drosophila* AMP expression is also regulated by nuclear hormone receptors, such as the ecdysone receptor (EcR; Rus et al., 2013). The ligand 20-hydroxyecdysone (20E) has been shown to be involved in pathogen-induced AMP expression in flies and in cell lines (Dimarcq et al., 1997; Meister & Richards, 1996), but a role in epidermal AMP gene expression has so far not been reported. In contrast to the regulation of human immune genes by the VDR, which directly targets the AMP gene regulatory regions, *Drosophila* EcR regulation seems to be indirect, via regulation of the pattern recognition receptor PGRP-LC and of other IMD pathway components (Rus et al., 2013). However, the role of VDR in human innate immunity also plays many other roles, both direct and indirect. The roles of nuclear hormone receptor signaling in innate immunity are likely much larger than our knowledge of today. The use of *Drosophila* should enable systematic analysis of individual hormone receptors and their functions.

6.2 Skin Microbiota

The skin of humans and the cuticle of insects are habitats of huge and diverse populations of microbiota. Many of these are probably just transient "guests," but many species can be recognized as human skin commensals. The composition of resident microbes in the skin may play important roles in both causing and preventing noninfectious skin diseases, such as psoriasis, atopic dermatitis, rosacea, and acne. As in other epithelia, commensal and symbiotic bacteria can serve as beneficial constituents by directly competing with and protect against growth of more pathogenic species. They also stimulate innate and adaptive immunity in the host, thereby strengthening both barrier functions and responses that prevent infections. However, genetic predisposition, injuries, and other causes of altered barrier integrity may promote pathogenic growth of normally nonpathogenic species or drive skin microbiota to initiate or amplify human skin disorders (Belkaid & Segre, 2014).

Analysis of bacterial composition of human skin has revealed four dominating phyla: *Actinobacteria*, *Firmicutes*, *Bacteroidetes*, and *Proteobacteria* (Grice et al., 2009). These phyla also dominate the human inner mucosal surfaces, but the relative proportions differ considerably. Depending on local differences in skin physiology (moist, dry, or sebaceous), certain species are dominating in different areas, such as *Propionibacterium* spp. at sebaceous sites, *Staphylococcus* spp. and *Corynebacterium* spp. in moist areas, while *Malassezia* fungal species dominate dry areas of the body (Findley et al., 2013).

Analysis of the *Drosophila* microbiota has to a large extent focused on the analysis of the gut (Broderick & Lemaitre, 2012). *Drosophila* laboratory strains were also found to carry several bacterial species in their guts that can be considered as human skin commensals, but that also can cause severe skin infections, such as *Staphylococcus* spp., *Streptococcus* spp., *Corynebacterium* spp., and *Neisseria* spp. (Dantoft et al., 2016). It suggests that *Drosophila* may harbor a gut flora that includes a broad range of human commensals and opportunistic species. Although this may primarily reflect the close physical interaction between humans and flies in laboratory settings, it also suggests that a broader range of host–microbe interaction studies may be conducted in flies than previously anticipated.

6.3 Wound Healing and Immunity

Wound healing and related problems are very common in medical clinics. Much research is focused on human cell cultures and mouse models. But tissue damage is a process that involves the entire organism, and whole-animal models are needed for a comprehensive understanding. Wound healing, tissue repair, and regeneration are intimately coupled to activation of immune responses. This will prevent infection and fight invading pathogens, but is also involved in local and systemic signals that induce tissue repair or replacement. *Drosophila* is a good model for many aspects of wound repair including local immune responses that have been shown to be evolutionarily conserved.

An important part of the wound healing process is the reepithelialization and recreation of barrier functions. This process differs in an important aspect between humans and *Drosophila*, as human epidermis contains stem cells that proliferate and migrate to the wound site to heal the wounds, while *Drosophila* epidermis do not have proliferating cells (Harden, 2005) and reepithelialization has to occur by other mechanisms. In the *Drosophila* embryo, an actin–myosin cable closes the hole like a purse string, and the actin cytoskeleton is also important in larval wound healing (Razzell et al., 2011). In *Drosophila* adult skin, diploid epithelial cells undergo polyploidization and cell fusion to create large cells that can grow in size, spread, and heal the wound (Losick et al., 2013). Although the mechanisms that replace the lost cell mass is different, considerable conservation in activation of signals and cellular activities during wound healing has been demonstrated between *Drosophila* and humans, as reviewed in Davis and Engstrom (2012), Lee and Miura (2014), Munoz-Soriano et al. (2014), Razzell et al. (2011), and Stramer and Dionne (2014), and only a few examples will be given here. In both *Drosophila* and mouse, transcription factors of the Grainy head (GRH) family have been shown to activate genes involved in cross-linking processes and in scab formation at the wound site (Harden, 2005) as well as in cuticle formation in flies and corneum stratum formation in humans (Mace et al., 2005; Ting et al., 2005; Wang & Samakovlis, 2012). The c-Jun N-terminal kinase (JNK) cascade is activated at the wound site and is necessary for tissue repair in both flies and human (Angel et al., 2001; Ramet, Lanot, et al., 2002). The damaged cells produce ROS locally, such as the release of hydrogen peroxide by calcium flashes and DUOX activation, which subsequently triggers recruitment of inflammatory cells both in zebra fish and in flies (Razzell et al., 2013). As mentioned earlier, immune responses are also triggered at the wound site, as revealed by local expression of AMPs in *Drosophila* (Önfelt Tingvall et al., 2001) and in humans (Mangoni et al., 2016). Finally, the tissue damage activates systemic responses where the *Drosophila* system provides a good model to follow the inter organ communication and its consequences at the whole organism level (Lee & Miura, 2014). These examples indicate the impact research in *Drosophila* has had for our general understanding of epithelial repair processes. The high level of conservation underscores the usefulness of this model in future studies of skin integrity, barrier functions, and wound healing.

7. CONCLUDING REMARKS

Here we have reviewed the innate immune system of *Drosophila* and human with a focus on the gut, the respiratory tract, and the skin. There is no doubt that many key components of innate immunity have been well conserved during evolution, including microbial recognition, intracellular signaling pathways, and effector mechanisms. Future studies of epithelial immunity using the *Drosophila* model are in fact likely to identify many more components and processes that are evolutionarily ancient. In addition to increasing our present knowledge, such findings may have great relevance for understanding the underlying mechanisms of human disease. A deepened collaboration between researchers active in the *Drosophila* field and medical scientists should likely promote scientific breakthroughs of medical importance in this area.

The accessibility to next-generation sequencing in clinical practice has opened up a new avenue for molecular understanding of human disease with great clinical relevance. In fact, it is now possible to perform whole genome sequencing of a patient's DNA and to obtain a list of candidate genes in a few days. However, the functional validation of candidate genes and linking them to disease is still a huge undertaking. The *Drosophila* model with its versatile genetic toolbox provides excellent possibilities to unravel the specific roles of candidate genes emanating from such human diagnostic projects.

In human medical research of innate immunity the field is now ready to turn the detailed knowledge on innate immunity into therapeutic approaches. There are many attempts to boost impaired immune pathways or to block excessive inflammation by targeted approaches. Another approach is to induce effector mechanisms, such as AMP expression or activation of autophagy. The *Drosophila* model can provide a fast-track to screen for novel compounds directed toward specific receptors or pathways. In particular, this strategy could be very useful if coupled to large chemical libraries consisting of already approved drugs, which will shorten the time from experimental setup to clinical use by many years.

ACKNOWLEDGMENTS

P.B. was supported by Swedish Research Council, the Swedish Heart and Lung-foundation, the Groschinsky foundation, Swedish Society for Physicians, Scandinavian Society for Antimicrobial Chemotherapy, and Karolinska Institutet1 and Y.E. was supported by the Swedish Cancer Society and Stockholm University.

REFERENCES

Abnave, P., Mottola, G., Gimenez, G., Boucherit, N., Trouplin, V., Torre, C., et al. (2014). Screening in planarians identifies MORN2 as a key component in LC3-associated phagocytosis and resistance to bacterial infection. *Cell Host & Microbe*, *16*, 338–350.

Adams, M. D., Celniker, S. E., Holt, R. A., Evans, C. A., Gocayne, J. D., Amanatides, P. G., et al. (2000). The genome sequence of Drosophila melanogaster. *Science*, *287*, 2185–2195.

Agaisse, H., Petersen, U. M., Boutros, M., Mathey-Prevot, B., & Perrimon, N. (2003). Signaling role of hemocytes in *Drosophila* JAK/STAT-dependent response to septic injury. *Developmental Cell*, *5*, 441–450.

Agerberth, B., Gunne, H., Odeberg, J., Kogner, P., Boman, H. G., & Gudmundsson, G. H. (1995). FALL-39, a putative human peptide antibiotic, is cysteine-free and expressed in bone marrow and testis. *Proceedings of the National Academy of Sciences of the United States of America*, *92*, 195–199.

Aggrawal, K., & Silverman, N. (2007). Peptidoglycan recognition in *Drosophila*. *Biochemical Society Transactions*, *35*, 1496–1500.
Ahlund, M. K., Ryden, P., Sjostedt, A., & Stoven, S. (2010). Directed screen of *Francisella novicida* virulence determinants using *Drosophila melanogaster*. *Infection and Immunity*, *78*, 3118–3128.
Akhouayri, I., Turc, C., Royet, J., & Charroux, B. (2011). Toll-8/Tollo negatively regulates antimicrobial response in the *Drosophila* respiratory epithelium. *PLoS Pathogens*, *7*, e1002319.
Amoyel, M., Anderson, A. M., & Bach, E. A. (2014). JAK/STAT pathway dysregulation in tumors: A *Drosophila* perspective. *Seminars in Cell & Developmental Biology*, *28*, 96–103.
Anderson, K. V., & Nusslein-Volhard, C. (1984). Information for the dorsal–ventral pattern of the *Drosophila* embryo is stored as maternal mRNA. *Nature*, *311*, 223–227.
Angel, P., Szabowski, A., & Schorpp-Kistner, M. (2001). Function and regulation of AP-1 subunits in skin physiology and pathology. *Oncogene*, *20*, 2413–2423.
Ayres, J. S., & Schneider, D. S. (2012). Tolerance of infections. *Annual Review of Immunology*, *30*, 271–294.
Ayyaz, A., Li, H., & Jasper, H. (2015). Haemocytes control stem cell activity in the *Drosophila* intestine. *Nature Cell Biology*, *17*, 736–748.
Bae, Y. S., Choi, M. K., & Lee, W. J. (2010). Dual oxidase in mucosal immunity and host-microbe homeostasis. *Trends in Immunology*, *31*, 278–287.
Becker, T., Loch, G., Beyer, M., Zinke, I., Aschenbrenner, A. C., Carrera, P., et al. (2010). FOXO-dependent regulation of innate immune homeostasis. *Nature*, *463*, 369–373.
Belkaid, Y., & Segre, J. A. (2014). Dialogue between skin microbiota and immunity. *Science*, *346*, 954–959.
Ben-Ami, R., Watson, C. C., Lewis, R. E., Albert, N. D., Arias, C. A., Raad, I. I., et al. (2013). Drosophila melanogaster as a model to explore the effects of methicillin-resistant *Staphylococcus aureus* strain type on virulence and response to linezolid treatment. *Microbial Pathogenesis*, *55*, 16–20.
Benjamin, J. L., Sumpter, R., Jr., Levine, B., & Hooper, L. V. (2013). Intestinal epithelial autophagy is essential for host defense against invasive bacteria. *Cell Host & Microbe*, *13*, 723–734.
Bidla, G., Dushay, M. S., & Theopold, U. (2007). Crystal cell rupture after injury in *Drosophila* requires the JNK pathway, small GTPases and the TNF homolog Eiger. *Journal of Cell Science*, *120*, 1209–1215.
Boman, H. G., Nilsson, I., & Rasmuson, B. (1972). Inducible antibacterial defence system in *Drosophila*. *Nature*, *237*, 232–235.
Bond, D., & Foley, E. (2009). A quantitative RNAi screen for JNK modifiers identifies Pvr as a novel regulator of *Drosophila* immune signaling. *PLoS Pathogens*, *5*, e1000655.
Bonnay, F., Nguyen, X. H., Cohen-Berros, E., Troxler, L., Batsche, E., Camonis, J., et al. (2014). Akirin specifies NF-kappaB selectivity of *Drosophila* innate immune response via chromatin remodeling. *The EMBO Journal*, *33*, 2349–2362.
Bosco-Drayon, V., Poidevin, M., Boneca, I. G., Narbonne-Reveau, K., Royet, J., & Charroux, B. (2012). Peptidoglycan sensing by the receptor PGRP-LE in the *Drosophila* gut induces immune responses to infectious bacteria and tolerance to microbiota. *Cell Host & Microbe*, *12*, 153–165.
Boutros, M., Agaisse, H., & Perrimon, N. (2002). Sequential activation of signaling pathways during innate immune responses in *Drosophila*. *Developmental Cell*, *3*, 711–722.
Brandt, S. M., Dionne, M. S., Khush, R. S., Pham, L. N., Vigdal, T. J., & Schneider, D. S. (2004). Secreted bacterial effectors and host-produced eiger/TNF drive death in a *Salmonella*-infected fruit fly. *PLoS Biology*, *2*, e418.
Britton, R. A., & Young, V. B. (2014). Role of the intestinal microbiota in resistance to colonization by *Clostridium difficile*. *Gastroenterology*, *146*, 1547–1553.

Broderick, N. A., & Lemaitre, B. (2012). Gut-associated microbes of *Drosophila melanogaster. Gut Microbes*, *3*, 307–321.

Broides, A., Sagi, O., Pinsk, V., Levy, J., & Yerushalmi, B. (2016). Subclinical intestinal inflammation in chronic granulomatous disease patients. *Immunologic Research*, *64*, 155–159.

Buchon, N., Broderick, N. A., Chakrabarti, S., & Lemaitre, B. (2009). Invasive and indigenous microbiota impact intestinal stem cell activity through multiple pathways in *Drosophila. Genes & Development*, *23*, 2333–2344.

Buchon, N., Broderick, N. A., & Lemaitre, B. (2013). Gut homeostasis in a microbial world: Insights from *Drosophila melanogaster. Nature Reviews. Microbiology*, *11*, 615–626.

Buchon, N., Broderick, N. A., Poidevin, M., Pradervand, S., & Lemaitre, B. (2009). *Drosophila* intestinal response to bacterial infection: Activation of host defense and stem cell proliferation. *Cell Host & Microbe*, *5*, 200–211.

Buchon, N., Osman, D., David, F. P., Fang, H. Y., Boquete, J. P., Deplancke, B., et al. (2013). Morphological and molecular characterization of adult midgut compartmentalization in *Drosophila. Cell Reports*, *3*, 1725–1738.

Buchon, N., Silverman, N., & Cherry, S. (2014). Immunity in *Drosophila melanogaster*—From microbial recognition to whole-organism physiology. *Nature Reviews Immunology*, *14*, 796–810.

Canfora, E. E., Jocken, J. W., & Blaak, E. E. (2015). Short-chain fatty acids in control of body weight and insulin sensitivity. *Nature Reviews. Endocrinology*, *11*, 577–591.

Cao, X. (2015). Self-regulation and cross-regulation of pattern-recognition receptor signalling in health and disease. *Nature Reviews Immunology*, *16*, 35–50.

Cash, H. L., Whitham, C. V., & Hooper, L. V. (2006). Refolding, purification, and characterization of human and murine RegIII proteins expressed in *Escherichia coli. Protein Expression and Purification*, *48*, 151–159.

Castonguay-Vanier, J., Vial, L., Tremblay, J., & Deziel, E. (2010). *Drosophila melanogaster* as a model host for the *Burkholderia cepacia* complex. *PLoS One*, *5*, e11467.

Chen, J., Xie, C., Tian, L., Hong, L., Wu, X., & Han, J. (2010). Participation of the p38 pathway in *Drosophila* host defense against pathogenic bacteria and fungi. *Proceedings of the National Academy of Sciences of the United States of America*, *107*, 20774–20779.

Choe, K. M., Werner, T., Stoven, S., Hultmark, D., & Anderson, K. V. (2002). Requirement for a peptidoglycan recognition protein (PGRP) in Relish activation and antibacterial immune responses in *Drosophila. Science*, *296*, 359–362.

Chu, H., Pazgier, M., Jung, G., Nuccio, S. P., Castillo, P. A., de Jong, M. F., et al. (2012). Human alpha-defensin 6 promotes mucosal innate immunity through self-assembled peptide nanonets. *Science*, *337*, 477–481.

Clark, R. I., Salazar, A., Yamada, R., Fitz-Gibbon, S., Morselli, M., Alcaraz, J., et al. (2015). Distinct shifts in microbiota composition during *Drosophila* aging impair intestinal function and drive mortality. *Cell Reports*, *12*, 1656–1667.

Clemmons, A. W., Lindsay, S. A., & Wasserman, S. A. (2015). An effector peptide family required for *Drosophila* toll-mediated immunity. *PLoS Pathogens*, *11*, e1004876.

Cole, A. M., Wang, W., Waring, A. J., & Lehrer, R. I. (2004). Retrocyclins: Using past as prologue. *Current Protein & Peptide Science*, *5*, 373–381.

Cronin, S. J., Nehme, N. T., Limmer, S., Liegeois, S., Pospisilik, J. A., Schramek, D., et al. (2009). Genome-wide RNAi screen identifies genes involved in intestinal pathogenic bacterial infection. *Science*, *325*, 340–343.

Dai, L., Aye Thu, C., Liu, X. Y., Xi, J., & Cheung, P. C. (2012). TAK1, more than just innate immunity. *IUBMB Life*, *64*, 825–834.

Dantoft, W., Davis, M. M., Lindvall, J. M., Tang, X., Uvell, H., Junell, A., et al. (2013). The Oct1 homolog Nubbin is a repressor of NF-kappaB-dependent immune gene expression that increases the tolerance to gut microbiota. *BMC Biology*, *11*, 99.

Dantoft, W., Lundin, D., Esfahani, S. S., & Engstrom, Y. (2016). The POU/Oct transcription factor pdm1/nub is necessary for a beneficial gut microbiota and normal lifespan of *Drosophila*. *Journal of Innate Immunity*, *8*, 412–426.
D'Argenio, D. A., Gallagher, L. A., Berg, C. A., & Manoil, C. (2001). Drosophila as a model host for *Pseudomonas aeruginosa* infection. *Journal of Bacteriology*, *183*, 1466–1471.
Davis, M. M., Alvarez, F. J., Ryman, K., Holm, A. A., Ljungdahl, P. O., & Engstrom, Y. (2011). Wild-type *Drosophila melanogaster* as a model host to analyze nitrogen source dependent virulence of *Candida albicans*. *PLoS One*, *6*, e27434.
Davis, M. M., & Engstrom, Y. (2012). Immune response in the barrier epithelia: Lessons from the fruit fly *Drosophila melanogaster*. *Journal of Innate Immunity*, *4*, 273–283.
Davis, M. M., Primrose, D. A., & Hodgetts, R. B. (2008). A member of the p38 mitogen-activated protein kinase family is responsible for transcriptional induction of Dopa decarboxylase in the epidermis of *Drosophila melanogaster* during the innate immune response. *Molecular and Cellular Biology*, *28*, 4883–4895.
De Gregorio, E., Spellman, P. T., Tzou, P., Rubin, G. M., & Lemaitre, B. (2002). The Toll and Imd pathways are the major regulators of the immune response in *Drosophila*. *The EMBO Journal*, *21*, 2568–2579.
de Luca, A., Smeekens, S. P., Casagrande, A., Iannitti, R., Conway, K. L., Gresnigt, M. S., et al. (2014). IL-1 receptor blockade restores autophagy and reduces inflammation in chronic granulomatous disease in mice and in humans. *Proceedings of the National Academy of Sciences of the United States of America*, *111*, 3526–3531.
Delaney, J. R., Stoven, S., Uvell, H., Anderson, K. V., Engstrom, Y., & Mlodzik, M. (2006). Cooperative control of *Drosophila* immune responses by the JNK and NF-kappaB signaling pathways. *The EMBO Journal*, *25*, 3068–3077.
Dimarcq, J. L., Imler, J. L., Lanot, R., Ezekowitz, R. A., Hoffmann, J. A., Janeway, C. A., et al. (1997). Treatment of l(2)mbn *Drosophila* tumorous blood cells with the steroid hormone ecdysone amplifies the inducibility of antimicrobial peptide gene expression. *Insect Biochemistry and Molecular Biology*, *27*, 877–886.
Dionne, M. S., Ghori, N., & Schneider, D. S. (2003). *Drosophila melanogaster* is a genetically tractable model host for *Mycobacterium marinum*. *Infection and Immunity*, *71*, 3540–3550.
Dionne, M. S., Pham, L. N., Shirasu-Hiza, M., & Schneider, D. S. (2006). Akt and FOXO dysregulation contribute to infection-induced wasting in Drosophila. *Current Biology*, *16*, 1977–1985.
Donia, M. S., & Fischbach, M. A. (2015). HUMAN MICROBIOTA. Small molecules from the human microbiota. *Science*, *349*, 1254766.
Dushay, M. S., Åsling, B., & Hultmark, D. (1996). Origins of immunity: Relish, a compound Rel-like gene in the antibacterial defense of *Drosophila*. *Proceedings of the National Academy of Sciences of the United States of America*, *93*, 10343–10347.
Dutta, D., Dobson, A. J., Houtz, P. L., Glasser, C., Revah, J., Korzelius, J., et al. (2015). Regional cell-specific transcriptome mapping reveals regulatory complexity in the adult *Drosophila* midgut. *Cell Reports*, *12*, 346–358.
Engström, Y., Kadalayil, L., Sun, S. C., Samakovlis, C., Hultmark, D., & Faye, I. (1993). kappa B-like motifs regulate the induction of immune genes in *Drosophila*. *Journal of Molecular Biology*, *232*, 327–333.
Erkosar, B., & Leulier, F. (2014). Transient adult microbiota, gut homeostasis and longevity: Novel insights from the *Drosophila* model. *FEBS Letters*, *588*, 4250–4257.
Erturk-Hasdemir, D., Broemer, M., Leulier, F., Lane, W. S., Paquette, N., Hwang, D., et al. (2009). Two roles for the *Drosophila* IKK complex in the activation of Relish and the induction of antimicrobial peptide genes. *Proceedings of the National Academy of Sciences of the United States of America*, *106*, 9779–9784.

Esfahani, S. S., & Engstrom, Y. (2010). Activation of an innate immune response in large numbers of permeabilized *Drosophila* embryos. *Developmental and Comparative Immunology, 35*, 263–266.
Felmy, B., Songhet, P., Slack, E. M., Muller, A. J., Kremer, M., Van Maele, L., et al. (2013). NADPH oxidase deficient mice develop colitis and bacteremia upon infection with normally avirulent, TTSS-1- and TTSS-2-deficient *Salmonella* Typhimurium. *PLoS One, 8*, e77204.
Ferrandon, D., Jung, A. C., Criqui, M., Lemaitre, B., Uttenweiler-Joseph, S., Michaut, L., et al. (1998). A *drosomycin-GFP* reporter transgene reveals a local immune response in *Drosophila* that is not dependent on the Toll pathway. *The EMBO Journal, 17*, 1217–1227.
Findley, K., Oh, J., Yang, J., Conlan, S., Deming, C., Meyer, J. A., et al. (2013). Topographic diversity of fungal and bacterial communities in human skin. *Nature, 498*, 367–370.
Frosali, S., Pagliari, D., Gambassi, G., Landolfi, R., Pandolfi, F., & Cianci, R. (2015). How the intricate interaction among toll-like receptors, microbiota, and intestinal immunity can influence gastrointestinal pathology. *Journal of Immunology Research, 2015*, 489821.
Gallo, R. L., & Hooper, L. V. (2012). Epithelial antimicrobial defence of the skin and intestine. *Nature Reviews. Immunology, 12*, 503–516.
Ganz, T. (2003). Defensins: Antimicrobial peptides of innate immunity. *Nature Reviews. Immunology, 3*, 710–720.
Ganz, T., Selsted, M. E., Szklarek, D., Harwig, S. S., Daher, K., Bainton, D. F., et al. (1985). Defensins. Natural peptide antibiotics of human neutrophils. *The Journal of Clinical Investigation, 76*, 1427–1435.
Gay, N. J., & Keith, F. J. (1991). Drosophila toll and IL-1 receptor. *Nature, 351*, 355–356.
Gendrin, M., Zaidman-Remy, A., Broderick, N. A., Paredes, J., Poidevin, M., Roussel, A., et al. (2013). Functional analysis of PGRP-LA in *Drosophila* immunity. *PLoS One, 8*, e69742.
Glittenberg, M. T., Kounatidis, I., Christensen, D., Kostov, M., Kimber, S., Roberts, I., et al. (2011). Pathogen and host factors are needed to provoke a systemic host response to gastrointestinal infection of Drosophila larvae by *Candida albicans*. *Disease Models & Mechanisms, 4*, 515–525.
Glocker, E. O., Kotlarz, D., Boztug, K., Gertz, E. M., Schaffer, A. A., Noyan, F., et al. (2009). Inflammatory bowel disease and mutations affecting the interleukin-10 receptor. *The New England Journal of Medicine, 361*, 2033–2045.
Gobert, V., Gottar, M., Matskevich, A. A., Rutschmann, S., Royet, J., Belvin, M., et al. (2003). Dual activation of the *Drosophila* toll pathway by two pattern recognition receptors. *Science, 302*, 2126–2130.
Gold, K. S., & Bruckner, K. (2014). *Drosophila* as a model for the two myeloid blood cell systems in vertebrates. *Experimental Hematology, 42*, 717–727.
Gombart, A. F., Borregaard, N., & Koeffler, H. P. (2005). Human cathelicidin antimicrobial peptide (CAMP) gene is a direct target of the vitamin D receptor and is strongly up-regulated in myeloid cells by 1,25-dihydroxyvitamin D3. *FASEB Journal: Official Publication of the Federation of American Societies for Experimental Biology, 19*, 1067–1077.
Gordon, M. D., Dionne, M. S., Schneider, D. S., & Nusse, R. (2005). WntD is a feedback inhibitor of Dorsal/NF-kappaB in *Drosophila* development and immunity. *Nature, 437*, 746–749.
Goto, A., Matsushita, K., Gesellchen, V., El Chamy, L., Kuttenkeuler, D., Takeuchi, O., et al. (2008). Akirins are highly conserved nuclear proteins required for NF-kappaB-dependent gene expression in drosophila and mice. *Nature Immunology, 9*, 97–104.
Gottar, M., Gobert, V., Matskevich, A. A., Reichhart, J. M., Wang, C., Butt, T. M., et al. (2006). Dual detection of fungal infections in *Drosophila* via recognition of glucans and sensing of virulence factors. *Cell, 127*, 1425–1437.

Gottar, M., Gobert, V., Michel, T., Belvin, M., Duyk, G., Hoffmann, J. A., et al. (2002). The *Drosophila* immune response against Gram-negative bacteria is mediated by a peptidoglycan recognition protein. *Nature, 416*, 640–644.

Grasberger, H., Gao, J., Nagao-Kitamoto, H., Kitamoto, S., Zhang, M., Kamada, N., et al. (2015). Increased expression of DUOX2 is an epithelial response to mucosal dysbiosis required for immune homeostasis in mouse intestine. *Gastroenterology, 149*, 1849–1859.

Grice, E. A., Kong, H. H., Conlan, S., Deming, C. B., Davis, J., Young, A. C., et al. (2009). Topographical and temporal diversity of the human skin microbiome. *Science, 324*, 1190–1192.

Gudmundsson, G. H., Bergman, P., Andersson, J., Raqib, R., & Agerberth, B. (2010). Battle and balance at mucosal surfaces—The story of Shigella and antimicrobial peptides. *Biochemical and Biophysical Research Communications, 396*, 116–119.

Guo, H., Callaway, J. B., & Ting, J. P. (2015). Inflammasomes: Mechanism of action, role in disease, and therapeutics. *Nature Medicine, 21*, 677–687.

Guo, L., Karpac, J., Tran, S. L., & Jasper, H. (2014). PGRP-SC2 promotes gut immune homeostasis to limit commensal dysbiosis and extend lifespan. *Cell, 156*, 109–122.

Gur, C., Ibrahim, Y., Isaacson, B., Yamin, R., Abed, J., Gamliel, M., et al. (2015). Binding of the Fap2 protein of Fusobacterium nucleatum to human inhibitory receptor TIGIT protects tumors from immune cell attack. *Immunity, 42*, 344–355.

Ha, E. M., Lee, K. A., Park, S. H., Kim, S. H., Nam, H. J., Lee, H. Y., et al. (2009a). Regulation of DUOX by the Galphaq-phospholipase Cbeta-Ca2+ pathway in *Drosophila* gut immunity. *Developmental Cell, 16*, 386–397.

Ha, E. M., Lee, K. A., Seo, Y. Y., Kim, S. H., Lim, J. H., Oh, B. H., et al. (2009b). Coordination of multiple dual oxidase-regulatory pathways in responses to commensal and infectious microbes in *Drosophila* gut. *Nature Immunology, 10*, 949–957.

Hang, S., Purdy, A. E., Robins, W. P., Wang, Z., Mandal, M., Chang, S., et al. (2014). The acetate switch of an intestinal pathogen disrupts host insulin signaling and lipid metabolism. *Cell Host & Microbe, 16*, 592–604.

Hanratty, W. P., & Dearolf, C. R. (1993). The *Drosophila* tumorous-lethal hematopoietic oncogene is a dominant mutation in the hopscotch locus. *Molecular & General Genetics: MGG, 238*, 33–37.

Harden, N. (2005). Cell biology. Of grainy heads and broken skins. *Science, 308*, 364–365.

Hedengren, M., Asling, B., Dushay, M. S., Ando, I., Ekengren, S., Wihlborg, M., et al. (1999). Relish, a central factor in the control of humoral but not cellular immunity in *Drosophila*. *Molecular Cell, 4*, 827–837.

Heguy, A., Baldari, C. T., Macchia, G., Telford, J. L., & Melli, M. (1992). Amino acids conserved in interleukin-1 receptors (IL-1Rs) and the *Drosophila* toll protein are essential for IL-1R signal transduction. *The Journal of Biological Chemistry, 267*, 2605–2609.

Helenius, I. T., Haake, R. J., Kwon, Y. J., Hu, J. A., Krupinski, T., Casalino-Matsuda, S. M., et al. (2016). Identification of *Drosophila* Zfh2 as a mediator of hypercapnic immune regulation by a genome-wide RNA interference screen. *Journal of Immunology, 196*, 655–667.

Helenius, I. T., Nair, A., Bittar, H. E., Sznajder, J. I., Sporn, P. H., & Beitel, G. J. (2016). Focused screening identifies evoxine as a small molecule that counteracts CO_2-induced immune suppression. *Journal of Biomolecular Screening, 21*(4), 363–371.

Hepburn, L., Prajsnar, T. K., Klapholz, C., Moreno, P., Loynes, C. A., Ogryzko, N. V., et al. (2014). Innate immunity. A Spaetzle-like role for nerve growth factor beta in vertebrate immunity to *Staphylococcus aureus*. *Science, 346*, 641–646.

Hiemstra, P. S., McCray, P. B., Jr., & Bals, R. (2015). The innate immune function of airway epithelial cells in inflammatory lung disease. *The European Respiratory Journal, 45*, 1150–1162.

Honti, V., Csordas, G., Kurucz, E., Markus, R., & Ando, I. (2014). The cell-mediated immunity of *Drosophila melanogaster*: Hemocyte lineages, immune compartments, microanatomy and regulation. *Developmental and Comparative Immunology, 42*, 47–56.

Hornef, M. W., Normark, B. H., Vandewalle, A., & Normark, S. (2003). Intracellular recognition of lipopolysaccharide by toll-like receptor 4 in intestinal epithelial cells. *The Journal of Experimental Medicine, 198*, 1225–1235.

Horowitz, A., & Simons, M. (2008). Branching morphogenesis. *Circulation Research, 103*, 784–795.

Hultmark, D. (1993). Immune reactions in *Drosophila* and other insects: A model for innate immunity. *Trends in Genetics: TIG, 9*, 178–183.

Hultmark, D., Steiner, H., Rasmuson, T., & Boman, H. G. (1980). Insect immunity. *European Journal of Biochemistry/FEBS, 106*, 7–16.

Imler, J. L. (2014). Overview of *Drosophila* immunity: A historical perspective. *Developmental and Comparative Immunology, 42*, 3–15.

Ip, Y. T., Reach, M., Engström, Y., Kadalayil, L., Cai, H., Gonzalez-Crespo, S., et al. (1993). *Dif*, a dorsal-related gene that mediates an immune response in *Drosophila*. *Cell, 75*, 753–763.

Islam, D., Bandholtz, L., Nilsson, J., Wigzell, H., Christensson, B., Agerberth, B., et al. (2001). Downregulation of bactericidal peptides in enteric infections: A novel immune escape mechanism with bacterial DNA as a potential regulator. *Nature Medicine, 7*, 180–185.

Jaiswal, M., Sandoval, H., Zhang, K., Bayat, V., & Bellen, H. J. (2012). Probing mechanisms that underlie human neurodegenerative diseases in *Drosophila*. *Annual Review of Genetics, 46*, 371–396.

Jasper, H. (2015). Exploring the physiology and pathology of aging in the intestine of *Drosophila melanogaster*. *Invertebrate Reproduction and Development, 59*, 51–58.

Jiang, H., & Edgar, B. A. (2012). Intestinal stem cell function in *Drosophila* and mice. *Current Opinion in Genetics & Development, 22*, 354–360.

Jones, R. M., Desai, C., Darby, T. M., Luo, L., Wolfarth, A. A., Scharer, C. D., et al. (2015). Lactobacilli modulate epithelial cytoprotection through the Nrf2 pathway. *Cell Reports, 12*, 1217–1225.

Kabouridis, P. S., & Pachnis, V. (2015). Emerging roles of gut microbiota and the immune system in the development of the enteric nervous system. *The Journal of Clinical Investigation, 125*, 956–964.

Kallio, J., Leinonen, A., Ulvila, J., Valanne, S., Ezekowitz, R. A., & Ramet, M. (2005). Functional analysis of immune response genes in *Drosophila* identifies JNK pathway as a regulator of antimicrobial peptide gene expression in S2 cells. *Microbes and Infection/Institut Pasteur, 7*, 811–819.

Kallsen, K., Zehethofer, N., Abdelsadik, A., Lindner, B., Kabesch, M., Heine, H., et al. (2015). ORMDL deregulation increases stress responses and modulates repair pathways in *Drosophila* airways. *The Journal of Allergy and Clinical Immunology, 136*, 1105–1108.

Kaneko, T., Goldman, W. E., Mellroth, P., Steiner, H., Fukase, K., Kusumoto, S., et al. (2004). Monomeric and polymeric gram-negative peptidoglycan but not purified LPS stimulate the *Drosophila* IMD pathway. *Immunity, 20*, 637–649.

Kang, D., Liu, G., Lundstrom, A., Gelius, E., & Steiner, H. (1998). A peptidoglycan recognition protein in innate immunity conserved from insects to humans. *Proceedings of the National Academy of Sciences of the United States of America, 95*, 10078–10082.

Kappler, C., Meister, M., Lagueux, M., Gateff, E., Hoffmann, J. A., & Reichhart, J. M. (1993). Insect immunity. Two 17 bp repeats nesting a kappa B-related sequence confer inducibility to the diptericin gene and bind a polypeptide in bacteria-challenged *Drosophila*. *The EMBO Journal, 12*, 1561–1568.

Kawai, T., & Akira, S. (2011). Toll-like receptors and their crosstalk with other innate receptors in infection and immunity. *Immunity, 34*, 637–650.

Kim, S. H., & Lee, W. J. (2014). Role of DUOX in gut inflammation: Lessons from *Drosophila* model of gut-microbiota interactions. *Frontiers in Cellular and Infection Microbiology, 3*, 116.

Kim, T., Yoon, J., Cho, H., Lee, W. B., Kim, J., Song, Y. H., et al. (2005). Downregulation of lipopolysaccharide response in *Drosophila* by negative crosstalk between the AP1 and NF-kappaB signaling modules. *Nature Immunology, 6*, 211–218.

Kleino, A., & Silverman, N. (2014). The *Drosophila* IMD pathway in the activation of the humoral immune response. *Developmental and Comparative Immunology, 42*, 25–35.

Kovach, M. A., Ballinger, M. N., Newstead, M. W., Zeng, X., Bhan, U., Yu, F. S., et al. (2012). Cathelicidin-related antimicrobial peptide is required for effective lung mucosal immunity in Gram-negative bacterial pneumonia. *Journal of Immunology, 189*, 304–311.

Krauss-Etschmann, S., Bush, A., Bellusci, S., Brusselle, G. G., Dahlen, S. E., Dehmel, S., et al. (2013). Of flies, mice and men: A systematic approach to understanding the early life origins of chronic lung disease. *Thorax, 68*, 380–384.

Krautz, R., Arefin, B., & Theopold, U. (2014). Damage signals in the insect immune response. *Frontiers in Plant Science, 5*, 342.

Kuraishi, T., Kenmoku, H., & Kurata, S. (2015). From mouth to anus: Functional and structural relevance of enteric neurons in the *Drosophila melanogaster* gut. *Insect Biochemistry and Molecular Biology, 67*, 21–26.

Kurata, S. (2014). Peptidoglycan recognition proteins in *Drosophila* immunity. *Developmental and Comparative Immunology, 42*, 36–41.

Lambeth, J. D., & Neish, A. S. (2014). Nox enzymes and new thinking on reactive oxygen: A double-edged sword revisited. *Annual Review of Pathology, 9*, 119–145.

Lamiable, O., & Imler, J. L. (2014). Induced antiviral innate immunity in *Drosophila. Current Opinion in Microbiology, 20*, 62–68.

Lee, Y. C., Chiang, T. H., Chou, C. K., Tu, Y. K., Liao, W. C., Wu, M. S., et al. (2016). Association between helicobacter pylori eradication and gastric cancer incidence: A systematic review and meta-analysis. *Gastroenterology, 150*(5), 1113–1124.e5.

Lee, K. A., Kim, B., Bhin, J., Kim do, H., You, H., Kim, E. K., et al. (2015). Bacterial uracil modulates Drosophila DUOX-dependent gut immunity via Hedgehog-induced signaling endosomes. *Cell Host & Microbe, 17*, 191–204.

Lee, K. A., Kim, S. H., Kim, E. K., Ha, E. M., You, H., Kim, B., et al. (2013). Bacterial-derived uracil as a modulator of mucosal immunity and gut-microbe homeostasis in *Drosophila. Cell, 153*, 797–811.

Lee, K. A., & Lee, W. J. (2014). *Drosophila* as a model for intestinal dysbiosis and chronic inflammatory diseases. *Developmental and Comparative Immunology, 42*, 102–110.

Lee, W. J., Lee, J. D., Kravchenko, V. V., Ulevitch, R. J., & Brey, P. T. (1996). Purification and molecular cloning of an inducible gram-negative bacteria-binding protein from the silkworm, *Bombyx mori. Proceedings of the National Academy of Sciences of the United States of America, 93*, 7888–7893.

Lee, W. J., & Miura, M. (2014). Mechanisms of systemic wound response in *Drosophila. Current Topics in Developmental Biology, 108*, 153–183.

Leiding, J. W., & Holland, S. M. (1993). Chronic granulomatous disease. In R. A. Pagon, M. P. Adam, H. H. Ardinger, S. E. Wallace, A. Amemiya, L. J. H. Bean, et al. *GeneReviews(R)*. Seattle, WA: University of Washington.

Lemaitre, B., & Hoffmann, J. (2007). The host defense of *Drosophila melanogaster. Annual Review of Immunology, 25*, 697–743.

Lemaitre, B., Kromer-Metzger, E., Michaut, L., Nicolas, E., Meister, M., Georgel, P., et al. (1995). A recessive mutation, *immune deficiency (imd)*, defines two distinct control

pathways in the *Drosophila* host defense. *Proceedings of the National Academy of Sciences of the United States of America, 92*, 9465–9469.
Lemaitre, B., Meister, M., Govind, S., Georgel, P., Steward, R., Reichhart, J. M., et al. (1995). Functional analysis and regulation of nuclear import of dorsal during the immune response in *Drosophila*. *The EMBO Journal, 14*, 536–545.
Lemaitre, B., & Miguel-Aliaga, I. (2013). The digestive tract of *Drosophila melanogaster*. *Annual Review of Genetics, 47*, 377–404.
Lemaitre, B., Nicolas, E., Michaut, L., Reichhart, J. M., & Hoffmann, J. A. (1996). The dorsoventral regulatory gene cassette spatzle/Toll/cactus controls the potent antifungal response in *Drosophila* adults. *Cell, 86*, 973–983.
Lemaitre, B., Reichhart, J. M., & Hoffmann, J. A. (1997). *Drosophila* host defense: Differential induction of antimicrobial peptide genes after infection by various classes of microorganisms. *Proceedings of the National Academy of Sciences of the United States of America, 94*, 14614–14619.
Li, H., & Jasper, H. (2016). Gastrointestinal stem cells in health and disease: From flies to humans. *Disease Models & Mechanisms, 9*, 487–499.
Li, H. S., & Watowich, S. S. (2014). Innate immune regulation by STAT-mediated transcriptional mechanisms. *Immunological Reviews, 261*, 84–101.
Liehl, P., Blight, M., Vodovar, N., Boccard, F., & Lemaitre, B. (2006). Prevalence of local immune response against oral infection in a *Drosophila/Pseudomonas* infection model. *PLoS Pathogens, 2*, e56.
Lindsay, S. A., & Wasserman, S. A. (2014). Conventional and non-conventional *Drosophila* Toll signaling. *Developmental and Comparative Immunology, 42*, 16–24.
Lionakis, M. S., & Kontoyiannis, D. P. (2005). Fruit flies as a minihost model for studying drug activity and virulence in *Aspergillus*. *Medical Mycology, 43*(Suppl. 1), S111–S114.
Lionakis, M. S., Lim, J. K., Lee, C. C., & Murphy, P. M. (2011). Organ-specific innate immune responses in a mouse model of invasive candidiasis. *Journal of Innate Immunity, 3*, 180–199.
Liu, P. T., Stenger, S., Li, H., Wenzel, L., Tan, B. H., Krutzik, S. R., et al. (2006). Toll-like receptor triggering of a vitamin D-mediated human antimicrobial response. *Science, 311*, 1770–1773.
Liu, J. Z., van Sommeren, S., Huang, H., Ng, S. C., Alberts, R., Takahashi, A., et al. (2015). Association analyses identify 38 susceptibility loci for inflammatory bowel disease and highlight shared genetic risk across populations. *Nature Genetics, 47*, 979–986.
Liu, B., Zheng, Y., Yin, F., Yu, J., Silverman, N., & Pan, D. (2016). Toll receptor-mediated hippo signaling controls innate immunity in *Drosophila*. *Cell, 164*, 406–419.
Lloyd, T. E., & Taylor, J. P. (2010). Flightless flies: *Drosophila* models of neuromuscular disease. *Annals of the New York Academy of Sciences, 1184*, E1–E20.
Longworth, M. S., Walker, J. A., Anderssen, E., Moon, N. S., Gladden, A., Heck, M. M., et al. (2012). A shared role for RBF1 and dCAP-D3 in the regulation of transcription with consequences for innate immunity. *PLoS Genetics, 8*, e1002618.
Loof, T. G., Schmidt, O., Herwald, H., & Theopold, U. (2011). Coagulation systems of invertebrates and vertebrates and their roles in innate immunity: The same side of two coins? *Journal of Innate Immunity, 3*, 34–40.
Loonen, L. M., Stolte, E. H., Jaklofsky, M. T., Meijerink, M., Dekker, J., van Baarlen, P., et al. (2014). REG3gamma-deficient mice have altered mucus distribution and increased mucosal inflammatory responses to the microbiota and enteric pathogens in the ileum. *Mucosal Immunology, 7*, 939–947.
Losick, V. P., Fox, D. T., & Spradling, A. C. (2013). Polyploidization and cell fusion contribute to wound healing in the adult *Drosophila* epithelium. *Current Biology, 23*, 2224–2232.

Louis, P., Hold, G. L., & Flint, H. J. (2014). The gut microbiota, bacterial metabolites and colorectal cancer. *Nature Reviews. Microbiology*, *12*, 661–672.

Lozupone, C. A., Stombaugh, J. I., Gordon, J. I., Jansson, J. K., & Knight, R. (2012). Diversity, stability and resilience of the human gut microbiota. *Nature*, *489*, 220–230.

Mabbott, N. A. (2015). A breakdown in communication? Understanding the effects of aging on the human small intestine epithelium. *Clinical Science (London)*, *129*, 529–531.

Mace, K. A., Pearson, J. C., & McGinnis, W. (2005). An epidermal barrier wound repair pathway in *Drosophila* is mediated by grainy head. *Science*, *308*, 381–385.

Macia, L., Tan, J., Vieira, A. T., Leach, K., Stanley, D., Luong, S., et al. (2015). Metabolite-sensing receptors GPR43 and GPR109A facilitate dietary fibre-induced gut homeostasis through regulation of the inflammasome. *Nature Communications*, *6*, 6734.

Maillet, F., Bischoff, V., Vignal, C., Hoffmann, J., & Royet, J. (2008). The *Drosophila* peptidoglycan recognition protein PGRP-LF blocks PGRP-LC and IMD/JNK pathway activation. *Cell Host & Microbe*, *3*, 293–303.

Mangoni, M. L., McDermott, A. M., & Zasloff, M. (2016). Antimicrobial peptides and wound healing: Biological and therapeutic considerations. *Experimental Dermatology*, *25*(3), 167–173.

Mansour, S. C., Pena, O. M., & Hancock, R. E. (2014). Host defense peptides: Front-line immunomodulators. *Trends in Immunology*, *35*, 443–450.

Marianes, A., & Spradling, A. C. (2013). Physiological and stem cell compartmentalization within the *Drosophila* midgut. *eLife*, *2*, e00886.

Martinon, F., Mayor, A., & Tschopp, J. (2009). The inflammasomes: Guardians of the body. *Annual Review of Immunology*, *27*, 229–265.

Medzhitov, R., Preston-Hurlburt, P., & Janeway, C. A., Jr. (1997). A human homologue of the Drosophila Toll protein signals activation of adaptive immunity. *Nature*, *388*, 394–397.

Meister, M., & Richards, G. (1996). Ecdysone and insect immunity: The maturation of the inducibility of the diptericin gene in *Drosophila* larvae. *Insect Biochemistry and Molecular Biology*, *26*, 155–160.

Mellroth, P., Karlsson, J., & Steiner, H. (2003). A scavenger function for a *Drosophila* peptidoglycan recognition protein. *The Journal of Biological Chemistry*, *278*, 7059–7064.

Merga, Y., Campbell, B. J., & Rhodes, J. M. (2014). Mucosal barrier, bacteria and inflammatory bowel disease: Possibilities for therapy. *Digestive Diseases*, *32*, 475–483.

Micchelli, C. A., & Perrimon, N. (2006). Evidence that stem cells reside in the adult *Drosophila* midgut epithelium. *Nature*, *439*, 475–479.

Michel, T., Reichhart, J. M., Hoffmann, J. A., & Royet, J. (2001). *Drosophila* Toll is activated by Gram-positive bacteria through a circulating peptidoglycan recognition protein. *Nature*, *414*, 756–759.

Miller, M., Beppu, A., Rosenthal, P., Pham, A., Das, S., Karta, M., et al. (2015). Fstl1 promotes asthmatic airway remodeling by inducing oncostatin M. *Journal of Immunology*, *195*, 3546–3556.

Mishima, Y., Quintin, J., Aimanianda, V., Kellenberger, C., Coste, F., Clavaud, C., et al. (2009). The N-terminal domain of *Drosophila* Gram-negative binding protein 3 (GNBP3) defines a novel family of fungal pattern recognition receptors. *The Journal of Biological Chemistry*, *284*, 28687–28697.

Moffatt, M. F., Kabesch, M., Liang, L., Dixon, A. L., Strachan, D., Heath, S., et al. (2007). Genetic variants regulating ORMDL3 expression contribute to the risk of childhood asthma. *Nature*, *448*, 470–473.

Mowat, A. M., & Agace, W. W. (2014). Regional specialization within the intestinal immune system. *Nature Reviews. Immunology*, *14*, 667–685.

Moy, R. H., Gold, B., Molleston, J. M., Schad, V., Yanger, K., Salzano, M. V., et al. (2014). Antiviral autophagy restrictsRift Valley fever virus infection and is conserved from flies to mammals. *Immunity, 40,* 51–65.
Mulcahy, H., Sibley, C. D., Surette, M. G., & Lewenza, S. (2011). Drosophila melanogaster as an animal model for the study of Pseudomonas aeruginosa biofilm infections in vivo. *PLoS Pathogens,* 7, e1002299.
Munoz-Soriano, V., Lopez-Domenech, S., & Paricio, N. (2014). Why mammalian wound-healing researchers may wish to turn to *Drosophila* as a model. *Experimental Dermatology, 23,* 538–542.
Myllymaki, H., & Ramet, M. (2014). JAK/STAT pathway in *Drosophila* immunity. *Scandinavian Journal of Immunology, 79,* 377–385.
Myllymaki, H., Valanne, S., & Ramet, M. (2014). The *Drosophila* imd signaling pathway. *Journal of Immunology, 192,* 3455–3462.
Nehme, N. T., Liegeois, S., Kele, B., Giammarinaro, P., Pradel, E., Hoffmann, J. A., et al. (2007). A model of bacterial intestinal infections in *Drosophila melanogaster. PLoS Pathogens, 3,* e173.
Oh, C. T., Moon, C., Choi, T. H., Kim, B. S., & Jang, J. (2013). *Mycobacterium marinum* infection in *Drosophila melanogaster* for antimycobacterial activity assessment. *The Journal of Antimicrobial Chemotherapy, 68,* 601–609.
Oh, C. T., Moon, C., Park, O. K., Kwon, S. H., & Jang, J. (2014). Novel drug combination for *Mycobacterium abscessus* disease therapy identified in a *Drosophila* infection model. *The Journal of Antimicrobial Chemotherapy, 69,* 1599–1607.
Ohlstein, B., & Spradling, A. (2006). The adult *Drosophila* posterior midgut is maintained by pluripotent stem cells. *Nature, 439,* 470–474.
Önfelt Tingvall, T., Roos, E., & Engström, Y. (2001). The imd gene is required for local *Cecropin* expression in *Drosophila* barrier epithelia. *EMBO Reports, 2,* 239–243.
Osman, D., Buchon, N., Chakrabarti, S., Huang, Y. T., Su, W. C., Poidevin, M., et al. (2012). Autocrine and paracrine unpaired signalling regulate intestinal stem cell maintenance and division. *Journal of Cell Science, 125,* 5944–5949.
Panayidou, S., Ioannidou, E., & Apidianakis, Y. (2014). Human pathogenic bacteria, fungi, and viruses in *Drosophila*: Disease modeling, lessons, and shortcomings. *Virulence, 5,* 253–269.
Parsons, B., & Foley, E. (2015). Cellular immune defenses of *Drosophila melanogaster. Developmental and Comparative Immunology, 58,* 95–101.
Persson, C., Oldenvi, S., & Steiner, H. (2007). Peptidoglycan recognition protein LF: A negative regulator of *Drosophila* immunity. *Insect Biochemistry and Molecular Biology, 37,* 1309–1316.
Petersen, U. M., Björklund, G., Ip, Y. T., & Engström, Y. (1995). The dorsal-related immunity factor, Dif, is a sequence-specific trans-activator of *Drosophila* Cecropin gene expression. *The EMBO Journal, 14,* 3146–3158.
Pitout, J. D., & Laupland, K. B. (2008). Extended-spectrum beta-lactamase-producing Enterobacteriaceae: An emerging public-health concern. *The Lancet Infectious Diseases, 8,* 159–166.
Poltorak, A., Smirnova, I., He, X., Liu, M. Y., Van Huffel, C., McNally, O., et al. (1998). Genetic and physical mapping of the Lps locus: Identification of the toll-4 receptor as a candidate gene in the critical region. *Blood Cells, Molecules & Diseases, 24,* 340–355.
Ramet, M., Lanot, R., Zachary, D., & Manfruelli, P. (2002). JNK signaling pathway is required for efficient wound healing in Drosophila. *Developmental Biology, 241,* 145–156.
Ramet, M., Manfruelli, P., Pearson, A., Mathey-Prevot, B., & Ezekowitz, R. A. (2002). Functional genomic analysis of phagocytosis and identification of a *Drosophila* receptor for *E. coli. Nature, 416,* 644–648.

Raqib, R., Sarker, P., Bergman, P., Ara, G., Lindh, M., Sack, D. A., et al. (2006). Improved outcome in shigellosis associated with butyrate induction of an endogenous peptide antibiotic. *Proceedings of the National Academy of Sciences of the United States of America*, *103*, 9178–9183.

Razzell, W., Evans, I. R., Martin, P., & Wood, W. (2013). Calcium flashes orchestrate the wound inflammatory response through DUOX activation and hydrogen peroxide release. *Current Biology*, *23*, 424–429.

Razzell, W., Wood, W., & Martin, P. (2011). Swatting flies: Modelling wound healing and inflammation in *Drosophila*. *Disease Models & Mechanisms*, *4*, 569–574.

Reichhart, J. M., Georgel, P., Meister, M., Lemaitre, B., Kappler, C., & Hoffmann, J. A. (1993). Expression and nuclear translocation of the rel/NF-kappa B-related morphogen dorsal during the immune response of *Drosophila*. *Comptes rendus de l'Academie des sciences. Serie III, Sciences de la vie*, *316*, 1218–1224.

Reichhart, J. M., Meister, M., Dimarcq, J. L., Zachary, D., Hoffmann, D., Ruiz, C., et al. (1992). Insect immunity: Developmental and inducible activity of the *Drosophila* diptericin promoter. *The EMBO Journal*, *11*, 1469–1477.

Reiter, L. T., Potocki, L., Chien, S., Gribskov, M., & Bier, E. (2001). A systematic analysis of human disease-associated gene sequences in *Drosophila melanogaster*. *Genome Research*, *11*, 1114–1125.

Rera, M., Clark, R. I., & Walker, D. W. (2012). Intestinal barrier dysfunction links metabolic and inflammatory markers of aging to death in *Drosophila*. *Proceedings of the National Academy of Sciences of the United States of America*, *109*, 21528–21533.

Rock, F. L., Hardiman, G., Timans, J. C., Kastelein, R. A., & Bazan, J. F. (1998). A family of human receptors structurally related to *Drosophila* Toll. *Proceedings of the National Academy of Sciences of the United States of America*, *95*, 588–593.

Rodrigues-Sousa, T., Ladeirinha, A. F., Santiago, A. R., Carvalheiro, H., Raposo, B., Alarcao, A., et al. (2014). Deficient production of reactive oxygen species leads to severe chronic DSS-induced colitis in Ncf1/p47phox-mutant mice. *PLoS One*, *9*, e97532.

Roeder, T., Isermann, K., Kallsen, K., Uliczka, K., & Wagner, C. (2012). A *Drosophila* asthma model—What the fly tells us about inflammatory diseases of the lung. *Advances in Experimental Medicine and Biology*, *710*, 37–47.

Rosetto, M., Engström, Y., Baldari, C. T., Telford, J. L., & Hultmark, D. (1995). Signals from the IL-1 receptor homolog, Toll, can activate an immune response in a *Drosophila* hemocyte cell line. *Biochemical and Biophysical Research Communications*, *209*, 111–116.

Royet, J. (2011). Epithelial homeostasis and the underlying molecular mechanisms in the gut of the insect model *Drosophila melanogaster*. *Cellular and Molecular Life Sciences: CMLS*, *68*, 3651–3660.

Royet, J., & Charroux, B. (2013). Mechanisms and consequence of bacteria detection by the *Drosophila* gut epithelium. *Gut Microbes*, *4*, 259–263.

Royet, J., Gupta, D., & Dziarski, R. (2011). Peptidoglycan recognition proteins: Modulators of the microbiome and inflammation. *Nature Reviews. Immunology*, *11*, 837–851.

Rus, F., Flatt, T., Tong, M., Aggarwal, K., Okuda, K., Kleino, A., et al. (2013). Ecdysone triggered PGRP-LC expression controls *Drosophila* innate immunity. *The EMBO Journal*, *32*, 1626–1638.

Rutschmann, S., Jung, A. C., Hetru, C., Reichhart, J. M., Hoffmann, J. A., & Ferrandon, D. (2000). The Rel protein DIF mediates the antifungal but not the antibacterial host defense in *Drosophila*. *Immunity*, *12*, 569–580.

Ryu, J. H., Ha, E. M., Oh, C. T., Seol, J. H., Brey, P. T., Jin, I., et al. (2006). An essential complementary role of NF-kappaB pathway to microbicidal oxidants in *Drosophila* gut immunity. *The EMBO Journal*, *25*, 3693–3701.

Ryu, J. H., Nam, K. B., Oh, C. T., Nam, H. J., Kim, S. H., Yoon, J. H., et al. (2004). The homeobox gene Caudal regulates constitutive local expression of antimicrobial peptide genes in *Drosophila* epithelia. *Molecular and Cellular Biology, 24*, 172–185.
Salem, M., Ammitzboell, M., Nys, K., Seidelin, J. B., & Nielsen, O. H. (2015). ATG16L1: A multifunctional susceptibility factor in Crohn disease. *Autophagy, 11*, 585–594.
Salzman, N. H., Hung, K., Haribhai, D., Chu, H., Karlsson-Sjoberg, J., Amir, E., et al. (2010). Enteric defensins are essential regulators of intestinal microbial ecology. *Nature Immunology, 11*, 76–83.
Samakovlis, C., Hacohen, N., Manning, G., Sutherland, D. C., Guillemin, K., & Krasnow, M. A. (1996). Development of the *Drosophila* tracheal system occurs by a series of morphologically distinct but genetically coupled branching events. *Development, 122*, 1395–1407.
Samuelson, D. R., Welsh, D. A., & Shellito, J. E. (2015). Regulation of lung immunity and host defense by the intestinal microbiota. *Frontiers in Microbiology, 6*, 1085.
Sancho, D., & Reis e Sousa, C. (2012). Signaling by myeloid C-type lectin receptors in immunity and homeostasis. *Annual Review of Immunology, 30*, 491–529.
Sayem, M. A., Ahmad, S. M., Rekha, R. S., Sarker, P., Agerberth, B., Talukder, K. A., et al. (2011). Differential host immune responses to epidemic and endemic strains of *Shigella dysenteriae* type I. *Journal of Health, Population, and Nutrition, 29*, 429–437.
Schauber, J., Svanholm, C., Termen, S., Iffland, K., Menzel, T., Scheppach, W., et al. (2003). Expression of the cathelicidin LL-37 is modulated by short chain fatty acids in colonocytes: Relevance of signalling pathways. *Gut, 52*, 735–741.
Schneider, D. S., Hudson, K. L., Lin, T. Y., & Anderson, K. V. (1991). Dominant and recessive mutations define functional domains of Toll, a transmembrane protein required for dorsal-ventral polarity in the *Drosophila* embryo. *Genes & Development, 5*, 797–807.
Schuliga, M. (2015). NF-kappaB signaling in chronic inflammatory airway disease. *Biomolecules, 5*, 1266–1283.
Schuster, A. T., Homer, C. R., Kemp, J. R., Nickerson, K. P., Deutschman, E., Kim, Y., et al. (2015). Chromosome-associated protein D3 promotes bacterial clearance in human intestinal epithelial cells by repressing expression of amino acid transporters. *Gastroenterology, 148*(1405–1416), e1403.
Schwager, S., Agnoli, K., Kothe, M., Feldmann, F., Givskov, M., Carlier, A., et al. (2013). Identification of *Burkholderia cenocepacia* strain H111 virulence factors using non-mammalian infection hosts. *Infection and Immunity, 81*, 143–153.
Scott, M. G., Dullaghan, E., Mookherjee, N., Glavas, N., Waldbrook, M., Thompson, A., et al. (2007). An anti-infective peptide that selectively modulates the innate immune response. *Nature Biotechnology, 25*, 465–472.
Selsted, M. E., Brown, D. M., DeLange, R. J., & Lehrer, R. I. (1983). Primary structures of MCP-1 and MCP-2, natural peptide antibiotics of rabbit lung macrophages. *The Journal of Biological Chemistry, 258*, 14485–14489.
Selsted, M. E., Harwig, S. S., Ganz, T., Schilling, J. W., & Lehrer, R. I. (1985). Primary structures of three human neutrophil defensins. *The Journal of Clinical Investigation, 76*, 1436–1439.
Sen, R., & Baltimore, D. (1986). Inducibility of kappa immunoglobulin enhancer-binding protein Nf-kappa B by a posttranslational mechanism. *Cell, 47*, 921–928.
Shai, Y. (1999). Mechanism of the binding, insertion and destabilization of phospholipid bilayer membranes by alpha-helical antimicrobial and cell non-selective membrane-lytic peptides. *Biochimica et Biophysica Acta, 1462*, 55–70.
Shaukat, Z., Liu, D., & Gregory, S. (2015). Sterile inflammation in *Drosophila*. *Mediators of Inflammation, 2015*, 369286.

Shin, S. C., Kim, S. H., You, H., Kim, B., Kim, A. C., Lee, K. A., et al. (2011). *Drosophila* microbiome modulates host developmental and metabolic homeostasis via insulin signaling. *Science*, *334*, 670–674.
Sluss, H. K., Han, Z., Barrett, T., Goberdhan, D. C., Wilson, C., Davis, R. J., et al. (1996). A JNK signal transduction pathway that mediates morphogenesis and an immune response in *Drosophila*. *Genes & Development*, *10*, 2745–2758.
Sommer, F., & Backhed, F. (2013). The gut microbiota—Masters of host development and physiology. *Nature Reviews. Microbiology*, *11*, 227–238.
Sorrentino, R. P., Melk, J. P., & Govind, S. (2004). Genetic analysis of contributions of dorsal group and JAK-Stat92E pathway genes to larval hemocyte concentration and the egg encapsulation response in *Drosophila*. *Genetics*, *166*, 1343–1356.
Sperandio, B., Fischer, N., & Sansonetti, P. J. (2015). Mucosal physical and chemical innate barriers: Lessons from microbial evasion strategies. *Seminars in Immunology*, *27*, 111–118.
Sperandio, B., Regnault, B., Guo, J., Zhang, Z., Stanley, S. L., Jr., Sansonetti, P. J., et al. (2008). Virulent *Shigella flexneri* subverts the host innate immune response through manipulation of antimicrobial peptide gene expression. *The Journal of Experimental Medicine*, *205*, 1121–1132.
Steiner, H., Hultmark, D., Engstrom, A., Bennich, H., & Boman, H. G. (1981). Sequence and specificity of two antibacterial proteins involved in insect immunity. *Nature*, *292*, 246–248.
Steward, R. (1987). Dorsal, an embryonic polarity gene in *Drosophila*, is homologous to the vertebrate proto-oncogene, c-rel. *Science*, *238*, 692–694.
Stoesser, N., Mathers, A. J., Moore, C. E., Day, N. P., & Crook, D. W. (2016). Colistin resistance gene mcr-1 and pHNSHP45 plasmid in human isolates of *Escherichia coli* and *Klebsiella pneumoniae*. *The Lancet. Infectious Diseases*, *16*(3), 285–286.
Storelli, G., Defaye, A., Erkosar, B., Hols, P., Royet, J., & Leulier, F. (2011). *Lactobacillus plantarum* promotes *Drosophila* systemic growth by modulating hormonal signals through TOR-dependent nutrient sensing. *Cell Metabolism*, *14*, 403–414.
Stöven, S., Ando, I., Kadalayil, L., Engström, Y., & Hultmark, D. (2000). Activation of the *Drosophila* NF-kappaB factor Relish by rapid endoproteolytic cleavage. *EMBO Reports*, *1*, 347–352.
Stöven, S., Silverman, N., Junell, A., Hedengren-Olcott, M., Erturk, D., Engström, Y., et al. (2003). Caspase-mediated processing of the *Drosophila* NF-kappaB factor Relish. *Proceedings of the National Academy of Sciences of the United States of America*, *100*, 5991–5996.
Stramer, B. M., & Dionne, M. S. (2014). Unraveling tissue repair immune responses in flies. *Seminars in Immunology*, *26*, 310–314.
Sun, S. C., Lindstrom, I., Lee, J. Y., & Faye, I. (1991). Structure and expression of the attacin genes in *Hyalophora cecropia*. *European Journal of Biochemistry/FEBS*, *196*, 247–254.
Takahasi, K., Ochiai, M., Horiuchi, M., Kumeta, H., Ogura, K., Ashida, M., et al. (2009). Solution structure of the silkworm betaGRP/GNBP3 N-terminal domain reveals the mechanism for beta-1,3-glucan-specific recognition. *Proceedings of the National Academy of Sciences of the United States of America*, *106*, 11679–11684.
Takehana, A., Yano, T., Mita, S., Kotani, A., Oshima, Y., & Kurata, S. (2004). Peptidoglycan recognition protein (PGRP)-LE and PGRP-LC act synergistically in *Drosophila* immunity. *The EMBO Journal*, *23*, 4690–4700.
Takeuchi, O., Kawai, T., Sanjo, H., Copeland, N. G., Gilbert, D. J., Jenkins, N. A., et al. (1999). TLR6: A novel member of an expanding toll-like receptor family. *Gene*, *231*, 59–65.
Tang, H. (2009). Regulation and function of the melanization reaction in *Drosophila*. *Fly (Austin)*, *3*, 105–111.

Tartey, S., Matsushita, K., Imamura, T., Wakabayashi, A., Ori, D., Mino, T., et al. (2015). Essential function for the nuclear protein Akirin2 in B cell activation and humoral immune responses. *Journal of Immunology*, *195*, 519–527.
Tartey, S., Matsushita, K., Vandenbon, A., Ori, D., Imamura, T., Mino, T., et al. (2014). Akirin2 is critical for inducing inflammatory genes by bridging IkappaB-zeta and the SWI/SNF complex. *The EMBO Journal*, *33*, 2332–2348.
Theopold, U., Krautz, R., & Dushay, M. S. (2014). The *Drosophila* clotting system and its messages for mammals. *Developmental and Comparative Immunology*, *42*, 42–46.
Thomas, S., Fisher, K., Snowden, J., Danson, S., Brown, S., & Zeidler, M. (2015). Effect of methotrexate on JAK/STAT pathway activation in myeloproliferative neoplasms. *Lancet*, *385*(Suppl. 1), S98.
Thompson, G. R., 3rd, Albert, N., Hodge, G., Wilson, M. D., Sykes, J. E., Bays, D. J., et al. (2014). Phenotypic differences of *Cryptococcus* molecular types and their implications for virulence in a *Drosophila* model of infection. *Infection and Immunity*, *82*, 3058–3065.
Ting, S. B., Caddy, J., Hislop, N., Wilanowski, T., Auden, A., Zhao, L. L., et al. (2005). A homolog of *Drosophila* grainy head is essential for epidermal integrity in mice. *Science*, *308*, 411–413.
Tingvall, T. Ö., Roos, E., & Engström, Y. (2001). The GATA factor Serpent is required for the onset of the humoral immune response in *Drosophila* embryos. *Proceedings of the National Academy of Sciences of the United States of America*, *98*, 3884–3888.
Tzelepis, I., Kapsetaki, S. E., Panayidou, S., & Apidianakis, Y. (2013). *Drosophila melanogaster*: A first step and a stepping-stone to anti-infectives. *Current Opinion in Pharmacology*, *13*, 763–768.
Tzou, P., Ohresser, S., Ferrandon, D., Capovilla, M., Reichhart, J. M., Lemaitre, B., et al. (2000). Tissue-specific inducible expression of antimicrobial peptide genes in *Drosophila* surface epithelia. *Immunity*, *13*, 737–748.
Tzou, P., Reichhart, J. M., & Lemaitre, B. (2002). Constitutive expression of a single antimicrobial peptide can restore wild-type resistance to infection in immunodeficient *Drosophila* mutants. *Proceedings of the National Academy of Sciences of the United States of America*, *99*, 2152–2157.
Underhill, D. M., & Pearlman, E. (2015). Immune interactions with pathogenic and commensal fungi: A two-way street. *Immunity*, *43*, 845–858.
Uttenweiler-Joseph, S., Moniatte, M., Lagueux, M., Van Dorsselaer, A., Hoffmann, J. A., & Bulet, P. (1998). Differential display of peptides induced during the immune response of *Drosophila*: A matrix-assisted laser desorption ionization time-of-flight mass spectrometry study. *Proceedings of the National Academy of Sciences of the United States of America*, *95*, 11342–11347.
Uvell, H., & Engstrom, Y. (2007). A multilayered defense against infection: Combinatorial control of insect immune genes. *Trends in Genetics: TIG*, *23*, 342–349.
Vaishnava, S., Yamamoto, M., Severson, K. M., Ruhn, K. A., Yu, X., Koren, O., et al. (2011). The antibacterial lectin RegIIIgamma promotes the spatial segregation of microbiota and host in the intestine. *Science*, *334*, 255–258.
Valanne, S., Wang, J. H., & Ramet, M. (2011). The *Drosophila* Toll signaling pathway. *Journal of Immunology*, *186*, 649–656.
van de Veerdonk, F. L., & Dinarello, C. A. (2014). Deficient autophagy unravels the ROS paradox in chronic granulomatous disease. *Autophagy*, *10*, 1141–1142.
van der Flier, L. G., & Clevers, H. (2009). Stem cells, self-renewal, and differentiation in the intestinal epithelium. *Annual Review of Physiology*, *71*, 241–260.
Vanaja, S. K., Rathinam, V. A., & Fitzgerald, K. A. (2015). Mechanisms of inflammasome activation: Recent advances and novel insights. *Trends in Cell Biology*, *25*, 308–315.
Veillard, F., Troxler, L., & Reichhart, J. M. (2016). *Drosophila melanogaster* clip-domain serine proteases: Structure, function and regulation. *Biochimie*, *122*, 255–269.

Voronin, D., Cook, D. A., Steven, A., & Taylor, M. J. (2012). Autophagy regulates Wolbachia populations across diverse symbiotic associations. *Proceedings of the National Academy of Sciences of the United States of America*, *109*, E1638–E1646.
Wagner, C., Isermann, K., Fehrenbach, H., & Roeder, T. (2008). Molecular architecture of the fruit fly's airway epithelial immune system. *BMC Genomics*, *9*, 446.
Wagner, C., Isermann, K., & Roeder, T. (2009). Infection induces a survival program and local remodeling in the airway epithelium of the fly. *FASEB Journal: Official Publication of the Federation of American Societies for Experimental Biology*, *23*, 2045–2054.
Wang, T. T., Nestel, F. P., Bourdeau, V., Nagai, Y., Wang, Q., Liao, J., et al. (2004). Cutting edge: 1,25-dihydroxyvitamin D3 is a direct inducer of antimicrobial peptide gene expression. *Journal of Immunology*, *173*, 2909–2912.
Wang, S., & Samakovlis, C. (2012). Grainy head and its target genes in epithelial morphogenesis and wound healing. *Current Topics in Developmental Biology*, *98*, 35–63.
Watson, R. (2011). Europe launches 12 point plan to tackle antimicrobial resistance. *BMJ*, *343*, d7528.
Weber, A. N., Tauszig-Delamasure, S., Hoffmann, J. A., Lelievre, E., Gascan, H., Ray, K. P., et al. (2003). Binding of the *Drosophila* cytokine Spatzle to Toll is direct and establishes signaling. *Nature Immunology*, *4*, 794–800.
Werner, T., Liu, G., Kang, D., Ekengren, S., Steiner, H., & Hultmark, D. (2000). A family of peptidoglycan recognition proteins in the fruit fly *Drosophila melanogaster*. *Proceedings of the National Academy of Sciences of the United States of America*, *97*, 13772–13777.
Wu, K., Conly, J., Surette, M., Sibley, C., Elsayed, S., & Zhang, K. (2012). Assessment of virulence diversity of methicillin-resistant *Staphylococcus aureus* strains with a *Drosophila melanogaster* infection model. *BMC Microbiology*, *12*, 274.
Wu, H., Tremaroli, V., & Backhed, F. (2015). Linking microbiota to human diseases: A systems biology perspective. *Trends in Endocrinology and Metabolism*, *26*, 758–770.
Yano, T., Mita, S., Ohmori, H., Oshima, Y., Fujimoto, Y., Ueda, R., et al. (2008). Autophagic control of listeria through intracellular innate immune recognition in drosophila. *Nature Immunology*, *9*, 908–916.
Yoshida, H., Kinoshita, K., & Ashida, M. (1996). Purification of a peptidoglycan recognition protein from hemolymph of the silkworm, *Bombyx mori*. *The Journal of Biological Chemistry*, *271*, 13854–13860.
Zaidman-Remy, A., Herve, M., Poidevin, M., Pili-Floury, S., Kim, M. S., Blanot, D., et al. (2006). The *Drosophila* amidase PGRP-LB modulates the immune response to bacterial infection. *Immunity*, *24*, 463–473.
Zaidman-Remy, A., Regan, J. C., Brandao, A. S., & Jacinto, A. (2012). The *Drosophila* larva as a tool to study gut-associated macrophages: PI3K regulates a discrete hemocyte population at the proventriculus. *Developmental and Comparative Immunology*, *36*, 638–647.

CHAPTER THREE

Drosophila melanogaster as a Model of Muscle Degeneration Disorders

R.E. Kreipke*[,†], Y.V. Kwon*, H.R. Shcherbata[‡], H. Ruohola-Baker*[,†,1]
*University of Washington, School of Medicine, Seattle, WA, United States
[†]Institute for Stem Cell and Regenerative Medicine, University of Washington, School of Medicine, Seattle, WA, United States
[‡]Max Planck Research Group of Gene Expression and Signaling, Max Planck Institute for Biophysical Chemistry, Göttingen, Germany
[1]Corresponding author: e-mail address: hannele@uw.edu

Contents

Abstract

Drosophila melanogaster provides a powerful platform with which researchers can dissect complex genetic questions and biochemical pathways relevant to a vast array of human diseases and disorders. Of particular interest, much work has been done with flies to elucidate the molecular mechanisms underlying muscle degeneration diseases. The fly is particularly useful for modeling muscle degeneration disorders because there are no identified satellite muscle cells to repair adult muscle following injury. This allows for the identification of endogenous processes of muscle degeneration as discrete

Current Topics in Developmental Biology, Volume 121
ISSN 0070-2153
http://dx.doi.org/10.1016/bs.ctdb.2016.07.003

events, distinguishable from phenotypes due to the lack of stem cell-based regeneration. In this review, we will discuss the ways in which the fruit fly provides a powerful platform with which to study human muscle degeneration disorders.

1. INTRODUCTION

The fruit fly, *Drosophila melanogaster*, provides an ideal model organism with which to study a number of human diseases. The vast array of sophisticated genetic techniques are available to manipulate the *D. melanogaster* genome coupled with their rapid life cycle, and the large number of progeny that can be generated quickly makes them an efficient and cost-effective system to model the biological basis for disease. Additionally, a large number of proteins and pathways are conserved between flies and higher order vertebrates—approximately 75% of known human disease genes have fly homologues (Reiter, Potocki, Chien, Gribskov, & Bier, 2001).

Flies have proven to be an especially useful model for diseases involving the degeneration of muscles, given the stereotyped arrangement and accessibility of their muscles for high-resolution microscopy study, as well as functional and physiological studies. Of particular interest to studies of muscle is the fact that, unlike mammalian muscle cells, adult *Drosophila* muscles lack an active adult stem cell population. In mammalian systems, muscular satellite cells regulate the regeneration of muscle tissue following injury. This can make it difficult to differentiate between phenotypes related directly to degeneration of muscle tissue as opposed to failures of the endogenous population of stem cells to regenerate. The absence of a muscle stem cell population in the fly makes it ideal for elucidation of the endogenous molecular mechanisms that underlie muscle degeneration. In this review, we will discuss the insights into muscle degeneration disorders gained from the fruit fly.

2. *DROSOPHILA* MODELS FOR MUSCULAR DYSTROPHIES

2.1 Duchenne Muscular Dystrophy

Muscle cells, controlled by the excitation–contraction coupling from neurons in the neuromuscular junctions (NMJs), support our body by generating force that allows life-supporting components, oxygen, and nutrients to circulate throughout an organism. Muscles, the engines in the animal body, are carefully built and controlled to last the lifetime of an organism.

Unfortunately, this system breaks down during aging. The aging animal body experiences dramatic wasting of cellular capacities, including wasting of the muscle. In some diseases, for example, Duchenne muscular dystrophy (DMD), this age-dependent degeneration of muscle is dramatically accelerated.

DMD is an X-linked, lethal muscle-wasting disease affecting 1/3500 male births per year in the United States. DMD is caused mainly by mutations in the Dystrophin gene that codes for a large cytoskeletal component of a Dystrophin–Dystroglycan plasma membrane complex (DGC). The complex has an important role in connecting the extracellular matrix through muscle cell plasma membrane (sarcolemma) to the cytoskeleton (Fig. 1;

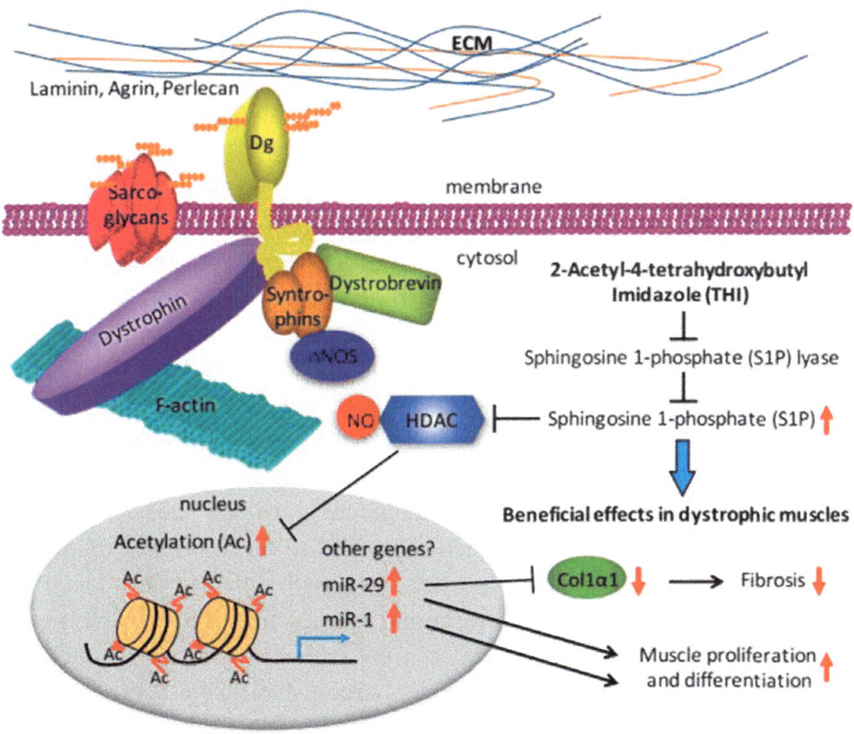

Fig. 1 Model of *Drosophila* DGC. The transmembrane protein Dystroglycan is a key component of the complex that via binding to Laminins, Perlecan, and Agrin connects the ECM to the actin cytoskeleton via the cytoplasmic protein, Dystrophin. The DGC also acts as a scaffold for many signaling pathways, including syntrophin-nitric oxide signaling that produces nitric oxide (NO). NO is involved in nitrosylation of histone deacetylases (HDACs), influencing gene expression.

Barresi & Campbell, 2006; Lapidos, Kakkar, & McNally, 2004; Li et al., 2014). Many signaling molecules are anchored to the sarcolemma via Dystrophin (Liu, Burkin, & Kaufman, 2008); hence the lack of Dystrophin in DMD patients results in mislocalization of key signaling components (Brenman, Chao, Xia, Aldape, & Bredt, 1995). In contrast to other muscle degeneration disorders which involve the pathologic interaction of multiple tissues or systems, in DMD, a mutation in the Dystrophin gene in muscle cells alone is sufficient for the pathogenesis of the disease in muscle, making it an interesting target of study in the field of muscle degeneration disorders.

Drosophila DGC is highly evolutionarily conserved and contains all major components present in the vertebrate DGC (Hoffman, Brown, & Kunkel, 1987; Koenig et al., 1987; Moore & Winder, 2010; Shcherbata et al., 2007; Yatsenko et al., 2007, 2009). Since many DMD phenotypes, including muscle wasting, could be modeled in *D. melanogaster* (Christoforou, Greer, Challoner, Charizanos, & Ray, 2008; Kucherenko, Marrone, Rishko, de Fatima Magliarelli, & Shcherbata, 2011; Mirouse, Christoforou, Fritsch, & Ray, 2009; Pilgram, Potikanond, Baines, Fradkin, & Noordermeer, 2010; Taghli-Lamallem, Jagla, Chamberlain, & Bodmer, 2014), the powerful *Drosophila* forward genetics was a viable approach to identify additional genes that interact with the disease causing Dystrophin mutations. Interestingly, it was found that stress induces muscle degeneration and accelerates age-dependent muscular dystrophy (Kucherenko et al., 2011; Mirouse et al., 2009). In vivo genetic interaction screen in aging dystrophic *Drosophila* identified genes that had not been previously shown to play a role in the development of muscular dystrophy and to interact with the DGC (Kucherenko et al., 2011). Mutations in many of these genes cause developmental and age-dependent morphological and heat-induced physiological defects in muscles and in the nervous tissue (Kucherenko et al., 2011; Marrone, Kucherenko, Wiek, Göpfert, & Shcherbata, 2011). Since most of them have human homologues that have been associated with different disorders, these genes can be potentially used as drug targets for muscular dystrophy treatment.

In addition to mutations in Dystrophin, mutations in other interacting proteins, for example, syntrophin and γ-sarcoglycan, have been shown to cause muscle degeneration phenotypes in *Drosophila* (McNally & Goldstein, 2012). Furthermore, recent data on gene editing have revealed *Drosophila* as a powerful system to identify and test potential gene-editing paradigms, not only in the context of Dystrophin mutations but also with γ-sarcoglycan (Gao et al., 2015).

2.2 Congenital Muscular Dystrophies

Dystroglycan is a major nonintegrin extracellular matrix receptor involved in cell adhesion, mechanoreception, and "inside-out" and "outside-in" signaling. While mutations in Dystrophin cause muscular dystrophies associated with muscle degeneration, abnormal functions in Dystroglycan in humans result in development of dystroglycanopathies or congenital muscular dystrophies (CMDs). CMD patients, in addition to muscle degeneration, exhibit morphological brain and eye defects, suffer from cognitive impairment and learning disability, and develop behavioral and neuropsychiatric disorders (Waite, Brown, & Blake, 2012). In particular, abnormal interaction of Dystroglycan and the ECM results in the cobblestone brain, which is a peculiar brain malformation, caused by neuron over-migration and formation of an extracortical layer that looks like a bumpy cobblestone surface (Devisme et al., 2012; Waite et al., 2012). There are three cobblestone lissencephaly categories: Walker–Warburg syndrome, Fukuyama congenital muscular dystrophy, and Finnish muscle eye brain disease. These disorders are extremely pleiotropic, which can be explained by the diverse nature of the mutations that affect Dystroglycan physiological functions (Moore & Winder, 2010; Inamori et al., 2012). Interestingly, Dystroglycan protein can be detected in virtually all tissues in mammals and in *Drosophila* (Dekkers et al., 2004; Ibraghimov-Beskrovnaya et al., 1993; Shcherbata et al., 2007) and there are multiple Dystroglycan gene transcripts listed in GenBank. However, it remains elusive how Dystroglycan is regulated at the transcriptional and posttranscriptional levels. Recently, a *Drosophila* CDM model was developed (Yatsenko, Marrone, & Shcherbata, 2014). It has been shown that *Drosophila* Dystroglycan misexpression mutants present all typical syndromes associated with dystroglycanopathy: muscle degeneration, cobblestone brain, and eye defects (Kucherenko et al., 2008, 2011; Marrone et al., 2011; Shcherbata et al., 2007; Yatsenko et al., 2014).

The precision of Dystroglycan expression is critical for proper nervous system development, since lower levels slow down neuronal stem cell division, while higher levels accelerate proliferation and perturb neuron differentiation (Yatsenko et al., 2014). In general, both lower and higher Dystroglycan levels cause defective brain morphogenesis. It has been shown that abnormal Dystroglycan levels affect the distribution of major cell adhesion proteins in differentiating neurons, altering their interactions with other neurons and the ECM, consequently causing anomalous brain tissue assembly. Interestingly, the intensity of the cobblestone-like brain phenotype is proportional to the amount and the duration of Dystroglycan protein misexpression. It implies that maintenance of Dystroglycan expression levels is key for proper tissue differentiation. In the brain, this is achieved via a novel perceptive–executive mechanism of gene expression regulation accomplished via miRNAs (Yatsenko et al., 2014). In future, *Drosophila* cobblestone lissencephaly model will allow a better understanding of the molecular basis of dystroglycanopathies.

3. MUSCLE DISEASE MODELS RELATED TO MOTOR NEURON DISORDERS

Motor neuron disorders comprise a heterogeneous group of disorders that involve the selective degeneration of the upper and/or lower motor neurons and include such disorders as spinal muscular atrophy (SMA) and amyotrophic lateral sclerosis (ALS). In these muscle degeneration diseases, it is the loss of innervation of muscle tissue that leads to progressive degeneration of the muscle fibers, rather than mutations within the muscles themselves. Depending on the disease and the severity, fatality can occur fewer than 5 years following diagnosis, usually as a result of respiratory failure (NINDS, 2012). Genetic studies in humans have provided a useful guide in terms of identifying therapeutic targets and mutations underlying the pathogenesis of these disorders. However, even with the identification of mutations leading to the development of these diseases, knowledge of the underlying disease-relevant biochemical pathways remains elusive. Thus, there remains a need for a model system that allows the elucidation of the basic biological mechanisms that these genes govern, thereby clarifying the pathway from mutation to pathogenesis.

There are several characteristics that make *D. melanogaster* ideal in this situation. The synapse between motor neurons and muscle, the NMJ, has been extensively studied and well characterized. Neurons leave the central nervous system of the fly, fasciculated into neuron bundles. When they reach the appropriate muscle fiber, they defasciculate, forming stereotyped synapses onto the muscle fibers. This well-established and regular pattern of innervation makes the fly NMJ the ideal system in which to study the effect of motor neuron degeneration on the corresponding muscle (Keshishian, Broadie, Chiba, & Bate, 1996). Furthermore, the fly NMJ is found close to the body wall, allowing easy access for high-resolution microscopy, as well as functional and physiological studies (Deshpande & Rodal, 2015).

This, in addition to their genetic manipulability, makes *Drosophila* an ideal platform to begin to elucidate the underlying pathogenesis of these complex genetic disorders. As a result, much has been learned about the etiology of these diseases. For the purpose of this review, we will focus on DMD, ALS, and SMA. While much has been done to elucidate the mechanisms underlying neurodegeneration in motor neuron disorders, for the purpose of this review, we will focus our scope on muscle degeneration phenotypes.

3.1 DMD: Hyperthermic Seizures

Interestingly, though it is not typically considered a motor neuron disorder, DMD patients experience a number of neuron-related symptoms, including life-threatening complications during and after general anesthesia, such as hyperthermia, muscle contracture, myolysis, hypercapnia, and metabolic acidosis. *Drosophila Dystrophin* mutants, similar to human DMD patients, have muscle function defects resulting in hyperthermic seizures of their indirect flight muscles. Specifically in a *Drosophila* model, Dystrophin and Dystroglycan have been shown to be localized postsynaptically at the NMJ and are required for appropriate homeostatic control of neurotransmitter release (Bogdanik et al., 2008; Dickman & Davis, 2009; Van der Plas et al., 2006). This indicates that the DGC may play two important roles in muscle tissue: that of muscle stabilization and mechanoreception and that of neurotransmission regulation. A long isoform of *Dystrophin* (dystrophin-like protein 2) was shown by Van der Plas et al. (2006), when deficient at the NMJ, to cause increased quantal content dependent upon *wnt*, a presynaptic receptor. The increase in quantal content was determined not to be due to increased Ca^{2+} sensor sensitivity, but more likely caused by altered modulation of presynaptic Ca^{2+} channel activity leading to presynaptic increase in Ca^{2+} influx causing an increased probability of neurotransmitter release. Interestingly, the Ca^{2+}-binding protein, Calmodulin, which binds to dystrophin in mammals, also rescued hyperthermic seizures, implying that Ca^{2+} levels are important for the dystrophic seizure activity. Furthermore, treatment of *Dystrophin* mutants with various Ca^{2+} channel blockers [Nifedipine (dihydropyridine channel), 2-APB (inositol 1,4,5-trisphosphate receptors, IP3R), and Ryanodine (Ryanodine receptors, RyR)] alleviated dystrophic seizures phenotype, demonstrating that Ca^{2+} released via IP3R and RyR activated channels plays a role in hyperthermic seizures (Marrone et al., 2011). Thus, the hyperthermic seizures are dependent upon neuronal input at the NMJ and the proteins involved in neurotransmitter receptor localization affect the phenotype, indicating that Dystrophin is not only required for muscle maintenance and neuron differentiation but also involved in neuron–muscle communication (Marrone et al., 2011; Pilgram, Potikanond, van der Plas, Fradkin, & Noordermeer, 2011).

Since Dystroglycan is a binding partner of Dys, it is expected that mutations in Dystroglycan would phenocopy the dystrophic seizure phenotype. Surprisingly, *Dystroglycan* loss-of-function mutants exhibit no seizing activity (Marrone et al., 2011). Even more, reduction of *Dystroglycan* can actually rescue dystrophic seizures, revealing novel dynamics between two components of the DGC and their interaction.

Apparently, in the absence of Dg, glutamate receptors are improperly localized causing insufficient muscle response (Marrone et al., 2011), while release of neurotransmitter at the NMJ results in depolarization of the muscle membrane causing release of Ca^{2+} from the SR required for muscle contraction. While Dystroglycan is involved in proper assembly of the neurotransmitter receptors, Dystrophin is involved in retrograde signaling. Upon its deficiency the muscle does not signal back to the neuron that keeps activating muscle contraction. Taken together, the data show that the DGC acts at the muscle side of the NMJ to regulate muscle cell homeostasis in response to neuronal signaling, which implies the DGC involvement in muscle–neuron communication (Fig. 2).

3.2 Amyotrophic Lateral Sclerosis

ALS is a progressive, upper, and lower motor neuron degenerative disease. Because both upper and lower motor neurons are lost, ALS patients gradually lose control of all voluntary motion. Most cases are rapid, and when the muscles in the chest wall and diaphragm become too weak to function, the result is typically death by respiratory failure, usually within 3–5 years of the onset of symptoms (NINDS, 2012). There are familial cases of ALS, but these account for only about 10% of all cases. Study of these familial cases has identified over 100 mutations in a single gene leading to ALS: the Cu/An superoxide dismutase gene (*SOD1*). *SOD1*-linked familial ALS resembles the sporadic disease closely, indicating that *SOD1* mutations likely have an important role in sporadic as well as in familial ALS (Rosen et al., 1993). However, the exact relationship between SOD1 protein activity and the progression of ALS has remained unclear. The development of fly mutants of SOD1 has yielded crucial information regarding the activity of this protein and how it may impact the pathogenesis of ALS.

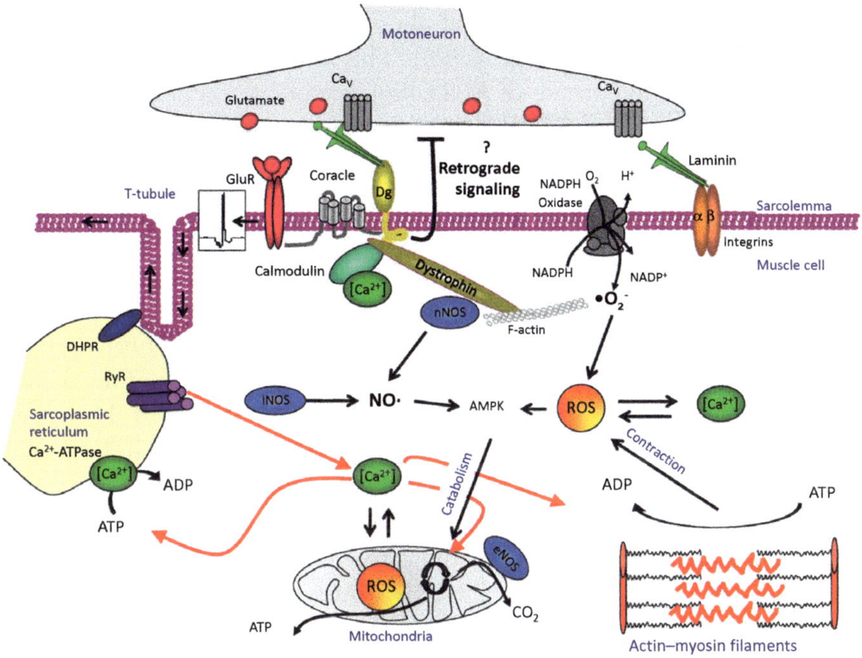

Fig. 2 At the NMJ, the DGC plays a role in cellular homeostasis. The state of Ca^{2+} homeostasis is altered in *Dystrophin* mutants, leading to an increase in mitochondrial function resulting in pathophysiology. The model suggests that homeostasis is dependent upon retrograde signaling that relies on proper glutamate receptor localization mediated by Dystroglycan binding to Dystrophin and Coracle. The loss of retrograde signaling control results in hyperexcitability of the motoneuron causing an increase in muscle contraction that is easily revealed at elevated temperatures. This in turn results in an increase of ROS that activates AMPK altering mitochondrial biogenesis. Other possible pathways for ROS generation include high levels of NO generated via iNOS and NAD(P)H oxidase.

SOD1 is a copper-dependent enzyme that catalyzes the conversion of toxic superoxide radicals to hydrogen peroxide and oxygen. *SOD1*-linked ALS mutations confer a dominant toxicity onto the function of the protein. Intriguingly, though it is ubiquitously expressed throughout most cells, mutations in SOD1 selectively lead to motor neuron degeneration. Fly models of SOD1 mutations have provided valuable insight into possible mechanisms governing the pathogenesis of ALS.

Due to its role in eliminating radicals, it was originally hypothesized that neuron degeneration was mediated by a loss of SOD1 function, leading to a lethal increase in damaging free radicals. However, expression of mutant SOD1 was shown to actually increase *Drosophila* lifespan and increases resistance to oxidative stress (Elia et al., 1999; Mockett, Radyuk, Benes, Orr, & Sohal, 2003; Parkes et al., 1998). This finding was critical in leading to the pro-oxidant hypothesis of ALS falling out of favor. Furthermore, evidence from fly models of SOD1 mutations has helped establish an aggregation model as a possibility in the pathogenesis of ALS. SOD1 overexpression in fly motor neurons results in motor deficits, defective neural circuit electrophysiology, and also elicited a stress response in surrounding glia (Watson, Lagow, Xu, Zhang, & Bonini, 2008). Importantly, it also revealed focal aggregation of SOD1 in motor neurons. Watson and colleagues reported visible protein accumulation of SOD1 in *Drosophila* motor neurons as early as 1 day after eclosion. As the flies aged, both the number of aggregates within cells and the number of cells with aggregates increased. However, these aggregates did not display the ubiquitination that would indicate that the proteins are being recognized as misfolded by the ubiquitin–proteasome or chaperon systems. This provided the first line of evidence in *Drosophila* that ALS may be join the ranks of other protein aggregation neurodegenerative disorders (Casci & Pandey, 2015; McGurk, Berson, & Bonini, 2015). Together, these crucial studies in flies suggest that protein aggregation, rather than toxic increase of radicals, may be an underlying cause for the pathogenesis of ALS. Additional fly studies have also elucidated more mechanisms of ALS-related pathophysiology and are discussed in more detail by Bellen and colleagues (Lin, Mao, & Bellen, 2017).

3.3 Spinal Muscular Atrophy

SMA is one of the most common autosomal recessive diseases and is considered to be one of the leading genetic causes of infant mortality, affecting 1 in 6000–10,000 live births. It is characterized by the selective degeneration of anterior horn lower motor neurons, which leads to muscle wasting (Nishimura et al., 2004). Tragically, though the cause of SMA has been linked to mutations in a single gene—the survival motor neuron (SMN) protein gene—there is no effective treatment (Grice, Praveen, Matera, & Liu, 2013). Furthermore, there is a dearth of knowledge of disease-relevant biochemical pathways and it remains unclear why the loss of a protein that is required for survival regardless of cell type would lead to selective motor neuron loss.

Drosophila provides an ideal model for SMA because there is a single *SMN* gene. This has allowed for the development of a number of mutant models of SMA in the fly, the severity of which correlates to the level of

SMN reduction (Chang et al., 2008; Lorson et al., 1998; Praveen, Wen, & Matera, 2012). Studies in these mutants have provided valuable insight into the requirements of SMN to maintain a healthy NMJ. When SMA levels are reduced, it results in abnormal motor behavior, disorganized synaptic boutons, and reduced postsynaptic currents at the NMJ (Lloyd & Taylor, 2010). It is possible to rescue these defects by expressing wild-type SMN in a tissue-specific manner. Intriguingly, SMN is required in both neurons and muscles for successful rescue, indicating that it is required in both cell types (Chang et al., 2008). Additionally, work in flies has provided information about the role of SMN in the mature neuromuscular system. It has been demonstrated that SMN colocalizes with sarcomeric actin, forming a complex with α-actinin at the sarcomeric Z-disc (Rajendra et al., 2007). This relationship is conserved in vertebrates, possibly indicating a muscle-specific role for SMN. Flies have also been used to demonstrate a potential role for SMN in RNA biogenesis, mRNA transport, and translational control (Fallini et al., 2011). These crucial insights from the fly begin to provide valuable insight into the biochemical pathways and processes that this protein may mediate, elucidating potential therapeutic targets.

4. *DROSOPHILA* MODELS OF CACHEXIA-LIKE WASTING

4.1 Modeling Muscle Wasting in *Drosophila*

Muscle wasting as a result of pathological conditions in other organ systems can also be modeled in *Drosophila*. Wasting is characterized by involuntary loss of body mass, manifested in particular by degeneration of skeletal muscle and adipose tissue. Wasting is not only a physiological response to extremely low energy intake, but it is also a part of a complex systemic disorder associated with many diseases, including cancers, chronic obstructive pulmonary disease, congestive heart failure, and chronic kidney disease (Argilés, Busquets, Stemmler, & López-Soriano, 2014; Fearon, Glass, & Guttridge, 2012; Mangner, Matsuo, Schuler, & Adams, 2013; Remels, Gosker, Langen, & Schols, 2013; Tsoli & Robertson, 2013). Strikingly, cancer cachexia, the wasting syndrome observed in cancer patients, affects more than 80% of advanced cancer patients and accounts for 20% of cancer deaths (Argilés et al., 2014; Lok, 2015). Although recent findings provide further insights into the molecular etiology of cachexia, the genetics and molecular basis of cachexia are far from understood. Unbiased forward genetics and organismal-level studies will greatly help to improve our understanding of this disease and identify novel therapeutic strategies for its treatment.

Recently, two research groups discovered that in *Drosophila*, induction of localized tumors in adult flies causes systemic organ wasting reminiscent of cachexia (Figueroa-Clarevega & Bilder, 2015; Kwon et al., 2015). Both groups utilized well-characterized tumor models to address how localized tumor growth affects the homeostasis of distant tissues. Yorkie is the transcriptional coactivator of the Hippo signaling pathway, which regulates organ growth and regeneration in many animals (Cannon et al., 2015; Hariharan, 2015; Johnson & Halder, 2014; Oh & Irvine, 2010; Pan, 2010; Yu, Zhao, & Guan, 2015). Interestingly, the mammalian ortholog of Yorkie, Yap1, is an oncogene, amplified in multiple types of cancers, including hepatocellular carcinoma, medulloblastoma, and esophageal squamous cell carcinoma (Johnson & Halder, 2014; Yu et al., 2015; Zender et al., 2006). Kwon et al. expressed an active form of Yorkie in intestinal stem cells and enteroblasts to induce a localized tumor growth in adult flies. Figueroa-Clarevega et al. utilized a classical transplantation technique for examining tissue growth (Hadorn, 1963). They transplanted eye imaginal disc tumors harboring a mutation in *scribble* (*scrib*) and expressing the oncogenic *Ras* allele (*scrib*/*Ras*^{V12}) into wild-type fly's hemocoel, the open body cavity of adults.

Various wasting phenotypes were observed in the flies with either intestinal tumors or disc tumors in hemocoel. In particular, these tumors induced wasting of muscle, ovary, and fat body, which are the tissues that preserve energy in the form of lipids, glycogen, and proteins. Interestingly, the wasting flies showed depletion of energy stores and hyperglycemia, which may be explained by a decrease in sugar uptake by these tissues during wasting. The hyperglycemia phenotype is reminiscent of the impaired glucose handling observed in both cancer patients and mouse cachexia models (Petruzzelli & Wagner, 2016; Yoshikawa, Noguchi, Doi, Makino, & Nomura, 2001). Strikingly, the growth of intestinal and transplanted disc tumors is not affected during wasting. Conceivably, these observations suggest that this aversive condition can be advantageous to the tumors by making more nutrients available for them to support growth. These systemic organ-wasting phenotypes are not caused by starvation. Strikingly, both studies discovered that the tumors express the secreted antagonist of insulin/IGF signaling, ImpL2, which inhibits Dilps by direct binding (Honegger et al., 2008; Fig. 3). Furthermore, suppression of *ImpL2* expression in tumors using a tissue-specific RNAi approach was sufficient to rescue the various wasting phenotypes, including muscle degeneration, observed in the flies with tumors (Figueroa-Clarevega & Bilder, 2015; Kwon et al., 2015).

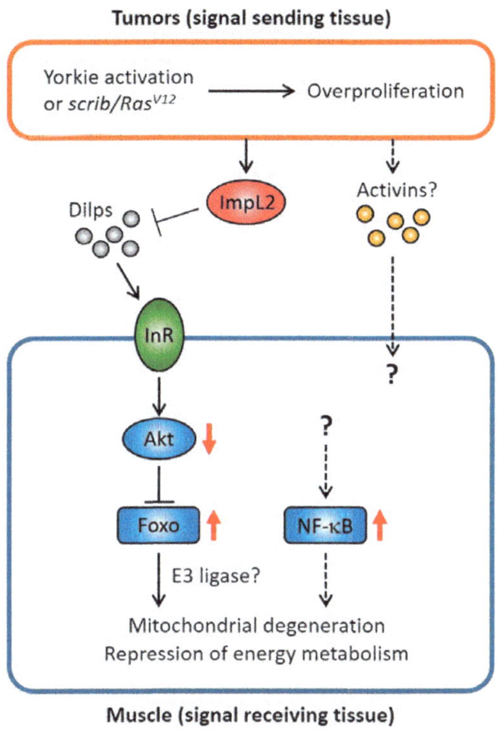

Fig. 3 Signaling pathways altered during muscle wasting. *ImpL2* expression is elevated in intestinal tumors induced by activation of Yki or transplanted *scrib/Ras*^{V12} disc tumors. ImpL2 causes systemic downregulation of insulin/IGF signaling by antagonizing Drosophila insulin-like peptides (Dilps). In muscles, Akt phosphorylation is reduced, leading to activation of Foxo. Additionally, activation of NF-kB signaling is observed, while the contribution of NF-kB activation to muscle wasting is not characterized.

4.2 Muscle-Wasting Phenotypes

Both studies demonstrate that the tumors in intestine or hemocoel greatly impair muscle function and metabolism (Figueroa-Clarevega & Bilder, 2015; Kwon et al., 2015). The flies with Yki-induced intestinal tumors showed a progressive climbing defect and a downturned wing phenotype (Kwon et al., 2015), which are general markers of muscle weakening/degeneration (Demontis & Perrimon, 2010; Greene et al., 2003). Moreover, both studies also reported abnormalities in muscle mitochondrial morphology. Figueroa-Clarevega and Bilder (2015) noted that the mitochondria in muscle were irregularly packed and displayed abnormal morphology. Electron microscopy of muscles from the flies with intestinal tumors revealed signs of mitochondrial degeneration: mitochondria were swollen, cristae were fragmented, and the electron density in the inner mitochondrial space was greatly reduced (Kwon et al., 2015). Strikingly, similar alterations in muscle mitochondrial morphology were observed in a mouse model of cancer cachexia (Fontes-Oliveira et al., 2013). Moreover, muscle ATP levels are greatly reduced in both *Drosophila* and mouse during wasting (Figueroa-Clarevega & Bilder, 2015; Fontes-Oliveira et al., 2013; Kwon et al., 2015), which can be explained by mitochondrial degeneration. Interestingly, *ImpL2* knockdown specifically in the tumors rescued the muscle dysfunction as well as the mitochondrial degeneration phenotypes (Figueroa-Clarevega & Bilder, 2015; Kwon et al., 2015).

Another striking phenotype observed in the flies with tumors is metabolic reprogramming in muscle. Kwon et al. (2015) thoroughly examined metabolic reprogramming in the wasting muscle. They discovered that expression of the genes in energy metabolism, comprising glycolysis, TCA cycle, and oxidative phosphorylation, is systematically downregulated. Consistently, the activities of hexokinase and pyruvate kinase, which catalyze two rate-limiting steps in glycolysis, are significantly reduced in muscle from the flies with intestinal tumors (Kwon et al., 2015). Metabolomic analysis of hemolymph showed a reduction in ATP, NADPH, and NADH levels, which are the main products of energy metabolism. Moreover, the repression of energy metabolism in muscle can contribute to the hyperglycemia phenotype by facilitating accumulation of sugar in hemolymph. RNA-seq analysis of the muscle transcriptome revealed alterations in additional metabolic pathways. In particular, the wasting muscles showed a systematic repression of the genes involved in glutathione metabolism and ascorbate metabolism, which play critical roles in the response to oxidative stress. It has been shown that oxidative stress contributes to the development of anorexia and cachexia in cancer patients (Mantovani, Madeddu, & Maccio, 2012). Therefore, repression of glutathione metabolism and ascorbate metabolism could greatly contribute to the wasting phenotypes in flies. The metabolic changes in the flies with tumors can be explained, at least in part, by the action of ImpL2 since *ImpL2* knockdown in the tumors significantly restored ATP levels in muscles and rescued hyperglycemia (Figueroa-Clarevega & Bilder, 2015; Kwon et al., 2015). Further investigation of metabolic reprogramming during wasting using state-of-the-art genetics and genomics tools will help to unveil the roles of these metabolic alterations in the wasting processes.

4.3 Signaling Pathways Altered in Wasting Muscles

Cachexia is a multifactorial disease, resulting from alternations in a combination of signaling pathways (Argilés et al., 2014; Fearon, Arends, & Baracos, 2013; Petruzzelli & Wagner, 2016; Tsoli & Robertson, 2013). The fly models of cachexia-like wasting provide a unique opportunity to delineate the signaling pathways involved in cachexia. IGF signaling plays a crucial role in muscle growth, and reduction of IGF signaling is observed in many wasting conditions (Costelli et al., 2006; Penna et al., 2010; Petruzzelli & Wagner, 2016; Rommel et al., 2001). Strikingly, studies in *Drosophila* discovered that systemic insulin/IGF signaling is significantly reduced in the flies with tumors in intestine or hemocoel (Figueroa-Clarevega & Bilder, 2015; Kwon et al., 2015). Phosphorylation of AKT, a marker of insulin/IGF signaling, is dramatically decreased in muscle of the flies with intestinal tumors (Fig. 3). Foxo is a transcription factor in the insulin/IGF signaling pathway; AKT directly inhibits Foxo by phosphorylation. In rodent models of wasting, Foxo plays an important role in muscle wasting by inducing the expression of two ubiquitin ligases, Muscle Ring Finger 1 (Murf1; also known as TRIM63) and Muscle Atrophy F-box protein (MAFbx, also known as Atrogin-1), which are involved in degradation of muscle proteins (Bodine et al., 2001). Notably, *Drosophila* lacks MuRF1, and expression of the *Drosophila* MAFBx/Atrogin ortholog is unaltered during muscle degeneration. Therefore, further scrutiny is required to address whether degradation of the sarcomere is a critical process during wasting in *Drosophila*, and searching for ubiquitin ligases involved in muscle degeneration in *Drosophila* may lead to the identification of novel catabolic mechanisms underlying muscle wasting. In *Drosophila*, 4E-BP mRNA levels are greatly increased during muscle wasting, an indication of increased Foxo activity (Fig. 3). Furthermore, PIP_3 levels are greatly reduced in fat body and ovary of tumor-bearing flies, which indicates a decrease in PI3-kinase activity (Figueroa-Clarevega & Bilder, 2015). Although severe reduction of food consumption can lead to the same observations, food consumption is not significantly altered in the flies with tumors. Strikingly, tumors in the intestine and hemocoel secrete ImpL2, which antagonizes insulin/IGF signaling by binding to *Drosophila* insulin-like peptides (Dilps). Specific knockdown of *ImpL2* in the tumors suppresses muscle wasting and increases insulin/IGF signaling in muscle and other distant tissues. *ImpL2* knockout further rescued systemic organ-wasting phenotypes induced by intestinal tumors, suggesting that an increase in insulin/IGF signaling is sufficient to rescue wasting of muscle, ovary, and fat body. Additionally, reduction of *ImpL2* expression in tumors by RNAi significantly, but only partially, rescued hyperglycemia phenotype observed in the flies bearing tumors (Figueroa-Clarevega & Bilder, 2015; Kwon et al., 2015).

Systemic inflammation is commonly observed in mouse models of cachexia and cancer patients (Fearon et al., 2013; Petruzzelli & Wagner, 2016; Tsoli & Robertson, 2013). It has been reported that nuclear factor NF-kB, which can be activated by proinflammatory cytokines including tumor necrosis factor α (Lawrence, 2009), is activated in both mouse models of cachexia and cancer patients (Argilés et al., 2014; Fearon et al., 2013). Interestingly, sequencing mRNA from the muscle in the flies with intestinal tumors showed that expression of NF-kB target genes are greatly increased during muscle wasting in *Drosophila* (Kwon et al., 2015). Another signaling pathway that might be altered in the flies with tumors is the transforming growth factor-β (TGF-β) signaling pathway, which is known to play an important role in muscle wasting. TGF-β family members myostatin and activin-β are negative regulators of muscle mass and antagonizing TGF-β signaling has been shown to suppress muscle wasting in mouse models of cancer cachexia (Chen et al., 2014; Zhou et al., 2010). Therefore, investigation of NF-kB signaling and TGF-β signaling during wasting in *Drosophila* will likely provide further insights into the molecular mechanisms by which these signaling pathways regulate the wasting process.

5. THERAPEUTIC POTENTIAL OF IDENTIFIED IN *DROSOPHILA* SCREENS FACTORS

5.1 Sphingosine-1-Phosphate Pathway

Multiple mutant screens or environment alterations have been tested to identify the critical modifiers for DMD in the *Drosophila* model, in hopes of ultimately helping to generate therapeutic approaches to ameliorate the disease. Among identified interesting candidates is a bioactive lipid, sphingosine-1-phosphate (S1P) and its pathway components (Kucherenko et al., 2008, 2011; Mosqueira, Willmann, Ruohola-Baker, & Khurana, 2010; Pantoja, Fischer, Ieronimakis, Reyes, & Ruohola-Baker, 2013; Pantoja & Ruohola-Baker, 2013). One of the suppressors of the dystrophic defect in *Drosophila* was a lipid phosphate phosphatase (wunen), which was initially identified as a gene required for proper germ cell migration in the *Drosophila* embryo (Kucherenko et al., 2008; Starz-Gaiano, Cho, Forbes, & Lehmann, 2001; Zhang, Zhang, Cheng, & Howard, 1996; Zhang, Zhang,

Purcell, Cheng, & Howard, 1997). Lipid phosphate phosphatases typically inactivate S1P by dephosphorylating the active molecule. Hence, mutations in lipid phosphate phosphatases increase the level of the active lipid, S1P. Rescue analysis revealed that human lipid phosphate phosphatase homologue, LPP3, rescues the wunen phenotype, acting as a functional homologue of *wunen* in germ cell migration (Burnett & Howard, 2003). Despite all the LPPs having the same broad substrate range in vitro, phosphatidic acid, S1P, and their derivatives (Pyne, Long, Ktistakis, & Pyne, 2005), only LPP3 knockout mice are embryonic lethal, indicating a central role for this enzyme in development (Brindley & Pilquil, 2009).

While the S1P pathway had been implicated in many anabolic processes in muscle, this was the first time it was implicated with therapeutic potential with DMD. Previous studies demonstrated that S1P was implicated in muscle repair and satellite cell proliferation and myoblast differentiation in vitro (Bruni & Donati, 2008; Nagata, Partridge, Matsuda, & Zammit, 2006; Rapizzi, Donati, Cencetti, Nincheri, & Bruni, 2008). In *Drosophila*, an increase of S1P suppressed dystrophic muscle defects, as assayed by muscle integrity, Projectin (the *Drosophila* Titin homologue) protein localization in sarcomeres, and fly movement during aging. Furthermore, mutations that increase intracellular S1P levels either biochemically or through the reduction of transport also suppress these dystrophic muscle phenotypes.

Importantly, pharmacological agents that elevate S1P in adult flies also show suppression of the muscular dystrophy phenotypes. For example, oral delivery of THI, an S1P lyase inhibitor, suppresses the dystrophic muscle phenotype in *Drosophila* (Pantoja et al., 2013). In mice, THI also shows partial amelioration of acute muscle injury, significantly increasing muscle fiber size and muscle force, while reducing DMD pathology of fibrosis and fat deposition. Furthermore, treatment with S1P promotes muscle repair in *mdx* mice (Ieronimakis et al., 2013; Loh et al., 2012; Pantoja et al., 2013). These data highlight the potential for using *Drosophila* as a drug-screening tool for DMD. Interestingly, there are already specific S1P receptor agonists (e.g., FTY720) that are currently FDA approved and are in clinical trials for other diseases. Therefore, future studies on S1P-based treatment of DMD and related myopathies should be fruitful. However, it is important to keep in mind that in mammalian organisms S1P has also shown to affect immune cell trafficking, adding complexity to the system (Kunkel, Maceyka, Milstien, & Spiegel, 2013). Since altered sphingolipid metabolism can affect muscle wasting in a muscular dystrophy context, it is also possible that the elevation of S1P could affect other muscle degeneration problems, such as age-dependent degeneration or wasting (Nishimura, Ocorr, Bodmer, & Cartry, 2011).

The identification of a common mode of action of S1P in *Drosophila* and mouse models remains an open area of investigation. S1P action can be mediated by a family of five G protein-coupled receptors (GPCR; Maceyka, Harikumar, Milstien, & Spiegel, 2012; Rosen, Gonzalez-Cabrera, Sanna, & Brown, 2009; Strub, Maceyka, Hait, Milstien, & Spiegel, 2010). However, more recently, S1P has also shown to have important intracellular targets, for example, histone deacetylases, HDAC1 and HDAC2 (Alvarez et al., 2010; Hait et al., 2009). Since *Drosophila* do not express GPCR homologues of known vertebrate S1P receptors, it was proposed that the common mode of function for S1P in dystrophic flies and mice is intracellular. Since muscular dystrophies correlate with increased HDAC2 levels and HDAC inhibitors are beneficial for the DMD disease (Colussi et al., 2008; Consalvi et al., 2013; Minetti et al., 2006), the S1P mode of action in DMD through HDACs was tested. Importantly, increasing nuclear S1P levels in *mdx* mice decreases HDAC activity, which, in turn, increases histone acetylation that results in upregulation of metabolic muscle genes and key microRNAs (Nguyen-Tran et al., 2014). It is, therefore, plausible that HDAC inhibition might be the ancestral function of S1P in muscle.

5.2 miRNAs in Muscular Dystrophies

Previously performed genetic screens in *Drosophila* allowed to determine multiple novel factors that can potentially act as modifiers of muscular dystrophy phenotypes. These screens also revealed a signaling function for the DGC to regulate miRNA expression (Marrone et al., 2011). miRNAs are a group of small noncoding RNAs that negatively regulate gene expression. Since their discovery only two decades ago, they have been implicated in virtually all biological processes. Even though we still do not fully understand the molecular mechanisms of miRNAs-based regulation of gene expression, nor do we know the full list of mRNAs each miRNA regulates, miRNAs have already developed as biomarkers and new therapeutic targets for muscular dystrophies (Cacchiarelli et al., 2011; Li et al., 2014). It has been shown that miRNAs not only play an important function in muscle development and integrity; several miRNAs have been identified as novel components that may be involved in DGC signaling and regulation (Cardinali et al., 2009; Chen et al., 2006; de la Garza-Rodea et al., 2014;

Shcherbata & Edeleva, 2013; Kim, Lee, Sivaprasad, Malhotra, & Dutta, 2006; Marrone & Shcherbata, 2011; van Rooij, Liu, & Olson, 2008). Even more, expression of multiple myo-specific and stress-responsive miRNAs has been shown to be dependent on the activity of the DGC complex (Adams et al., 2008; Cacchiarelli et al., 2010; Marrone, Edeleva, Kucherenko, Hsiao, & Shcherbata, 2012).

The DGC is known as a scaffold responsible for the localization of signaling proteins; various kinases, channels, and other enzymes have been shown to associate with the DGC (Adams et al., 2008; Pilgram et al., 2010). Particularly, DGC-coordinated neuronal nitric oxide synthase (nNOS) signaling plays a role in nitrosylation of histone deacetylases (HDACs; see Fig. 1). Upon nitrosylation, HDACs can no longer deacetylase histones, which has a global influence on gene expression. Furthermore, in mammals and *Drosophila*, the DGC–syntrophin–nNOS–HDAC signaling cascade is responsible for the direct regulation of a subset of myo-specific miRNAs (Adams et al., 2008; Cacchiarelli et al., 2010; Marrone et al., 2012). Further detailed analysis of DGC-dependent miRNAs might be useful to monitor the DMD pathological progression and develop miRNAs as novel biomarkers for DMD diagnosis and targets for therapeutics.

Interestingly, not only does the DGC play a role in miRNA regulation, but it is also regulated by miRNAs (Fiorillo et al., 2015; Yatsenko & Shcherbata, 2014; Yatsenko et al., 2014). The ECM receptor Dystroglycan is subjected to miRNA regulation in a tissue-specific manner. In *Drosophila*, miRNA-based regulation of Dystroglycan by *miR-9a* is essential for proper formation of muscle attachment sites in the developing embryo (Yatsenko & Shcherbata, 2014). During embryogenesis, both *miR-9a* and Dystroglycan have dynamic expression patterns that become mutually exclusive in the regions of muscle-tendon connections. Elimination of Dystroglycan from tendon precursor cells is required for accurate muscle-tendon matrix assembly and *miR-9a* ensures that Dystroglycan is not misexpressed in tendon precursors due to transcriptional noise, as these epidermal cells invaginate into and reside within the mesoderm. Since many other essential muscle differentiation genes are *miR-9a* predicted targets, *miR-9a* prevents noisy muscle gene expression in the muscle attachment precursor cells and acts as a guardian angel that reassures the fidelity of muscle attachment site formation (Yatsenko & Shcherbata, 2014).

Adjustment of Dystroglycan levels by miRNAs also takes place during neuron differentiation as neurons undergo complex rearrangements during development, migration, axon growth and projections, and synapse formation. In humans, the most profound neurodevelopmental disorder associated with Dystroglycan dysfunction is the cobblestone brain, which is an odd malformation caused by neuron over-migration and formation of cobblestone-like structures that outgrow the normal contour of the ECM-defined brain space (Devisme et al., 2012; Waite et al., 2012). Analysis of a recently developed *Drosophila* model for cobblestone lissencephaly revealed the *miR-310s* as an executive mechanism to buffer levels of the muscular dystrophy-associated ECM receptor Dystroglycan via its alternative *3′UTR* to establish the proper brain structure (Yatsenko et al., 2014). Since multiple aspects of neuron maturation involve rigorous rearrangements of cell shape and form, extracellular matrix proteins, including Dystroglycan, are involved in regulation of numerous aspects of neuron differentiation and neural tissue assembly (Marrone et al., 2011; Shcherbata et al., 2007). Dystroglycan is a mechanosensor that is activated by mechanical stress due to cell shape reorganization and after the completion of cell shape rearrangements; Dystroglycan levels have to be reestablished by miRNA action. Thus, studies in *Drosophila* show that miRNAs play fundamental roles in control of this critical protein expression.

6. CONCLUSIONS

Though seemingly far removed from humans and the diseases we suffer, *D. melanogaster* actually provides an ideal platform with which one can study the genetic basis of complex human diseases. A century's worth of close study has yielded a plethora of genetic techniques that have enabled researchers to elucidate the biochemical pathways in flies that would be too difficult or expensive to fully trace originally in higher order vertebrates. Because many of the molecular pathways involved in basic functions of the cell are well conserved between invertebrates and vertebrates, the foundation established in fly work has allowed the development of much more refined theories surrounding the pathogenesis of debilitating human diseases. Additionally, the high level of genetic tractability of the fly continues to allow for screens which will reveal novel enhancers and suppressors of disease-relevant pathways, possibly even leading to the development of novel therapeutics, as seen with DMD.

Here, we have demonstrated the powerful role that *D. melanogaster* has played in elucidating pathophysiological mechanisms of several muscle degenerative disorders. In many of these diseases, there is a striking similarity between the human muscle disease and the fly disease and many of the phenotypes can be recapitulated. This allows us to use the fly muscle as a model to study many diseases of wasting and degeneration. It is also possible that insight gained from one model may be used to inform progress made on other diseases. For example, as we have discussed, S1P can be beneficial in ameliorating muscle wasting in DMD. It is of interest to test that this understanding can be applied to other muscle-wasting diseases, such as cachexia, to provide additional therapeutic targets and pathways.

Since Morgan and colleagues first demonstrated the power of *D. melanogaster* as model for human genetics a century ago, much has been learned about the pathogenesis of human diseases. In the next decades of study, flies will continue to provide solid groundwork to underpin the development of new theories about the pathogenesis of and therapeutic targets for diseases that to date have no known cure or treatment.

REFERENCES

Adams, M. E., Tesch, Y., Percival, J. M., Albrecht, D. E., Conhaim, J. I., Anderson, K., et al. (2008). Differential targeting of nNOS and AQP4 to dystrophin-deficient sarcolemma by membrane-directed α-dystrobrevin. *Journal of Cell Science, 121*, 48–54.

Alvarez, S. E., Harikumar, K. B., Hait, N. C., Allegood, J., Strub, G. M., Kim, E. Y., et al. (2010). Sphingosine-1-phosphate is a missing cofactor for the E3 ubiquitin ligase TRAF2. *Nature, 465*, 1084–1088.

Argilés, J. M., Busquets, S., Stemmler, B., & López-Soriano, F. J. (2014). Cancer cachexia: Understanding the molecular basis. *Nature Reviews. Cancer, 14*, 754–762.

Barresi, R., & Campbell, K. P. (2006). Dystroglycan: From biosynthesis to pathogenesis of human disease. *Journal of Cell Science, 119*, 199–207.

Bodine, S. C., Latres, E., Baumhueter, S., Lai, V. K.-M., Nunez, L., Clarke, B. A., et al. (2001). Identification of ubiquitin ligases required for skeletal muscle atrophy. *Science, 294*, 1704–1708.

Bogdanik, L., Framery, B., Frölich, A., Franco, B., Mornet, D., Bockaert, J., et al. (2008). Muscle dystroglycan organizes the postsynapse and regulates presynaptic neurotransmitter release at the *Drosophila* neuromuscular junction. *PLoS One, 3*, e2084.

Brenman, J. E., Chao, D. S., Xia, H., Aldape, K., & Bredt, D. S. (1995). Nitric oxide synthase complexed with dystrophin and absent from skeletal muscle sarcolemma in Duchenne muscular dystrophy. *Cell, 82*, 743–752.

Brindley, D. N., & Pilquil, C. (2009). Lipid phosphate phosphatases and signaling. *Journal of Lipid Research, 50*, S225–S230.

Bruni, P., & Donati, C. (2008). Pleiotropic effects of sphingolipids in skeletal muscle. *Cellular and Molecular Life Sciences, 65*, 3725–3736.

Burnett, C., & Howard, K. (2003). Fly and mammalian lipid phosphate phosphatase isoforms differ in activity both in vitro and in vivo. *EMBO Reports, 4*, 793–799.

Cacchiarelli, D., Legnini, I., Martone, J., Cazzella, V., D'Amico, A., Bertini, E., et al. (2011). miRNAs as serum biomarkers for Duchenne muscular dystrophy. *EMBO Molecular Medicine, 3*(5), 258–265.

Cacchiarelli, D., Martone, J., Girardi, E., Cesana, M., Incitti, T., Morlando, M., et al. (2010). MicroRNAs involved in molecular circuitries relevant for the Duchenne muscular dystrophy pathogenesis are controlled by the dystrophin/nNOS pathway. *Cell Metabolism, 12*, 341–351.

Cannon, L., Yu, Z.-Y., Marciniec, T., Waardenberg, A. J., Iismaa, S. E., Nikolova-Krstevski, V., et al. (2015). Irreversible triggers for hypertrophic cardiomyopathy are established in the early postnatal period. *Journal of the American College of Cardiology, 65*, 560–569. http://dx.doi.org/10.1016/j.jacc.2014.10.069.

Cardinali, B., Castellani, L., Fasanaro, P., Basso, A., Alema, S., Martelli, F., et al. (2009). Microrna-221 and microrna-222 modulate differentiation and maturation of skeletal muscle cells. *PLoS ONE, 4*(10), e7607.

Casci, I., & Pandey, U. B. (2015). A fruitful endeavor: Modeling ALS in the fruit fly. *Brain Research, 1607*, 47–74.

Chang, H. C.-H., Dimlich, D. N., Yokokura, T., Mukherjee, A., Kankel, M. W., Sen, A., et al. (2008). Modeling spinal muscular atrophy in *Drosophila*. *PLoS One, 3*, e3209.

Chen, J. F., Mandel, E. M., Thomson, J. M., Wu, Q., Callis, T. E., Hammond, S. M., et al. (2006). The role of microRNA-1 and microRNA-133 in skeletal muscle proliferation and differentiation. *Nature Genetics, 38*(2), 228–233.

Chen, J. L., et al. (2014). Elevated expression of activins promotes muscle wasting and cachexia. *The FASEB Journal, 28*(4), 1711–1723.

Christoforou, C. P., Greer, C. E., Challoner, B. R., Charizanos, D., & Ray, R. P. (2008). The detached locus encodes *Drosophila* Dystrophin, which acts with other components of the Dystrophin Associated Protein Complex to influence intercellular signalling in developing wing veins. *Developmental Biology, 313*, 519–532.

Colussi, C., Mozzetta, C., Gurtner, A., Illi, B., Rosati, J., Straino, S., et al. (2008). HDAC2 blockade by nitric oxide and histone deacetylase inhibitors reveals a common target in Duchenne muscular dystrophy treatment. *Proceedings of the National Academy of Sciences of the United States of America, 105*, 19183–19187.

Consalvi, S., Mozzetta, C., Bettica, P., Germani, M., Fiorentini, F., Del Bene, F., et al. (2013). Preclinical studies in the mdx mouse model of duchenne muscular dystrophy with the histone deacetylase inhibitor givinostat. *Molecular Medicine, 19*, 79–87.

Costelli, P., Muscaritoli, M., Bossola, M., Penna, F., Reffo, P., Bonetto, A., et al. (2006). IGF-1 is downregulated in experimental cancer cachexia. *American Journal of Physiology. Regulatory, Integrative and Comparative Physiology, 291*, R674–R683.

Dekkers, L. C., van der Plas, M. C., van Loenen, P. B., den Dunnen, J. T., van Ommen, G.-J. B., Fradkin, L. G., et al. (2004). Embryonic expression patterns of the Drosophila dystrophin-associated glycoprotein complex orthologs. *Gene Expression Patterns, 4*, 153–159.

de la Garza-Rodea, A. S., Baldwin, D. M., Oskouian, B., Place, R. F., Bandhuvula, P., Kumar, A., et al. (2014). Sphingosine phosphate lyase regulates myogenic differentiation via S1P receptor-mediated effects on myogenic microRNA expression. *The FASEB Journal, 28*(1), 506–519.

Demontis, F., & Perrimon, N. (2010). FOXO/4E-BP signaling in *Drosophila* muscles regulates organism-wide proteostasis during aging. *Cell, 143*, 813–825.

Deshpande, M., & Rodal, A. A. (2015). The crossroads of synaptic growth signaling, membrane traffic and neurological disease: Insights from *Drosophila*. *Traffic, 17*, 87–101.

Devisme, L., Bouchet, C., Gonzalès, M., Alanio, E., Bazin, A., Bessières, B., et al. (2012). Cobblestone lissencephaly: Neuropathological subtypes and correlations with genes of dystroglycanopathies. *Brain, 135*, 469–482.

Dickman, D. K., & Davis, G. W. (2009). The schizophrenia susceptibility gene dysbindin controls synaptic homeostasis. *Science, 326*, 1127–1130.

Elia, A. J., Parkes, T. L., Kirby, K., St George-Hyslop, P., Boulianne, G.-H. L., Phillips, J. P., et al. (1999). Expression of human FALS SOD in motorneurons of *Drosophila*. *Free Radical Biology & Medicine, 26*, 1332–1338.

Fallini, C., Zhang, H., Su, Y., Silani, V., Singer, R. H., Rossoll, W., et al. (2011). The survival of motor neuron (SMN) protein interacts with the mRNA-binding protein HuD and regulates localization of poly (A) mRNA in primary motor neuron axons. *The Journal of Neuroscience, 31*, 3914–3925.

Fearon, K., Arends, J., & Baracos, V. (2013). Understanding the mechanisms and treatment options in cancer cachexia. *Nature Reviews. Clinical Oncology, 10*, 90–99.

Fearon, K. C. H., Glass, D. J., & Guttridge, D. C. (2012). Cancer cachexia: Mediators, signaling, and metabolic pathways. *Cell Metabolism, 16*, 153–166.

Figueroa-Clarevega, A., & Bilder, D. (2015). Malignant *Drosophila* tumors interrupt insulin signaling to induce cachexia-like wasting. *Developmental Cell, 33*, 47–55.

Fiorillo, A. A., Heier, C. R., Novak, J. S., Tully, C. B., Brown, K. J., Uaesoontrachoon, K., et al. (2015). TNF-α-induced microRNAs control dystrophin expression in Becker muscular dystrophy. *Cell Reports, 12*, 1678–1690.

Fontes-Oliveira, C. C., Busquets, S., Toledo, M., Penna, F., Aylwin, M. P., Sirisi, S., et al. (2013). Mitochondrial and sarcoplasmic reticulum abnormalities in cancer cachexia: Altered energetic efficiency? *Biochimica et Biophysica Acta (BBA) General Subject, 1830*, 2770–2778.

Gao, X., Shen, X., Dong, X., Ran, N., Han, G., Cao, L., et al. (2015). Peptide nucleic acid promotes systemic dystrophin expression and functional rescue in dystrophin-deficient mdx mice. *Molecular Therapy. Nucleic Acids, 4*, e255.

Greene, J. C., Whitworth, A. J., Kuo, I., Andrews, L. A., Feany, M. B., & Pallanck, L. J. (2003). Mitochondrial pathology and apoptotic muscle degeneration in *Drosophila* parkin mutants. *Proceedings of the National Academy of Sciences of the United States of America, 100*, 4078–4083.

Grice, S. J., Praveen, K., Matera, A. G., & Liu, J. L. (2013). *Spinal muscular atrophy*. Hauppauge, NY: Nova Science Publishers, Inc.

Hadorn, E. (1963). Differenzierungsleistungen wiederholt fragmentierter Teilstücke männlicher Genitalscheiben von *Drosophila melanogaster* nach Kultur in vivo. *Developmental Biology*, 7, 617–629.

Hait, N. C., Allegood, J., Maceyka, M., Strub, G. M., Harikumar, K. B., Singh, S. K., et al. (2009). Regulation of histone acetylation in the nucleus by sphingosine-1-phosphate. *Science, 325*, 1254–1257.

Hariharan, I. K. (2015). Organ size control: Lessons from *Drosophila*. *Developmental Cell, 34*, 255–265.

Hoffman, E. P., Brown, R. H., & Kunkel, L. M. (1987). Dystrophin: The protein product of the Duchenne muscular dystrophy locus. *Cell, 51*, 919–928.

Honegger, B., Galic, M., Köhler, K., Wittwer, F., Brogiolo, W., Hafen, E., et al. (2008). Imp-L2, a putative homolog of vertebrate IGF-binding protein 7, counteracts insulin signaling in *Drosophila* and is essential for starvation resistance. *Journal of Biology*, 7, 1.

Ibraghimov-Beskrovnaya, O., Milatovich, A., Ozcelik, T., Yang, B., Koepnick, K., Francke, U., et al. (1993). Human dystroglycan: Skeletal muscle cDNA, genomic structure, origin of tissue specific isoforms and chromosomal localization. *Human Molecular Genetics, 2*, 1651–1657.

Ieronimakis, N., Hays, A. L., Janebodin, K., Mahoney, W. M., Duffield, J. S., Majesky, M. W., et al. (2013). Coronary adventitial cells are linked to perivascular cardiac fibrosis via TGFβ1 signaling in the mdx mouse model of Duchenne muscular dystrophy. *Journal of Molecular and Cellular Cardiology, 63*, 122–134.

Inamori, K., Yoshida-Moriguchi, T., Hara, Y., Anderson, M. E., Yu, L., & Campbell, K. P. (2012). Dystroglycan function requires xylosyl-and glucuronyltransferase activities of LARGE. *Science, 335*, 93–96.

Johnson, R., & Halder, G. (2014). The two faces of Hippo: Targeting the Hippo pathway for regenerative medicine and cancer treatment. *Nature Reviews. Drug Discovery, 13*, 63–79.

Keshishian, H., Broadie, K., Chiba, A., & Bate, M. (1996). The *Drosophila* neuromuscular junction: A model system for studying synaptic development and function. *Annual Review of Neuroscience, 19*, 545–575.
Kim, H. K., Lee, Y. S., Sivaprasad, U., Malhotra, A., & Dutta, A. (2006). Muscle-specific microRNA miR-206 promotes muscle differentiation. *The Journal of Cell Biology, 174*(5), 677–687.
Koenig, M., Hoffman, E. P., Bertelson, C. J., Monaco, A. P., Feener, C., & Kunkel, L. M. (1987). Complete cloning of the Duchenne muscular dystrophy (DMD) cDNA and preliminary genomic organization of the DMD gene in normal and affected individuals. *Cell, 50*, 509–517.
Kucherenko, M. M., Marrone, A. K., Rishko, V. M., de Fatima Magliarelli, H., & Shcherbata, H. R. (2011). Stress and muscular dystrophy: A genetic screen for dystroglycan and dystrophin interactors in *Drosophila* identifies cellular stress response components. *Developmental Biology, 352*, 228–242.
Kucherenko, M. M., Pantoja, M., Yatsenko, A. S., Shcherbata, H. R., Fischer, K. A., Maksymiv, D. V., et al. (2008). Genetic modifier screens reveal new components that interact with the *Drosophila* dystroglycan-dystrophin complex. *PLoS One, 3*, e2418.
Kunkel, G. T., Maceyka, M., Milstien, S., & Spiegel, S. (2013). Targeting the sphingosine-1-phosphate axis in cancer, inflammation and beyond. *Nature Reviews Drug Discovery, 12*, 688–702.
Kwon, Y., Song, W., Droujinine, I. A., Hu, Y., Asara, J. M., & Perrimon, N. (2015). Systemic organ wasting induced by localized expression of the secreted insulin/IGF antagonist ImpL2. *Developmental Cell, 33*, 36–46.
Lapidos, K. A., Kakkar, R., & McNally, E. M. (2004). The dystrophin glycoprotein complex signaling strength and integrity for the sarcolemma. *Circulation Research, 94*, 1023–1031.
Lawrence, T. (2009). The nuclear factor NF-kappaB pathway in inflammation. *Cold Spring Harbor Perspectives in Biology, 1*, a001651.
Li, X., Li, Y., Zhao, L., Zhang, D., Yao, X., Zhang, H., et al. (2014). Circulating muscle-specific miRNAs in Duchenne muscular dystrophy patients. *Molecular Therapy Nucleic Acids, 3*, e177.
Lin, G., Mao, D., & Bellen, H. J. (2017). Amyotrophic lateral sclerosis pathogenesis converges on defects in protein homeostasis associated with TDP-43 mislocalization and proteasome-mediated degradation overload. *Current Topics in Developmental Biology, 121*, 111–171.
Liu, J., Burkin, D. J., & Kaufman, S. J. (2008). Increasing $\alpha7\beta1$-integrin promotes muscle cell proliferation, adhesion, and resistance to apoptosis without changing gene expression. *American Journal of Physiology—Cell Physiology, 294*, C627–C640.
Lloyd, T. E., & Taylor, J. P. (2010). Flightless flies: *Drosophila* models of neuromuscular disease. *Annals of the New York Academy of Sciences, 1184*, E1–E20.
Loh, K. C., Leong, W.-I., Carlson, M. E., Oskouian, B., Kumar, A., Fyrst, H., et al. (2012). Sphingosine-1-phosphate enhances satellite cell activation in dystrophic muscles through a S1PR2/STAT3 signaling pathway. *PLoS One, 7*, e37218.
Lok, C. (2015). Cachexia: The last illness. *Nature, 528*, 182.
Lorson, C. L., Strasswimmer, J., Yao, J.-M., Baleja, J. D., Hahnen, E., Wirth, B., et al. (1998). SMN oligomerization defect correlates with spinal muscular atrophy severity. *Nature Genetics, 19*, 63–66.
Maceyka, M., Harikumar, K. B., Milstien, S., & Spiegel, S. (2012). Sphingosine-1-phosphate signaling and its role in disease. *Trends in Cell Biology, 22*, 50–60.
Mangner, N., Matsuo, Y., Schuler, G., & Adams, V. (2013). Cachexia in chronic heart failure: Endocrine determinants and treatment perspectives. *Endocrine, 43*, 253–265.
Mantovani, G., Madeddu, C., & Maccio, A. (2012). Cachexia and oxidative stress in cancer: An innovative therapeutic management. *Current Pharmaceutical Design, 18*, 4813–4818.

Marrone, A. K., Edeleva, E. V., Kucherenko, M. M., Hsiao, N.-H., & Shcherbata, H. R. (2012). Dg-Dys-Syn1 signaling in Drosophila regulates the microRNA profile. *BMC Cell Biology*, *13*, 1.

Marrone, A. K., Kucherenko, M. M., Wiek, R., Göpfert, M. C., & Shcherbata, H. R. (2011). Hyperthermic seizures and aberrant cellular homeostasis in *Drosophila* dystrophic muscles. *Scientific Reports*, *1*, 47.

Marrone, A. K., & Shcherbata, H. R. (2011). Dystrophin orchestrates the epigenetic profile of muscle cells via miRNAs. *Frontiers in Genetics*, *2*, 64.

McGurk, L., Berson, A., & Bonini, N. M. (2015). *Drosophila* as an in vivo model for human neurodegenerative disease. *Genetics*, *201*, 377–402.

McNally, E. M., & Goldstein, J. A. (2012). Interplay between heart and skeletal muscle disease in heart failure the 2011 George E. Brown Memorial Lecture. *Circulation Research*, *110*, 749–754.

Minetti, G. C., Colussi, C., Adami, R., Serra, C., Mozzetta, C., Parente, V., et al. (2006). Functional and morphological recovery of dystrophic muscles in mice treated with deacetylase inhibitors. *Nature Medicine*, *12*, 1147–1150.

Mirouse, V., Christoforou, C. P., Fritsch, C., & Ray, R. P. (2009). Dystroglycan and perlecan provide a basal cue required for epithelial polarity during energetic stress. *Developmental Cell*, *16*, 83.

Mockett, R. J., Radyuk, S. N., Benes, J. J., Orr, W. C., & Sohal, R. S. (2003). Phenotypic effects of familial amyotrophic lateral sclerosis mutant Sod alleles in transgenic *Drosophila. Proceedings of the National Academy of Sciences of the United States of America*, *100*, 301–306.

Moore, C. J., & Winder, S. J. (2010). Dystroglycan versatility in cell adhesion: A tale of multiple motifs. *Cell Communication and Signaling*, *8*, 1.

Mosqueira, M., Willmann, G., Ruohola-Baker, H., & Khurana, T. S. (2010). Chronic hypoxia impairs muscle function in the *Drosophila* model of Duchenne's muscular dystrophy (DMD). *PLoS One*, *5*, e13450.

Nagata, Y., Partridge, T. A., Matsuda, R., & Zammit, P. S. (2006). Entry of muscle satellite cells into the cell cycle requires sphingolipid signaling. *The Journal of Cell Biology*, *174*, 245–253.

Nguyen-Tran, D.-H., Hait, N. C., Sperber, H., Qi, J., Fischer, K., Ieronimakis, N., et al. (2014). Molecular mechanism of sphingosine-1-phosphate action in Duchenne muscular dystrophy. *Disease Models & Mechanisms*, 7, 41–54.

NINDS. (2012). *Motor Neuron Diseases Fact Sheet*. Bethesda, MD: National Institutes of Health.

Nishimura, A. L., Mitne-Neto, M., Silva, H. C. A., Oliveira, J. R. M., Vainzof, M., & Zatz, M. (2004). A novel locus for late onset amyotrophic lateral sclerosis/motor neurone disease variant at 20q13. *Journal of Medical Genetics*, *41*, 315–320. http://dx.doi.org/10.1136/jmg.2003.013029.

Nishimura, M., Ocorr, K., Bodmer, R., & Cartry, J. (2011). *Drosophila* as a model to study cardiac aging. *Experimental Gerontology*, *46*, 326–330.

Oh, H., & Irvine, K. D. (2010). Yorkie: The final destination of Hippo signaling. *Trends in Cell Biology*, *20*, 410–417.

Pan, D. (2010). The hippo signaling pathway in development and cancer. *Developmental Cell*, *19*, 491–505.

Pantoja, M., Fischer, K. A., Ieronimakis, N., Reyes, M., & Ruohola-Baker, H. (2013). Genetic elevation of sphingosine 1-phosphate suppresses dystrophic muscle phenotypes in *Drosophila. Development*, *140*, 136–146.

Pantoja, M., & Ruohola-Baker, H. (2013). *Drosophila* as a starting point for developing therapeutics for the rare disease Duchenne Muscular Dystrophy. *Rare Diseases*, *1*, 136–146.

Parkes, T. L., Elia, A. J., Dickinson, D., Hilliker, A. J., Phillips, J. P., & Boulianne, G. L. (1998). Extension of *Drosophila* lifespan by overexpression of human SOD1 in motorneurons. *Nature Genetics*, *19*, 171–174.
Penna, F., Bonetto, A., Muscaritoli, M., Costamagna, D., Minero, V. G., Bonelli, G., et al. (2010). Muscle atrophy in experimental cancer cachexia: Is the IGF-1 signaling pathway involved? *International Journal of Cancer*, *127*, 1706–1717.
Petruzzelli, M., & Wagner, E. F. (2016). Mechanisms of metabolic dysfunction in cancer-associated cachexia. *Genes & Development*, *30*, 489–501.
Pilgram, G. S. K., Potikanond, S., Baines, R. A., Fradkin, L. G., & Noordermeer, J. N. (2010). The roles of the dystrophin-associated glycoprotein complex at the synapse. *Molecular Neurobiology*, *41*, 1–21.
Pilgram, G. S. K., Potikanond, S., van der Plas, M. C., Fradkin, L. G., & Noordermeer, J. N. (2011). The RhoGAP crossveinless-c interacts with Dystrophin and is required for synaptic homeostasis at the *Drosophila* neuromuscular junction. *The Journal of Neuroscience*, *31*, 492–500.
Praveen, K., Wen, Y., & Matera, A. G. (2012). A *Drosophila* model of spinal muscular atrophy uncouples snRNP biogenesis functions of survival motor neuron from locomotion and viability defects. *Cell Reports*, *1*, 624–631.
Pyne, S., Long, J. S., Ktistakis, N. T., & Pyne, N. J. (2005). Lipid phosphate phosphatases and lipid phosphate signalling. *Biochemical Society Transactions*, *33*, 1370–1374.
Rajendra, T. K., Gonsalvez, G. B., Walker, M. P., Shpargel, K. B., Salz, H. K., & Matera, A. G. (2007). A *Drosophila melanogaster* model of spinal muscular atrophy reveals a function for SMN in striated muscle. *The Journal of Cell Biology*, *176*, 831–841.
Rapizzi, E., Donati, C., Cencetti, F., Nincheri, P., & Bruni, P. (2008). Sphingosine 1-phosphate differentially regulates proliferation of C2C12 reserve cells and myoblasts. *Molecular and Cellular Biochemistry*, *314*, 193–199.
Reiter, L. T., Potocki, L., Chien, S., Gribskov, M., & Bier, E. (2001). A systematic analysis of human disease-associated gene sequences in *Drosophila melanogaster*. *Genome Research*, *11*, 1114–1125. http://dx.doi.org/10.1101/gr.169101.
Remels, A. H. V., Gosker, H. R., Langen, R. C. J., & Schols, A. M. W. J. (2013). The mechanisms of cachexia underlying muscle dysfunction in COPD. *Journal of Applied Physiology*, *114*, 1253–1262.
Rommel, C., Bodine, S. C., Clarke, B. A., Rossman, R., Nunez, L., Stitt, T. N., et al. (2001). Mediation of IGF-1-induced skeletal myotube hypertrophy by PI (3) K/Akt/mTOR and PI (3) K/Akt/GSK3 pathways. *Nature Cell Biology*, *3*, 1009–1013.
Rosen, H., Gonzalez-Cabrera, P. J., Sanna, M. G., & Brown, S. (2009). Sphingosine 1-phosphate receptor signaling. *Annual Review of Biochemistry*, *78*, 743–768.
Rosen, D. R., Siddique, T., Patterson, D., Figlewicz, D. A., Sapp, P., Hentati, A., et al. (1993). Mutations in Cu/Zn superoxide dismutase gene are associated with familial amyotrophic lateral sclerosis. *Nature*, *362*, 59–62.
Shcherbata, H. R., & Edeleva, E. (2013). Stress-induced ECM alteration modulates cellular microRNAs that feedback to readjust the extracellular environment and cell behavior. *Frontiers in Genetics*, *4*, 305.
Shcherbata, H. R., Yatsenko, A. S., Patterson, L., Sood, V. D., Nudel, U., Yaffe, D., et al. (2007). Dissecting muscle and neuronal disorders in a *Drosophila* model of muscular dystrophy. *The EMBO Journal*, *26*, 481–493.
Starz-Gaiano, M., Cho, N. K., Forbes, A., & Lehmann, R. (2001). Spatially restricted activity of a *Drosophila* lipid phosphatase guides migrating germ cells. *Development*, *128*, 983–991.
Strub, G. M., Maceyka, M., Hait, N. C., Milstien, S., & Spiegel, S. (2010). Extracellular and intracellular actions of sphingosine-1-phosphate. In *Sphingolipids as signaling and regulatory molecules* (pp. 141–155). Medford, MA: Springer Science & Business Media.

Taghli-Lamallem, O., Jagla, K., Chamberlain, J. S., & Bodmer, R. (2014). Mechanical and non-mechanical functions of Dystrophin can prevent cardiac abnormalities in *Drosophila*. *Experimental Gerontology*, *49*, 26–34.
Tsoli, M., & Robertson, G. (2013). Cancer cachexia: Malignant inflammation, tumorkines, and metabolic mayhem. *Trends in Endocrinology and Metabolism*, *24*, 174–183.
Van der Plas, M. C., Pilgram, G. S. K., Plomp, J. J., de Jong, A., Fradkin, L. G., & Noordermeer, J. N. (2006). Dystrophin is required for appropriate retrograde control of neurotransmitter release at the *Drosophila* neuromuscular junction. *The Journal of Neuroscience*, *26*, 333–344.
van Rooij, E., Liu, N., & Olson, E. N. (2008). MicroRNAs flex their muscles. *Trends in Genetics*, *24*(4), 159–166.
Waite, A., Brown, S. C., & Blake, D. J. (2012). The dystrophin–glycoprotein complex in brain development and disease. *Trends in Neurosciences*, *35*, 487–496.
Watson, M. R., Lagow, R. D., Xu, K., Zhang, B., & Bonini, N. M. (2008). A drosophila model for amyotrophic lateral sclerosis reveals motor neuron damage by human SOD1. *The Journal of Biological Chemistry*, *283*, 24972–24981.
Yatsenko, A. S., Gray, E. E., Shcherbata, H. R., Patterson, L. B., Sood, V. D., Kucherenko, M. M., et al. (2007). A putative Src homology 3 domain binding motif but not the C-terminal dystrophin WW domain binding motif is required for dystroglycan function in cellular polarity in Drosophila. *The Journal of Biological Chemistry*, *282*, 15159–15169.
Yatsenko, A. S., Kucherenko, M. M., Pantoja, M., Fischer, K. A., Madeoy, J., Deng, W. M., et al. (2009). The conserved WW-domain binding sites in Dystroglycan C-terminus are essential but partially redundant for Dystroglycan function. *BMC Developmental Biology*, *9*, 1.
Yatsenko, A. S., Marrone, A. K., & Shcherbata, H. R. (2014). miRNA-based buffering of the cobblestone–lissencephaly-associated extracellular matrix receptor dystroglycan via its alternative 3′-UTR. *Nature Communications*, *5*, 4906.
Yatsenko, A. S., & Shcherbata, H. R. (2014). *Drosophila* miR-9a targets the ECM receptor dystroglycan to canalize myotendinous junction formation. *Developmental Cell*, *28*, 335–348. http://dx.doi.org/10.1016/j.devcel.2014.01.004.
Yoshikawa, T., Noguchi, Y., Doi, C., Makino, T., & Nomura, K. (2001). Insulin resistance in patients with cancer: Relationships with tumor site, tumor stage, body-weight loss, acute-phase response, and energy expenditure. *Nutrition*, *17*, 590–593.
Yu, F.-X., Zhao, B., & Guan, K.-L. (2015). Hippo pathway in organ size control, tissue homeostasis, and cancer. *Cell*, *163*, 811–828.
Zender, L., Spector, M. S., Xue, W., Flemming, P., Cordon-Cardo, C., Silke, J., et al. (2006). Identification and validation of oncogenes in liver cancer using an integrative oncogenomic approach. *Cell*, *125*, 1253–1267.
Zhang, N., Zhang, J., Cheng, Y., & Howard, K. (1996). Identification and genetic analysis of wunen, a gene guiding *Drosophila melanogaster* germ cell migration. *Genetics*, *143*, 1231–1241.
Zhang, N., Zhang, J., Purcell, K. J., Cheng, Y., & Howard, K. (1997). The *Drosophila* protein Wunen repels migrating germ cells. *Nature*, *385*(6611), 64–67.
Zhou, X., et al. (2010). Reversal of cancer cachexia and muscle wasting by ActRIIB antagonism leads to prolonged survival. *Cell*, *142*(4), 531–543.

CHAPTER FOUR

Amyotrophic Lateral Sclerosis Pathogenesis Converges on Defects in Protein Homeostasis Associated with TDP-43 Mislocalization and Proteasome-Mediated Degradation Overload

G. Lin[*,†,1], D. Mao[†,‡,1], H.J. Bellen[*,†,‡,§,2]
[*]Baylor College of Medicine, Houston, TX, United States
[†]Jan and Dan Duncan Neurological Research Institute, Texas Children's Hospital, Houston, TX, United States
[‡]Program in Developmental Biology, Baylor College of Medicine, Houston, TX, United States
[§]Howard Hughes Medical Institute, Baylor College of Medicine, Houston, TX, United States
[2]Corresponding author: e-mail address: hbellen@bcm.edu

Contents

Abstract

Amyotrophic lateral sclerosis (ALS) is a fatal neurodegenerative disorder that affects upper and/or lower motor neurons. It usually affects people between the ages of 40–70. The average life expectancy is about 3–5 years after diagnosis and there is no

[1] These authors contributed equally to this work.

Current Topics in Developmental Biology, Volume 121
ISSN 0070-2153
http://dx.doi.org/10.1016/bs.ctdb.2016.07.004

effective cure available. Identification of variants in more than 20 different loci has provided insight into the pathogenic molecular mechanisms mediating disease pathogenesis. In this review, we focus on seven ALS-causing genes: TDP-43, FUS, C9orf72, VCP, UBQLN2, VAPB and SOD-1, which encompass about 90% of the variants causing familial ALS. We examine the biological functions of these genes to assess how these pathogenic variants contribute to ALS pathogenesis by integrating findings from studies in *Drosophila melanogaster* and mammals. Additionally, we highlight the functional and genetic connections between these loci. Altogether, this review reveals that the majority of biological studies converge on defects in proteostasis due to the mislocalization of TDP-43 and/or altering the function of specific proteins mediating or modulating proteasomal degradation.

1. INTRODUCTION

Amyotrophic lateral sclerosis (ALS), also known as Lou Gehrig's disease, is a fatal neurodegenerative disorder characterized by progressive degeneration of upper and lower motor neurons with a prevalence of 4–6 per 100,000 in the population worldwide (Majoor-Krakauer, Willems, & Hofman, 2003) and a lifelong risk of 1 in 600 individuals (Gonzalez-Perez et al., 2012). The incidence of this disease increases with age and is higher in males than females (Alonso, Logroscino, Jick, & Hernan, 2009; Fang et al., 2009). ALS was first described by Charles Bell in 1824 and Jean-Martin Charcot first linked the clinical symptoms with the neurological pathology in 1869 (Rowland & Shneider, 2001). Charcot introduced the term amyotrophic lateral sclerosis, where amyotrophic means "no muscle nourishment" in Greek (Rowland & Shneider, 2001). The most characteristic clinical feature of ALS is the progressive motor weakness and atrophy (Peters, Ghasemi, & Brown, 2015).

There are two types of ALS, familial ALS (FALS) and sporadic ALS (SALS) (Gros-Louis, Gaspar, & Rouleau, 2006). About 10% of patients with a positive family history are classified as FALS, whereas the remainder are SALS (Mitchell & Borasio, 2007). Pathologically, protein and RNA aggregates are widely observed in motor neurons of postmortem ALS brain and spinal cord (Peters, Ghasemi, et al., 2015). These aggregates usually contain ubiquitin and TDP-43 (Blokhuis, Groen, Koppers, van den Berg, & Pasterkamp, 2013). Hence, defects in RNA metabolism and an imbalance in proteostasis are thought to play an important role in ALS pathogenesis. However, it is unclear how an imbalance in proteostasis develops and how this relates to ALS pathogenesis. The genes identified in FALS provide an opportunity to dissect the pathogenic mechanisms.

In this review, we focus on seven major ALS-causing genes: *TARDBP* (*TDP-43*), *FUS*, *C9orf72*, *VCP*, *UBQLN2*, *VAPB*, and *SOD-1*. *Drosophila* has been used to study the mechanisms associated with the function of these loci in neurodegenerative diseases. Table 1 summarizes techniques that are frequently used in *Drosophila* to define phenotypes associated with neurodegeneration. Please refer to the references therein for a detailed description of these methods. All the genes discussed in this review, except *C9orf72*, have *Drosophila* homologs and have been explored in this model organism. For simplicity, we use the human nomenclature rather than the fly

Table 1 Assays Frequently Used to Characterize Neurodegenerative Defects in *Drosophila*

Defects	Tissue	Assay	References
Survival	Whole animal	Life span	Linford, Bilgir, Ro, and Pletcher (2013)
		Lethal phase	
		Eclosion rate	
Neuronal defects	Photoreceptor morphology	Transmission electron microscopy	Wolff (2011)
		Immunofluorescence	Walther and Pichaud (2006)
	Photoreceptor function	Electroretinogram	Dolph, Nair, and Raghu (2011)
	Neuromuscular junction (NMJ)	Giant fiber assay (Adult)	Allen and Godenschwege (2010)
		NMJ morphology (Larva)	Andlauer and Sigrist (2012a, 2012b, 2012c)
		NMJ electrophysiology (Larva)	Imlach and McCabe (2009)
Locomotor defects	Behavioral assays	Negative geotaxis assay (Adult)	Ali, Escala, Ruan, and Zhai (2011)
		Flight assay (Adult)	Babcock and Ganetzky (2014)
		Crawling assay (Larva)	Nichols, Becnel, and Pandey (2012)

nomenclature (*TDP-43* in flies is *TAR DNA-binding protein-43 homolog* (*TBPH*); *FUS* is *cabeza* (*caz*); *VCP* is *Transitional endoplasmic reticulum ATPase* (*TER94*); *UBQLN2* is *Ubiquilin* (*Ubqn*); and *VAPB* is *VAMP-associated protein-33-1* (*Vap-33-1*)). Findings from vertebrate and *Drosophila* studies suggest that variants in these seven loci cause progressive defects in protein homeostasis that converge on the inability of neurons to cope with protein degradation. We propose a model in which four factors contribute to the proteinopathy. First, in C9orf72-related ALS, the presence of a hexanucleotide repeat expansion (HRE), (GGGGCC)*n*, in the noncoding intron of the *C9orf72* gene (DeJesus-Hernandez et al., 2011; Renton et al., 2011) causes the accumulation of abortive transcripts that form foci containing RNAs and proteins (Donnelly et al., 2013; Haeusler et al., 2014). In addition, dipeptide proteins repeat (DPRs) synthesized via an ATG-independent translation mechanism has been proposed to mediate HRE toxicity. Moreover, the HRE-containing transcripts or the arginine-containing DPRs recruit various proteins, including proteins required for nuclear import and export (Freibaum et al., 2015; Zhang, Donnelly, et al., 2015). These in turn lead to cytoplasmic mislocalization of TDP-43 (Chew et al., 2015; Cooper-Knock et al., 2015; Zhang, Donnelly, et al., 2015). Second, the aggregated and mislocalized TDP-43 lead to the inclusion of cryptic exons in primary mRNA transcripts (Ling, Pletnikova, Troncoso, & Wong, 2015), thereby producing aberrant mRNA and truncated proteins (Ling et al., 2015). Third, these aberrant proteins and aggregates are typically removed by proteasome degradation or autophagy (Wong & Cuervo, 2010). However, ALS-causing variants in genes implicated in proteasome degradation like VCP, UBQLN2, and VAPB, impair this mechanism and cause ALS (Deng et al., 2011; Johnson et al., 2010; Nishimura et al., 2004). Finally, the accumulation of RNA and protein aggregates lead to progressive degenerative phenotypes (Blokhuis et al., 2013; Ling, Polymenidou, & Cleveland, 2013). Table 2 summarizes many of the key papers discussed in this review. We realize that this model is an oversimplification of the available data, but it provides a useful framework to integrate many data sets that will be expanded upon below.

2. CURRENT CHALLENGES IN ALS RESEARCH

In the past two decades, many of the breakthroughs in ALS research are associated with the identification of novel ALS pathogenic variants and the characterization of the respective genes (Ajroud-Driss & Siddique, 2015).

Table 2 Key Papers Discussed in This Review

Gene	Genetic Manipulation	TDP-43 Phenotype	Proteostasis	Species	References
TDP-43	FTD and ALS patients	Identification of hyperphosphorylated TDP-43 aggregates	Hyperphosphorylated TDP-43 aggregates	Human	Arai et al. (2010), Hasegawa et al. (2008), and Neumann et al. (2009)
		Identification of hyperphosphorylated, ubiquitinated, and cleaved 25-kDa TDP-43	N.D.[a]	Human	Neumann et al. (2006)
	TDP-43 deleted mouse embryonic stem cells; ALS patients	Loss of TDP-43 causes the incorporation of cryptic exons into mRNAs, disrupts mRNA translation, and may lead to protein aggregation	N.D.	Human; mammalian cell line	Ling et al. (2015)
	Ectopic expression of wild-type TDP-43	Nuclear localization	No protein aggregation	Mammalian cell line	Zhang et al. (2009), Arai et al. (2010), and Yang et al. (2010)
		Formation of cytosolic dTDP-43-positive aggregates	Protein aggregates	*Drosophila*	Lin, Cheng, and Shen (2011)

Continued

Table 2 Key Papers Discussed in This Review—cont'd

Gene	Genetic Manipulation	TDP-43 Phenotype	Proteostasis	Species	References
	Ectopic expression of C-terminal disease-related 25-kDa or 35-kDa fragments of TDP-43	Cytoplasmic localized TDP-43	Protein aggregations	Mammalian cell line	Zhang et al. (2009), Arai et al. (2010), and Yang et al. (2010)
TDP-43	Ectopic expression of wild-type, ΔNLS, disease related (G287S, G348C, N390D,A315T, or A382T), or RNA-binding deficient TDP-43	Only ΔNLS leads to a cytoplasmic localization of TDP-43	N.D.	Mammalian cell line and *Drosophila*	Voigt et al. (2010)
	Ectopic expression of wild-type or disease-related (A315T) TDP-43	TDP-43 and A315T aggregate in axons of larval eye imaginal discs	N.D.	*Drosophila*	Estes et al. (2011)
	Ectopic expression of disease-related (A315T) TDP-43	Neurons with ubiquitin pathology show loss of nuclear TDP-43 staining	Ubiquitin-positive aggregations in frontal cortex and spinal motor neurons	Mouse	Wegorzewska, Bell, Cairns, Miller, and Baloh (2009)

	Ectopic expression of TDP-43 with Ser 409/410 mutated to phosphomimetic aspartic acids	Reduces TDP-43-positive aggregates	Reduces TDP-43-positive aggregates	Mammalian cell line	Brady, Meng, Zheng, Mao, and Hu (2011) and Li, Yeh, Chiu, Tang, and Tu (2011)
TDP-43	Overexpression of a chaperone, CG14207	Reduces TDP-43-positive aggregates and TDP-43 toxicity	Reduces TDP-43-positive aggregates and TDP-43 toxicity	*Drosophila*	Gregory, Barros, Meehan, Dobson, and Luheshi (2012)
	Overexpression of a chaperone, Hsp104	Suppresses TDP-43 toxicity, aggregation, and mislocalization	Suppresses TDP-43 toxicity, aggregation, and mislocalization	Yeast	Jackrel et al. (2014)
	FALS patient or mammalian cell lines ectopically express wild-type or disease-related FUS (R521G)	Wild-type FUS localized to nucleus, whereas cytoplasmic localization is observed in disease-related (R521G) FUS	Cytoplasmic aggregates	Human and mammalian cell line	Kwiatkowski et al. (2009) and Vance et al. (2009)
FUS[b]	Ectopic expression of wild-type or disease-related FUS (R518K)	Wild-type FUS is localized to nucleus, whereas cytoplasmic localization is observed in disease-related (R518K) FUS; FUS genetically interacts with TDP-43 in a disease mutation-dependent manner to cause eye degeneration	N.D.	*Drosophila*	Lanson et al. (2011)

Continued

Table 2 Key Papers Discussed in This Review—cont'd

Gene	Genetic Manipulation	TDP-43 Phenotype	Proteostasis	Species	References
	Ectopic expression of wild-type or phosphorylation deficient FUS	DNA damage induces FUS phosphorylation and cytoplasmic localization FUS	N.D.	Mammalian cell lines	Deng, Holler, et al. (2014)
	Ectopic expression of wild-type or phosphorylation deficient FUS	C-terminal phosphorylation on Tyr526 impairs FUS nucleotransport machinary	N.D.	Mammalian cell lines	Darovic et al. (2015)
FUS[b]	Ectopic expression of wild-type or argine methylation deficient FUS	Argine methylation of FUS is required for its interaction with transportin1 and nuclear transport	N.D.	Mammalian cell lines; *Drosophila*	Dormann et al. (2012), Scaramuzzino et al. (2013), and Jackel et al. (2015)
	shRNA knockdown of FUS	Misspliced Tau protein and formation of Tau aggregates	N.D.	Mammalian cell lines	Orozco et al. (2012)
	Motorneuron expression of wild-type or disease-related FUS (R518K or R521C)	Wild-type FUS localized to nucleus, whereas cytoplasmic localization is observed in disease-related (R518K or R521C) FUS	N.D.	*Drosophila*	Daigle et al. (2013)

	FALS patients	N.D.	Formation of RNA foci in nuclei	Human	DeJesus-Hernandez et al. (2011)
		Cytoplasmic TDP-43 aggregates	Cytoplasmic TDP-43 aggregates	Human	Stewart et al. (2012) and Rohrer et al. (2015)
		Cytoplasmic TDP-43, p62 and Ubiquilin-2-positive aggregates	Cytoplasmic TDP-43, p62, and Ubiquilin-2-positive aggregates	Human	Fratta et al. (2013)
C9orf72	CNS samples from C9orf72 ALS patients	TDP-43 mislocalization	Formation of RNA foci	Human	Cooper-Knock et al. (2015)
		N.D.	Detection of DPR aggregates in the cerebellum of C9orf72 ALS patients	Human	Ash et al. (2013), Mori et al. (2013), and Zu et al. (2013)
	CNS samples from C9orf72 ALS patients	N.D.	DPR aggregates are rarely present in the spinal cords or motor neurons of C9orf72 ALS patients	Human	Gomez-Deza et al. (2015)
	Neuronal and glial knockout of C9orf72	No obvious TDP-43 mislocalization	No protein aggregation	Mouse	Koppers et al. (2015)
	iPSC-derived from C9orf72 ALS patients	N.D.	Intranuclear RNA foci that contain HRE and HRE-binding protein	Human	Donnelly et al. (2013)

Continued

Table 2 Key Papers Discussed in This Review—cont'd

Gene	Genetic Manipulation	TDP-43 Phenotype	Proteostasis	Species	References
		N.D.	HRE forms G-quadruplex that impairs RNA polymerase II activity and leads to the accumulation of short incomplete transcripts that bind nucleolar proteins to form aggregates	Human	Haeusler et al. (2014)
C9orf72	Transgenetic *Drosophila* expressing HRE; iPSC-derived from C9orf72 ALS patients	Mislocalization of TDP-43 in the cytoplasm is observed in mutant *Drosophila* as well as iPSCs	Formation of RanGAP1 and ubiquitin-containing aggregates in the cytoplasm	*Drosophila* and Human	Zhang, Donnelly, et al. (2015)
		N.D.	Impairment of nuclear import machinery and formation of RNA foci in the nucleus	*Drosophila* and Human	Freibaum et al. (2015)
	Transgenetic *Drosophila* expressing HRE	N.D.	Formation of DPR aggregates correlate with neurodegeneration	*Drosophila*	Mizielinska et al. (2014), Freibaum et al. (2015), and Tran et al. (2015)
	Transgenetic mice expressing HRE	Phosphorylated TDP-43 cytoplasmic inclusions are observed	Formation of intranuclear RNA foci and DPR aggregates	Mouse	Chew et al. (2015)
	Transgenetic yeast expressing HRE	N.D.	Formation of arginine-rich DPRs are toxic in	Yeast	Jovicic et al. (2015)

	IBMPFD patients with VCP variant R155H (brain)	Accumulations of insoluble, phosphorylated TDP-43	Accumulations of TDP-43 colocalized with ubiquitin	Human	Neumann et al. (2007)
	IBMPFD patients with VCP variant R155H N387H (muscle)	TDP-43 is present as large inclusions in muscle cytoplasm	TDP-43 inclusions colocalize with ubiquitin	Human	Weihl et al. (2008)
	IBMPFD patients with VCP variant R155H (brain)	TDP-43 redistributes to cytoplasmic inclusions from the nucleus	N.D.	Human	Johnson et al. (2010)
VCP	Frontal cortex of IBMPFD patients	TDP-43 redistributes to cytoplasm and some nuclei show dense inclusions	N.D.	Human	Ritson et al. (2010)
	Cold-sensitive cdc48-1 (yeast ortholog of VCP) mutant strain	N.D.	Required in ubiquitin–proteasome degradation	Yeast	Dai and Li (2001)
	siRNA knockdown of VCP	N.D.	Accumulation of high-molecular-weight conjugates of ubiquitin	*Drosophila* S2 and mammalian cell line	Wojcik, Yano, and DeMartino (2004)

Continued

Table 2 Key Papers Discussed in This Review—cont'd

Gene	Genetic Manipulation	TDP-43 Phenotype	Proteostasis	Species	References
VCP	siRNA knockdown or ectopic expression of IBMPFD mutant VCP (R155H, A232E, or E587Q)	TDP-43 mislocalizes to cytoplasm	Protein aggregation caused by autophagy degradation defect	Mouse skeletal muscle and mammalian cell line	Ju et al. (2009)
	siRNA knockdown of VCP and expression of dominant-negative or disease-associated VCP variants (R155H or A232E)	N.D.	Accumulation of immature autophagic vesicles	Mammalian cell line	Tresse et al. (2010)
	Transgenic mice expressing VCP/p97 mutations (R155H or A232E)	Cytoplasmic accumulation of TDP-43	TDP-43 accumulations are also ubiquitin positive	Mouse	Custer, Neumann, Lu, Wright, and Taylor (2010)
	Mouse primary cortical neurons expressing mutant VCP (R155H or A232E)	Cytoplasmic accumulation of TDP-43	N.D.	Mouse	Ritson et al. (2010)
	Knock in mice expressing VCP/p97 variant R155H	Cytoplasmic accumulation of TDP-43	Ubiquitin-positive inclusion bodies, and increased LC3-II staining	Mouse	Nalbandian et al. (2013)
		TDP-43 positive cytosolic inclusions	Ubiquitin and SQSTM1/p62 aggregates	Mouse	Yin et al. (2012)

	Myoblasts derived from patients with IBMPFD carrying R155H or R155S mutations	N.D.	Accumulation of numerous, large LAMP-1- and LAMP-2-positive vacuoles and accumulation of LC3-II	Human	Tresse et al. (2010)
VCP	Expression of human Cdc48 variant (D592N)	N.D.	Impaired 20S proteasome binding and proteolytic communication	*Thermoplasma acidophilum*	Barthelme, Jauregui, and Sauer (2015)
	Patients with UBQLN2 variant (P497L)	TDP-43-immunoreactive inclusions	Ubiquilin-2-positive inclusions	Human	Fahed et al. (2014)
	Patients with UBQLN2 variants (P497H P506T)	Inclusions that are positive for TDP-43	Inclusions positive for Ubiquilin-2, ubiquitin, and p62	Human	Deng et al. (2011)
	Patient with UBQLN2 (T487I) variant	Frequent colocalization of TDP-43 with Ubiquilin2	Colocalization of Ubiquilin-2 with ubiquitin in all neuronal inclusions	Human	Williams et al. (2012)
UBQLN1/2	Endogenous Ubiquilin	N.D.	Ubiquilin is present in autophagosomes and binds to the autophagosome marker LC3	Mouse and mammalian cell line	Rothenberg et al. (2010)
	siRNA knockdown	N.D.	Stabilizes ERAD substrates	Mammalian cell line	Kim, Kim, Yoon, and Yoon (2008) and Lim et al. (2009)

Continued

Table 2 Key Papers Discussed in This Review—cont'd

Gene	Genetic Manipulation	TDP-43 Phenotype	Proteostasis	Species	References
	siRNA knockdown of UBQLN1 or UBQLN2	N.D.	Inhibit autophagosome formation and delay the delivery of autophagosomes to lysosomes	Mammalian cell line	N'Diaye et al. (2009) and Rothenberg et al. (2010)
	siRNA knockdown of UBQLN	N.D.	Induction of the ER stress reporter, accumulation of polyubiquited proteins	*C. elegans*	Lim et al. (2009)
UBQLN1/2	Overexpression or siRNA knockdown of UBQLN2	N.D.	Accumulation of ERAD substrates	Mammalian cell line	Xia et al. (2014)
	Ectopic expression of Flag-hUbiquilin-1 (E54D)	Reduces the level of endogenous TDP-43	N.D.	Mammalian cell line	Gonzalez-Perez et al. (2012)
	Coexpression of Flag-hUbiquilin-1 (E54D) and TDP-43-EGFP	Redistributes TDP-43 to the cytoplasm	N.D.	Mammalian cell line	Gonzalez-Perez et al. (2012)
	Ectopic expression of UAS-HA-hUBQLN1 (E54D)	Reduces the level of endogenous TDP-43	N.D.	*Drosophila*	Gonzalez-Perez et al. (2012)

	Coexpression of WT or variant (P497H) Ubiquilin-2 with C-terminal fragment of TDP43	TDP43 inclusions are colocalized with wild-type and mutant (P497H) Ubiquilin-2	N.D.	Mammalian cell line	Deng et al. (2011)
	Ectopic expression of UBQLN 2 variants (P497H or P506T)	N.D.	Ubiquitin-mediated impairment of proteasomal degradation	Mammalian cell line	Deng et al. (2011)
	Ectopic expression of UBQLN variants (P497H, P497S, P506T, P509S, or P525S)	N.D.	Weaker binding to the S5a subunit of the proteasome	Mammalian cell line	Chang and Monteiro (2015)
UBQLN1/2	Overexpression of UBQLN-1/2	N.D.	Suppress starvation-induced cell death	Mammalian cell line	N'Diaye et al. (2009)
	Overexpression of UBQLN-1/2	N.D.	Reduce starvation-induced cell death caused by overexpression of presenilin-2 proteins	Mammalian cell line	Rothenberg et al. (2010)
	Overexpression of UBQLN-1	Potentiates TDP-43 aggregate formation in the cytoplasm	Recruits TDP-43 to detergent-resistant cytoplasmic aggregates that colocalize with autophagosomal markers	Mammalian cell line	Kim et al. (2009)

Continued

Table 2 Key Papers Discussed in This Review—cont'd

Gene	Genetic Manipulation	TDP-43 Phenotype	Proteostasis	Species	References
UBQLN1/2	Coexpression of HA-TDP-43 and Myc-Ubiquilin-1	Physically interacts with TDP-43 through the C-terminal UBA domain of Ubiquilin-1	N.D.	Yeast and mammalian cell line	Kim et al. (2009)
	Coexpression of human UBQLN and TDP-43	Reduced eclosion rate as well as shortened life span caused by TDP-43 overexpression	N.D.	*Drosophila*	Hanson, Kim, Wassarman, and Tibbetts (2010)
VAPB	VAPB null mutation	N.D.	Aberrant ER expansion, accumulation of membrane proteins on the ER and alternatively spliced XBP-1	*Drosophila*	Moustaqim-Barrette et al. (2013)
	Ectopic expression of VAPB (mammalian P56S or *Drosophila* P58S) variant	N.D.	Formation of protein aggregates, ER stress, and traps endogenous wild-type VAPB proteins	*Drosophila* and mammalian cell line	Teuling et al. (2007) and Tsuda et al. (2008)
	VAPB (P56S) transgenic mice	N.D.	Mutant VAPB proteins accumulate in large puncta and significant increase in ER stress	Mouse	Aliaga et al. (2013)

VAPB	Knock-in mice carrying the VAPB (P56S) variant	N.D.	Accumulation of ubiquitinated proteins in cytoplasmic inclusions that colocalize with autophagosome markers	Mouse	Larroquette et al. (2015)
SOD-1	SOD-1 patients	N.D.	SOD-1 patients show elevated ER stress	Human	Atkin et al. (2008) and Ilieva et al. (2007)
	Transgenetic mouse lines expressing disease-related SOD-1 (G93A; G85R, or G37R)	N.D.	Formation of SOD-1 mutant containing protein aggregates in all transgenic lines	Mouse	Bruijn et al. (1997, 1998), Cleveland (1999) (review), Gurney et al. (1994), Julien and Kriz (2006) (review), and Wong et al. (1995)
	Transgenetic *Drosophila* lines expressing disease-related SOD-1 variants (A4V or G85R)	N.D.	Accumulation of SOD-1 aggregates with age in motor neurons	*Drosophila*	Watson, Lagow, Xu, Zhang, and Bonini (2008)
	Ectopic expression of wild-type or disease-related SOD-1 (A4V; G85R, or G93A)	N.D.	Mutant but not wild-type SOD-1 forms in cultured cells promote ER stress	Mammalian cell lines	Nishitoh et al. (2008), Oh, Shin, Yuan, and Kang (2008), and Tobisawa et al. (2003)
	Transgenetic mouse lines expressing disease-related SOD-1 (G93A or G85R)	N.D.	Transgenic mice expressing mutant SOD-1 exhibits elevated ER stress	Mouse	Atkin et al. (2006), Kikuchi et al. (2006), Saxena, Cabuy, and Caroni (2009), Tobisawa et al. (2003), and Wate et al. (2005)

Continued

Table 2 Key Papers Discussed in This Review—cont'd

Gene	Genetic Manipulation	TDP-43 Phenotype	Proteostasis	Species	References
	Transgenetic mouse lines expressing disease-related SOD-1 (G85R)	N.D.	Remove one copy of PERK in SOD-1 (G85R) mice accelerates aggregate formation, shorten life span, and enhances degeneration phenotypes	Mouse	Wang, Popko, and Roos (2011)
SOD-1	Transgenetic mouse lines expressing disease-related SOD-1 (G85R)	N.D.	Genetic or pharmacological activation of PERK ameliorates mutant SOD-1-induced ALS phenotypes	Mouse	Wang, Popko, and Roos (2014), Das et al. (2015), and Wang, Popko, Tixier, and Roos (2014)
		N.D.	Inactivation of the IRE1 pathway reduces the accumulation of mutant SOD-1 in spinal cord, prolongs life span and alleviates degeneration phenotypes	Mouse	Hetz et al. (2009)

[a]Not determined.
[b]Since FUS and TDP-43 may function in the same pathway, we summarize FUS phenotypes in this section.

However, most of the isolated genes are poorly characterized, and relatively little is known about their in vivo function. Yeast as well as animal models like mice, flies, and worms are valuable to help define the function of these proteins. These studies are typically based on loss-of-function alleles as well as expression of mutant alleles in the null or wild-type background. Results from these studies are very important as they may affect the design of therapeutic strategies. For example, if the disease-causing variant results in loss of function, then an overexpression strategy may be advised; but if the variant results in gain of function, then suppressing gene function may be advised. In addition to examine individual gene functions, animal models also allow the elucidation of genetic interactions between disease-causing genes in controlled experiments. Delineating the genetic interactions between different disease-causing genes allows elucidation of the pathogenic mechanisms. Here, we will summarize the salient features that contributed to the model outlined earlier and describe some of the contribution of *Drosophila* research to the field. We only cover seven of the ALS-causing loci because they cover the vast majority of the FALS cases and because they have been studied to some extent in *Drosophila.*

3. RNA PROCESSING, SPLICING, RNA FOCI, AND PROTEIN AGGREGATION

A general pathological feature of neurodegenerative diseases is the presence of insoluble protein aggregates that are often abnormally phosphorylated and ubiquitinated (Ling et al., 2013). A fundamental question is how the formation of these insoluble protein aggregates contributes to the pathogenesis of ALS. Interestingly, several of the ALS genes identified so far encode DNA/RNA-binding proteins, including *TDP-43*, *FUS*, and *hnRNPs* (Peters, Ghasemi, et al., 2015). More recently, HRE in *C9orf72* was found to cause RNA foci and affect RNA metabolism (Donnelly et al., 2013; Haeusler et al., 2014). Therefore, the formation of protein aggregates abolishing normal RNA splicing may play an important role in ALS pathogenesis. In this section, we will discuss the connection between protein aggregates, RNA metabolism, and ALS pathogenesis with a special emphasis on *TDP-43*, *FUS*, and *C9orf72*.

3.1 TDP-43

TAR DNA-binding protein (TDP-43) is a 43 kDa RNA/DNA-binding protein encoded by *TARDBP* gene that is evolutionarily conserved

(Buratti et al., 2001; Ou, Wu, Harrich, Garcia-Martinez, & Gaynor, 1995). TDP-43 is often found in pathological inclusions in spinal cord motor neurons, hippocampal, and frontal cortex neurons as well as glial cells in SALS and FALS patients that are not caused by SOD-1 variants (Blokhuis et al., 2013). Variants in *TARDBP* (also known as ALS10) that affect TDP-43 have been reported in ~4% of FALS and a few SALS cases (Renton, Chio, & Traynor, 2014). TDP-43 is normally localized to the nucleus but in ALS patients, the protein is present in the cytoplasm and excluded from the nucleus (Neumann et al., 2006). These observations suggest that TDP-43 plays an important role in the development of ALS.

Loss of TDP-43 in flies causes semilethality (Feiguin et al., 2009). Motor neuronal expression of either human or *Drosophila* TDP-43 in the TDP-43 null mutant prolongs life span, suggesting a functional conservation of TDP-43 between *Drosophila* and human (Feiguin et al., 2009). In addition, loss of TDP-43 in fly larvae reduced motility and decreased boutons at the neuromuscular junctions (NMJs) (Feiguin et al., 2009). Moreover, neuronal overexpression of the wild-type human transgene causes a decrease in bouton and branch number associated with protein aggregates (Li et al., 2010), suggesting that loss and gain of TDP-43 function affects NMJ morphology.

In 90% of all ALS patients, TDP-43 is absent from the nucleus and present in ubiquitinated cytoplasmic aggregates (Ling et al., 2013). Ectopic expression of a C-terminal fragment TDP-43 in cell lines causes protein aggregations (Zhang, Gui, et al., 2009), suggesting that TDP-43 is physically prone to aggregate formation. Moreover, overexpression of the human wild-type or mutant TDP-43 in *Drosophila* or mice leads to locomotor defects (Estes et al., 2011; Li et al., 2010; Lin et al., 2011; Voigt et al., 2010; Wegorzewska et al., 2009). These data indicate that TDP-43 aggregates may play an important role in ALS pathology. A predominant marker of TDP-43 aggregates is the presence of hyperphosphorylated TDP-43 as well as the presence of the N-terminal truncated 25 kDa TDP-43 (will be referred to as TDP-25) in cytoplasmic inclusions (Arai et al., 2010; Neumann et al., 2006). TDP-43 is hyperphosphorylated at several serine residues, including S409/410 (Hasegawa et al., 2008; Neumann et al., 2009). Mutating the S409/410 to phosphomimetic aspartic acid residues greatly reduces protein aggregation (Brady et al., 2011; Li et al., 2011), suggesting that the hyperphosphorylation on these two sites alleviates TDP-43 proteinopathy. TDP-25 forms cytoplasmic protein aggregates when expressed in the cell lines (Yang et al., 2010). Similarly, expression of TDP-25 in *Drosophila* eyes causes a rough eye phenotype suggesting that the toxicity is associated with TDP-25 (Gregory et al., 2012). However, its toxicity is milder than the wild-type human TDP-43 (Gregory et al., 2012). Moreover, the authors show that TDP-25 is more susceptible to protein degradation (Gregory et al., 2012). This may contribute to the reduction of its toxicity. Furthermore, coexpression of chaperones in yeast as well as *Drosophila* suppresses TDP-43 aggregation, cytoplasmic localization, and toxicity (Estes et al., 2011; Gregory et al., 2012; Jackrel et al., 2014), suggesting that promoting protein refolding and thereby reducing the burden of proteasomal degradation is beneficial to ALS.

To assess the correlation between TDP-43 nuclear localization and degeneration, Miguel, Frebourg, Campion, and Lecourtois (2011) expressed wild type or mutant human TDP-43 that lacks either the nuclear localization sequence (NLS) or nuclear export sequence (NES) in *Drosophila* eyes and examined their ability to cause protein aggregations and eye degeneration. Only mutant TDP-43 lacking NLS caused severe eye degeneration. Interestingly, human TDP-43 proteins expressed in adult fly neurons are abnormally phosphorylated on the disease-specific sites, Ser409/Ser410 (Miguel et al., 2011). This study highlights the importance of TDP-43 mislocalization in the induction of TDP-43 proteinopathy. Note that TDP-43 aggregation is not commonly observed in mouse models of TDP-43 proteinopathy (Choksi et al., 2014; Gendron & Petrucelli, 2011). Hence, *Drosophila* provides a platform to study TDP-43 proteinopathy.

TDP-43 is a member of the heterogeneous nuclear ribonucleoprotein (hnRNP) family of proteins (Buratti & Baralle, 2008) and affects pre-mRNA splicing, transcription, mRNA stability, and mRNA transport (Buratti & Baralle, 2008). TDP-43 and other members of the hnRNP family bind RNA, but the RNA-binding specificity of hRNPs varies substantially, ranging from nonspecific to highly defined RNA sequences. TDP-43 specifically binds to a long UG-repeat sequence in RNA (Ayala et al., 2005; Buratti et al., 2001) via two RNA recognition motifs (Lukavsky et al., 2013) and regulates alternative splicing of mRNA including several human disease-related genes (Bose, Wang, Hung, Tarn, & Shen, 2008; Buratti & Baralle, 2001; Buratti et al., 2001; Mercado, Ayala, Romano, Buratti, & Baralle, 2005). In an in vitro splicing assay, the *Drosophila TDP-43* shows comparable ability to recognize and splice an UG-rich substrate with human TDP-43 (Ayala et al., 2005), suggesting that the pre-mRNA splicing activity of TDP-43 is conserved in *Drosophila*. Neuronal expression of a human TDP-43 lacking RNA-binding activity in *Drosophila* reverses life span reductions and locomotor defects caused by the expression of human TDP-43 harboring ALS-linked pathogenic variants (Voigt et al., 2010), indicating that the RNA-binding activity of TDP-43 may be important for ALS pathology.

Cytoplasmic inclusion and nuclear exclusion of TDP-43 are a hallmark of ALS. Studies in cells suggest that TDP-43 is an important regulator of pre-mRNA splicing. However, how TDP-43 leads to the formation of aggregates in ALS patients and how its pre-mRNA splicing activity is involved in ALS pathogenesis remains to be established. A recent publication by Ling et al. (2015) provides an important hint on how these two phenomena maybe interconnected. By using CreER-inducible *TDP-43* knockout mouse embryonic stem cells and high-resolution RNA-sequencing (RNA-seq) technology, Ling et al. (2015) found that nonconserved cryptic exons were not spliced out with loss of *TDP-43*. The incorporation of cryptic exons in mature mRNA disrupts translation and produces numerous aberrant proteins (Fig. 1A). Interestingly, suppression of the incorporation of cryptic exons prevents cell death in *TDP-43* knockout cells. These data suggest that the chronic production of aberrant proteins may overload the protein degradation machinery.

These findings raise the question whether there is a few or a group of conserved TDP-43-binding exons that are responsible for the disease. Interestingly, while all TDP-43 targets share the UG-rich repeat consistent with its binding preference in vitro, TDP-43 is associated with a subset of cryptic exons that is totally different in mice vs human (Ling et al., 2015). Taken together, these data indicate that there is probably not a specific TDP-43 target responsible for the degenerative phenotypes. Rather, loss of *TDP-43* may affect numerous genes containing UG-rich repeats and the incorporation of cryptic exons produces numerous aberrant proteins, eventually leading to the formation of protein aggregates that overload the protein removal machinery.

3.2 FUS

FUS (fused in sarcoma) was first reported to be an oncoprotein in liposarcomas associated with a *t*(12; 16) chromosomal translocation (Crozat, Aman, Mandahl, & Ron, 1993; Rabbitts, Forster, Larson, & Nathan, 1993). It is a member of a protein family containing a DNA/RNA-binding domain (Tan & Manley, 2009). FUS regulates RNA processing, pre-mRNA splicing and mRNA trafficking (Deng, Gao, & Jankovic, 2014). Variants in *FUS* (also known as ALS6) lead to the formation

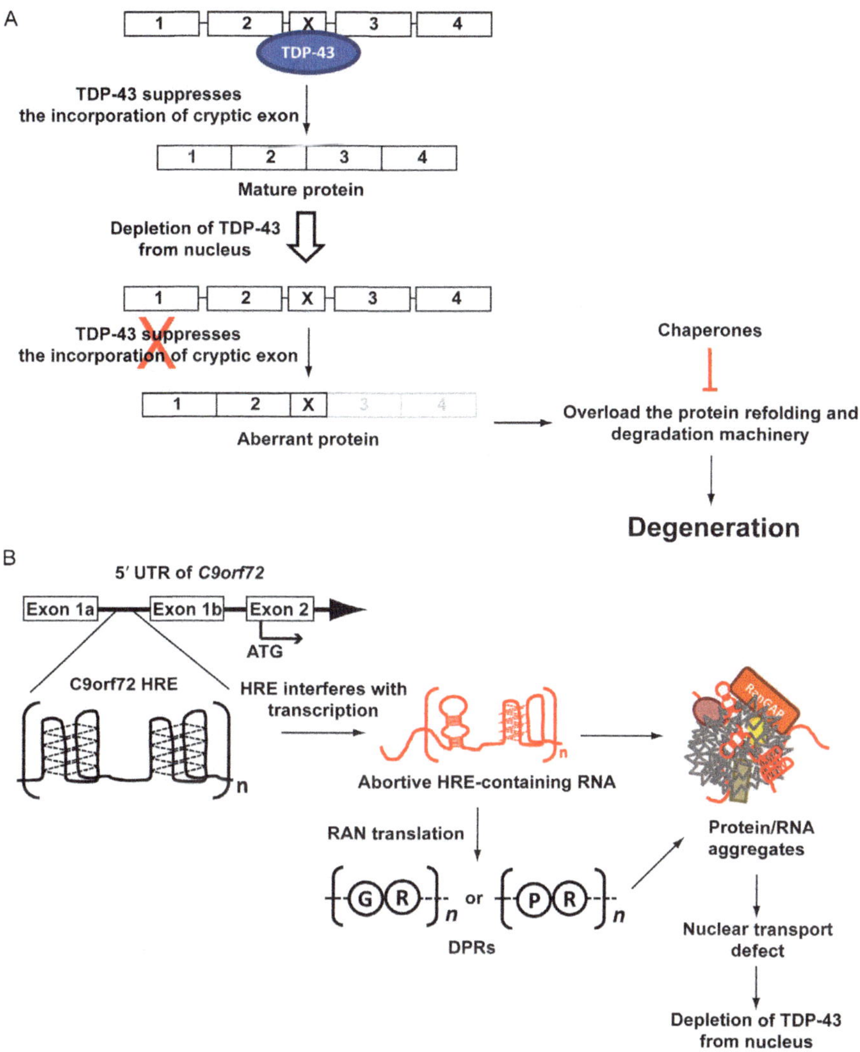

Fig. 1 A model for the involvement of RNA metabolism in the formation of protein/RNA aggregates and degeneration. (A) Variants in TDP-43 or C9orf72 cause TDP-43 exclusion from the nucleus. This disrupts the function of TDP-43 to suppress the incorporation of the cryptic exon to mRNAs and leads to the synthesis of misfolded proteins that overload the protein refolding and degradation machinery. This results in a chronic accumulation of protein aggregates and degeneration. The enhancement of chaperone function may relieve protein aggregation and suppress degeneration. (B) The insertion of HREs in the 5′ noncoding intron of C9orf72 gene interfere with transcription and lead to the generation of HRE-containing abortive RNAs. These aberrant RNAs result in the formation of protein/RNA aggregates or DPRs via RAN translation which in turn cause defects in the nuclear transport machinery and promote the accumulation of nuclear proteins including TDP-43 in the cytoplasm.

of FUS-positive inclusions in motor neurons in 4–5% of all FALS (Kwiatkowski et al., 2009; Vance et al., 2009) and 1% of SALS patients (Belzil et al., 2009; Deng, Gao, et al., 2014; Rademakers et al., 2010). The homolog of FUS in *Drosophila* corresponds to *caz* (Stolow & Haynes, 1995). Loss of *FUS* in *Drosophila* causes reduced eclosion rates, a shorter life span, and defects in locomotion (Wang, Brent, Tomlinson, Shneider, & McCabe, 2011; Xia et al., 2012). Neuronal expression of fly or human *FUS* cDNA rescues these defects, suggesting a conservation of function (Wang, Brent, et al., 2011). Mice with a genetic knockout of *FUS* die 16 hours after birth (Hicks et al., 2000), but in an independent knockout line with a different genetic background is viable with only a reduction in body weight and loss of fertility in males (Kuroda et al., 2000). Nevertheless, neither of them displays obvious neurological defects, suggesting that *Drosophila* is a better model to study FUS-related ALS.

The pathology associated with FUS pathogenic variants is characterized by the presence of FUS immunoreactive inclusion in the cytoplasm of neurons and glia (Deng, Gao, et al., 2014). In addition to ALS, FUS-containing inclusions have also been detected in disorders including frontotemporal dementia (FTD) and polyglutamine diseases, such as Huntington's chorea, spinocerebellar ataxia types 1–3 and dentatorubral–pallidoluysian atrophy (Deng, Gao, et al., 2014). FUS contains a NLS and is predominantly localized to the nucleus. Expression of the ALS-linked variants affecting the NLS (R518K, R521C, R521H, R524S, and R525L) in *Drosophila* eyes causes age- and dosage-dependent degenerations (Lanson et al., 2011). Interestingly, removal of the nuclear export sequence (NES) from these mutants rescues the eye degeneration phenotypes (Lanson et al., 2011). These findings support the argument that cytoplasmic localization of FUS plays an important role in the development of the degenerative phenotypes.

Similar to TDP-43, several FUS modifications, including phosphorylation and methylation, have been associated with the FUS-containing inclusions. Phosphorylation of the N-terminus of FUS by the DNA-dependent protein kinase (DNA-PK) contributes to the formation of inclusions (Deng, Holler, et al., 2014) and phosphorylation of the C-terminal tyrosine 526 impairs its nuclear import (Darovic et al., 2015). Arginine methylation is a common posttranslational modification modulating its affinity to transportin 1 and controls nuclear transport (Dormann & Haass, 2013). FUS is methylated at a site near the NLS by arginine methyltransferase 1 and 8 - (PRMT1 and PRMT8) (Dormann et al., 2012; Scaramuzzino et al., 2013). In *Drosophila* photoreceptors, a reduction of PRMT protein, Dart1, leads to a decrease in methylation of the ectopically expressed human FUS P525L and promotes degeneration caused by FUS P525L (Jackel et al., 2015). This finding suggests that arginine methylation of FUS promotes the toxicity of ALS-linked FUS variants.

FUS dysfunction alters the splicing of several genes encoding proteins forming pathological inclusions, suggesting that the pathogenic variants create stresses on protein degradation. For example, loss of FUS causes the aggregation of mis-spliced Tau protein in neurons (Orozco & Edbauer, 2013; Orozco et al., 2012). Indeed, ectopic expression of human FUS carrying ALS-causing variants (R518K or R521C) in *Drosophila* causes degeneration in eyes, brains, and motor neurons (Daigle et al., 2013). The introduction of point mutations that disrupt the RNA-binding ability of these FUS transgene constructs prevents the incorporation of FUS in these inclusions and reduces degeneration (Daigle et al., 2013). These data suggest that FUS RNA-binding activity may contribute to the formation of protein aggregates and cause degeneration. In addition, FUS is also present in the Drosha complex, an RNase III enzyme required for microRNA biogenesis (Gregory et al., 2004), suggesting that loss of FUS might disrupt microRNA biogenesis and contribute to the pathogenesis of ALS6. Unlike TDP-43 which shows a high binding specificity to the UG-rich sequence, FUS shows little to no binding preference to specific RNA sequences (Wang, Schwartz, & Cech, 2015). Hence, FUS dysfunction may result in a broader pathology than TDP-43 dysfunction.

TDP-43 and FUS not only share similar functions in pre-mRNA splicing, but are also physically and functionally interconnected, as suggested by genetic and biochemical studies in *Drosophila* and cell culture (Wang, Brent, et al., 2011). Neuronal expression of FUS in a TDP-43 null background suppresses lethality, locomotion defects, and the reduced bouton number caused by loss of TDP-43 (Wang, Brent, et al., 2011). Moreover, coexpression of either wild type or the ALS-causing TDP-43 variant (M337V) enhances eye degeneration caused by either wild type or variant (R521H) FUS (Lanson et al., 2011). These findings collectively argue that FUS and TDP-43 may function in the same pathway.

In summary, the formation of cytoplasmic FUS aggregates is likely to be important in the pathogenesis of *FUS* variant-linked ALS. The role of FUS in the pathogenesis may resemble that of TDP-43. Loss of *FUS* may cause aberrant splicing of its target genes that eventually leads to a burden of aberrant proteins that are ineffectively removed in neurons and cause degeneration.

3.3 C9orf72

The HRE, (GGGGCC)*n*, in the noncoding region of *C9orf72* was recently linked to ALS and FTD (DeJesus-Hernandez et al., 2011; Renton et al., 2011). It accounts for 40% of FALS and 8% of SALS (DeJesus-Hernandez et al., 2011; Majounie et al., 2012; Renton et al., 2011). Cytoplasmic aggregates that contain TDP-43, FUS, and Tau are a common observation in motor neurons in *C9orf72*-associated ALS cases (Fratta et al., 2013; Rohrer et al., 2015; Stewart et al., 2012), implicating that there is an important connection between *C9orf72* HRE and protein cytoplasmic aggregates in ALS pathophysiology. In addition to ALS, HRE insertion in the *C9orf72* gene has also been widely associated with FTD. FTD is a type of dementia that results in a progressive degeneration of the frontal and/or temporal lobes of the brain. Although ALS and FTD are distinct disorders, accumulating evidence suggest that they may share common neurodegenerative pathways and may be part of a spectrum. Indeed, TDP-43 or FUS-containing protein aggregates are usually observed in FTD. In addition, ALS-causing genetic loci such as C9orf72 and VCP have also been connected with FTD (Ferrari, Kapogiannis, Huey, & Momeni, 2013).

Three models may underlie the pathogenesis of *C9orf72*-mediated ALS: (1) HRE interferes with C9orf72 protein level and loss C9orf72 protein mediates pathogenesis. (2) HRE forms a toxic RNA structure that binds and sequesters RNA-binding proteins (RBPs). (3) HRE is translated into toxic DPRs via repeat-associated and non-ATG translation (RAN translation, or RANT). Both models (2) and (3) lead to protein aggregates that deplete regulators of nucleocytoplasmic transport and contribute to degeneration. The data related to C9orf72 loss of expression in pathogenesis is ambiguous. Most studies agree that *C9orf72* mRNA and protein expression is lower in brain tissue of C9-ALS patients or in induced pluripotent stem (iPS) neurons derived from C9-ALS/FTD patients (Ciura et al., 2013; Donnelly et al., 2013; Waite et al., 2014). Moreover, loss of the *Caenorhabditis elegans* C9orf72 ortholog, *alfa-1*, causes age-dependent locomotion defects resulting in paralysis as well as the specific degeneration of GABAergic motor neurons (Therrien, Rouleau, Dion, & Parker, 2013). In addition, *C9orf72* knockdown in zebrafish embryos causes axon degeneration of motor neurons (Ciura et al., 2013). However, in a rare homozygous *C9orf72* patient, the clinical or pathological features are not worse than the dominant cases (Fratta et al., 2013). Whole body knockout of C9orf72 causes a robust immunological defect, but no motor defects were observed (Atanasio et al., 2016; Jiang et al., 2016; Sudria-Lopez et al., 2016). In addition, selective deletion of *C9orf72* in neurons and glia in mice causes reduced body weight but does not induce motor neuron loss, or affect survival, and no mislocalization of TDP-43 was observed (Koppers et al., 2015). Hence, loss of *C9orf72* may not play a key role in ALS pathogenesis.

C9-ALS/FTD patients exhibit RNA foci in the motor cortex and spinal cord (Donnelly et al., 2013). Antisense oligonucleotides targeting HRE suppress HRE RNA foci-induced cytotoxicity in C9orf72 iPSC neurons (Donnelly et al., 2013), suggesting that HRE RNA underlies the pathogenesis of HRE-linked ALS. The HRE RNA or the single-stranded DNA formed during transcription has a high tendency to form a G-quadruplex, a stack of hydrogen-bonded guanine tetramers (Haeusler et al., 2014). During transcription, the single-stranded G-quadruplex DNA impairs RNA polymerase II elongation and leads to the accumulation of abortive transcripts (the RNA G-quadruplexes). The abortive transcripts bind and sequester RBPs (Donnelly et al., 2013), which further lead to aggregation of ribonucleoproteins and nucleolar stress (Haeusler et al., 2014). Nineteen RBPs were identified that exhibit a higher binding affinity to HREs in a protein–array analysis (Donnelly et al., 2013); among them RanGAP1 (*Drosophila*: RanGAP) was identified as a suppressor of HRE-mediated toxicity from an RNAi screen in *Drosophila* using eye degeneration as a phenotypic readout (Zhang, Donnelly, et al., 2015). RanGAP (Ran GTPase-activating protein) is a master regulator of nucleocytoplasmic transport, suggesting that this process is involved in HRE-induced pathogenesis. Indeed, either knockdown of Exportin proteins or overexpression of Importin proteins that promotes nuclear import suppresses HRE-mediated eye degeneration. In addition, RanGAP/RanGAP1 was observed to interact and colocalize with RNA foci in *Drosophila* S2 cells and/or iPSCs derived from *C9orf72*-linked ALS patients. Interestingly, feeding the HRE-expressing flies antisense oligonucleotides or small inhibitors that target HRE suppresses nucleocytoplasmic transport defects as well as the degenerative phenotypes in *Drosophila* eyes, providing further support that nucleocytoplasmic transport mediates HRE-induced toxicity. Moreover, TDP-43 is observed to be mislocalized to the cytoplasm in HRE-expressing flies, iPS neurons, transgenic mice that express HRE (Chew et al., 2015; Zhang, Donnelly, et al., 2015) as well as tissue samples from *C9orf72* patients (Cooper-Knock et al., 2015), suggesting that TDP-43 mislocalization may play a role in *C9orf72*-linked ALS. Taken together, these data support a model in which the HRE binds and recruits RBPs such as RanGAP to the RNA foci, leading to their mislocalization and loss of function. These defects in nucleocytoplasmic transportation cause cytoplasmic mislocalization and accumulation of nuclear proteins including TDP-43 that contributes to the pathogenesis.

The third model argues that DPR expressed by the repeat-associated non-ATG (RAN) translation of HRE causes neuronal toxicity (Cleary & Ranum, 2013; Zu et al., 2011). RAN translation can occur in all three reading frames independent of an ATG start codon, generating six different types of DPRs in total. However, since poly(GP) can be translated from either direction, only five types, poly-GA, GP, GR, PA, and PR, are generated. These DPRs are highly prone to aggregate formation and have been shown to accumulate in the cerebellum of *C9orf72* patients (Ash et al., 2013; Mori et al., 2013; Zu et al., 2013). However, recent observations suggest that DPRs are rarely present in the spinal cords or motor neurons of *C9orf72*-linked ALS patients (Gomez-Deza et al., 2015). Hence, the neurotoxicity of the DPRs is still under debate. In *Drosophila*, ectopic expression of HRE in *Drosophila* eyes under control of the GAL4/UAS system causes degeneration (Mizielinska et al., 2014). Introduction of stop codons that block RAN translation in all six reading frames without affecting the circular dichromatic property of the HRE RNA suppresses the HRE-induced eye degeneration, suggesting that DPRs, but not the HRE RNA, are toxic. Moreover, Mizielinska et al. (2014) generated "protein-only" constructs using alternative codons to individually express four major forms of DPRs, poly(GR), poly(PR), poly(PA), or poly(GA) encoded by the HRE. Only the arginine-containing DPRs, poly(GR) and poly(PR), caused eye degeneration, indicating that some DPRs but not the G-quadruplex RNA lead to eye degeneration (Mizielinska et al., 2014). Freibaum et al. (2015) confirmed these results in an independent line of HRE-expressing transgenic flies. Moreover, they found that expression of HRE cause locomotion defects and reduced bouton numbers at NMJs in larvae (Freibaum et al., 2015). They further performed a genetic modifier screen that covers ~80% of the *Drosophila* genome to identify genetic loci whose partial loss of function modified the HRE-induced eye degeneration. They discovered that loss of proteins that play a role in nuclear import either enhance or suppress HRE-induced eye degeneration, suggesting that nuclear import is compromised. Moreover, they identify a nuclear export protein as a suppressor of HRE-induced eye degeneration, suggesting that nuclear export is also important. In agreement with this, nuclear RNA is accumulated in mammalian cell lines and in iPSCs derived from C9 patients. In addition,

Freibaum et al. (2015) also observed morphological disruption of nuclear envelope and RNA foci in the nuclei of *Drosophila* salivary glands. As low levels of DPRs were observed in this study, the DPRs may be involved in *C9orf72* toxicity. The DPR toxicity model is further supported recently by Jovicic et al. (2015). They reported that arginine-rich DPRs, expressed via alternative codons that express the DPR but do not form G-quadruplex, are toxic in yeast. Through genome-wide complementary screens, they also found an enrichment of modifiers functioning in nucleocytoplasmic transport (Jovicic et al., 2015). Moreover, an independent study in which HRE is expressed in *Drosophila* supports the DPR-toxicity model (Tran et al., 2015). Taken together, these studies converge on an effect of HRE toxicity via disrupting nucleocytoplasmic transport (Freibaum et al., 2015; Jovicic et al., 2015; Zhang, Donnelly, et al., 2015). However, it is still not obvious whether the toxicity of HRE is derived from the G-quadruplex, DPRs, or both. Zhang, Donnelly, et al. (2015) and Zhang, Gui, et al. (2015) did not detect DPRs, poly(GR) or poly(GP), in HRE-expressing flies, arguing that G-quadruplex is toxic. Freibaum et al. (2015) observed DPRs as well as RNA foci in HRE-expressing flies. They support the DPR toxicity model but do not rule out an involvement of G-quadruplex toxicity. Both Jovicic et al. (2015) and Tran et al. (2015) argued that it is the DPR not G-quadruplex that cause toxicity in yeast or fly, respectively.

The above findings suggest that C9orf72 loss of expression may not play an important role in ALS, indicating that the first model is unlikely. Rather, the HRE seems to be a key player in C9orf72 pathogenesis. The second model argues that HRE sequences disrupt RNA polymerase activity, leading to the formation of toxic RNA foci containing short abortive transcripts (Fig. 1B). These HRE-containing transcripts lead to the formation of protein aggregates containing proteins such as RanGAP that disrupts nuclear import machinery. One likely target is TDP-43. The aberrantly localized TDP-43 causes abnormal incorporation of cryptic exons and leads to a proteinopathy. The third model argues that arginine-containing DPRs synthesized via RAN translation leads to degeneration. The expression of the arginine-containing DPRs causes impairment in nuclear import and export and may contribute to the accumulation of protein in the cytoplasm and the formation of nuclear RNA foci, respectively. The latter are probably not toxic. It is important to note that mouse models that carry an HRE have been generated. The sense, antisense RNA foci as well as DPRs were observed in neuronal tissues including motor cortex and spinal cord (O'Rourke et al., 2015; Peters, Cabrera, et al., 2015). However, no obvious behavioral and histological defects that indicate neurodegeneration has been observed in these mice after a 2-year period. Interestingly, the authors did not observe a TDP-43 cytoplasmic mislocalization. These data suggest that the role of the RNA foci and DPRs in C9orf72-linked degeneration is still not resolved.

4. PROTEOSTASIS DEFICIENCY IN ALS

Protein aggregation and formation of intracellular inclusions are central features of many neurodegenerative diseases. It is interesting that at least three genes that have been implicated in protein quality control at the ER are linked to ALS. These genes correspond to VCP (Johnson et al., 2010), UBQLN2 (Deng et al., 2011), and VAPB (Nishimura et al., 2004).

Accumulation of unfolded or misfolded proteins in the ER lead to ER stress as well as unfolded protein response (UPR) (Ron & Walter, 2007). UPR activation attenuates protein translation, promotes protein degradation, as well as upregulates the expression of chaperones that are targeted to the ER to refold proteins. If the proteins cannot be properly folded, the ER-associated protein degradation (ERAD) pathway is activated (Vembar & Brodsky, 2008). Misfolded proteins are ubiquitinated and degraded by the proteasome machinery or autophagy. Defects in ERAD lead to protein aggregations that are often observed in many neurodegenerative diseases, including ALS (Ling et al., 2013). Indeed, one of the most common observations in the cell body of motor neurons is protein aggregation in the cytoplasm (Al-Chalabi et al., 2012). Hence, a better understanding of mechanisms affecting proteostasis will provide better insight into ALS pathology.

4.1 VCP

Valosin-containing protein (VCP)/p97 belongs to the hexameric AAA (ATPase associated with diverse cellular activities) protein family. Proteins of this family generally consume energy from ATP hydrolysis to structurally remodel their targets (Erzberger & Berger, 2006). VCP is conserved across many species and named Cdc48 in yeast and Ter94 in flies, respectively. Variants in *VCP* were first found to associate with IBMPFD (inclusion body myopathy associated with Paget disease of bone (PDB) and FTD), a type of inclusion body myopathy (Watts et al., 2004). IBMPFD is a dominant autosomal progressive disorder (Kimonis et al., 2000). It is characterized by an adult-onset proximal and distal muscle weakness, early onset PDB, and FTD (Kovach et al., 2001). Moreover, TDP-43 and ubiquitin-positive cytoplasmic inclusions in brains and muscles are observed in IBMPFD patients regardless of the affected residues of VCP or clinical symptoms (Ju et al., 2009; Neumann et al., 2007; Weihl et al., 2008). Interestingly, variants in the same protein domain of VCP can cause IBMPFD as well as FALS (Johnson et al., 2010). Variants in *VCP* are responsible for 2% of FALS (also known as ALS14) cases and are inherited in a dominant manner (Johnson et al., 2010; Miller et al., 2012). Moreover, TDP-43-positive cytoplasmic inclusions are observed in the spinal cord in these patients (Johnson et al., 2010). In this section, we will discuss how variants in *VCP* cause TDP-43 proteinopathy and contribute to degeneration.

Loss of VCP in *Drosophila* is cell lethal (Leon & McKearin, 1999) and loss of VCP in mice leads to lethality prior to implantation (Muller, Deinhardt, Rosewell, Warren, & Shima, 2007). VCP has been shown to affect numerous cellular process including membrane fusion (Latterich, Frohlich, & Schekman, 1995; Rabouille, Levine, Peters, & Warren, 1995), cell cycle progression (Cao, Nakajima, Meyer, & Zheng, 2003), and transcriptional activation (Rape et al., 2001). In addition, VCP has also been shown to regulate several proteostasis pathways including proteasome degradation (Dai & Li, 2001), ERAD (Jarosch, Geiss-Friedlander, Meusser, Walter, & Sommer, 2002; Ye, Meyer, & Rapoport, 2001), and autophagy (Ju et al., 2009). VCP consists of an N-terminal regulatory domain, two ATPase domains (D1 and D2), and a C-terminal HbYX motif (Brunger & DeLaBarre, 2003; DeLaBarre & Brunger, 2003; Stolz, Hilt, Buchberger, & Wolf, 2011). The N-terminal domain is responsible for the recognition of substrates. This interaction is achieved either through a direct binding to polyubiquitin chains of substrates (Dai & Li, 2001) or through cofactors that mediate the interaction with polyubiquitinated targets (Meyer, Wang, & Warren, 2002). The D1 and D2 ATPase domains hydrolyze ATP to generate energy for VCP to function as a segregase to structurally remodel ubiquitinated proteins. This facilitates the extraction of aberrant proteins from ER, plasma membranes, or protein aggregates for proteasome degradation (Buchberger, Schindelin, & Hanzelmann, 2015). The HbYX motif together with a pore-2 loop at the bottom of the D2 domain mediates the docking of aberrant proteins into the proteasome and facilitates the transfer of unfolded protein to the proteasome for degradation (Barthelme & Sauer, 2012, 2013). Variants in the pore-2 loop of VCP impair 20S proteasome binding and have been linked to FALS (Abramzon et al., 2012; Barthelme et al., 2015; Johnson et al., 2010). These data suggest that VCP plays a role to remodel aberrant proteins and promote the degradation of the aberrant proteins via the proteasome system.

Interestingly, most of the variants identified in ALS or IBMPFD patients localize to the N-terminal domain or the interface between N-terminal domain and D1 domain, suggesting that substrate interaction plays a key role in the pathogenesis of VCP-linked diseases, including ALS. To date, more than 30 cofactors have been identified and most of them bind to the N-terminal domain of VCP (Barthelme & Sauer, 2016). Loss of VCP in yeast causes accumulation of ubiquitinated proteins (Dai & Li, 2001). Similarly, in *Drosophila* and human cells, suppression of VCP expression by RNAi leads to an upregulation of ubiquitinated proteins (Wojcik et al., 2004) supporting the observation from in vitro assays that VCP loss of function causes defects in proteostasis. The ectopic expression of mutant VCP with disease variants in cell lines disrupts ERAD and leads to accumulation of ubiquitinated proteins, indicating a defect in proteostasis which resembles the defect in patients with VCP variants (Ritson et al., 2010). Knock-in of the VCP(R155H/+) variant in mice, which account for 50% of IBMPFD cases, causes an accumulation of cytoplasmic TDP-43 and ubiquitin-positive inclusions (Custer et al., 2010; Nalbandian et al., 2013; Yin et al., 2012). Hence, both loss of VCP and the expression of R155H VCP variant lead to similar proteostasis defects, suggesting that VCP loss of function may be a key. Indeed, VCP variants have been shown to decrease protein stability and/or impair VCP binding with cofactors (Zhang, Gui, et al., 2015). In summary, disease-associated VCP variants may impair the interaction with cofactors that regulate degradation of ubiquitinated proteins, which in turn leads to the formation of protein aggregates that are likely to play a role in the degeneration.

The relationship between VCP and TDP-43 has been explored in *Drosophila* (Ritson et al., 2010). Ectopic expression of mutant VCPs that contain disease-causing variants, R152H or A229E, in *Drosophila* eyes leads to a severe photoreceptor loss when compared to wild-type VCP expression, which only causes a subtle defect (Ritson et al., 2010). Expression of these mutant VCPs in the *Drosophila* central nervous system (CNS) leads to a reduced life span. Moreover, loss of TDP-43 suppresses mutant VCP-induced photoreceptor loss, whereas coexpression of wild-type TDP-43 with mutant VCP enhances photoreceptor loss, suggesting that phenotypes mediated by expression of pathogenic VCP are mediated through TDP-43 in *Drosophila*. Interestingly, TDP-43 without the NES has a minor effect on VCP-mediated degenerative phenotypes, while TDP-43 without a NLS enhances VCP-mediated photoreceptor loss (Ritson et al., 2010). These observations suggest that the nuclear exclusion of TDP-43 and/or the formation of TDP-43 cytoplasmic inclusions play a key role in the pathogenesis of VCP-related diseases. In addition, the redistribution of TDP-43 is also observed in cultured primary neurons or brain cortex expressing VCP (R155H) (Nalbandian et al., 2013; Ritson et al., 2010). Consistently, knock-in of the VCP (R155H/+) variant in mice causes motor neuron loss and muscle denervation at 24–27 months. Furthermore, there is also an accumulation of cytoplasmic TDP-43 and ubiquitin-positive inclusions in hippocampi as well as spinal cords in VCP(R155H/+) mice (Nalbandian et al., 2013; Yin et al., 2012). These data suggest that TDP-43 mislocalization and TDP-43 pathology play a key role in the pathogenesis of VCP-linked ALS.

VCP has also been shown to interact genetically with *FUS* in *Drosophila*. Reduced levels of FUS in *Drosophila* lead to a rough-eye phenotype, locomotion defects, as well as a reduction of the bouton numbers and branch length in NMJ (Azuma et al., 2014). Overexpression of VCP suppresses these phenotypes while loss of VCP exacerbates the photoreceptor defects (Azuma et al., 2014). In addition, overexpression of VCP promotes FUS nuclear localization, suggesting that VCP promotes nuclear translocation of FUS or suppresses FUS degradation (Azuma et al., 2014).

Although most data relate VCP to defects in the proteasomal degradation pathway, the accumulation of LC3 and p62 has also been observed in patients with VCP R155H variant, suggesting defects in autophagy (Ju et al., 2009; Tresse et al., 2010). Both LC3 and p62 are key components of autophagy (Xie & Klionsky, 2007). Indeed, both loss of VCP and expression of VCP carrying disease-causing variants impair autophagosome maturation in cultured cells, indicating that VCP regulates autophagy pathway (Ju et al., 2009; Tresse et al., 2010). In *Drosophila*, HDAC6, a microtubule-associated deacetylase that binds to polyubiquitinated proteins and VCP, has been shown to mediate a compensatory autophagic degradation upon proteasome inhibition (Pandey, Batlevi, Baehrecke, & Taylor, 2007; Pandey, Nie, et al., 2007; Seigneurin-Berny et al., 2001). Overexpression of VCP (R155H) disrupts autophagic degradation but not proteasome degradation in cultured cells (Tresse et al., 2010). HDAC6 may modulate the role of VCP between these two functions. Interestingly, suppression of autophagy in cultured cells causes cytoplasmic TDP-43 accumulation (Ju et al., 2009), suggesting that cytoplasmic mislocalized TDP-43 can be degraded through autophagy.

In summary, VCP extracts proteins from membranes and promotes protein refolding and degradation via the proteasomal and/or autophagic degradation pathway (Fig. 2A). Protein cofactors such as HDAC6 play an important role in modulating this process. Variants in VCP may disrupt its binding to cofactors and block aberrant protein refolding and degradation, leading to an accumulation of misfolded proteins. Moreover, VCP loss of function impairs nuclear localization of TDP-43, leading to the accumulation of TDP-43 in cytoplasm which in turn promotes the formation of protein aggregates, possibly further exacerbating the proteostasis defects (Fig. 2B).

4.2 UBQLN2

UBQLN2 belongs to an evolutionarily conserved protein family that functions in proteasomal degradation (Mah, Perry, Smith, & Monteiro, 2000; Wu, Wang, Zheleznyak, & Brown, 1999). There are four homologs in mammals, *UBQLN1–4*, and a single homolog in *Drosophila*. Human Ubiquilin-1 and -2 share 79% homology and are conserved in domain structure, indicating that they might share a similar molecular function. Ubiquilin-1 is expressed ubiquitously, whereas Ubiquilin-2 is mainly expressed in muscles and nervous system (Wu et al., 1999; Zhang & Saunders, 2009). Variants in *UBQLN1* have been linked to Alzheimer's disease (AD) and Brown–Vialetto–Van Laere syndrome (BVVLS) (Bertram et al., 2005; Gonzalez-Perez et al., 2012). BVVLS is a rare neurological disorder characterized by a progressive loss of hearing and lower or upper motor neuron degeneration. It is also associated with TDP-43-containing protein aggregates in cultured neurons that express the mutant protein (Gonzalez-Perez et al., 2012; Sathasivam, 2008). Ubiquilin-1 has also been shown to bind to polyubiquitinated presenilin-1, a protease that regulates amyloid formation in AD, and targets it to the proteasome (Haapasalo et al., 2010). Variants in *UBQLN2* cause a dominantly inherited, X-linked form of ALS (also known as ALS15) and FTD (Deng et al., 2011; Fahed et al., 2014; Gellera et al., 2013; Synofzik et al., 2012; Williams et al., 2012). Expression of disease-causing variants of Ubiqulin-2 in cell lines disrupts proteasomal degradation and causes accumulation of ubiquitinated substrates (Deng et al., 2011). Hence, defects in proteostasis play a role in the pathogenesis of diseases associated with *UBQLN* variants.

Ubiquilin-1 and -2 are cytosolic proteins that control proteasomal degradation and ERAD (Walters, Kleijnen, Goh, Wagner, & Howley, 2002; Zhang & Saunders, 2009). Ubiquilin-2 contains an N-terminal UBL

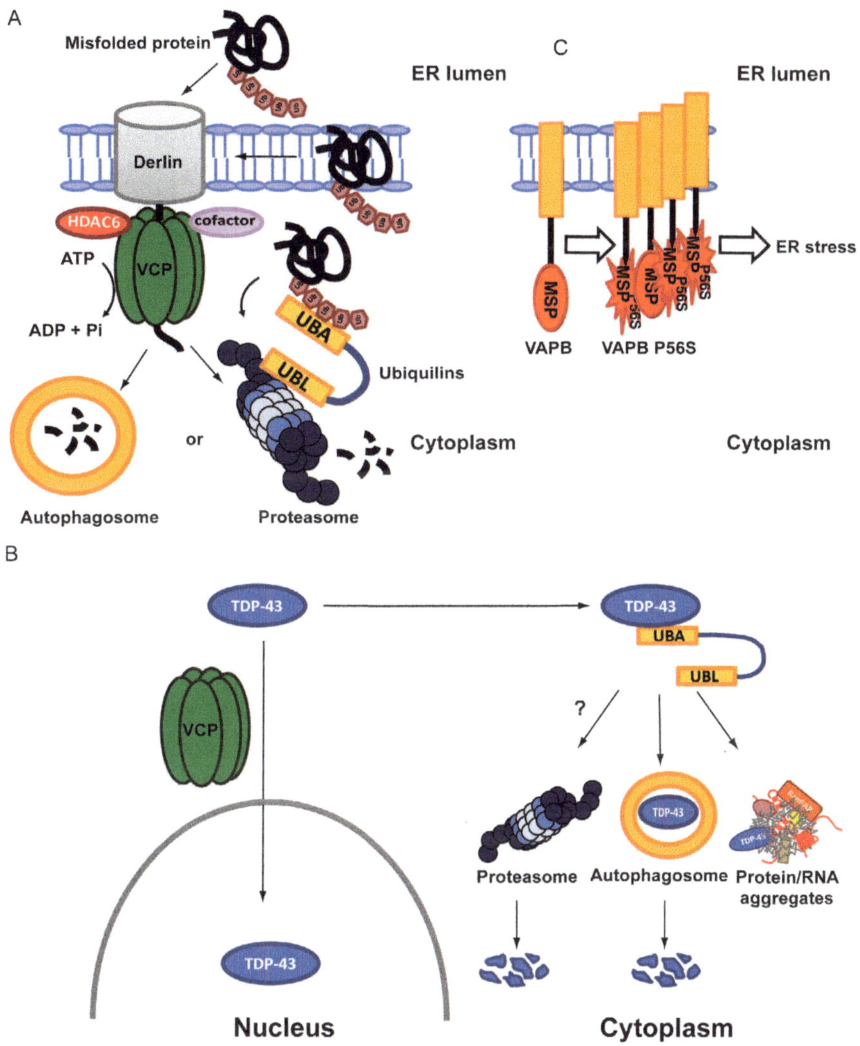

Fig. 2 A model for the involvement of the proteasome degradation system in the formation of protein aggregates and degeneration. (A) VCP facilitates the unfolding and translocation of misfolded proteins from membranes or protein aggregates and consumes ATP. The translocation machinery (e.g., Derlin) is required for extracting proteins from the membrane or the ER lumen. Cofactors, such as HDAC6, modulate VCP function in protein degradation through proteasomes or autophagosomes. UBQLNs bind to ubiquitinated proteins from the ER or cytoplasm through their UBA domain. This facilitates the delivery of target proteins to proteasome for degradation. (B) In addition to their function to regulate protein degradation through the proteasome degradation system, VCP and Ubiquilins also play a role to control TDP-43 localization. VCP can affect TDP-43 localization to the nucleus. UBQLNs bind to cytoplasmic TDP-43 and deliver it to

(Continued)

(ubiquitin-like) domain followed by STI1-like motifs, PXX repeats, and a C-terminal UBA (ubiquitin-associated) domain. The UBL domain mediates the interaction of Ubiquilins with the proteasome while the UBA domain mediates its interaction with polyubiquitinated proteins (Kleijnen, Alarcon, & Howley, 2003; Walters et al., 2002). This domain organization is characteristic of proteins that are involved in facilitating the presentation of polyubiquitinated proteins to the proteasomes for degradation (Elsasser & Finley, 2005; Miller & Gordon, 2005). Hence, it has been proposed that variants of Ubiquilin-1 and -2 may disrupt its binding to the proteasome or polyubiquitinated proteins. This in turn may impair the shuttling of polyubiquitinated proteins to the proteasomal degradation machinery and causes a chronic accumulation of protein aggregates. However, most of the disease-causing variants identified in patients localize to the PXX repeats, of which the molecular function is poorly characterized (Deng et al., 2011; Fahed et al., 2014; Gellera et al., 2013). Ectopic expression of Ubiquilin-2 containing human variants in the PXX repeats in a neuronal cell line leads to an aggregation of a proteasome reporter substrate in the cytoplasm (Deng et al., 2011). These disease mutants show a reduced binding affinity to the proteasome, whereas their interaction with polyubiquitinated proteins is not affected (Chang & Monteiro, 2015). The data suggest that the PXX repeats may specifically affect the interaction of Ubiquilin with the proteasome. Hence, the ALS-causing variants in *UBQLN2* may disrupt this interaction and impair the delivery of polyubiquitinated proteins to the proteasome and therefore causes accumulation of ubiquitinated proteins.

The molecular mechanism related to how Ubiquilins control ERAD has been investigated in cultured cells. Overexpression of Ubiquilins promotes the degradation of ERAD substrates whereas loss of Ubiquilins or overexpression of mutant forms leads to substrate accumulation (Kim et al., 2008; Xia et al., 2014), indicating that Ubiquilins are required for ERAD. It is still unclear how Ubiquilins regulate ERAD and current observations suggest that Ubiquilins may bind to ER stress response proteins like Erasin

Fig. 2—Cont'd either the proteasomes or autophagosomes for degradation. Moreover, UBQLNs also transfer TDP-43 to aggregates. Collectively, these mechanisms lead to TDP-43 nuclear exclusion and cause the incorporation of cryptic exon sequences in mRNAs. This results in the accumulation of aberrant proteins. (C) VAPB is localized to the ER through its C-terminal transmembrane domain. The N-terminal MSP domain is localized to the cytoplasm. The disease-causing variant (P56S) disrupts the proper function of VAPB and leads to the formation of protein aggregates, traps endogenous wild-type VAPB proteins, induces ER stress and degeneration.

(Lim et al., 2009), Herp (Kim et al., 2008), or UBXD8 (Xia et al., 2014), to facilitate the recognition of misfolded ER proteins for proteasome degradation. Erasin is an ER membrane-associated protein that functions as a platform to recruit VCP and Ubiquilin to facilitate ERAD (Liang et al., 2006; Lim et al., 2009). Suppression of the expression of Erasin or Ubiquilin causes an accumulation of ERAD substrates (Lim et al., 2009). In *C. elegans*, loss of Erasin or Ubiquilin leads to the accumulation of polyubiquited proteins, activation of ER stress, and a reduced life span (Lim et al., 2009), providing in vivo evidence that Erasin and Ubiquilin play a key role in ERAD. Moreover, Erasin binds to Derlin, an ER membrane protein that may form a channel that allows the export of misfolded proteins or extract misfolded protein from the ER in a VCP-dependent manner (Lilley & Ploegh, 2004; Mehnert, Sommer, & Jarosch, 2014). Hence, Erasin has been proposed to recruit Ubiquilin and promotes the shuttling of misfolded proteins to the proteasomal degradation machinery (Lim et al., 2009). In addition to Erasin, Herp, and UBXD8 have also been reported to bind to Ubiquilin (Kim et al., 2008; Xia et al., 2014). Disrupting the interaction between Ubiquilin and Herp or UBXD8 reduces ERAD (Kim et al., 2008; Xia et al., 2014). Pathogenic variants of *UBQLN2* impair its interaction with UBXD8 and disrupt ERAD (Xia et al., 2014). Hence, Ubiquilin dysfunction may impair ERAD, leading to the accumulation of misfolded proteins that in turn contribute to neuronal degeneration in ALS.

In addition to the proteasomal degradation machinery and ERAD, evidence from cultured cell lines suggests that Ubiquilins may also play a role in autophagy (N'Diaye et al., 2009; Rothenberg et al., 2010). Ubiquilin-1 and -2 have been shown to localize to autophagosomes in HeLa cells as well as mouse brain and liver tissue. Loss of Ubiquilins reduces autophagosome formation and maturation (Rothenberg et al., 2010). Overexpression of Ubiquilin-1 or -2 suppresses starvation-induced cell death, and removing key proteins of the autophagy machinery, ATG5 and ATG7, abolish this protective effect (N'Diaye et al., 2009). These findings indicate that Ubiquilin may promote autophagy to suppress starvation-induced cell death. Moreover, Ubiquilin-1 physically interacts with LC3, a key component in autophagosomes. Deleting the UBA domain of Ubiquilin-1 abolishes this interaction as well as the protective effect in starvation-induced cell death (N'Diaye et al., 2009). Therefore, it is proposed that Ubiquilins may function in the delivery of ubiquitinated cargo to autophagosome for degradation (Fig. 2A). However, these data have not yet been validated in vivo and the mechanism is not understood.

UBQLN2 also interacts with *TDP-43*. Cytoplasmic inclusions that contain TDP-43 are often observed in ALS patients with UBQLN2 variants (Fahed et al., 2014; Williams et al., 2012) and Ubiquilin-2-positive aggregates in other ALS patient samples also colocalize with TDP-43 inclusions (Deng et al., 2011; Williams et al., 2012). In addition, Ubiquilin-1 physically interacts with TDP-43 in yeast and the C-terminal UBA domain of Ubiquilin-1 is necessary and sufficient for this interaction (Kim et al., 2009). Overexpression of Ubiquilin-1 in HeLa cells potentiates TDP-43 aggregate formation in the cytoplasm (Kim et al., 2009), suggesting that Ubiquilin-1 promotes TDP-43 aggregation. Moreover, the reduced eclosion rate as well as shortened life span caused by TDP-43 overexpression is exacerbated upon coexpressing human Ubiquilin in *Drosophila* (Hanson et al., 2010), suggesting that Ubiquilins enhance TDP-43 toxicity. In addition, the same lab has also documented that overexpression of Ubiquilin-1 reduces the detergent soluble form of TDP-43 in *Drosophila* and in cultured cells, consistent with the hypothesis that insoluble TDP-43 is toxic (Hanson et al., 2010). Furthermore, overexpression of Ubiquilin-1 with disease variant (E54D) in both *Drosophila* as well as HEK293 cells also reduces the soluble form of TDP-43 and redistributes TDP-43 from the nucleus to the cytoplasm (Gonzalez-Perez et al., 2012). Given that Ubiquilin physically interacts with TDP-43, it is proposed that Ubiquilin modulates the level of soluble TDP-43 either by coaggregation with TDP-43 or by delivering it to proteasome or autophagosome for degradation (Fig. 2B). Hence, defects in Ubiquilin pathway disturb the proper function of TDP-43 in the nucleus and may lead to the neurodegeneration.

4.3 VAPB

VAPB (VAMP-associated protein B) belongs to the VAMP-associated protein family. It is highly conserved across species and named *Vap-33-1* in *Drosophila*, *Vpr-1* in *C. elegans*, and *Scs2p* in *Saccharomyces cerevisiae* (Nikawa, Murakami, Esumi, & Hosaka, 1995; Pennetta, Hiesinger, Fabian-Fine, Meinertzhagen, & Bellen, 2002; Weir, Klip, & Trimble, 1998). We will use VAPB for all species. There are three homologs of VAPs in mammals, VAPA, VAPB, and VAPC (Nishimura, Hayashi, Inada, & Tanaka, 1999). Both VAPA and VAPB contain an N-terminal major sperm proteins (MSP) domain that shows homology to MSP in *C. elegans*, a coiled-coil domain, and a C-terminal transmembrane (TM) domain (Nishimura et al., 1999). Multiple variants in VAPB (P56S, T46I, A145V, S160Δ, and V234I) have been identified in patients with ALS (also known as ALS8) or spinal muscular atrophy (Chen et al., 2010; Kabashi et al., 2013; Nishimura et al., 2004; van Blitterswijk et al., 2012). Among them, the P56S variant is by far the most prevalent form that cosegregates with disease in ALS patients.

VAPB plays a role in several cellular processes and functions both cell autonomously and nonautonomously. The cell autonomous function of VAPs (VAPA and VAPB) has been associated with its localization to the ER (Kagiwada, Hosaka, Murata, Nikawa, & Takatsuki, 1998; Tsuda et al., 2008). Both VAPA and VAPB are required for the transfer of oxysterols and ceramides from the ER to Golgi via interactions with lipid-binding proteins, such as OSBP (oxysterol-binding protein) and CERT (ceramide transfer protein) (Kawano, Kumagai, Nishijima, & Hanada, 2006; Loewen, Roy, & Levine, 2003; Sanhueza et al., 2015). The disease-causing variant, P56S, disrupts these interactions in *Drosophila* and may also do so in mammalian cells (Moustaqim-Barrette et al., 2013; Teuling et al., 2007) and loss of VAPs affects the structure of ER and Golgi (Moustaqim-Barrette et al., 2013; Peretti, Dahan, Shimoni, Hirschberg, & Lev, 2008). In *Drosophila*, overexpression of specific OSBPs in neuronal tissues suppresses the early lethality and locomotor defect in *VAPB (P58S)* (equivalent to the P56S in human) mutant flies (Moustaqim-Barrette et al., 2013). These observations suggest that promoting lipid transfer can partially suppress the phenotypes associated with *VAPB (P58S)* in flies.

VAPB has also been shown to function in a cell nonautonomous fashion, as the MSP domain of VAPB is cleaved and secreted from cells (Charng, Yamamoto, & Bellen, 2014; Tsuda et al., 2008). In *Drosophila* wing discs, ectopically expressing VAPB (P58S) results in protein aggregation and failure of MSP secretion (Tsuda et al., 2008). In *C. elegans*, secreted MSP binds to the Ephrin A4 receptor as well as Lar-like protein–tyrosine–phosphatase (CLR) and Roundabout (Robo) present on muscles (Han et al., 2012). The binding of MSP to these receptors is required to retain cytoskeletal organization in muscles as well as proper mitochondrial morphology, which relies on actin and myosins (Han et al., 2012). Loss of VAPB in the nervous system causes a severe muscle mitochondria defect in flies and worms. Secreted MSP is also detected in human serum and cerebrospinal fluid (Deidda et al., 2014; Tsuda et al., 2008), and the MSP level is reduced in patients with SALS (Deidda et al., 2014), indicating that loss of MSP secretion may play a role in ALS.

Loss of VAPB in *Drosophila* leads to pupal lethality. Introducing one copy of wild-type VAPB or VAPB (P58S) in VAPB null flies rescues the lethality (Moustaqim-Barrette et al., 2013). However, VAPB (P58S) flies exhibit severely reduced life span with age-dependent climbing defects and reduced flying ability (Moustaqim-Barrette et al., 2013). Heterozygous VAPB (P58S/+) flies have a normal life span as well as locomotion (Moustaqim-Barrette et al., 2013), indicating that the P58S allele is a partial loss of function. Interestingly, the VAPB protein level is reduced in motor neurons derived from fibroblasts of ALS patients carrying the VAPB (P56S) variant (Mitne-Neto et al., 2011). Similarly, the VAPB level is decreased in spinal cords of SOD-1 (G93A) transgenic mice as well as SALS patients (Anagnostou et al., 2010; Teuling et al., 2007). These data suggest that the VAPB (P56S) is a partial loss-of-function variant and that loss of VAPB may play a role in FALS and SALS.

VAPB knockout mice are viable and display a mild defect in motor activity starting at 18 months of age (Kabashi et al., 2013). Homozygous VAPB (P56S) knock-in mice start to develop defects at 9–11 months age, including cytoplasmic inclusions, locomotion defects as well as muscle denervation (Larroquette et al., 2015). The VAPB (P56S/+) heterozygous mice display similar but milder phenotypes (Larroquette et al., 2015), consistent with the dominant inheritance pattern of VAPB (P56S) (Nishimura et al., 2004). The data indicate that the VAPB (P56S) variant causes both a loss of function and/or a dominant-negative effect. The later may be due to the observation that the mutant VAPB protein preferentially binds to wild-type protein and this hypothesis is supported by in vivo data (Teuling et al., 2007). This hypothesis is also consistent with the observations in transgenic mice (Aliaga et al., 2013; Tudor et al., 2010) and *Drosophila* expressing various VAPB variants (Chen et al., 2010; Forrest et al., 2013; Sanhueza, Zechini, Gillespie, & Pennetta, 2013; Tsuda et al., 2008).

The most striking phenotypes associated with VAPB studies are defects in ER quality control and protein aggregation. Studies in mammalian cells and *Drosophila* have shown that VAPB is associated with ER. In *Drosophila*, loss of VAPB causes aberrant ER expansion, accumulation of membrane proteins on the ER, and alternatively spliced XBP-1, which are typically associated with ER stress (Moustaqim-Barrette et al., 2013; Tsuda et al., 2008). Similarly, expression of VAPB (P58S) also leads to formation of protein aggregates, ER stress, and traps endogenous wild-type VAPB proteins (Teuling et al., 2007; Tsuda et al., 2008). Knock-in mice carrying the *VAPB (P56S)* variant develop an accumulation of ubiquitinated proteins in cytoplasmic inclusions in motor neurons prior to the onset of the defects in locomotion. Those aggregates also partially colocalize with autophagosome markers (Larroquette et al., 2015). These data suggest that defects in VAPB function lead to ER stress and protein aggregation. Mutant VAPB proteins accumulate in large puncta in both corticospinal motor neurons and spinal motor neurons in VAPB (P56S) transgenic mice (Aliaga et al., 2013). VAPB (P56S) leads to a significant increase in ER stress and proapoptotic factors expression (Aliaga et al., 2013). Therefore, VAPB (P56S) protein accumulates and the failure of protein aggregate clearance may affect neuronal survival.

In summary, VAPB has pleiotropic functions in ER quality control, lipid transfer from ER to Golgi, and mitochondrial dynamics in muscles (Charng et al., 2014). Disease-causing variants in VAPB disrupt lipid transfer, cause protein aggregations, and trap wild-type VAPB protein. The accumulation of VAPB aggregates correlates with ER stress (Fig. 2C). However, the induced ER stress is not sufficient to clear all the aggregates, which may eventually lead to neuronal degeneration. These results indicate that a defect in protein clearance combined with reduced MSP secretion may play a significant role in ALS pathology.

5. SOD-1 AND PROTEINOPATHY IN ALS

Copper zinc superoxide dismutase, SOD-1, is a ubiquitously expressed and highly conserved enzyme that plays a role in scavenging superoxide radicals. It was the first gene to be associated with FALS (Rosen et al., 1993) and is present in 10–20% of all FALS cases (also known as ALS1) (Renton et al., 2014). Two major hypotheses may underlie SOD-1 toxicity: (1) a loss of dismutase activity and/or (2) toxicity caused by protein aggregation.

A reduction of dismutase activity and an increase in superoxides were detected in tissue extracts from ALS SOD-1 patients (Bowling et al., 1995; Cohen, Kohen, Lavon, Abramsky, & Steiner, 1996; Deng et al., 1993; Moumen, Nouvelot, Duval, Lechevalier, & Viader, 1997; Rosen et al., 1993). It was proposed that some ALS-related SOD-1 variants disrupt dimerization and reduce its dismutase activity. This in turn leads to the elevation of oxidative stress that causes protein misfolding and aggregation and contributes to degeneration. The dismutase activity of SOD-1 decreases with age in ALS patients but not in normal controls (Fiszman, Borodinsky, Ricart, Sanz, & Sica, 1999). In contrast, loss of dismutase activity does not correlate with age of onset or aggressiveness of the disease in SOD-1 patients (Fiszman et al., 1999; Ratovitski et al., 1999) and many question a role for the enzymatic function in the pathogenesis of the disease.

Interestingly, loss of SOD-1 in *Drosophila* causes a mild (~10%) reduction of life span, a severe loss of fertility, hypersensitivity to oxidative stress (Phillips, Campbell, Michaud, Charbonneau, & Hilliker, 1989), and retinal degeneration (Phillips et al., 1995). In mice, five-independent SOD-1 knockout mice have been established (Ho et al., 1998; Huang et al., 1997; Matzuk, Dionne, Guo, Kumar, & Lebovitz, 1998; Reaume et al., 1996; Yoshida, Maulik, Engelman, Ho, & Das, 2000) and all these mutants exhibit mild defects, including a progressive loss of locomotion, denervation at NMJs, and muscle atrophy (Reaume et al., 1996; Saccon, Bunton-Stasyshyn, Fisher, & Fratta, 2013), symptoms that are all typically associated with ALS. In addition, an increased vulnerability to oxidative stress has also been observed in two mice models (Ho et al., 1998; Reaume et al., 1996). However, no obvious motor neuron loss or accumulation of ubiquitinated protein inclusions has been documented in these mice. The above data suggest that loss of dismutase activity in ALS-linked SOD-1 variants may play a role in the pathogenesis of the disease. This loss may create a sensitized background to oxidative stress which may promote the accumulation of misfolded proteins and protein aggregates. However, the lack of motor neuron death and paralysis suggests that the SOD-1 loss of dismutase activity may not cause ALS, but may promote the pathogenesis.

The formation of SOD-1-containing insoluble inclusions in the affected motor neurons is a hallmark of SOD-1-related FALS cases and has been considered a cause of motor neuron death (Bruijn et al., 1998). SOD-1 mutants that carry human ALS-linked variants are susceptible to partial unfolding and prone to the formation of aggregates (Tiwari & Hayward, 2005). Transgenic mice overexpressing ALS-related SOD-1 variants exhibit large cytoplasmic SOD-1- and ubiquitin-positive inclusions in motor neurons, a shorter life span, hind limb paralysis, muscle atrophy, motor neuron loss, and axon degeneration (Bruijn et al., 1997; Cleveland, 1999; Gurney et al., 1994; Julien & Kriz, 2006; Wong et al., 1995). However, the dismutase activity is either unchanged relative to the number of copies or even elevated in at least two transgenic strains, SOD-1 (G93A) and SOD-1 (G37R) (Bruijn et al., 1997; Deng et al., 2006; Saccon et al., 2013; Subramaniam et al., 2002), suggesting that the loss of dismutase activity discussed previously may not be implicated. In *Drosophila*, overexpression of human wild type or ALS-linked (A4V, G85R) SOD-1 in motor neurons causes a progressive climbing defect and abnormal synaptic transmission (Watson et al., 2008). However, overexpression of *Drosophila* wild-type SOD-1 does not cause any defects (Watson et al., 2008), suggesting that a gain of the dismutase activity does not cause neuronal toxicity. In addition, Watson et al. (2008) found that the degenerative phenotypes are accompanied by the focal accumulation of human wild-type and mutant SOD-1 proteins in motor neurons (Watson et al., 2008), suggesting that chronic accumulation of misfolded SOD-1 in the ubiquitin-positive inclusions contributes to the degeneration. This model is supported by the fact that infusion of oligonucleotides complementary to human SOD-1 mRNA prolongs survival in rats expressing human SOD-1 (G93A) (Smith et al., 2006). Clinical trials that involve the introduction of these oligonucleotides to SOD-1 FALS patients are under evaluation (Miller et al., 2013). Interestingly, TDP-43 proteinopathy is absent from SOD-1-positive inclusions in SOD-1-related ALS cases (Blokhuis et al., 2013; Mackenzie et al., 2007), implicating that SOD-1 inclusions affect degeneration via a distinct molecular pathway.

Extensive datasets suggest that ER stress may play an important role in SOD-1 FALS. Indeed, overexpression of the mutant but not wild-type SOD-1 forms in cultured cells promotes ER stress (Nishitoh et al., 2008; Oh et al., 2008; Tobisawa et al., 2003). Similarly, transgenic mice that express mutant SOD-1 (Atkin et al., 2006; Kikuchi et al., 2006; Saxena et al., 2009; Tobisawa et al., 2003; Wate et al., 2005) and motor neurons of SOD-1 patients exhibit an elevated level of ER stress (Atkin et al., 2008; Ilieva et al., 2007). The ER stress responses promote the expression of chaperones that are targeted to the ER to refold misfolded proteins (Gardner, Pincus, Gotthardt, Gallagher, & Walter, 2013; Ron & Walter, 2007). The ER stress response is activated through three sensor pathways, IRE1, PERK, or ATF6, which activate distinct downstream effectors that promote chaperone expression to facilitate protein refolding (Ron & Walter, 2007).

To assess the role of the PERK pathway, Wang, Popko, et al. (2011) removed one copy of *PERK* in SOD-1 (G85R) mice and noted that this accelerated aggregate formation, shortened life span, and enhanced the degeneration phenotypes (Wang, Popko, et al., 2011). These data suggest that activation of the PERK pathway provides a neuroprotective effect. Indeed, removal of one copy of GADD34 (growth arrest and DNA damage-inducible protein), a downstream suppressor of PERK, enhanced PERK signaling, and reduced aggregate formation, prolonged life span, and alleviated the degenerative phenotypes (Wang, Popko, et al., 2014). Similarly,

Guanabenz or Sephin1, drugs which inhibit GADD34, and hence promote PERK signaling, ameliorated the mutant SOD-1-induced ALS phenotypes in mice (Das et al., 2015; Wang, Popko, et al., 2014), further suggesting that activation of PERK is beneficial in ALS. However, a neuronal knockout of XBP-1 in SOD-1 (G85R), a downstream component of the IRE1 pathway, reduced the accumulation of mutant SOD-1 in spinal cord, prolonged life span, and alleviated degeneration phenotypes (Hetz et al., 2009). Hence, activation of the PERK1 pathway as well as inactivation of the IRE1 pathway is beneficial. The latter data were attributed to the observation that basal autophagy activity is enhanced in these mice and may compensate for the ERAD defect (Hetz et al., 2009). In summary, variants in SOD-1 cause misfolding of the protein and the formation of protein aggregates. However, activation of the ER stress and promotion of ERAD and/or basal autophagy may be beneficial to the progression of the disease.

6. PRION-LIKE PROTEIN TOXICITY AND ALS

The general concept of prion diseases is that an infectious misfolded protein, such as a misfolded prion isoform, PrP^{sc}, forms aggregates through self-aggregation and propagates through binding to wild type or normal protein, such as prion (PrP^{C}), to produce a misfolded, pathogenic version (Colby & Prusiner, 2011; Prusiner, 1982). In addition, the misfolded protein has the capacity to be transmitted from one cell to another. Therefore, the disease can spread from the site of onset to surrounding tissues. ALS shows a number of similarities with prion diseases. (1) All ALS patients have misfolded protein-containing aggregates in their affected neurons (Blokhuis et al., 2013). (2) In cultured neuronal cells, misfolded forms of SOD-1 or TDP-43 can self-propagate within the cells and be transmitted to neighboring cells via cellular connections (Johnson et al., 2009; Munch, O'Brien, & Bertolotti, 2011). (3) Studies of tissue samples from postmortem ALS patients suggest that lower motor neuron degeneration is a focal process that gradually spreads to distal sites (Lee & Kim, 2015; Ravits, Laurie, Fan, & Moore, 2007), similar to the spreading of the prion diseases. (4) ALS-causing genes, including TDP43, FUS, and hnRNPA1 have a prion-like domain (Kim et al., 2013), which may contribute to the spread.

SOD-1 aggregates in patients or mice exhibit a fibrillary structure enriched in β-sheets, a feature characteristic for a template to transform a normal protein into a misfolded protein which elongates into fibrils

(Furukawa, Kaneko, Yamanaka, O'Halloran, & Nukina, 2008). This hypothesis has been supported experimentally as protein extracts from spinal cords of SOD-1 (G93A) mice are able to induce wild-type SOD-1 to form amyloid-like fibrils (Chia et al., 2010; Grad et al., 2011). Moreover, in vitro data suggest that aggregates composed of mutated SOD-1 can be efficiently transferred from a sick cell to a healthy cell in culture (Munch et al., 2011). These data argue that ALS-related SOD-1 mutants might use a prion-like mechanism to spread the disease.

TDP-43 harbors a predicted prion-like domain (Lee & Kim, 2015). The predicted prion domain of TDP-43 contains a glycine-rich domain in the C-terminal region (Lee & Kim, 2015). Expression of the full-length human wild-type and ALS mutant TDP-43 in *Drosophila* eyes causes an eye degeneration (Estes et al., 2011; Hanson et al., 2010; Li et al., 2010; Ritson et al., 2010). Interestingly, the severity of eye degeneration correlates with the level of a truncated 25 kDa C-terminal fragment that contains the prion-like domain (Ritson et al., 2010) and the 25 kDa C-terminal domain of TDP-43 is present in the cytoplasmic inclusions of ALS patients (Neumann et al., 2006). These data argue that there is a relationship between the formation of TDP-43-containing aggregates, the prion-like mechanism, and the degeneration phenotype.

The predicted prion-like domain of FUS is localized at the very N-terminal part that is SYGQ-rich (Lee & Kim, 2015), required for FUS self-assembly, and formation of cytoplasmic FUS-containing inclusions (Yang, Gal, Chen, & Zhu, 2014). In *Drosophila*, ectopic expression of either wild-type or human ALS-linked variants in FUS in the eyes causes degeneration (Chen et al., 2011; Lanson et al., 2011). However, the role of the prion-like domain alone in the induction of degeneration in *Drosophila* has not been tested. Finally, another prion-like domain containing RBP, hnRNPA1, has recently been associated with ALS and contains a prion-like domain (Alberti, Halfmann, King, Kapila, & Lindquist, 2009; Kim et al., 2013). The above data argue that a prion-like disease propagation mechanism may play a role in the pathogenesis of FALS and SALS.

7. CONCLUSION AND FUTURE PERSPECTIVE

In summary, we propose that variants in seven genes (*TARDBP*, *FUS*, *C9ofr72*, *VCP*, *UBQLN2*, *VAPB*, and *SOD-1*) that cause ALS converge on chronic defects in RNA and protein homeostasis. Mutations in these proteins lead to the aberrant localization and/or degradation of TDP-43.

Variants in TDP-43 cause the exclusion of TDP-43 from the nucleus and lead to the incorporation of cryptic exons in the mRNAs of thousands of genes, causing the formation of misfolded proteins that overload the protein refolding and degradation machinery and leads to protein aggregates. Similarly, FUS variants have been shown to cause aberrant splicing of numerous genes. The presence of the HRE in *C9orf72* causes protein aggregates that impair nucleocytoplasmic transport, promoting TDP-43 mislocalization to the cytoplasm, which in turn promotes TDP-43 proteinopathy. This model is further supported by the observation that variants in VCP or UBQLNs that affect protein refolding, ubiquitination, and degradation via the proteasome degradation pathway also lead to mislocalization of TDP-43. Finally, variants in VAPB and SOD-1 also cause protein aggregates, induce ER stress, and affect ERAD function even though TDP-43 pathology may not play a role in the latter.

ACKNOWLEDGMENTS

We thank Karen Schulze, Ke Zhang, Shinya Yamamoto, Mumine Senturk, and Hsiao-Tuan Chao for their critical reading of the manuscript. This work was supported by Target ALS, the Robert A. and Renee E. Belfer Family Foundation, and the Huffington Foundation. H.J.B. is an Investigator of the Howard Hughes Medical Institute.

REFERENCES

Abramzon, Y., Johnson, J. O., Scholz, S. W., Taylor, J. P., Brunetti, M., Calvo, A., et al. (2012). Valosin-containing protein (VCP) mutations in sporadic amyotrophic lateral sclerosis. *Neurobiology of Aging*, *33*(9), 2231.e1–2231.e6.

Ajroud-Driss, S., & Siddique, T. (2015). Sporadic and hereditary amyotrophic lateral sclerosis (ALS). *Biochimica et Biophysica Acta*, *1852*(4), 679–684.

Alberti, S., Halfmann, R., King, O., Kapila, A., & Lindquist, S. (2009). A systematic survey identifies prions and illuminates sequence features of prionogenic proteins. *Cell*, *137*(1), 146–158.

Al-Chalabi, A., Jones, A., Troakes, C., King, A., Al-Sarraj, S., & van den Berg, L. H. (2012). The genetics and neuropathology of amyotrophic lateral sclerosis. *Acta Neuropathologica*, *124*(3), 339–352.

Ali, Y. O., Escala, W., Ruan, K., & Zhai, R. G. (2011). Assaying locomotor, learning, and memory deficits in Drosophila models of neurodegeneration. *Journal of Visualized Experiments*, (49), pii: 2504.

Aliaga, L., Lai, C., Yu, J., Chub, N., Shim, H., Sun, L., et al. (2013). Amyotrophic lateral sclerosis-related VAPB P56S mutation differentially affects the function and survival of corticospinal and spinal motor neurons. *Human Molecular Genetics*, *22*(21), 4293–4305.

Allen, M. J., & Godenschwege, T. A. (2010). Electrophysiological recordings from the Drosophila giant fiber system (GFS). *Cold Spring Harbor Protocols*. *2010*(7). http://dx.doi.org/10.1101/pdb.prot5453. pdb prot5453.

Alonso, A., Logroscino, G., Jick, S. S., & Hernan, M. A. (2009). Incidence and lifetime risk of motor neuron disease in the United Kingdom: A population-based study. *European Journal of Neurology*, *16*(6), 745–751.

Anagnostou, G., Akbar, M. T., Paul, P., Angelinetta, C., Steiner, T. J., & de Belleroche, J. (2010). Vesicle associated membrane protein B (VAPB) is decreased in ALS spinal cord. *Neurobiology of Aging*, *31*(6), 969–985.
Andlauer, T. F., & Sigrist, S. J. (2012a). In vivo imaging of Drosophila larval neuromuscular junctions to study synapse assembly. *Cold Spring Harbor Protocols*, *2012*(4), 407–413.
Andlauer, T. F., & Sigrist, S. J. (2012b). In vivo imaging of the Drosophila larval neuromuscular junction. *Cold Spring Harbor Protocols*, *2012*(4), 481–489.
Andlauer, T. F., & Sigrist, S. J. (2012c). Quantitative analysis of Drosophila larval neuromuscular junction morphology. *Cold Spring Harbor Protocols*, *2012*(4), 490–493.
Arai, T., Hasegawa, M., Nonoka, T., Kametani, F., Yamashita, M., Hosokawa, M., et al. (2010). Phosphorylated and cleaved TDP-43 in ALS, FTLD and other neurodegenerative disorders and in cellular models of TDP-43 proteinopathy. *Neuropathology*, *30*(2), 170–181.
Ash, P. E., Bieniek, K. F., Gendron, T. F., Caulfield, T., Lin, W. L., Dejesus-Hernandez, M., et al. (2013). Unconventional translation of C9ORF72 GGGGCC expansion generates insoluble polypeptides specific to c9FTD/ALS. *Neuron*, *77*(4), 639–646.
Atanasio, A., Decman, V., White, D., Ramos, M., Ikiz, B., Lee, H. C., et al. (2016). C9orf72 ablation causes immune dysregulation characterized by leukocyte expansion, autoantibody production, and glomerulonephropathy in mice. *Scientific Reports*, *6*, 23204.
Atkin, J. D., Farg, M. A., Turner, B. J., Tomas, D., Lysaght, J. A., Nunan, J., et al. (2006). Induction of the unfolded protein response in familial amyotrophic lateral sclerosis and association of protein-disulfide isomerase with superoxide dismutase 1. *Journal of Biological Chemistry*, *281*(40), 30152–30165.
Atkin, J. D., Farg, M. A., Walker, A. K., McLean, C., Tomas, D., & Horne, M. K. (2008). Endoplasmic reticulum stress and induction of the unfolded protein response in human sporadic amyotrophic lateral sclerosis. *Neurobiology of Disease*, *30*(3), 400–407.
Ayala, Y. M., Pantano, S., D'Ambrogio, A., Buratti, E., Brindisi, A., Marchetti, C., et al. (2005). Human, Drosophila, and C. elegans TDP43: Nucleic acid binding properties and splicing regulatory function. *Journal of Molecular Biology*, *348*(3), 575–588.
Azuma, Y., Tokuda, T., Shimamura, M., Kyotani, A., Sasayama, H., Yoshida, T., et al. (2014). Identification of ter94, Drosophila VCP, as a strong modulator of motor neuron degeneration induced by knockdown of Caz, Drosophila FUS. *Human Molecular Genetics*, *23*(13), 3467–3480.
Babcock, D. T., & Ganetzky, B. (2014). An improved method for accurate and rapid measurement of flight performance in Drosophila. *Journal of Visualized Experiments*, *84*, e51223.
Barthelme, D., Jauregui, R., & Sauer, R. T. (2015). An ALS disease mutation in Cdc48/p97 impairs 20S proteasome binding and proteolytic communication. *Protein Science*, *24*(9), 1521–1527.
Barthelme, D., & Sauer, R. T. (2012). Identification of the Cdc48*20S proteasome as an ancient AAA+ proteolytic machine. *Science*, *337*(6096), 843–846.
Barthelme, D., & Sauer, R. T. (2013). Bipartite determinants mediate an evolutionarily conserved interaction between Cdc48 and the 20S peptidase. *Proceedings of the National Academy of Sciences of the United States of America*, *110*(9), 3327–3332.
Barthelme, D., & Sauer, R. T. (2016). Origin and functional evolution of the Cdc48/p97/VCP AAA+ protein unfolding and remodeling machine. *Journal of Molecular Biology*, *428*(9 Pt B), 1861–1869.
Belzil, V. V., Valdmanis, P. N., Dion, P. A., Daoud, H., Kabashi, E., Noreau, A., et al. (2009). Mutations in FUS cause FALS and SALS in French and French Canadian populations. *Neurology*, *73*(15), 1176–1179.

Bertram, L., Hiltunen, M., Parkinson, M., Ingelsson, M., Lange, C., Ramasamy, K., et al. (2005). Family-based association between Alzheimer's disease and variants in UBQLN1. *The New England Journal of Medicine*, *352*(9), 884–894.

Blokhuis, A. M., Groen, E. J., Koppers, M., van den Berg, L. H., & Pasterkamp, R. J. (2013). Protein aggregation in amyotrophic lateral sclerosis. *Acta Neuropathologica*, *125*(6), 777–794.

Bose, J. K., Wang, I. F., Hung, L., Tarn, W. Y., & Shen, C. K. (2008). TDP-43 overexpression enhances exon 7 inclusion during the survival of motor neuron pre-mRNA splicing. *Journal of Biological Chemistry*, *283*(43), 28852–28859.

Bowling, A. C., Barkowski, E. E., McKenna-Yasek, D., Sapp, P., Horvitz, H. R., Beal, M. F., et al. (1995). Superoxide dismutase concentration and activity in familial amyotrophic lateral sclerosis. *Journal of Neurochemistry*, *64*(5), 2366–2369.

Brady, O. A., Meng, P., Zheng, Y., Mao, Y., & Hu, F. (2011). Regulation of TDP-43 aggregation by phosphorylation and p62/SQSTM1. *Journal of Neurochemistry*, *116*(2), 248–259.

Bruijn, L. I., Becher, M. W., Lee, M. K., Anderson, K. L., Jenkins, N. A., Copeland, N. G., et al. (1997). ALS-linked SOD1 mutant G85R mediates damage to astrocytes and promotes rapidly progressive disease with SOD1-containing inclusions. *Neuron*, *18*(2), 327–338.

Bruijn, L. I., Houseweart, M. K., Kato, S., Anderson, K. L., Anderson, S. D., Ohama, E., et al. (1998). Aggregation and motor neuron toxicity of an ALS-linked SOD1 mutant independent from wild-type SOD1. *Science*, *281*(5384), 1851–1854.

Brunger, A. T., & DeLaBarre, B. (2003). NSF and p97/VCP: Similar at first, different at last. *FEBS Letters*, *555*(1), 126–133.

Buchberger, A., Schindelin, H., & Hanzelmann, P. (2015). Control of p97 function by cofactor binding. *FEBS Letters*, *589*(19 Pt. A), 2578–2589.

Buratti, E., & Baralle, F. E. (2001). Characterization and functional implications of the RNA binding properties of nuclear factor TDP-43, a novel splicing regulator of CFTR exon 9. *Journal of Biological Chemistry*, *276*(39), 36337–36343.

Buratti, E., & Baralle, F. E. (2008). Multiple roles of TDP-43 in gene expression, splicing regulation, and human disease. *Frontiers in Bioscience*, *13*, 867–878.

Buratti, E., Dork, T., Zuccato, E., Pagani, F., Romano, M., & Baralle, F. E. (2001). Nuclear factor TDP-43 and SR proteins promote in vitro and in vivo CFTR exon 9 skipping. *EMBO Journal*, *20*(7), 1774–1784. http://dx.doi.org/10.1093/emboj/20.7.1774.

Cao, K., Nakajima, R., Meyer, H. H., & Zheng, Y. (2003). The AAA-ATPase Cdc48/p97 regulates spindle disassembly at the end of mitosis. *Cell*, *115*(3), 355–367.

Chang, L., & Monteiro, M. J. (2015). Defective proteasome delivery of polyubiquitinated proteins by ubiquilin-2 proteins containing ALS mutations. *PLoS One*, *10*(6), e0130162.

Charng, W. L., Yamamoto, S., & Bellen, H. J. (2014). Shared mechanisms between Drosophila peripheral nervous system development and human neurodegenerative diseases. *Current Opinion in Neurobiology*, *27*, 158–164.

Chen, H. J., Anagnostou, G., Chai, A., Withers, J., Morris, A., Adhikaree, J., et al. (2010). Characterization of the properties of a novel mutation in VAPB in familial amyotrophic lateral sclerosis. *Journal of Biological Chemistry*, *285*(51), 40266–40281.

Chen, Y., Yang, M., Deng, J., Chen, X., Ye, Y., Zhu, L., et al. (2011). Expression of human FUS protein in Drosophila leads to progressive neurodegeneration. *Protein & Cell*, *2*(6), 477–486.

Chew, J., Gendron, T. F., Prudencio, M., Sasaguri, H., Zhang, Y. J., Castanedes-Casey, M., et al. (2015). Neurodegeneration. C9ORF72 repeat expansions in mice cause TDP-43 pathology, neuronal loss, and behavioral deficits. *Science*, *348*(6239), 1151–1154.

Chia, R., Tattum, M. H., Jones, S., Collinge, J., Fisher, E. M., & Jackson, G. S. (2010). Superoxide dismutase 1 and tgSOD1 mouse spinal cord seed fibrils, suggesting a propagative cell death mechanism in amyotrophic lateral sclerosis. *PLoS One*, *5*(5), e10627.

Choksi, D. K., Roy, B., Chatterjee, S., Yusuff, T., Bakhoum, M. F., Sengupta, U., et al. (2014). TDP-43 phosphorylation by casein kinase Iepsilon promotes oligomerization and enhances toxicity in vivo. *Human Molecular Genetics*, *23*(4), 1025–1035.
Ciura, S., Lattante, S., Le Ber, I., Latouche, M., Tostivint, H., Brice, A., et al. (2013). Loss of function of C9orf72 causes motor deficits in a zebrafish model of amyotrophic lateral sclerosis. *Annals of Neurology*, *74*(2), 180–187.
Cleary, J. D., & Ranum, L. P. (2013). Repeat-associated non-ATG (RAN) translation in neurological disease. *Human Molecular Genetics*, *22*(R1), R45–R51.
Cleveland, D. W. (1999). From Charcot to SOD1: Mechanisms of selective motor neuron death in ALS. *Neuron*, *24*(3), 515–520.
Cohen, O., Kohen, R., Lavon, E., Abramsky, O., & Steiner, I. (1996). Serum Cu/Zn superoxide dismutase activity is reduced in sporadic amyotrophic lateral sclerosis patients. *Journal of the Neurological Sciences*, *143*(1–2), 118–120.
Colby, D. W., & Prusiner, S. B. (2011). Prions. *Cold Spring Harbor Perspectives in Biology*, *3*(1), a006833.
Cooper-Knock, J., Higginbottom, A., Stopford, M. J., Highley, J. R., Ince, P. G., Wharton, S. B., et al. (2015). Antisense RNA foci in the motor neurons of C9ORF72-ALS patients are associated with TDP-43 proteinopathy. *Acta Neuropathologica*, *130*(1), 63–75.
Crozat, A., Aman, P., Mandahl, N., & Ron, D. (1993). Fusion of CHOP to a novel RNA-binding protein in human myxoid liposarcoma. *Nature*, *363*(6430), 640–644.
Custer, S. K., Neumann, M., Lu, H., Wright, A. C., & Taylor, J. P. (2010). Transgenic mice expressing mutant forms VCP/p97 recapitulate the full spectrum of IBMPFD including degeneration in muscle, brain and bone. *Human Molecular Genetics*, *19*(9), 1741–1755.
Dai, R. M., & Li, C. C. (2001). Valosin-containing protein is a multi-ubiquitin chain-targeting factor required in ubiquitin-proteasome degradation. *Nature Cell Biology*, *3*(8), 740–744.
Daigle, J. G., Lanson, N. A., Jr., Smith, R. B., Casci, I., Maltare, A., Monaghan, J., et al. (2013). RNA-binding ability of FUS regulates neurodegeneration, cytoplasmic mislocalization and incorporation into stress granules associated with FUS carrying ALS-linked mutations. *Human Molecular Genetics*, *22*(6), 1193–1205.
Darovic, S., Prpar Mihevc, S., Zupunski, V., Guncar, G., Stalekar, M., Lee, Y. B., et al. (2015). Phosphorylation of C-terminal tyrosine residue 526 in FUS impairs its nuclear import. *Journal of Cell Science*, *128*(22), 4151–4159.
Das, I., Krzyzosiak, A., Schneider, K., Wrabetz, L., D'Antonio, M., Barry, N., et al. (2015). Preventing proteostasis diseases by selective inhibition of a phosphatase regulatory subunit. *Science*, *348*(6231), 239–242.
Deidda, I., Galizzi, G., Passantino, R., Cascio, C., Russo, D., Colletti, T., et al. (2014). Expression of vesicle-associated membrane-protein-associated protein B cleavage products in peripheral blood leukocytes and cerebrospinal fluid of patients with sporadic amyotrophic lateral sclerosis. *European Journal of Neurology*, *21*(3), 478–485.
DeJesus-Hernandez, M., Mackenzie, I. R., Boeve, B. F., Boxer, A. L., Baker, M., Rutherford, N. J., et al. (2011). Expanded GGGGCC hexanucleotide repeat in noncoding region of C9ORF72 causes chromosome 9p-linked FTD and ALS. *Neuron*, *72*(2), 245–256.
DeLaBarre, B., & Brunger, A. T. (2003). Complete structure of p97/valosin-containing protein reveals communication between nucleotide domains. *Nature Structural Biology*, *10*(10), 856–863.
Deng, H. X., Chen, W., Hong, S. T., Boycott, K. M., Gorrie, G. H., Siddique, N., et al. (2011). Mutations in UBQLN2 cause dominant X-linked juvenile and adult-onset ALS and ALS/dementia. *Nature*, *477*(7363), 211–215.
Deng, H., Gao, K., & Jankovic, J. (2014). The role of FUS gene variants in neurodegenerative diseases. *Nature Reviews. Neurology*, *10*(6), 337–348.

Deng, H. X., Hentati, A., Tainer, J. A., Iqbal, Z., Cayabyab, A., Hung, W. Y., et al. (1993). Amyotrophic lateral sclerosis and structural defects in Cu, Zn superoxide dismutase. *Science*, *261*(5124), 1047–1051.

Deng, Q., Holler, C. J., Taylor, G., Hudson, K. F., Watkins, W., Gearing, M., et al. (2014). FUS is phosphorylated by DNA-PK and accumulates in the cytoplasm after DNA damage. *Journal of Neuroscience*, *34*(23), 7802–7813.

Deng, H. X., Shi, Y., Furukawa, Y., Zhai, H., Fu, R., Liu, E., et al. (2006). Conversion to the amyotrophic lateral sclerosis phenotype is associated with intermolecular linked insoluble aggregates of SOD1 in mitochondria. *Proceedings of the National Academy of Sciences of the United States of America*, *103*(18), 7142–7147.

Dolph, P., Nair, A., & Raghu, P. (2011). Electroretinogram recordings of Drosophila. *Cold Spring Harbor Protocols*, *2011*(1), pdb.prot5549. http://dx.doi.org/10.1101/pdb.prot5549.

Donnelly, C. J., Zhang, P. W., Pham, J. T., Haeusler, A. R., Mistry, N. A., Vidensky, S., et al. (2013). RNA toxicity from the ALS/FTD C9ORF72 expansion is mitigated by antisense intervention. *Neuron*, *80*(2), 415–428.

Dormann, D., & Haass, C. (2013). Fused in sarcoma (FUS): An oncogene goes awry in neurodegeneration. *Molecular and Cellular Neurosciences*, *56*, 475–486.

Dormann, D., Madl, T., Valori, C. F., Bentmann, E., Tahirovic, S., Abou-Ajram, C., et al. (2012). Arginine methylation next to the PY-NLS modulates transportin binding and nuclear import of FUS. *EMBO Journal*, *31*(22), 4258–4275.

Elsasser, S., & Finley, D. (2005). Delivery of ubiquitinated substrates to protein-unfolding machines. *Nature Cell Biology*, 7(8), 742–749.

Erzberger, J. P., & Berger, J. M. (2006). Evolutionary relationships and structural mechanisms of AAA+ proteins. *Annual Review of Biophysics and Biomolecular Structure*, *35*, 93–114.

Estes, P. S., Boehringer, A., Zwick, R., Tang, J. E., Grigsby, B., & Zarnescu, D. C. (2011). Wild-type and A315T mutant TDP-43 exert differential neurotoxicity in a Drosophila model of ALS. *Human Molecular Genetics*, *20*(12), 2308–2321.

Fahed, A. C., McDonough, B., Gouvion, C. M., Newell, K. L., Dure, L. S., Bebin, M., et al. (2014). UBQLN2 mutation causing heterogeneous X-linked dominant neurodegeneration. *Annals of Neurology*, *75*(5), 793–798.

Fang, F., Valdimarsdottir, U., Bellocco, R., Ronnevi, L. O., Sparen, P., Fall, K., et al. (2009). Amyotrophic lateral sclerosis in Sweden, 1991–2005. *Archives of Neurology*, *66*(4), 515–519.

Feiguin, F., Godena, V. K., Romano, G., D'Ambrogio, A., Klima, R., & Baralle, F. E. (2009). Depletion of TDP-43 affects Drosophila motoneurons terminal synapsis and locomotive behavior. *FEBS Letters*, *583*(10), 1586–1592.

Ferrari, R., Kapogiannis, D., Huey, E. D., & Momeni, P. (2013). FTD and ALS: A tale of two diseases. *Current Alzheimer Research*, *8*(3), 273–294.

Fiszman, M. L., Borodinsky, L. N., Ricart, K. C., Sanz, O. P., & Sica, R. E. (1999). Cu/Zn superoxide dismutase activity at different ages in sporadic amyotrophic lateral sclerosis. *Journal of the Neurological Sciences*, *162*(1), 34–37.

Forrest, S., Chai, A., Sanhueza, M., Marescotti, M., Parry, K., Georgiev, A., et al. (2013). Increased levels of phosphoinositides cause neurodegeneration in a Drosophila model of amyotrophic lateral sclerosis. *Human Molecular Genetics*, *22*(13), 2689–2704.

Fratta, P., Poulter, M., Lashley, T., Rohrer, J. D., Polke, J. M., Beck, J., et al. (2013). Homozygosity for the C9orf72 GGGGCC repeat expansion in frontotemporal dementia. *Acta Neuropathologica*, *126*(3), 401–409.

Freibaum, B. D., Lu, Y., Lopez-Gonzalez, R., Kim, N. C., Almeida, S., Lee, K. H., et al. (2015). GGGGCC repeat expansion in C9orf72 compromises nucleocytoplasmic transport. *Nature*, *525*(7567), 129–133.

Furukawa, Y., Kaneko, K., Yamanaka, K., O'Halloran, T. V., & Nukina, N. (2008). Complete loss of post-translational modifications triggers fibrillar aggregation of SOD1 in the familial form of amyotrophic lateral sclerosis. *Journal of Biological Chemistry, 283*(35), 24167–24176.

Gardner, B. M., Pincus, D., Gotthardt, K., Gallagher, C. M., & Walter, P. (2013). Endoplasmic reticulum stress sensing in the unfolded protein response. *Cold Spring Harbor Perspectives in Biology, 5*(3), a013169.

Gellera, C., Tiloca, C., Del Bo, R., Corrado, L., Pensato, V., Agostini, J., et al. (2013). Ubiquilin 2 mutations in Italian patients with amyotrophic lateral sclerosis and frontotemporal dementia. *Journal of Neurology, Neurosurgery, and Psychiatry, 84*(2), 183–187.

Gendron, T. F., & Petrucelli, L. (2011). Rodent models of TDP-43 proteinopathy: Investigating the mechanisms of TDP-43-mediated neurodegeneration. *Journal of Molecular Neuroscience, 45*(3), 486–499.

Gomez-Deza, J., Lee, Y. B., Troakes, C., Nolan, M., Al-Sarraj, S., Gallo, J. M., et al. (2015). Dipeptide repeat protein inclusions are rare in the spinal cord and almost absent from motor neurons in C9ORF72 mutant amyotrophic lateral sclerosis and are unlikely to cause their degeneration. *Acta Neuropathologica Communications, 3*, 38.

Gonzalez-Perez, P., Lu, Y., Chian, R. J., Sapp, P. C., Tanzi, R. E., Bertram, L., et al. (2012). Association of UBQLN1 mutation with Brown–Vialetto–Van Laere syndrome but not typical ALS. *Neurobiology of Disease, 48*(3), 391–398.

Grad, L. I., Guest, W. C., Yanai, A., Pokrishevsky, E., O'Neill, M. A., Gibbs, E., et al. (2011). Intermolecular transmission of superoxide dismutase 1 misfolding in living cells. *Proceedings of the National Academy of Sciences of the United States of America, 108*(39), 16398–16403.

Gregory, J. M., Barros, T. P., Meehan, S., Dobson, C. M., & Luheshi, L. M. (2012). The aggregation and neurotoxicity of TDP-43 and its ALS-associated 25 kDa fragment are differentially affected by molecular chaperones in Drosophila. *PLoS One*, 7(2), e31899.

Gregory, R. I., Yan, K. P., Amuthan, G., Chendrimada, T., Doratotaj, B., Cooch, N., et al. (2004). The microprocessor complex mediates the genesis of microRNAs. *Nature, 432*(7014), 235–240.

Gros-Louis, F., Gaspar, C., & Rouleau, G. A. (2006). Genetics of familial and sporadic amyotrophic lateral sclerosis. *Biochimica et Biophysica Acta, 1762*(11–12), 956–972.

Gurney, M. E., Pu, H., Chiu, A. Y., Dal Canto, M. C., Polchow, C. Y., Alexander, D. D., et al. (1994). Motor neuron degeneration in mice that express a human Cu, Zn superoxide dismutase mutation. *Science, 264*(5166), 1772–1775.

Haapasalo, A., Viswanathan, J., Bertram, L., Soininen, H., Tanzi, R. E., & Hiltunen, M. (2010). Emerging role of Alzheimer's disease-associated ubiquilin-1 in protein aggregation. *Biochemical Society Transactions, 38*(Pt 1), 150–155.

Haeusler, A. R., Donnelly, C. J., Periz, G., Simko, E. A., Shaw, P. G., Kim, M. S., et al. (2014). C9orf72 nucleotide repeat structures initiate molecular cascades of disease. *Nature, 507*(7491), 195–200.

Han, S. M., Tsuda, H., Yang, Y., Vibbert, J., Cottee, P., Lee, S. J., et al. (2012). Secreted VAPB/ALS8 major sperm protein domains modulate mitochondrial localization and morphology via growth cone guidance receptors. *Developmental Cell, 22*(2), 348–362.

Hanson, K. A., Kim, S. H., Wassarman, D. A., & Tibbetts, R. S. (2010). Ubiquilin modifies TDP-43 toxicity in a Drosophila model of amyotrophic lateral sclerosis (ALS). *Journal of Biological Chemistry, 285*(15), 11068–11072.

Hasegawa, M., Arai, T., Nonaka, T., Kametani, F., Yoshida, M., Hashizume, Y., et al. (2008). Phosphorylated TDP-43 in frontotemporal lobar degeneration and amyotrophic lateral sclerosis. *Annals of Neurology, 64*(1), 60–70.

Hetz, C., Thielen, P., Matus, S., Nassif, M., Court, F., Kiffin, R., et al. (2009). XBP-1 deficiency in the nervous system protects against amyotrophic lateral sclerosis by increasing autophagy. *Genes & Development, 23*(19), 2294–2306.

Hicks, G. G., Singh, N., Nashabi, A., Mai, S., Bozek, G., Klewes, L., et al. (2000). Fus deficiency in mice results in defective B-lymphocyte development and activation, high levels of chromosomal instability and perinatal death. *Nature Genetics*, *24*(2), 175–179.

Ho, Y. S., Gargano, M., Cao, J., Bronson, R. T., Heimler, I., & Hutz, R. J. (1998). Reduced fertility in female mice lacking copper-zinc superoxide dismutase. *Journal of Biological Chemistry*, *273*(13), 7765–7769.

Huang, T. T., Yasunami, M., Carlson, E. J., Gillespie, A. M., Reaume, A. G., Hoffman, E. K., et al. (1997). Superoxide-mediated cytotoxicity in superoxide dismutase-deficient fetal fibroblasts. *Archives of Biochemistry and Biophysics*, *344*(2), 424–432.

Ilieva, E. V., Ayala, V., Jove, M., Dalfo, E., Cacabelos, D., Povedano, M., et al. (2007). Oxidative and endoplasmic reticulum stress interplay in sporadic amyotrophic lateral sclerosis. *Brain*, *130*(Pt. 12), 3111–3123.

Imlach, W., & McCabe, B. D. (2009). Electrophysiological methods for recording synaptic potentials from the NMJ of Drosophila larvae. *Journal of Visualized Experiments*, *24*(24), pii: 1109.

Jackel, S., Summerer, A. K., Thommes, C. M., Pan, X., Voigt, A., Schulz, J. B., et al. (2015). Nuclear import factor transportin and arginine methyltransferase 1 modify FUS neurotoxicity in Drosophila. *Neurobiology of Disease*, *74*, 76–88.

Jackrel, M. E., DeSantis, M. E., Martinez, B. A., Castellano, L. M., Stewart, R. M., Caldwell, K. A., et al. (2014). Potentiated Hsp104 variants antagonize diverse proteotoxic misfolding events. *Cell*, *156*(1–2), 170–182.

Jarosch, E., Geiss-Friedlander, R., Meusser, B., Walter, J., & Sommer, T. (2002). Protein dislocation from the endoplasmic reticulum—Pulling out the suspect. *Traffic*, *3*(8), 530–536.

Jiang, J., Zhu, Q., Gendron, T. F., Saberi, S., McAlonis-Downes, M., Seelman, A., et al. (2016). Gain of toxicity from ALS/FTD-linked repeat expansions in C9ORF72 is alleviated by antisense oligonucleotides targeting GGGGCC-containing RNAs. *Neuron*, *90*(3), 535–550.

Johnson, J. O., Mandrioli, J., Benatar, M., Abramzon, Y., Van Deerlin, V. M., Trojanowski, J. Q., et al. (2010). Exome sequencing reveals VCP mutations as a cause of familial ALS. *Neuron*, *68*(5), 857–864.

Johnson, B. S., Snead, D., Lee, J. J., McCaffery, J. M., Shorter, J., & Gitler, A. D. (2009). TDP-43 is intrinsically aggregation-prone, and amyotrophic lateral sclerosis-linked mutations accelerate aggregation and increase toxicity. *Journal of Biological Chemistry*, *284*(30), 20329–20339.

Jovicic, A., Mertens, J., Boeynaems, S., Bogaert, E., Chai, N., Yamada, S. B., et al. (2015). Modifiers of C9orf72 dipeptide repeat toxicity connect nucleocytoplasmic transport defects to FTD/ALS. *Nature Neuroscience*, *18*(9), 1226–1229.

Ju, J. S., Fuentealba, R. A., Miller, S. E., Jackson, E., Piwnica-Worms, D., Baloh, R. H., et al. (2009). Valosin-containing protein (VCP) is required for autophagy and is disrupted in VCP disease. *Journal of Cell Biology*, *187*(6), 875–888.

Julien, J. P., & Kriz, J. (2006). Transgenic mouse models of amyotrophic lateral sclerosis. *Biochimica et Biophysica Acta*, *1762*(11–12), 1013–1024.

Kabashi, E., El Oussini, H., Bercier, V., Gros-Louis, F., Valdmanis, P. N., McDearmid, J., et al. (2013). Investigating the contribution of VAPB/ALS8 loss of function in amyotrophic lateral sclerosis. *Human Molecular Genetics*, *22*(12), 2350–2360.

Kagiwada, S., Hosaka, K., Murata, M., Nikawa, J., & Takatsuki, A. (1998). The Saccharomyces cerevisiae SCS2 gene product, a homolog of a synaptobrevin-associated protein, is an integral membrane protein of the endoplasmic reticulum and is required for inositol metabolism. *Journal of Bacteriology*, *180*(7), 1700–1708.

Kawano, M., Kumagai, K., Nishijima, M., & Hanada, K. (2006). Efficient trafficking of ceramide from the endoplasmic reticulum to the Golgi apparatus requires a VAMP-associated protein-interacting FFAT motif of CERT. *Journal of Biological Chemistry, 281*(40), 30279–30288.

Kikuchi, H., Almer, G., Yamashita, S., Guegan, C., Nagai, M., Xu, Z., et al. (2006). Spinal cord endoplasmic reticulum stress associated with a microsomal accumulation of mutant superoxide dismutase-1 in an ALS model. *Proceedings of the National Academy of Sciences of the United States of America, 103*(15), 6025–6030.

Kim, H. J., Kim, N. C., Wang, Y. D., Scarborough, E. A., Moore, J., Diaz, Z., et al. (2013). Mutations in prion-like domains in hnRNPA2B1 and hnRNPA1 cause multisystem proteinopathy and ALS. *Nature, 495*(7442), 467–473.

Kim, T. Y., Kim, E., Yoon, S. K., & Yoon, J. B. (2008). Herp enhances ER-associated protein degradation by recruiting ubiquilins. *Biochemical and Biophysical Research Communications, 369*(2), 741–746.

Kim, S. H., Shi, Y., Hanson, K. A., Williams, L. M., Sakasai, R., Bowler, M. J., et al. (2009). Potentiation of amyotrophic lateral sclerosis (ALS)-associated TDP-43 aggregation by the proteasome-targeting factor, ubiquilin 1. *Journal of Biological Chemistry, 284*(12), 8083–8092.

Kimonis, V. E., Kovach, M. J., Waggoner, B., Leal, S., Salam, A., Rimer, L., et al. (2000). Clinical and molecular studies in a unique family with autosomal dominant limb-girdle muscular dystrophy and Paget disease of bone. *Genetics in Medicine, 2*(4), 232–241.

Kleijnen, M. F., Alarcon, R. M., & Howley, P. M. (2003). The ubiquitin-associated domain of hPLIC-2 interacts with the proteasome. *Molecular Biology of the Cell, 14*(9), 3868–3875.

Koppers, M., Blokhuis, A. M., Westeneng, H. J., Terpstra, M. L., Zundel, C. A., Vieira de Sa, R., et al. (2015). C9orf72 ablation in mice does not cause motor neuron degeneration or motor deficits. *Annals of Neurology, 78*(3), 426–438.

Kovach, M. J., Waggoner, B., Leal, S. M., Gelber, D., Khardori, R., Levenstien, M. A., et al. (2001). Clinical delineation and localization to chromosome 9p13.3-p12 of a unique dominant disorder in four families: Hereditary inclusion body myopathy, Paget disease of bone, and frontotemporal dementia. *Molecular Genetics and Metabolism, 74*(4), 458–475.

Kuroda, M., Sok, J., Webb, L., Baechtold, H., Urano, F., Yin, Y., et al. (2000). Male sterility and enhanced radiation sensitivity in TLS(-/-) mice. *EMBO Journal, 19*(3), 453–462.

Kwiatkowski, T. J., Jr., Bosco, D. A., Leclerc, A. L., Tamrazian, E., Vanderburg, C. R., Russ, C., et al. (2009). Mutations in the FUS/TLS gene on chromosome 16 cause familial amyotrophic lateral sclerosis. *Science, 323*(5918), 1205–1208.

Lanson, N. A., Jr., Maltare, A., King, H., Smith, R., Kim, J. H., Taylor, J. P., et al. (2011). A Drosophila model of FUS-related neurodegeneration reveals genetic interaction between FUS and TDP-43. *Human Molecular Genetics, 20*(13), 2510–2523.

Larroquette, F., Seto, L., Gaub, P. L., Kamal, B., Wallis, D., Lariviere, R., et al. (2015). Vapb/amyotrophic lateral sclerosis 8 knock-in mice display slowly progressive motor behavior defects accompanying ER stress and autophagic response. *Human Molecular Genetics, 24*(22), 6515–6529.

Latterich, M., Frohlich, K. U., & Schekman, R. (1995). Membrane fusion and the cell cycle: Cdc48p participates in the fusion of ER membranes. *Cell, 82*(6), 885–893.

Lee, S., & Kim, H. J. (2015). Prion-like mechanism in amyotrophic lateral sclerosis: Are protein aggregates the key? *Experimental Neurobiology, 24*(1), 1–7.

Leon, A., & McKearin, D. (1999). Identification of TER94, an AAA ATPase protein, as a Bam-dependent component of the Drosophila fusome. *Molecular Biology of the Cell, 10*(11), 3825–3834.

Li, Y., Ray, P., Rao, E. J., Shi, C., Guo, W., Chen, X., et al. (2010). A Drosophila model for TDP-43 proteinopathy. *Proceedings of the National Academy of Sciences of the United States of America*, *107*(7), 3169–3174.
Li, H. Y., Yeh, P. A., Chiu, H. C., Tang, C. Y., & Tu, B. P. (2011). Hyperphosphorylation as a defense mechanism to reduce TDP-43 aggregation. *PLoS One*, *6*(8), e23075.
Liang, J., Yin, C., Doong, H., Fang, S., Peterhoff, C., Nixon, R. A., et al. (2006). Characterization of erasin (UBXD2): A new ER protein that promotes ER-associated protein degradation. *Journal of Cell Science*, *119*(Pt. 19), 4011–4024.
Lilley, B. N., & Ploegh, H. L. (2004). A membrane protein required for dislocation of misfolded proteins from the ER. *Nature*, *429*(6994), 834–840.
Lim, P. J., Danner, R., Liang, J., Doong, H., Harman, C., Srinivasan, D., et al. (2009). Ubiquilin and p97/VCP bind erasin, forming a complex involved in ERAD. *Journal of Cell Biology*, *187*(2), 201–217.
Lin, M. J., Cheng, C. W., & Shen, C. K. (2011). Neuronal function and dysfunction of Drosophila dTDP. *PLoS One*, *6*(6), e20371.
Linford, N. J., Bilgir, C., Ro, J., & Pletcher, S. D. (2013). Measurement of lifespan in Drosophila melanogaster. *Journal of Visualized Experiments*, (71), pii: 50068.
Ling, J. P., Pletnikova, O., Troncoso, J. C., & Wong, P. C. (2015). TDP-43 repression of nonconserved cryptic exons is compromised in ALS-FTD. *Science*, *349*(6248), 650–655.
Ling, S. C., Polymenidou, M., & Cleveland, D. W. (2013). Converging mechanisms in ALS and FTD: Disrupted RNA and protein homeostasis. *Neuron*, *79*(3), 416–438.
Loewen, C. J., Roy, A., & Levine, T. P. (2003). A conserved ER targeting motif in three families of lipid binding proteins and in Opi1p binds VAP. *EMBO Journal*, *22*(9), 2025–2035.
Lukavsky, P. J., Daujotyte, D., Tollervey, J. R., Ule, J., Stuani, C., Buratti, E., et al. (2013). Molecular basis of UG-rich RNA recognition by the human splicing factor TDP-43. *Nature Structural & Molecular Biology*, *20*(12), 1443–1449.
Mackenzie, I. R., Bigio, E. H., Ince, P. G., Geser, F., Neumann, M., Cairns, N. J., et al. (2007). Pathological TDP-43 distinguishes sporadic amyotrophic lateral sclerosis from amyotrophic lateral sclerosis with SOD1 mutations. *Annals of Neurology*, *61*(5), 427–434.
Mah, A. L., Perry, G., Smith, M. A., & Monteiro, M. J. (2000). Identification of ubiquilin, a novel presenilin interactor that increases presenilin protein accumulation. *Journal of Cell Biology*, *151*(4), 847–862.
Majoor-Krakauer, D., Willems, P. J., & Hofman, A. (2003). Genetic epidemiology of amyotrophic lateral sclerosis. *Clinical Genetics*, *63*(2), 83–101.
Majounie, E., Renton, A. E., Mok, K., Dopper, E. G., Waite, A., Rollinson, S., et al. (2012). Frequency of the C9orf72 hexanucleotide repeat expansion in patients with amyotrophic lateral sclerosis and frontotemporal dementia: A cross-sectional study. *Lancet Neurology*, *11*(4), 323–330.
Matzuk, M. M., Dionne, L., Guo, Q., Kumar, T. R., & Lebovitz, R. M. (1998). Ovarian function in superoxide dismutase 1 and 2 knockout mice. *Endocrinology*, *139*(9), 4008–4011.
Mehnert, M., Sommer, T., & Jarosch, E. (2014). Der1 promotes movement of misfolded proteins through the endoplasmic reticulum membrane. *Nature Cell Biology*, *16*(1), 77–86.
Mercado, P. A., Ayala, Y. M., Romano, M., Buratti, E., & Baralle, F. E. (2005). Depletion of TDP 43 overrides the need for exonic and intronic splicing enhancers in the human apoA-II gene. *Nucleic Acids Research*, *33*(18), 6000–6010.
Meyer, H. H., Wang, Y., & Warren, G. (2002). Direct binding of ubiquitin conjugates by the mammalian p97 adaptor complexes, p47 and Ufd1-Npl4. *EMBO Journal*, *21*(21), 5645–5652.

Miguel, L., Frebourg, T., Campion, D., & Lecourtois, M. (2011). Both cytoplasmic and nuclear accumulations of the protein are neurotoxic in Drosophila models of TDP-43 proteinopathies. *Neurobiology of Disease*, *41*(2), 398–406.
Miller, J., & Gordon, C. (2005). The regulation of proteasome degradation by multi-ubiquitin chain binding proteins. *FEBS Letters*, *579*(15), 3224–3230.
Miller, T. M., Pestronk, A., David, W., Rothstein, J., Simpson, E., Appel, S. H., et al. (2013). An antisense oligonucleotide against SOD1 delivered intrathecally for patients with SOD1 familial amyotrophic lateral sclerosis: A phase 1, randomised, first-in-man study. *Lancet Neurology*, *12*(5), 435–442.
Miller, J. W., Smith, B. N., Topp, S. D., Al-Chalabi, A., Shaw, C. E., & Vance, C. (2012). Mutation analysis of VCP in British familial and sporadic amyotrophic lateral sclerosis patients. *Neurobiology of Aging*, *33*(11). 2721 e2721–2722.
Mitchell, J. D., & Borasio, G. D. (2007). Amyotrophic lateral sclerosis. *Lancet*, *369*(9578), 2031–2041.
Mitne-Neto, M., Machado-Costa, M., Marchetto, M. C., Bengtson, M. H., Joazeiro, C. A., Tsuda, H., et al. (2011). Downregulation of VAPB expression in motor neurons derived from induced pluripotent stem cells of ALS8 patients. *Human Molecular Genetics*, *20*(18), 3642–3652.
Mizielinska, S., Gronke, S., Niccoli, T., Ridler, C. E., Clayton, E. L., Devoy, A., et al. (2014). C9orf72 repeat expansions cause neurodegeneration in Drosophila through arginine-rich proteins. *Science*, *345*(6201), 1192–1194.
Mori, K., Weng, S. M., Arzberger, T., May, S., Rentzsch, K., Kremmer, E., et al. (2013). The C9orf72 GGGGCC repeat is translated into aggregating dipeptide-repeat proteins in FTLD/ALS. *Science*, *339*(6125), 1335–1338.
Moumen, R., Nouvelot, A., Duval, D., Lechevalier, B., & Viader, F. (1997). Plasma superoxide dismutase and glutathione peroxidase activity in sporadic amyotrophic lateral sclerosis. *Journal of the Neurological Sciences*, *151*(1), 35–39.
Moustaqim-Barrette, A., Lin, Y. Q., Pradhan, S., Neely, G. G., Bellen, H. J., & Tsuda, H. (2013). The amyotrophic lateral sclerosis 8 protein, VAP, is required for ER protein quality control. *Human Molecular Genetics*, *23*(8), 1975–1989.
Muller, J. M., Deinhardt, K., Rosewell, I., Warren, G., & Shima, D. T. (2007). Targeted deletion of p97 (VCP/CDC48) in mouse results in early embryonic lethality. *Biochemical and Biophysical Research Communications*, *354*(2), 459–465.
Munch, C., O'Brien, J., & Bertolotti, A. (2011). Prion-like propagation of mutant superoxide dismutase-1 misfolding in neuronal cells. *Proceedings of the National Academy of Sciences of the United States of America*, *108*(9), 3548–3553.
Nalbandian, A., Llewellyn, K. J., Badadani, M., Yin, H. Z., Nguyen, C., Katheria, V., et al. (2013). A progressive translational mouse model of human valosin-containing protein disease: The VCP(R155H/+) mouse. *Muscle & Nerve*, *47*(2), 260–270.
N'Diaye, E. N., Kajihara, K. K., Hsieh, I., Morisaki, H., Debnath, J., & Brown, E. J. (2009). PLIC proteins or ubiquilins regulate autophagy-dependent cell survival during nutrient starvation. *EMBO Reports*, *10*(2), 173–179.
Neumann, M., Kwong, L. K., Lee, E. B., Kremmer, E., Flatley, A., Xu, Y., et al. (2009). Phosphorylation of S409/410 of TDP-43 is a consistent feature in all sporadic and familial forms of TDP-43 proteinopathies. *Acta Neuropathologica*, *117*(2), 137–149.
Neumann, M., Mackenzie, I. R., Cairns, N. J., Boyer, P. J., Markesbery, W. R., Smith, C. D., et al. (2007). TDP-43 in the ubiquitin pathology of frontotemporal dementia with VCP gene mutations. *Journal of Neuropathology and Experimental Neurology*, *66*(2), 152–157.
Neumann, M., Sampathu, D. M., Kwong, L. K., Truax, A. C., Micsenyi, M. C., Chou, T. T., et al. (2006). Ubiquitinated TDP-43 in frontotemporal lobar degeneration and amyotrophic lateral sclerosis. *Science*, *314*(5796), 130–133.

Nichols, C. D., Becnel, J., & Pandey, U. B. (2012). Methods to assay Drosophila behavior. *Journal of Visualized Experiments,* (61), pii: 3795.

Nikawa, J., Murakami, A., Esumi, E., & Hosaka, K. (1995). Cloning and sequence of the SCS2 gene, which can suppress the defect of INO1 expression in an inositol auxotrophic mutant of Saccharomyces cerevisiae. *Journal of Biochemistry, 118*(1), 39–45.

Nishimura, Y., Hayashi, M., Inada, H., & Tanaka, T. (1999). Molecular cloning and characterization of mammalian homologues of vesicle-associated membrane protein-associated (VAMP-associated) proteins. *Biochemical and Biophysical Research Communications, 254*(1), 21–26.

Nishimura, A. L., Mitne-Neto, M., Silva, H. C., Richieri-Costa, A., Middleton, S., Cascio, D., et al. (2004). A mutation in the vesicle-trafficking protein VAPB causes late-onset spinal muscular atrophy and amyotrophic lateral sclerosis. *American Journal of Human Genetics, 75*(5), 822–831.

Nishitoh, H., Kadowaki, H., Nagai, A., Maruyama, T., Yokota, T., Fukutomi, H., et al. (2008). ALS-linked mutant SOD1 induces ER stress- and ASK1-dependent motor neuron death by targeting Derlin-1. *Genes & Development, 22*(11), 1451–1464.

Oh, Y. K., Shin, K. S., Yuan, J., & Kang, S. J. (2008). Superoxide dismutase 1 mutants related to amyotrophic lateral sclerosis induce endoplasmic stress in neuro2a cells. *Journal of Neurochemistry, 104*(4), 993–1005.

O'Rourke, J. G., Bogdanik, L., Muhammad, A. K., Gendron, T. F., Kim, K. J., Austin, A., et al. (2015). C9orf72 BAC transgenic mice display typical pathologic features of ALS/FTD. *Neuron, 88*(5), 892–901.

Orozco, D., & Edbauer, D. (2013). FUS-mediated alternative splicing in the nervous system: Consequences for ALS and FTLD. *Journal of Molecular Medicine (Berlin, Germany), 91*(12), 1343–1354.

Orozco, D., Tahirovic, S., Rentzsch, K., Schwenk, B. M., Haass, C., & Edbauer, D. (2012). Loss of fused in sarcoma (FUS) promotes pathological Tau splicing. *EMBO Reports, 13*(8), 759–764.

Ou, S. H., Wu, F., Harrich, D., Garcia-Martinez, L. F., & Gaynor, R. B. (1995). Cloning and characterization of a novel cellular protein, TDP-43, that binds to human immunodeficiency virus type 1 TAR DNA sequence motifs. *Journal of Virology, 69*(6), 3584–3596.

Pandey, U. B., Batlevi, Y., Baehrecke, E. H., & Taylor, J. P. (2007). HDAC6 at the intersection of autophagy, the ubiquitin-proteasome system and neurodegeneration. *Autophagy, 3*(6), 643–645.

Pandey, U. B., Nie, Z., Batlevi, Y., McCray, B. A., Ritson, G. P., Nedelsky, N. B., et al. (2007). HDAC6 rescues neurodegeneration and provides an essential link between autophagy and the UPS. *Nature, 447*(7146), 859–863.

Pennetta, G., Hiesinger, P. R., Fabian-Fine, R., Meinertzhagen, I. A., & Bellen, H. J. (2002). Drosophila VAP-33A directs bouton formation at neuromuscular junctions in a dosage-dependent manner. *Neuron, 35*(2), 291–306.

Peretti, D., Dahan, N., Shimoni, E., Hirschberg, K., & Lev, S. (2008). Coordinated lipid transfer between the endoplasmic reticulum and the Golgi complex requires the VAP proteins and is essential for Golgi-mediated transport. *Molecular Biology of the Cell, 19*(9), 3871–3884.

Peters, O. M., Cabrera, G. T., Tran, H., Gendron, T. F., McKeon, J. E., Metterville, J., et al. (2015). Human C9ORF72 hexanucleotide expansion reproduces RNA foci and dipeptide repeat proteins but not neurodegeneration in BAC transgenic mice. *Neuron, 88*(5), 902–909.

Peters, O. M., Ghasemi, M., & Brown, R. H., Jr. (2015). Emerging mechanisms of molecular pathology in ALS. *Journal of Clinical Investigation, 125*(5), 1767–1779.

Phillips, J. P., Campbell, S. D., Michaud, D., Charbonneau, M., & Hilliker, A. J. (1989). Null mutation of copper/zinc superoxide dismutase in Drosophila confers hypersensitivity to

paraquat and reduced longevity. *Proceedings of the National Academy of Sciences of the United States of America, 86*(8), 2761–2765.
Phillips, J. P., Tainer, J. A., Getzoff, E. D., Boulianne, G. L., Kirby, K., & Hilliker, A. J. (1995). Subunit-destabilizing mutations in Drosophila copper/zinc superoxide dismutase: Neuropathology and a model of dimer dysequilibrium. *Proceedings of the National Academy of Sciences of the United States of America, 92*(19), 8574–8578.
Prusiner, S. B. (1982). Novel proteinaceous infectious particles cause scrapie. *Science, 216*(4542), 136–144.
Rabbitts, T. H., Forster, A., Larson, R., & Nathan, P. (1993). Fusion of the dominant negative transcription regulator CHOP with a novel gene FUS by translocation t(12;16) in malignant liposarcoma. *Nature Genetics, 4*(2), 175–180.
Rabouille, C., Levine, T. P., Peters, J. M., & Warren, G. (1995). An NSF-like ATPase, p97, and NSF mediate cisternal regrowth from mitotic Golgi fragments. *Cell, 82*(6), 905–914.
Rademakers, R., Stewart, H., Dejesus-Hernandez, M., Krieger, C., Graff-Radford, N., Fabros, M., et al. (2010). Fus gene mutations in familial and sporadic amyotrophic lateral sclerosis. *Muscle & Nerve, 42*(2), 170–176.
Rape, M., Hoppe, T., Gorr, I., Kalocay, M., Richly, H., & Jentsch, S. (2001). Mobilization of processed, membrane-tethered SPT23 transcription factor by CDC48(UFD1/NPL4), a ubiquitin-selective chaperone. *Cell, 107*(5), 667–677.
Ratovitski, T., Corson, L. B., Strain, J., Wong, P., Cleveland, D. W., Culotta, V. C., et al. (1999). Variation in the biochemical/biophysical properties of mutant superoxide dismutase 1 enzymes and the rate of disease progression in familial amyotrophic lateral sclerosis kindreds. *Human Molecular Genetics, 8*(8), 1451–1460.
Ravits, J., Laurie, P., Fan, Y., & Moore, D. H. (2007). Implications of ALS focality: Rostral-caudal distribution of lower motor neuron loss postmortem. *Neurology, 68*(19), 1576–1582.
Reaume, A. G., Elliott, J. L., Hoffman, E. K., Kowall, N. W., Ferrante, R. J., Siwek, D. F., et al. (1996). Motor neurons in Cu/Zn superoxide dismutase-deficient mice develop normally but exhibit enhanced cell death after axonal injury. *Nature Genetics, 13*(1), 43–47.
Renton, A. E., Chio, A., & Traynor, B. J. (2014). State of play in amyotrophic lateral sclerosis genetics. *Nature Neuroscience, 17*(1), 17–23.
Renton, A. E., Majounie, E., Waite, A., Simon-Sanchez, J., Rollinson, S., Gibbs, J. R., et al. (2011). A hexanucleotide repeat expansion in C9ORF72 is the cause of chromosome 9p21-linked ALS-FTD. *Neuron, 72*(2), 257–268.
Ritson, G. P., Custer, S. K., Freibaum, B. D., Guinto, J. B., Geffel, D., Moore, J., et al. (2010). TDP-43 mediates degeneration in a novel Drosophila model of disease caused by mutations in VCP/p97. *Journal of Neuroscience, 30*(22), 7729–7739.
Rohrer, J. D., Isaacs, A. M., Mizielinska, S., Mead, S., Lashley, T., Wray, S., et al. (2015). C9orf72 expansions in frontotemporal dementia and amyotrophic lateral sclerosis. *Lancet Neurology, 14*(3), 291–301.
Ron, D., & Walter, P. (2007). Signal integration in the endoplasmic reticulum unfolded protein response. *Nature Reviews. Molecular Cell Biology, 8*(7), 519–529.
Rosen, D. R., Siddique, T., Patterson, D., Figlewicz, D. A., Sapp, P., Hentati, A., et al. (1993). Mutations in Cu/Zn superoxide dismutase gene are associated with familial amyotrophic lateral sclerosis. *Nature, 362*(6415), 59–62.
Rothenberg, C., Srinivasan, D., Mah, L., Kaushik, S., Peterhoff, C. M., Ugolino, J., et al. (2010). Ubiquilin functions in autophagy and is degraded by chaperone-mediated autophagy. *Human Molecular Genetics, 19*(16), 3219–3232.
Rowland, L. P., & Shneider, N. A. (2001). Amyotrophic lateral sclerosis. *The New England Journal of Medicine, 344*(22), 1688–1700.
Saccon, R. A., Bunton-Stasyshyn, R. K., Fisher, E. M., & Fratta, P. (2013). Is SOD1 loss of function involved in amyotrophic lateral sclerosis? *Brain, 136*(Pt. 8), 2342–2358.

Sanhueza, M., Chai, A., Smith, C., McCray, B. A., Simpson, T. I., Taylor, J. P., et al. (2015). Network analyses reveal novel aspects of ALS pathogenesis. *PLoS Genetics*, *11*(3), e1005107.

Sanhueza, M., Zechini, L., Gillespie, T., & Pennetta, G. (2013). Gain-of-function mutations in the ALS8 causative gene VAPB have detrimental effects on neurons and muscles. *Biology Open*, *3*(1), 59–71.

Sathasivam, S. (2008). Brown–Vialetto–Van Laere syndrome. *Orphanet Journal of Rare Diseases*, *3*, 9.

Saxena, S., Cabuy, E., & Caroni, P. (2009). A role for motoneuron subtype-selective ER stress in disease manifestations of FALS mice. *Nature Neuroscience*, *12*(5), 627–636.

Scaramuzzino, C., Monaghan, J., Milioto, C., Lanson, N. A., Jr., Maltare, A., Aggarwal, T., et al. (2013). Protein arginine methyltransferase 1 and 8 interact with FUS to modify its sub-cellular distribution and toxicity in vitro and in vivo. *PLoS One*, *8*(4), e61576.

Seigneurin-Berny, D., Verdel, A., Curtet, S., Lemercier, C., Garin, J., Rousseaux, S., et al. (2001). Identification of components of the murine histone deacetylase 6 complex: Link between acetylation and ubiquitination signaling pathways. *Molecular and Cellular Biology*, *21*(23), 8035–8044.

Smith, R. A., Miller, T. M., Yamanaka, K., Monia, B. P., Condon, T. P., Hung, G., et al. (2006). Antisense oligonucleotide therapy for neurodegenerative disease. *Journal of Clinical Investigation*, *116*(8), 2290–2296.

Stewart, H., Rutherford, N. J., Briemberg, H., Krieger, C., Cashman, N., Fabros, M., et al. (2012). Clinical and pathological features of amyotrophic lateral sclerosis caused by mutation in the C9ORF72 gene on chromosome 9p. *Acta Neuropathologica*, *123*(3), 409–417.

Stolow, D. T., & Haynes, S. R. (1995). Cabeza, a Drosophila gene encoding a novel RNA binding protein, shares homology with EWS and TLS, two genes involved in human sarcoma formation. *Nucleic Acids Research*, *23*(5), 835–843.

Stolz, A., Hilt, W., Buchberger, A., & Wolf, D. H. (2011). Cdc48: A power machine in protein degradation. *Trends in Biochemical Sciences*, *36*(10), 515–523.

Subramaniam, J. R., Lyons, W. E., Liu, J., Bartnikas, T. B., Rothstein, J., Price, D. L., et al. (2002). Mutant SOD1 causes motor neuron disease independent of copper chaperone-mediated copper loading. *Nature Neuroscience*, *5*(4), 301–307.

Sudria-Lopez, E., Koppers, M., de Wit, M., van der Meer, C., Westeneng, H. J., Zundel, C. A., et al. (2016). Full ablation of C9orf72 in mice causes immune system-related pathology and neoplastic events but no motor neuron defects. *Acta Neuropathologica*, *132*(1), 145–147.

Synofzik, M., Maetzler, W., Grehl, T., Prudlo, J., Vom Hagen, J. M., Haack, T., et al. (2012). Screening in ALS and FTD patients reveals 3 novel UBQLN2 mutations outside the PXX domain and a pure FTD phenotype. *Neurobiology of Aging*, *33*. 2949 e2913–2947.

Tan, A. Y., & Manley, J. L. (2009). The TET family of proteins: Functions and roles in disease. *Journal of Molecular Cell Biology*, *1*(2), 82–92.

Teuling, E., Ahmed, S., Haasdijk, E., Demmers, J., Steinmetz, M. O., Akhmanova, A., et al. (2007). Motor neuron disease-associated mutant vesicle-associated membrane protein-associated protein (VAP) B recruits wild-type VAPs into endoplasmic reticulum-derived tubular aggregates. *Journal of Neuroscience*, *27*(36), 9801–9815.

Therrien, M., Rouleau, G. A., Dion, P. A., & Parker, J. A. (2013). Deletion of C9ORF72 results in motor neuron degeneration and stress sensitivity in C. elegans. *PLoS One*, *8*(12), e83450.

Tiwari, A., & Hayward, L. J. (2005). Mutant SOD1 instability: Implications for toxicity in amyotrophic lateral sclerosis. *Neurodegenerative Diseases*, *2*(3–4), 115–127.

Tobisawa, S., Hozumi, Y., Arawaka, S., Koyama, S., Wada, M., Nagai, M., et al. (2003). Mutant SOD1 linked to familial amyotrophic lateral sclerosis, but not wild-type SOD1, induces ER stress in COS7 cells and transgenic mice. *Biochemical and Biophysical Research Communications*, *303*(2), 496–503.

Tran, H., Almeida, S., Moore, J., Gendron, T. F., Chalasani, U., Lu, Y., et al. (2015). Differential toxicity of nuclear RNA foci versus dipeptide repeat proteins in a Drosophila model of C9ORF72 FTD/ALS. *Neuron*, *87*(6), 1207–1214.

Tresse, E., Salomons, F. A., Vesa, J., Bott, L. C., Kimonis, V., Yao, T. P., et al. (2010). VCP/p97 is essential for maturation of ubiquitin-containing autophagosomes and this function is impaired by mutations that cause IBMPFD. *Autophagy*, *6*(2), 217–227.

Tsuda, H., Han, S. M., Yang, Y., Tong, C., Lin, Y. Q., Mohan, K., et al. (2008). The amyotrophic lateral sclerosis 8 protein VAPB is cleaved, secreted, and acts as a ligand for Eph receptors. *Cell*, *133*(6), 963–977.

Tudor, E. L., Galtrey, C. M., Perkinton, M. S., Lau, K. F., De Vos, K. J., Mitchell, J. C., et al. (2010). Amyotrophic lateral sclerosis mutant vesicle-associated membrane protein-associated protein-B transgenic mice develop TAR-DNA-binding protein-43 pathology. *Neuroscience*, *167*(3), 774–785.

van Blitterswijk, M., van Es, M. A., Koppers, M., van Rheenen, W., Medic, J., Schelhaas, H. J., et al. (2012). VAPB and C9orf72 mutations in 1 familial amyotrophic lateral sclerosis patient. *Neurobiology of Aging*, *33*(12). 2950 e2951–2954.

Vance, C., Rogelj, B., Hortobagyi, T., De Vos, K. J., Nishimura, A. L., Sreedharan, J., et al. (2009). Mutations in FUS, an RNA processing protein, cause familial amyotrophic lateral sclerosis type 6. *Science*, *323*(5918), 1208–1211.

Vembar, S. S., & Brodsky, J. L. (2008). One step at a time: Endoplasmic reticulum-associated degradation. *Nature Reviews. Molecular Cell Biology*, *9*(12), 944–957.

Voigt, A., Herholz, D., Fiesel, F. C., Kaur, K., Muller, D., Karsten, P., et al. (2010). TDP-43-mediated neuron loss in vivo requires RNA-binding activity. *PLoS One*, *5*(8), e12247.

Waite, A. J., Baumer, D., East, S., Neal, J., Morris, H. R., Ansorge, O., et al. (2014). Reduced C9orf72 protein levels in frontal cortex of amyotrophic lateral sclerosis and frontotemporal degeneration brain with the C9ORF72 hexanucleotide repeat expansion. *Neurobiology of Aging*, *35*(7), 1779.e5–1779.e13.

Walters, K. J., Kleijnen, M. F., Goh, A. M., Wagner, G., & Howley, P. M. (2002). Structural studies of the interaction between ubiquitin family proteins and proteasome subunit S5a. *Biochemistry*, *41*(6), 1767–1777.

Walther, R. F., & Pichaud, F. (2006). Immunofluorescent staining and imaging of the pupal and adult Drosophila visual system. *Nature Protocols*, *1*(6), 2635–2642.

Wang, J. W., Brent, J. R., Tomlinson, A., Shneider, N. A., & McCabe, B. D. (2011). The ALS-associated proteins FUS and TDP-43 function together to affect Drosophila locomotion and life span. *Journal of Clinical Investigation*, *121*(10), 4118–4126.

Wang, L., Popko, B., & Roos, R. P. (2011). The unfolded protein response in familial amyotrophic lateral sclerosis. *Human Molecular Genetics*, *20*(5), 1008–1015.

Wang, L., Popko, B., & Roos, R. P. (2014a). An enhanced integrated stress response ameliorates mutant SOD1-induced ALS. *Human Molecular Genetics*, *23*(10), 2629–2638.

Wang, L., Popko, B., Tixier, E., & Roos, R. P. (2014b). Guanabenz, which enhances the unfolded protein response, ameliorates mutant SOD1-induced amyotrophic lateral sclerosis. *Neurobiology of Disease*, *71*, 317–324.

Wang, X., Schwartz, J. C., & Cech, T. R. (2015). Nucleic acid-binding specificity of human FUS protein. *Nucleic Acids Research*, *43*(15), 7535–7543.

Wate, R., Ito, H., Zhang, J. H., Ohnishi, S., Nakano, S., & Kusaka, H. (2005). Expression of an endoplasmic reticulum-resident chaperone, glucose-regulated stress protein 78, in the spinal cord of a mouse model of amyotrophic lateral sclerosis. *Acta Neuropathologica*, *110*(6), 557–562.

Watson, M. R., Lagow, R. D., Xu, K., Zhang, B., & Bonini, N. M. (2008). A drosophila model for amyotrophic lateral sclerosis reveals motor neuron damage by human SOD1. *Journal of Biological Chemistry*, *283*(36), 24972–24981.

Watts, G. D., Wymer, J., Kovach, M. J., Mehta, S. G., Mumm, S., Darvish, D., et al. (2004). Inclusion body myopathy associated with Paget disease of bone and frontotemporal

dementia is caused by mutant valosin-containing protein. *Nature Genetics*, *36*(4), 377–381.
Wegorzewska, I., Bell, S., Cairns, N. J., Miller, T. M., & Baloh, R. H. (2009). TDP-43 mutant transgenic mice develop features of ALS and frontotemporal lobar degeneration. *Proceedings of the National Academy of Sciences of the United States of America*, *106*(44), 18809–18814.
Weihl, C. C., Temiz, P., Miller, S. E., Watts, G., Smith, C., Forman, M., et al. (2008). TDP-43 accumulation in inclusion body myopathy muscle suggests a common pathogenic mechanism with frontotemporal dementia. *Journal of Neurology, Neurosurgery, and Psychiatry*, *79*(10), 1186–1189.
Weir, M. L., Klip, A., & Trimble, W. S. (1998). Identification of a human homologue of the vesicle-associated membrane protein (VAMP)-associated protein of 33 kDa (VAP-33): A broadly expressed protein that binds to VAMP. *Biochemical Journal*, *333*(Pt. 2), 247–251.
Williams, K. L., Warraich, S. T., Yang, S., Solski, J. A., Fernando, R., Rouleau, G. A., et al. (2012). UBQLN2/ubiquilin 2 mutation and pathology in familial amyotrophic lateral sclerosis. *Neurobiology of Aging*, *33*(10), 2527.e3–2527.10.
Wojcik, C., Yano, M., & DeMartino, G. N. (2004). RNA interference of valosin-containing protein (VCP/p97) reveals multiple cellular roles linked to ubiquitin/proteasome-dependent proteolysis. *Journal of Cell Science*, *117*(Pt. 2), 281–292.
Wolff, T. (2011). Preparation of Drosophila eye specimens for transmission electron microscopy. *Cold Spring Harbor Protocols*, *2011*(11), 1386–1388.
Wong, E., & Cuervo, A. M. (2010). Integration of clearance mechanisms: The proteasome and autophagy. *Cold Spring Harbor Perspectives in Biology*, *2*(12), a006734.
Wong, P. C., Pardo, C. A., Borchelt, D. R., Lee, M. K., Copeland, N. G., Jenkins, N. A., et al. (1995). An adverse property of a familial ALS-linked SOD1 mutation causes motor neuron disease characterized by vacuolar degeneration of mitochondria. *Neuron*, *14*(6), 1105–1116.
Wu, A. L., Wang, J., Zheleznyak, A., & Brown, E. J. (1999). Ubiquitin-related proteins regulate interaction of vimentin intermediate filaments with the plasma membrane. *Molecular Cell*, *4*(4), 619–625.
Xia, R., Liu, Y., Yang, L., Gal, J., Zhu, H., & Jia, J. (2012). Motor neuron apoptosis and neuromuscular junction perturbation are prominent features in a Drosophila model of Fus-mediated ALS. *Molecular Neurodegeneration*, 7, 10.
Xia, Y., Yan, L. H., Huang, B., Liu, M., Liu, X., & Huang, C. (2014). Pathogenic mutation of UBQLN2 impairs its interaction with UBXD8 and disrupts endoplasmic reticulum-associated protein degradation. *Journal of Neurochemistry*, *129*(1), 99–106.
Xie, Z., & Klionsky, D. J. (2007). Autophagosome formation: Core machinery and adaptations. *Nature Cell Biology*, *9*(10), 1102–1109.
Yang, L., Gal, J., Chen, J., & Zhu, H. (2014). Self-assembled FUS binds active chromatin and regulates gene transcription. *Proceedings of the National Academy of Sciences of the United States of America*, *111*(50), 17809–17814.
Yang, C., Tan, W., Whittle, C., Qiu, L., Cao, L., Akbarian, S., et al. (2010). The C-terminal TDP-43 fragments have a high aggregation propensity and harm neurons by a dominant-negative mechanism. *PLoS One*, *5*(12), e15878.
Ye, Y., Meyer, H. H., & Rapoport, T. A. (2001). The AAA ATPase Cdc48/p97 and its partners transport proteins from the ER into the cytosol. *Nature*, *414*(6864), 652–656.
Yin, H. Z., Nalbandian, A., Hsu, C. I., Li, S., Llewellyn, K. J., Mozaffar, T., et al. (2012). Slow development of ALS-like spinal cord pathology in mutant valosin-containing protein gene knock-in mice. *Cell Death & Disease*, *3*, e374.
Yoshida, T., Maulik, N., Engelman, R. M., Ho, Y. S., & Das, D. K. (2000). Targeted disruption of the mouse Sod I gene makes the hearts vulnerable to ischemic reperfusion injury. *Circulation Research*, *86*(3), 264–269.

Zhang, K., Donnelly, C. J., Haeusler, A. R., Grima, J. C., Machamer, J. B., Steinwald, P., et al. (2015). The C9orf72 repeat expansion disrupts nucleocytoplasmic transport. *Nature*, *525*(7567), 56–61.

Zhang, X., Gui, L., Zhang, X., Bulfer, S. L., Sanghez, V., Wong, D. E., et al. (2015). Altered cofactor regulation with disease-associated p97/VCP mutations. *Proceedings of the National Academy of Sciences of the United States of America*, *112*(14), E1705–E1714.

Zhang, C., & Saunders, A. J. (2009). An emerging role for Ubiquilin 1 in regulating protein quality control system and in disease pathogenesis. *Discovery Medicine*, *8*(40), 18–22.

Zhang, Y. J., Xu, Y. F., Cook, C., Gendron, T. F., Roettges, P., Link, C. D., et al. (2009). Aberrant cleavage of TDP-43 enhances aggregation and cellular toxicity. *Proceedings of the National Academy of Sciences of the United States of America*, *106*(18), 7607–7612.

Zu, T., Gibbens, B., Doty, N. S., Gomes-Pereira, M., Huguet, A., Stone, M. D., et al. (2011). Non-ATG-initiated translation directed by microsatellite expansions. *Proceedings of the National Academy of Sciences of the United States of America*, *108*(1), 260–265.

Zu, T., Liu, Y., Banez-Coronel, M., Reid, T., Pletnikova, O., Lewis, J., et al. (2013). RAN proteins and RNA foci from antisense transcripts in C9ORF72 ALS and frontotemporal dementia. *Proceedings of the National Academy of Sciences of the United States of America*, *110*(51), E4968–E4977.

CHAPTER FIVE

Mechanisms of Parkinson's Disease: Lessons from *Drosophila*

V.L. Hewitt, A.J. Whitworth[1]
Medical Research Council Mitochondrial Biology Unit, Cambridge, United Kingdom
[1]Corresponding author: e-mail address: a.whitworth@mrc-mbu.cam.ac.uk

Contents

Abstract

The power of *Drosophila* genetics has attracted attention in tackling important biomedical challenges such as the understanding and prevention of neurodegenerative diseases. Parkinson's disease (PD) is the most common neurodegenerative movement disorder which results from the relentless degeneration of midbrain dopaminergic neurons. Over the past two decades tremendous advances have been made in identifying genes responsible for inherited forms of PD. The ease of genetic manipulation in *Drosophila* has spurred the development of numerous models of PD, including expression of human genes carrying pathogenic mutations or the targeted mutation of conserved

Current Topics in Developmental Biology, Volume 121
ISSN 0070-2153
http://dx.doi.org/10.1016/bs.ctdb.2016.07.005

orthologs. The genetic and cellular analysis of these models is beginning to reveal fundamental insights into the pathogenic mechanisms. Numerous pathways and processes are disrupted in these models but some common themes are emerging. These often implicate aberrant synaptic function, protein aggregation, autophagy, oxidative stress, and mitochondrial dysfunction. Moreover, an impressive list of small molecule compounds have been identified as effective in reversing pathogenic phenotypes, paving the way to explore these for therapeutic interventions.

1. INTRODUCTION

Parkinson's disease (PD) affects ~1% of the retirement-age population. It is characterized by the progressive degeneration of dopaminergic (DA) neurons in the substantia nigra, which disrupts motor control causing symptoms such as resting tremor, bradykinesia, and postural instability. Pathological examination often reveals the presence of intraneuronal inclusions called Lewy bodies. Dopamine replacement by levodopa (L-DOPA) remains the mainstay therapeutic treatment but the effects diminish over time and there are currently no effective disease-modifying therapies.

The majority of cases are sporadic, implicating a combination of genetic susceptibility and environmental factors in the pathogenic cause. While the precise pathologic mechanisms remain unclear, prevailing hypotheses include aberrant protein degradation, mitochondrial dysfunction, calcium imbalance, and inflammation (Gupta, Dawson, & Dawson, 2008). In addition, oxidative stress is a prominent and common feature in many neurodegenerative diseases including PD and may represent a convergent toxic event leading to neuronal cell death.

Genetic analyses have identified a considerable number of single-gene mutations responsible for familial forms of PD (Hernandez, Reed, & Singleton, 2016). While inherited PD cases are relatively rare, many are clinically indistinguishable from the more common sporadic form, supporting the assumption of a shared pathogenic mechanism between inherited and sporadic cases. In addition, genome-wide associations studies (GWAS) have identified a large number of susceptibility loci for sporadic PD (Hernandez et al., 2016). The underlying genetic component(s) of the GWAS risk loci still need to be determined in most cases, thus the majority of efforts in modeling PD are still focused on genes linked to rare Mendelian forms of PD. Generating animal models that recapitulate the genetic lesions leading to disease are essential tools on the path to fully elucidating the disease biology. Although many genetic models have been generated in rodents few recapitulate the cardinal features of PD—progressive loss of nigral DA neurons and locomotor disturbances—and none are currently held as an ideal model (Blesa & Przedborski, 2014). Hence, there is still a pressing need to develop additional animal models suitable to investigate the basic pathogenic mechanisms. One model system that has proven to offer some insights is the fruit fly *Drosophila*.

2. *DROSOPHILA* AS A MODEL SYSTEM FOR PD

Drosophila are a preeminent genetic model organism for investigating a very broad range of biological questions, and the high degree of functional conservation with vertebrates makes them an excellent system to advance our understanding of neurological disease mechanisms. Both mutational and transgenic approaches have been used to create *Drosophila* models of neurological disease. Mutation of the fly homolog of a disease gene is useful for recessive traits, indicating genetic loss of function, whereas the transgenic approaches are more often used for mutations conferring a dominant, toxic gain-of-function mechanism. One of the greatest advantages of using *Drosophila* to understand disease mechanisms is the ability to conduct genetic modifier screens to identify genes that can suppress or enhance the disease model phenotypes. This approach has the potential to identify cellular factors that impinge on the pathogenic processes. Moreover, suppressors identified from such screens represent potential targets for therapeutic intervention.

When developing a model system of PD, the integrity of DA neurons and locomotor behaviors is obvious salient features for comparison to the human syndrome. Although the anatomical arrangement of DA neurons in the adult *Drosophila* brain differs significantly from the vertebrate brain, recent work has highlighted the ancient functional homology of these circuits (Strausfeld & Hirth, 2013), at least some of which are known to regulate locomotor behaviors (Riemensperger et al., 2011). This functional homology gives us confidence that the analysis of DA neuron integrity in *Drosophila* has at least some relevance to the DA neurons that are preferentially susceptible in PD.

Phenotypes arising from these genetic models in tissues not directly affected in the human syndrome can provide fundamental insights into the basic function of disease-related genes. For example, analysis of the highly metabolic flight muscle has greatly informed our understanding of

Table 1 Genes Linked to Parkinson's Disease

Locus	Chrom.	Gene	Inheritance	Mechanism (?)	Fly Homology (%)
PARK1/4	4q21	SNCA	AD	Protein aggregation, autophagy, synaptic vesicle dynamics	No homolog
PARK2	6q26	PARKIN	AR	Mitochondria, autophagy	49
PARK6	1p36	PINK1	AR	Mitochondria, autophagy	32
PARK7	1p36	DJ-1	AR	Ox. stress, mitochondria	56
PARK8	12q12	LRRK2	AD	Vesicles dynamics, autophagy	26
PARK9	1p36	ATP13A2	AR	Lysosome	No homolog
PARK14	22q13	PLA2G6	AR	Lipids, mitochondria	50
PARK15	22q12	FBXO7	AR	Mitochondria, autophagy	59
PARK16	1q32	RAB7L1	Risk locus	Vesicles dynamics, autophagy	51
PARK17	16q13	VPS35	AD	Vesicles dynamics, autophagy	61
PARK18	3q27	EIF4G1	AD	Translation	36
PARK19	1p32	DNAJC6	AR	Chaperone, vesicles dynamics	?
PARK20	21q22	SYNJ1	AR	Vesicles dynamics	55
Unassigned	1q22	GBA	Risk locus	Lysosome	32

the function of PD genes such as *PINK1* and *parkin* (see later for details). Other more obvious model systems, such as the larval neuromuscular junction (NMJ) and the adult visual system have yielded important electrophysiological findings, but so too have some more surprising tissues such as developing spermatids or oocytes.

Finally, as an animal model *Drosophila* also offers the opportunity to study the impact of genetic or environmental (e.g., toxin) influence on aging and age-related processes, and vice versa, to study the impact of aging on a cell biological process of interest. With the typical life span of wild-type flies around 2–3 months, this represents a convenient model system for the analysis of neurodegenerative processes across the complete life span which can encompass the presymptomatic phase of disease progression, important for understanding early, upstream events.

The existence of *Drosophila* homologs of most of the genes implicated in heritable forms of PD (Table 1) implies that the pathways regulated by these genes are also likely to be conserved in *Drosophila*. Arguably the biggest advantage that the *Drosophila* models can bring to the field are genetic analyses that reveal new insights into the pathogenic mechanism and potentially protective pathways. In this review we present a brief overview of the key features various genetic models of PD in *Drosophila* but focus more on recent work elucidating the functions of these genes and the implications not only for the pathogenic process but also for therapeutic targets.

3. DOMINANT TRAITS

3.1 α-Synuclein Models

SNCA encodes α-synuclein, an abundant neuronal protein of unclear function that has a propensity to self-aggregate. The presence of α-synuclein in Lewy body inclusions propelled α-synuclein aggregation into center stage in the pathogenesis of PD. *Drosophila* have no homolog of *SNCA*, but pathogenic missense mutations and gene duplication/triplication implicate a toxic gain of function which lend itself well to transgenic modeling in *Drosophila*.

Numerous studies have reported that expression of α-synuclein causes a variety of phenotypes including progressive neurodegeneration in DA neurons or retina, decline in locomotor ability, the accumulation of proteinaceous inclusions, and progressive degeneration of retinal tissue, although there is some discrepancy over the strength of α-synuclein phenotypes, which is discussed in detail elsewhere (Navarro et al., 2014; Whitworth, 2011). Early studies directly addressed the aggregation toxicity hypothesis. Genetic or pharmacological induction of the chaperone HSP70, as well as upregulation of proteasome-dependent degradation, abrogated the α-synuclein toxicity without apparently influencing inclusion formation (Auluck & Bonini, 2002; Auluck, Chan, Trojanowski, Lee, & Bonini, 2002; Lee et al., 2009). Moreover, the expression of phosphorylation site variants revealed that phospho-Ser129 increased toxicity and decreased aggregation, while phospho-Tyr125 had the opposite effects (Chen &

Feany, 2005; Chen et al., 2009). Overall, these findings suggest that an increase in α-synuclein aggregates correlate with decreased cellular toxicity and support recent in vitro evidence (Cremades et al., 2012; Karpinar et al., 2009) that smaller oligomeric species and not the large aggregates are the primary toxic species. Crucially, these findings challenge the rationale of current therapeutic strategies aimed at preventing inclusion formation as these strategies may even augment disease progression.

In addressing the pathogenic mechanism of preaggregate α-synuclein, several recent studies have implicated intracellular vesicle trafficking in α-synuclein pathogenesis, in particular by the small GTPase Rab proteins. Genetic and biochemical screens have identified functional interaction between α-synuclein and Rab1, Rab8a, and Rab11, which have all been shown to ameliorate toxicity in *Drosophila* (Breda et al., 2015; Cooper et al., 2006; Yin et al., 2014). Further studies will be needed to determine the precise cause of toxicity, but it is notable that both LRRK2 and PINK1 (see later) have also recently been functionally linked to Rab8 (Lai et al., 2015; Steger et al., 2016).

3.2 LRRK2

Mutations in *LRRK2* are a common cause of heritable PD and a robust GWAS risk locus, making *LRRK2* a target of intensive analysis. *LRRK2* encodes a large, complex, multidomain protein comprising 2527 amino acids with both kinase and GTPase activity. The complexity of LRRK2 means that its mutation has the potential to cause a multitude of pathogenic effects and perhaps unsurprisingly no consensus has yet emerged over a likely pathogenic mechanism. The most obvious target is its native enzymatic activity. While the physiological function of LRRK2 is still unclear, pathogenic mutations are generally considered to either increase kinase activity (e.g., the most prevalent G2019S mutation) or reduce GTPase activity (Esteves, Swerdlow, & Cardoso, 2014).

The *Drosophila* genome contains a single homolog of *LRRK2*, called *Lrrk*, which appears to lack certain N-terminal repeat sequences, though valuable insights may still be gained from studying this ancestral gene. Homozygous *Lrrk* mutants are viable with a normal life span and overall appear grossly normal. Several studies report no effect on DA neuron survival, but mild locomotor deficits have been measured (Lee, Kim, Lee, & Chung, 2007). Unexpectedly, female mutants have severely reduced fecundity, which appears to result from aberrant apoptosis in follicle cells (Dodson, Leung, Lone, Lizzio, & Guo, 2014; Lee et al., 2007).

Initial links to regulating protein translation came from observations that *Lrrk* genetically interacted with a negative regulator of the eIF4E protein translation machinery, *4E-BP* (Imai et al., 2008). 4E-BP is a downstream effector of TOR signaling, and its activity is important for survival under a wide variety of stresses including starvation, oxidative stress, unfolded protein stress, and immune challenge. While direct phosphorylation of 4E-BP1 by LRRK2 seems unlikely (Kumar et al., 2010; Trancikova et al., 2012), overexpression of *4E-BP* protected flies against DA neuron loss and locomotor deficits caused by pathogenic LRRK2 mutations (Imai et al., 2008). Subsequent studies presented additional evidence for either direct (Martin et al., 2014) or indirect regulation of translation (Gehrke, Imai, Sokol, & Lu, 2010). Martin and colleagues reported that LRRK2 phosphorylates several ribosomal proteins with manipulation specifically of a phosphorylation site on RPS15 providing protection against mutant LRRK2 in vivo and in vitro (Martin et al., 2014). Overall, the results from this study suggested that an increase in bulk translation may contribute to LRRK2 toxicity. This hypothesis is partly supported by evidence that an inhibitor of the eIF4E/4G interaction, 4EGI-1, could also prevent toxicity from LRRK2 G2019S (Chuang, Lu, Wang, & Chang, 2014). Evidence for an indirect effect on translation comes partly from observations in *Drosophila* that LRRK2 interacts with Ago-1 of the RNA-induced silencing complex inhibiting let-7 and miR-184*, which in turn regulate the translation of transcription factors E2F1 and DP (Gehrke et al., 2010). Together these findings have highlighted regulated protein translation as a therapeutic avenue to explore further.

Drosophila genetics has also linked LRRK2 to various endocytic processes, which have obvious implications in neurodegeneration. LRRK2 appears to phosphorylate Endophilin A, a central component of synaptic vesicle endocytosis, affecting its membrane binding and membrane remodeling properties. Both loss- and gain-of-function (G2019S) perturbed synaptic vesicle recycling in the fly NMJ (Matta et al., 2012). Intriguingly, Endophilin A has a close functional association with Synaptojanin (Verstreken et al., 2003), which was recently linked to recessive PD. LRRK2 G2019S has also been reported to cause functional decline and neurodegeneration in the fly visual system (Hindle et al., 2013). These tractable systems offer an excellent opportunity to explore the possible connection between LRRK2, Endophilin A, and Synaptojanin.

Another link to intracellular vesicle trafficking has identified a role for LRRK2/Lrrk in regulating Rab7-dependent endolysosomal maturation. *Drosophila* Lrrk was shown to physically interact with Rab5, Rab7, and Rab9 (in late endosomes and lysosomes) but did not colocalize with Rab5 in the early endosomes (Dodson et al., 2014; Dodson, Zhang, Jiang, Chen, & Guo, 2012). Loss of *Lrrk* resulted in distorted Rab7-positive late endosomal compartments and abnormal lysosome positioning (Dodson et al., 2012, 2014). Furthermore, expression of the pathogenic LRRK2 G2019S mutation also caused abnormal lysosome trafficking in flies, which was corroborated by similar results in cultured rat neurons (MacLeod et al., 2013). In addition, overexpression of Rab7 could rescue the LRRK2 G2019S premature mortality in flies, while deficiency of the putative *RAB7L1* homolog, *lightoid/Rab32*, results in DA neuron loss (MacLeod et al., 2013). These lysosomal trafficking defects have been linked to retromer and Vps35 function (see later), and Vps35 overexpression can rescue DA neuron loss and life span in LRRK2 mutant flies (Linhart et al., 2014; MacLeod et al., 2013). These findings together suggest that pathogenic LRRK2 mutations could be disrupting endosomal trafficking through altering interactions with these endosomal proteins. This is likely to have broad downstream effects resulting from aberrant degradation, but mistrafficking of signaling components can lead to aberrant signaling as recently shown for Notch (Imai et al., 2015).

An emerging area in which LRRK2 pathology may have a broad impact on intracellular trafficking and neuronal survival is in axonal transport. LRRK2 appears to bind microtubules and affect their acetylation state (Caesar et al., 2013; Law et al., 2014). Mutant LRRK2 shows enhanced microtubule binding and causes excess deacetylation. This caused an inhibition of axonal transport in *Drosophila* and mammalian neurons, and climbing deficits in flies (Godena et al., 2014). Interestingly, axonal transport was inhibited by LRRK2 mutations R1441C and Y1699G that affect the GTPase activity, but not the G2019S mutation. The mechanisms by which this occurs are not known, but pharmacological or genetic inhibition of the deacetylase enzymes HDAC6 and SIRT2 restored axonal transport (Godena et al., 2014). If this process were proven to contribute to the pathogenic process, such inhibitors would represent possible therapeutics.

3.3 Vps35

Several dominant mutations in *VPS35* have been identified, however, only D620N is currently confirmed as causing late-onset PD. VPS35 is a central component of the retromer complex that mediates the recycling of transmembrane cargo proteins from endosomes back to the *trans*-Golgi network or plasma membrane (Seaman, 2012). Cargo-selective sorting is critical for regulating the proper subcellular destination of endosomal proteins. Although VPS35 D620N is dominantly inherited, there is a general consensus that it has impaired function compared to wild type without overt dominant toxicity (Dhungel et al., 2015; McGough et al., 2014). Recent fly models also support this view as expression of various VPS35 variants show very limited toxicity (Malik, Godena, & Whitworth, 2015), although very mild loss of DA neurons was observed in another study (Tsika et al., 2014). Moreover, genetic rescue experiments indicate that Vps35 variants can only partially rescue *vps35* null mutants (Malik et al., 2015). These results suggest that the pathogenicity derives from haploinsufficiency. There is currently no consensus on the pathogenic mechanism as cycling membrane compartments can have pleiotropic effects including effects on synaptic and mitochondrial function (see later) but may also impact on autophagy degradation of α-synuclein (Miura et al., 2014).

3.4 GBA

One of the most significant recent discoveries in PD genetics has revealed that heterozygous mutations in *GBA* confer a significant risk of developing PD. *GBA* encodes the lysosomal protein glucocerebrosidase (GCase), and homozygous mutations lead to the lysosomal storage disorder Gaucher's disease. Current evidence indicates that dominant mutations in GCase, most commonly N370S and L444P, cause ER stress and induce the unfolded protein stress response. Moreover, lack of lysosomal GCase can lead to the accumulation of α-synuclein and its aggregates. Both of these features have been recapitulated in recently described *Drosophila* models of *GBA*.

Drosophila has two closely related homologs of *GBA*, *Gba1a* and *Gba1b*, although expression of *Gba1a* appears to be restricted to the gut (FlyBase). RNAi knockdown or loss-of-function mutants show a consistent pattern of shortened life span, progressive loss of climbing ability, and some degree of neurodegeneration, although the effect on DA neurons is disputed (Davis et al., 2016; Maor et al., 2013, 2016; Suzuki et al., 2015). Expression of human GCase N370S or L444P also led to motor deficits and neurodegeneration with some DA neuron loss (Maor et al., 2013; Suzuki et al., 2013). In the majority of these models there is clear evidence for ER stress induction and aberrant protein turnover, in particular, the degradation of transgenic α-synuclein was inhibited, although the effect on α-synuclein was inconsistent.

Encouragingly, feeding flies expressing mutant *GBA* the pharmacological chaperone ambroxol alleviated the ER stress, prevented DA neurodegeneration, and restored climbing ability (Maor et al., 2016; Suzuki et al., 2013). Treatment with ambroxol or other chaperones (e.g., isofagomine) is thought to both alleviate the effects of ER stress and promote lysosomal activity and α-synuclein turnover, thus providing a double benefit as a possible therapeutic.

4. RECESSIVE TRAITS

Most of the autosomal recessive PD genes are conserved in *Drosophila* (Table 1), which invites the power of classical genetic analysis in *Drosophila* to explore the basic biological functions of these evolutionarily conserved genes. This approach has provided fundamental insights into their function and provided compelling evidence for pathogenic mechanisms.

4.1 parkin

Mutations in *parkin* are the commonest cause of autosomal recessive parkinsonism. *parkin* encodes an E3 ubiquitin ligase which provoked the general hypothesis that it would regulate substrates prone to aggregation. However, it was the analysis of *Drosophila parkin* mutants that provided compelling evidence that Parkin plays an important role in mitochondrial homeostasis. Flies lacking *parkin* display reduced longevity, DA neuron degeneration, motor deficits stemming from apoptotic degeneration of the musculature, and male sterility (Greene et al., 2003; Whitworth et al., 2005). Ultrastructural studies revealed that profound loss of mitochondrial integrity was the earliest to manifest, during muscle degeneration and also apparent in spermatids and DA neurons (Greene et al., 2003). Mitochondrial defects now represent a conserved feature in all organisms with *parkin* mutations including humans (Winklhofer, 2014). Together these observations established that mitochondrial dysfunction is an important contributing factor to DA neuron death in patients with altered Parkin function.

4.2 DJ-1

DJ-1 encodes a highly conserved small protein of unclear function with homology to proteases, kinases, and small heat shock proteins. *Drosophila* has two genes highly homologous to human *DJ-1*. *DJ-1β* is ubiquitously expressed, whereas *DJ-1α* expression is largely restricted to testes (Meulener et al., 2005). Loss of one or both of the *DJ-1* genes is fully viable, fertile and display no evidence of DA neuron loss. However, *DJ-1β* mutants and *DJ-1α;DJ-1β* double mutants display a striking sensitivity to oxidative stress-inducing agents, including paraquat and rotenone (Meulener et al., 2005; Park et al., 2005). Like vertebrate DJ-1, fly DJ-1β is modified by sulfonification of a conserved cysteine residue upon oxidative stress (Meulener, Xu, Thompson, Ischiropoulos, & Bonini, 2006). C104 appears to be crucial for DJ-1β's protection from oxidative stress.

An important challenge of future work will be to discern between the myriad biological functions ascribed to this protein family to define the mechanism by which loss of DJ-1 function protects against neuron loss, but it seems most likely that the major contribution of DJ-1 is in the response to oxidative stress to maintain mitochondrial integrity. In support of this, feeding *DJ-1β* mutant flies various dietary antioxidants, including vitamin C, melatonin, and vitamin E, significantly extends life span of these mutants (Lavara-Culebras, Munoz-Soriano, Gomez-Pastor, Matallana, & Paricio, 2010), indicating the potential therapeutic benefit of antioxidants.

4.3 PINK1

Mutations in *PINK1*, which encodes a mitochondrially targeted serine–threonine kinase, are a rare cause of recessive parkinsonism but have provided some significant insights into possible pathogenic mechanisms. A major role for PINK1 in maintaining mitochondrial integrity was established again with mutational analysis of the *Drosophila* ortholog, *Pink1* (Clark et al., 2006; Park et al., 2006). Loss of *Pink1* causes degeneration of flight muscles, defective spermatid formation, loss of DA neurons, and locomotor deficits. These phenotypes were associated with profound mitochondrial disruptions.

These phenotypes bore a striking similarity with those previously described for *parkin* mutants, and genetic interaction studies unequivocally linked Pink1 and Parkin. Essentially, *Pink1;parkin* double mutants are phenotypically indistinguishable from the respective single mutants, consistent with the two mutations affecting a common pathway, and genetic epistasis experiments showed that Parkin acts downstream from Pink1 (Clark et al., 2006; Park et al., 2006). These findings provided compelling evidence that Pink1 and Parkin act in a common pathway that regulates mitochondrial homeostasis.

4.4 PLA2G6

Mutations in *PLA2G6* have been linked to a range of complex neurodegenerative disorders including a complex form of dystonia-parkinsonism. Since *PLA2G6* encodes a calcium-independent phospholipase enzyme that selectively hydrolyses glycerophospholipids to release free fatty acids, it was hypothesized that PLA2G6 may function in the repair of oxidatively damaged cardiolipin, a mitochondria-enriched phospholipid. *Drosophila* mutants in *iPLA2-VIA*, the homolog of *PLA2G6*, show reduced survival, locomotor deficits, and sensitivity to oxidative stress (Kinghorn et al., 2015; Malhotra et al., 2009). These phenotypes were accompanied by a number of mitochondrial abnormalities, including mitochondrial respiratory chain dysfunction, reduced ATP synthesis, and abnormal mitochondrial morphology (Kinghorn et al., 2015). While no evidence was found for oxidized cardiolipin, loss of iPLA2-VIA caused increased lipid peroxidation levels. Furthermore, administration of deuterated polyunsaturated fatty acids, which inhibit lipid peroxidation, was able to partially rescue the locomotor deficits in aged *iPLA2-VIA* mutant flies. Encouragingly, similar treatment of fibroblasts from patients with PLA2G6 mutations was able to restore mitochondrial membrane potential deficits seen in these cells (Kinghorn et al., 2015). These initial findings demonstrate that loss of PLA2G6 leads to lipid peroxidation and mitochondrial dysfunction, and again indicate another contributor to mitochondrial dysfunction in PD.

4.5 FBXO7

Mutations in *FBXO7* cause a severe form of autosomal recessive early-onset Parkinson's disease similar to that caused by mutations in *PINK1* or *parkin*. *FBXO7* encodes an F-box domain-containing protein, which often act as SCF-type (Skp1–Cul1–F-box) E3 ubiquitin ligase complexes. Recently, a combined in vivo/in vitro approach revealed that Fbxo7 acts in PINK1/Parkin mitochondrial quality control pathway (Burchell et al., 2013). In mammalian cells Fbxo7 was shown to directly bind Parkin and facilitates its translocation to mitochondria during mitophagy, whereas PD-causing mutations in Fbxo7 interfered with this. Moreover, ectopic expression of Fbxo7 in *Drosophila* potently rescued many *parkin* mutant phenotypes, while PD mutations were unable to. One surprising outcome of this study was the observation that, in contrast to *parkin* mutants, ectopic expression of Fbxo7 did not provide any rescue for *Pink1* mutants. It is hard to immediately reconcile these differential genetic interactions with the idea of a linear PINK1–Parkin pathway, although with time more detailed genetic analyses will likely resolve this. Nevertheless, the link between another familial PD gene and mitophagy further emphasizes the importance of this process in PD pathogenesis.

5. FUNCTIONS OF THE PINK1/PARKIN PATHWAY

5.1 Mitochondrial Dynamics

A recent wave of reports has begun to reveal new insights into the process in which PINK1 and Parkin normally function, and how their dysfunction may lead to neuronal death. These findings indicate an important function of the PINK1/Parkin pathway is to regulate mitochondrial dynamics, help segregate damaged or dysfunctional units, and promote their degradation by autophagy. This process likely acts as part of a quality control mechanism to recognize terminally damaged mitochondria and to safely degrade them to prevent increased production of reactive oxygen species (ROS) and potentially catastrophic rupture and release of proapoptotic factors.

Initial clues to how PINK1 and Parkin influence mitochondrial integrity in vivo converged on mitochondrial dynamics. Mitochondria are highly dynamic organelles that undergo frequent fission and fusion events controlled by evolutionarily conserved proteins (Chen & Chan, 2009). These include the profission factors dynamin-related protein 1 (Drp1) and Fis1, and the profusion factors optic atrophy 1 (Opa1) and mitofusins (Mfn1/2). While mitochondrial morphology is grossly affected in *Pink1/parkin* mutants tissues (Clark et al., 2006; Greene et al., 2003; Park et al., 2006), *Pink1* and *parkin* showed strong genetic interaction with components of the mitochondrial fission/fusion machinery (Deng, Dodson, Huang, & Guo, 2008; Poole et al., 2008; Yang et al., 2008). Genetic manipulations that promote mitochondrial fission, by either overexpression of *Drp1* or reduction of *Opa1* and *Marf* (homologous to Mfn1/2), significantly rescued many of the *Pink1/parkin* mutant defects including flight and climbing ability, muscle integrity, and mitochondrial morphology (Deng et al., 2008; Poole et al., 2008; Yang et al., 2008). These findings indicate that the PINK1/Parkin pathway promotes mitochondrial fission and/or inhibits mitochondrial fusion. This hypothesis gained weight with the subsequent identification that Parkin promotes the ubiquitylation of Marf/Mitofusins, first in flies (Ziviani, Tao, & Whitworth, 2010) and subsequently validated in mammals (Tanaka et al., 2010).

The central role of mitochondrial dynamics in *Pink1/parkin* mechanisms is further highlighted by evidence that overexpression of the mitochondrial SUMO/ubiquitin E3 ligase MUL1 can act in a parallel pathway to promote mitochondria fission and suppress many of the *Pink1/parkin* mutant phenotypes (Yun et al., 2014). Although the interaction was proposed to occur by MUL1-regulating Marf levels via ubiquitylation, this may also occur via MUL1's known regulation of Drp1 (Braschi, Zunino, & McBride, 2009; Karbowski, Neutzner, & Youle, 2007).

In addition to modulating mitochondrial fission–fusion dynamics, PINK1 and Parkin have been shown to affect mitochondrial axonal transport. First, PINK1 was found to physically interact with the Miro/Milton complex (Weihofen, Thomas, Ostaszewski, Cookson, & Selkoe, 2009). Miro is a mitochondrial Rho GTPase that binds the Kinesin-adapter protein Milton, linking mitochondria to Kinesin motors for microtubule-mediated transport in neurons (Schwarz, 2013). Recent work has shown that mitochondrial PINK1 can phosphorylate Miro on several sites which promotes its ubiquitylation by Parkin (Liu et al., 2012; Wang et al., 2011). This leads to the proteasome-dependent degradation of Miro, release of Kinesin and Milton from the mitochondrial surface, and the consequent arrest of mitochondrial motility in neurons (Wang et al., 2011). Similar to the inhibition of fusion, this process is proposed to help in the removal of malfunctioning mitochondria (see later). One attractive model is that the degradation of Miro and Mitofusin are both important as an early step to quarantine-damaged mitochondria and prevent them from fusing with other healthy mitochondria, thereby mixing their damaged components in with the healthy population. Moving mitochondria are much more likely to fuse than stationary mitochondria and therefore, together with the degradation of Mitofusin, the loss of Miro will produce a population of stationary, fragmented mitochondria that are ready for engulfment by autophagosomes. Supporting the hypothesis that this mechanism promotes neuronal survival, recent work in *Drosophila* has shown that expression of Miro that lacks the PINK1 phosphorylation sites results in loss of DA neurons (Tsai et al., 2014).

5.2 Mitophagy

Concurrent to these genetic interaction studies, several important studies revealed that the mitochondrial fission–fusion cycle acts to regulate the sequestration and destruction of damaged organelles via autophagy (Twig et al., 2008) and that this is dependent on PINK1 and Parkin (Narendra et al., 2010; Narendra, Tanaka, Suen, & Youle, 2008). This process, known as mitophagy, has become a dominant model in the field and a wealth of studies have revealed many of the molecular details (Pickrell & Youle, 2015). Mitophagy offers an explanation for the variety of mitochondrial defects that have been documented in PINK1 and Parkin-deficient cell models, including decreased membrane potential, deficits in the electron transport chain complexes, reduced ATP synthesis, decreased mitochondrial DNA synthesis, and aberrant mitochondrial calcium efflux (Gandhi et al., 2009; Gegg, Cooper, Schapira, & Taanman, 2009). These pleiotropic phenotypes could derive from the accumulation of damaged mitochondria in the absence of a functional mitochondrial quality control system. Selective mitophagy would also explain the protective effects of PINK1 and Parkin overexpression from exposure to mitochondrial toxins (Haque et al., 2008; Paterna, Leng, Weber, Feldon, & Bueler, 2007; Rosen et al., 2006). The abundant mitochondrial DNA mutational load of substantia nigra neurons (Bender et al., 2006) would also help account for the selective vulnerability of this population of cells to the loss of a mitochondrial quality control system.

While there is considerable debate surrounding the nature of the wholesale degradation of entire organelles or indeed the physiological stimulus, another mechanism has been proposed that offers a more piecemeal process of mitochondrial quality control—the formation of cargo-selective mitochondria-derived vesicles (MDVs) (Sugiura, McLelland, Fon, & McBride, 2014). The precise function of these MDVs is currently uncertain but evidence implicates them in the selective transport of oxidized mitochondrial components for degradation in lysosomes. Importantly, three factors linked to PD have recently been implicated in the formation and trafficking of MDVs: PINK1, Parkin, and Vps35 (McLelland, Soubannier, Chen, McBride, & Fon, 2014). Genetic studies in *Drosophila* have revealed a striking genetic interaction between *vps35* and *parkin*, providing strong support for a common pathway. Double heterozygotes confer age-related locomotor deficits, loss of DA neurons, and shortened life span, whereas Vps35 overexpression was able to rescue several *parkin* mutant phenotypes (Malik et al., 2015). Surprisingly though, *vps35* did not appear to genetically interact with *Pink1*. Ultimately, these genetic interactions support the idea of a common role of these components in the pathogenesis of PD, likely via some derivation of a mitochondrial quality control process. Given the previous links between PINK1/Parkin and mitochondrial dynamics, it is interesting to note that *VPS35* mutations have recently been shown to cause mitochondrial dysfunction by affecting the fission–fusion machinery (Tang et al., 2015; Wang et al., 2016).

A key issue for the field is to address whether mitophagy and/or MDV turnover occurs under truly physiological conditions. Compelling evidence has again emerged from *Drosophila* models using an innovative quantitative mass spectrometry approach to monitor mitochondrial protein turnover (Vincow et al., 2013). Working on the premise that under physiologically constant conditions mitochondrial components would be degraded and replaced at a constant rate to maintain steady-state levels, Pallanck and colleagues devised a method to monitor the accumulation of heavy isotope-labeled neuronal mitochondrial proteins in *Pink1* and *parkin* mutants compared to wild-type flies. In *Pink1* and *parkin* mutants the half-life of many mitochondrial proteins was significantly increased, consistent with a reduction in turnover rates and similar to autophagy-defective mutants. Moreover, this phenomenon was not random but showed an intriguing selectivity for respiratory chain components, and in particular for membrane-bound subunits. These observations offer the best evidence to date that Pink1/Parkin regulate the turnover of mitochondrial components under physiological conditions, although much still needs to be explored here.

5.3 Complex I

An alternative function for PINK1 has also been proposed that appears to be independent of parkin. For some time, a variety of functional deficits have been noted in multiple models of *PINK1* deficiency. Given the long-standing association of complex I (CI) deficiency with PD, it was notable that CI deficits were frequently observed upon loss of PINK1 (Morais et al., 2009). The relative contribution of CI deficits to *Drosophila Pink1* phenotypes was tested directly with transgenic expression of yeast *NDI1* which bypasses CI by transferring electrons directly from NADH to ubiquinone (Vilain et al., 2012). Expression of NDI1 was able to rescue several phenotypes of *Pink1* mutants, such as climbing ability, male sterility, and mitochondrial morphological and functional abnormalities. Interestingly other phenotypes were not rescued, including flight ability and flight muscle integrity indicating that pathology is not caused by CI deficiency in these tissues. Nevertheless, these data support the idea that CI deficits may underlie at least some of the *Pink1* phenotypes. In contrast, NDI1 expression was unable to rescue any phenotypes of *parkin* mutants (Vilain et al., 2012).

These findings gave rise to hypothesis that this function of PINK1 is separate from its role with Parkin.

To elucidate the molecular mechanism, phosphoproteomics revealed a particular CI subunit, NDUFA10 (called ND42 in flies), lacked phosphorylation at Ser-250 in *Pink1*$^{-/-}$ cells (Morais et al., 2014). Although this phosphorylation appears to not occur directly via PINK1, compelling evidence supports this modification in the regulation of CI activity. Expression of a phosphomimetic version of ND42/NDUFA10 specifically rescued phenotypes in multiple *PINK1*-deficient systems, while an S250A mutant version of ND42/NDUFA10 that is incapable of being phosphorylated was unable to confer rescue. However, this specific pattern was not observed in another report (Pogson et al., 2014).

While further studies are needed to clarify the functional relationship between PINK1 and NDUFA10 in the regulation of CI, there is growing evidence that many manipulations that can rescue *Pink1* mutants also promote CI activity. In addition to *NDUFA10/ND42* and *NDI1* described earlier these manipulations include overexpression of the *NDUFA10/ND42* chaperone *sicily* (Pogson et al., 2014), the mitochondrial Hsp90-like chaperone *TRAP1* (Costa, Loh, & Martins, 2013; Zhang et al., 2013), the receptor tyrosine kinase *Ret* (Klein et al., 2014), the mitochondrial fission factor *drp1* (Liu et al., 2011), deoxyribonucleoside kinase *dNK* (Tufi et al., 2014), or dietary treatment with vitamin K_2 (Vos et al., 2012). The diverse functions of these modifiers would suggest that a more general defect underlies the CI deficiency observed in *PINK1*-deficient models. For example, reduced CI activity in *Pink1* mutants may be due to a general destabilization of CI. Assembly is a particular challenge for such a large, multisubunit complex and occurs in a stepwise process that is highly regulated by many factors (Lazarou, Thorburn, Ryan, & McKenzie, 2009). Even its association with other electron transport chain complexes in supercomplexes affects CI's stability (Acin-Perez et al., 2004). Indeed, there is evidence for reduced complex stability in *Pink1* mutants, though this may not be specific to CI (Amo, Saiki, Sawayama, Sato, & Hattori, 2014; Liu et al., 2011; Tufi et al., 2014). One possibility is that PINK1 influences CI stability by directly promoting the assembly of CI, which may be regulated by (phosphorylated) NDUFA10.

Although the specific benefit of restoring or bypassing CI activity to *Pink1* mutants argues for a Parkin-independent function of PINK1, it was surprising that overexpression of *parkin* did not rescue the CI deficiency in *Pink1* mutants given that Parkin overexpression rescues all other *Pink1* phenotypes (Pogson et al., 2014). This suggests that CI deficiency alone cannot fully account for those *Pink1* phenotypes and questions whether CI deficiency is an important mediator of pathogenesis. Interestingly, a recent study showed that expression of yeast NDI1 generates a "protective" ROS signal via overreduction of the CoQ pool and reverse electron transfer through CI (Scialo et al., 2016). It will be interesting to determine whether other manipulations attributed to restoring CI activity also confer neuroprotection via this mechanism. Clearly further studies in animal models are needed to clarify the full spectrum of cellular defects in *Pink1* and *parkin* mutants, their related or independent nature and their relative importance to pathologic mechanisms.

6. CONVERGENT THERAPEUTIC APPROACHES

A vast body of evidence links many neurodegenerative diseases, directly or indirectly, with oxidative stress (Sayre, Perry, & Smith, 2008). An array of endogenous processes act to maintain the balance in redox signaling and antioxidant protection; however, the effects of compromised antioxidant defenses are most obviously threatening to highly energy-demanding, postmitotic tissues such as the nervous system. All of the *Drosophila* models of PD to date have exhibited some sensitivity to oxidative stress. Consequently, numerous transgenic or pharmacologic manipulations that boost antioxidant mechanisms have proven effective in these models.

Transgenic expression of enzymes that catalytically eliminate oxidizing species, such as Cu/Zn superoxide dismutase or peroxidredoxin-3, and others that more indirectly combat oxidative damage, such as glutathione *S*-transferases (GstS1, GstO2), methionine sulfoxide reductase, metal chelating, and metal responsive factors, have all suppressed phenotypes in various PD models (Angeles et al., 2014; Kim & Yim, 2013; Munoz-Soriano & Paricio, 2011). Moreover, the therapeutic potential for small molecule antioxidants has been explored including glutathione, *s*-methyl L-cysteine, *N*-acetylcysteine, ascorbic acid, polyphenols, allyl disulfide and sulforaphane, and dietary zinc; all had positive effects in several different models (Casani, Gomez-Pastor, Matallana, & Paricio, 2013; Filograna et al., 2016; Munoz-Soriano & Paricio, 2011). The master regulators of the antioxidant response, the Keap1–Nrf2 pathway, which controls expression of many of the factors mentioned earlier (O'Connell & Hayes, 2015), have also been shown to be neuroprotective by genetic or chemical activation

(Barone, Sykiotis, & Bohmann, 2011; Trinh et al., 2010; Wang, Liu, Shan, Xia, & Liu, 2015).

While all of these findings support oxidative stress as a contributing factor and an attractive therapeutic target, the normal metabolism of dopamine itself is known to produce ROS and harmful dopaquinone conjugates (Hattori et al., 2009). A number of groups have utilized the tractability of the *Drosophila* models to address the contribution of dopamine to selective neuronal vulnerability. It is hypothesized that promoting vesicular packaging or reducing the cytosolic dopamine will reduce the potential toxicity. Consistent with this, depletion of cytosolic dopamine by knockdown of the rate-limiting enzyme *tyrosine hydroxylase* (*pale* in *Drosophila*) or overexpression of the *vesicular monoamine transporter*, to accelerate dopamine packaging was protective in *α-synuclein* or *parkin* models (Park, Schulz, & Lee, 2007; Sang et al., 2007). Furthermore, these manipulations were also protective against rotenone toxicity (Bayersdorfer, Voigt, Schneuwly, & Botella, 2010; Lawal et al., 2010) supporting the potential impact of dopamine metabolism in sporadic PD.

Beyond oxidative stress, but converging with mechanisms relating to mitophagy, another major therapeutic target for numerous neurodegenerative diseases is the target of rapamycin (mTOR) signaling pathway; a master regulator of cellular responses to nutrient availability, sensing inputs from multiple signals including insulin receptors and AMPK among others (Laplante & Sabatini, 2012). Two of the major processes regulated by mTOR signaling are autophagy and protein translation, hence, the mTOR signaling pathway represents an attractive node for therapeutic targeting, especially given that there are highly effective chemical inhibitors. In addition to the link between LRRK2 and protein translation, *Pink1* and *parkin* mutant phenotypes are also alleviated by downregulation of protein translation via mTOR targets 4E-BP or S6K (Liu & Lu, 2010; Tain et al., 2009). Similarly, besides the impact of PINK1/parkin mutations on mitophagy, stimulating autophagy is broadly considered to help clear aggregation-prone proteins such as α-synuclein (Buttner et al., 2014) but may also be beneficial for LRRK2 mutations (Dodson et al., 2014).

As we learn more about the pathogenic mechanism(s) underlying the disease more putative therapeutic interventions will come to light. Current evidence indicates that there is no single pathogenic insult, decreasing the likelihood of a single "magic bullet" disease-modifying therapy. More likely any interventions will need to consider multiple toxic events. The findings discussed here have already highlighted a few key regulators of important protective mechanisms. The availability of relevant *Drosophila* models for PD provides an excellent opportunity for early stage drug testing to reduce the cost and risk associated with developing such a therapeutic, prior to further testing in mammalian preclinical models.

7. CONCLUDING REMARKS

Given the remarkable degree of genetic, molecular, and cell biological conservation between flies and mammals, *Drosophila* remains a valid model system in which to address novel biological questions in the future, including those relevant to human health. Work aimed at an understanding of the genes involved in heritable forms of PD has already made significant contributions to our understanding of the causes of this debilitating disease. Moreover, these studies have begun to define small molecule compounds that could potentially impinge on the pathways implicated in PD. *Drosophila* models are also amenable to medium-throughput screening of small molecule libraries in the search for better compounds that are able to combat pathogenesis.

While the *Drosophila* system has enormous potential for studies aimed at a mechanistic understanding of PD, it is also imperative that we have realistic expectations from these models. For example, some of the phenotypes presented by the models do not precisely mirror the phenotypes of humans with mutations in the corresponding genes. *Drosophila* are still a relatively simple model organism, far less complex than humans; hence, it is understandable that some aspects that may be relevant to our understanding of a particular human disease may not be evident in *Drosophila*. However, to date we have only just begun to tap the insights that modeling PD in *Drosophila* can potentially provide. Perhaps the greatest contribution that these models can give to the field is in identifying the basic aspects of the underlying disease mechanisms. As knowledge continues to grow in understanding the genetic and environmental causes of the disease, *Drosophila* models will be well placed to provide answers.

ACKNOWLEDGMENTS

We apologize to those authors whose work we have not cited due to space constraints. Work in the Whitworth lab is funded by MRC core funding (MC-A070-5PSB0), and an ERC Starting Grant (no. 309742). V.L.H. is funded by an EMBO Long-Term Fellowship.

REFERENCES

Acin-Perez, R., Bayona-Bafaluy, M. P., Fernandez-Silva, P., Moreno-Loshuertos, R., Perez-Martos, A., Bruno, C., et al. (2004). Respiratory complex III is required to maintain complex I in mammalian mitochondria. *Molecular Cell, 13*, 805–815.

Amo, T., Saiki, S., Sawayama, T., Sato, S., & Hattori, N. (2014). Detailed analysis of mitochondrial respiratory chain defects caused by loss of PINK1. *Neuroscience Letters, 580*, 37–40.

Angeles, D. C., Ho, P., Chua, L. L., Wang, C., Yap, Y. W., Ng, C., et al. (2014). Thiol peroxidases ameliorate LRRK2 mutant-induced mitochondrial and dopaminergic neuronal degeneration in Drosophila. *Human Molecular Genetics, 23*, 3157–3165.

Auluck, P. K., & Bonini, N. M. (2002). Pharmacological prevention of Parkinson disease in Drosophila. *Nature Medicine, 8*, 1185–1186.

Auluck, P. K., Chan, H. Y., Trojanowski, J. Q., Lee, V. M., & Bonini, N. M. (2002). Chaperone suppression of alpha-synuclein toxicity in a Drosophila model for Parkinson's disease. *Science (New York, N.Y.), 295*, 865–868.

Barone, M. C., Sykiotis, G. P., & Bohmann, D. (2011). Genetic activation of Nrf2 signaling is sufficient to ameliorate neurodegenerative phenotypes in a Drosophila model of Parkinson's disease. *Disease Models & Mechanisms, 4*, 701–707.

Bayersdorfer, F., Voigt, A., Schneuwly, S., & Botella, J. A. (2010). Dopamine-dependent neurodegeneration in Drosophila models of familial and sporadic Parkinson's disease. *Neurobiology of Disease, 40*, 113–119.

Bender, A., Krishnan, K. J., Morris, C. M., Taylor, G. A., Reeve, A. K., Perry, R. H., et al. (2006). High levels of mitochondrial DNA deletions in substantia nigra neurons in aging and Parkinson disease. *Nature Genetics, 38*, 515–517.

Blesa, J., & Przedborski, S. (2014). Parkinson's disease: Animal models and dopaminergic cell vulnerability. *Frontiers in Neuroanatomy, 8*, 155.

Braschi, E., Zunino, R., & McBride, H. M. (2009). MAPL is a new mitochondrial SUMO E3 ligase that regulates mitochondrial fission. *EMBO Reports, 10*, 748–754.

Breda, C., Nugent, M. L., Estranero, J. G., Kyriacou, C. P., Outeiro, T. F., Steinert, J. R., et al. (2015). Rab11 modulates alpha-synuclein-mediated defects in synaptic transmission and behaviour. *Human Molecular Genetics, 24*, 1077–1091.

Burchell, V. S., Nelson, D. E., Sanchez-Martinez, A., Delgado-Camprubi, M., Ivatt, R. M., Pogson, J. H., et al. (2013). The Parkinson's disease-linked proteins Fbxo7 and Parkin interact to mediate mitophagy. *Nature Neuroscience, 16*, 1257–1265.

Buttner, S., Broeskamp, F., Sommer, C., Markaki, M., Habernig, L., Alavian-Ghavanini, A., et al. (2014). Spermidine protects against alpha-synuclein neurotoxicity. *Cell Cycle, 13*, 3903–3908.

Caesar, M., Zach, S., Carlson, C. B., Brockmann, K., Gasser, T., & Gillardon, F. (2013). Leucine-rich repeat kinase 2 functionally interacts with microtubules and kinase-dependently modulates cell migration. *Neurobiology of Disease, 54*, 280–288.

Casani, S., Gomez-Pastor, R., Matallana, E., & Paricio, N. (2013). Antioxidant compound supplementation prevents oxidative damage in a Drosophila model of Parkinson's disease. *Free Radical Biology & Medicine, 61*, 151–160.

Chen, H., & Chan, D. C. (2009). Mitochondrial dynamics—Fusion, fission, movement, and mitophagy—In neurodegenerative diseases. *Human Molecular Genetics, 18*, R169–R176.

Chen, L., & Feany, M. B. (2005). Alpha-synuclein phosphorylation controls neurotoxicity and inclusion formation in a Drosophila model of Parkinson disease. *Nature Neuroscience, 8*, 657–663.

Chen, L., Periquet, M., Wang, X., Negro, A., McLean, P. J., Hyman, B. T., et al. (2009). Tyrosine and serine phosphorylation of alpha-synuclein have opposing effects on neurotoxicity and soluble oligomer formation. *The Journal of Clinical Investigation, 119*, 3257–3265.

Chuang, C. L., Lu, Y. N., Wang, H. C., & Chang, H. Y. (2014). Genetic dissection reveals that Akt is the critical kinase downstream of LRRK2 to phosphorylate and inhibit FOXO1, and promotes neuron survival. *Human Molecular Genetics*, *23*, 5649–5658.

Clark, I. E., Dodson, M. W., Jiang, C., Cao, J. H., Huh, J. R., Seol, J. H., et al. (2006). Drosophila pink1 is required for mitochondrial function and interacts genetically with parkin. *Nature*, *441*, 1162–1166.

Cooper, A. A., Gitler, A. D., Cashikar, A., Haynes, C. M., Hill, K. J., Bhullar, B., et al. (2006). Alpha-synuclein blocks ER-golgi traffic and Rab1 rescues neuron loss in Parkinson's models. *Science (New York, N.Y.)*, *313*, 324–328.

Costa, A. C., Loh, S. H., & Martins, L. M. (2013). Drosophila Trap1 protects against mitochondrial dysfunction in a PINK1/parkin model of Parkinson's disease. *Cell Death & Disease*, *4*, e467.

Cremades, N., Cohen, S. I., Deas, E., Abramov, A. Y., Chen, A. Y., Orte, A., et al. (2012). Direct observation of the interconversion of normal and toxic forms of alpha-synuclein. *Cell*, *149*, 1048–1059.

Davis, M. Y., Trinh, K., Thomas, R. E., Yu, S., Germanos, A. A., Whitley, B. N., et al. (2016). Glucocerebrosidase deficiency in Drosophila results in alpha-synuclein-independent protein aggregation and neurodegeneration. *PLoS Genetics*, *12*, e1005944.

Deng, H., Dodson, M. W., Huang, H., & Guo, M. (2008). The Parkinson's disease genes pink1 and parkin promote mitochondrial fission and/or inhibit fusion in Drosophila. *Proceedings of the National Academy of Sciences of the United States of America*, *105*, 14503–14508.

Dhungel, N., Eleuteri, S., Li, L. B., Kramer, N. J., Chartron, J. W., Spencer, B., et al. (2015). Parkinson's disease genes VPS35 and EIF4G1 interact genetically and converge on alpha-synuclein. *Neuron*, *85*, 76–87.

Dodson, M. W., Leung, L. K., Lone, M., Lizzio, M. A., & Guo, M. (2014). Novel ethyl methanesulfonate (EMS)-induced null alleles of the Drosophila homolog of LRRK2 reveal a crucial role in endolysosomal functions and autophagy in vivo. *Disease Models & Mechanisms*, 7, 1351–1363.

Dodson, M. W., Zhang, T., Jiang, C., Chen, S., & Guo, M. (2012). Roles of the Drosophila LRRK2 homolog in Rab7-dependent lysosomal positioning. *Human Molecular Genetics*, *21*, 1350–1363.

Esteves, A. R., Swerdlow, R. H., & Cardoso, S. M. (2014). LRRK2, a puzzling protein: Insights into Parkinson's disease pathogenesis. *Experimental Neurology*, *261*, 206–216.

Filograna, R., Godena, V. K., Sanchez-Martinez, A., Ferrari, E., Casella, L., Beltramini, M., et al. (2016). Superoxide dismutase (SOD)-mimetic M40403 is protective in cell and fly models of paraquat toxicity: Implications for Parkinson Disease. *The Journal of Biological Chemistry*, *291*, 9257–9267.

Gandhi, S., Wood-Kaczmar, A., Yao, Z., Plun-Favreau, H., Deas, E., Klupsch, K., et al. (2009). PINK1-associated Parkinson's disease is caused by neuronal vulnerability to calcium-induced cell death. *Molecular Cell*, *33*, 627–638.

Gegg, M. E., Cooper, J. M., Schapira, A. H., & Taanman, J. W. (2009). Silencing of PINK1 expression affects mitochondrial DNA and oxidative phosphorylation in dopaminergic cells. *PloS One*, *4*, e4756.

Gehrke, S., Imai, Y., Sokol, N., & Lu, B. (2010). Pathogenic LRRK2 negatively regulates microRNA-mediated translational repression. *Nature*, *466*, 637–641.

Godena, V. K., Brookes-Hocking, N., Moller, A., Shaw, G., Oswald, M., Sancho, R. M., et al. (2014). Increasing microtubule acetylation rescues axonal transport and locomotor deficits caused by LRRK2 Roc-COR domain mutations. *Nature Communications*, *5*, 5245.

Greene, J. C., Whitworth, A. J., Kuo, I., Andrews, L. A., Feany, M. B., & Pallanck, L. J. (2003). Mitochondrial pathology and apoptotic muscle degeneration in Drosophila

parkin mutants. *Proceedings of the National Academy of Sciences of the United States of America, 100*, 4078–4083.
Gupta, A., Dawson, V. L., & Dawson, T. M. (2008). What causes cell death in Parkinson's disease? *Annals of Neurology, 64*(Suppl. 2), S3–S15.
Haque, M. E., Thomas, K. J., D'Souza, C., Callaghan, S., Kitada, T., Slack, R. S., et al. (2008). Cytoplasmic Pink1 activity protects neurons from dopaminergic neurotoxin MPTP. *Proceedings of the National Academy of Sciences of the United States of America, 105*, 1716–1721.
Hattori, N., Wanga, M., Taka, H., Fujimura, T., Yoritaka, A., Kubo, S., et al. (2009). Toxic effects of dopamine metabolism in Parkinson's disease. *Parkinsonism & Related Disorders, 15*(Suppl. 1), S35–S38.
Hernandez, D. G., Reed, X., & Singleton, A. B. (2016). Genetics in Parkinson disease: Mendelian versus non-Mendelian inheritance. *Journal of Neurochemistry*. http://dx.doi.org/10.1111/jnc.13593. [Epub ahead of print].
Hindle, S., Afsari, F., Stark, M., Middleton, C. A., Evans, G. J., Sweeney, S. T., et al. (2013). Dopaminergic expression of the Parkinsonian gene LRRK2-G2019S leads to non-autonomous visual neurodegeneration, accelerated by increased neural demands for energy. *Human Molecular Genetics, 22*, 2129–2140.
Imai, Y., Gehrke, S., Wang, H. Q., Takahashi, R., Hasegawa, K., Oota, E., et al. (2008). Phosphorylation of 4E-BP by LRRK2 affects the maintenance of dopaminergic neurons in Drosophila. *The EMBO Journal, 27*, 2432–2443.
Imai, Y., Kobayashi, Y., Inoshita, T., Meng, H., Arano, T., Uemura, K., et al. (2015). The Parkinson's disease-associated protein kinase LRRK2 modulates notch signaling through the endosomal pathway. *PLoS Genetics, 11*, e1005503.
Karbowski, M., Neutzner, A., & Youle, R. J. (2007). The mitochondrial E3 ubiquitin ligase MARCH5 is required for Drp1 dependent mitochondrial division. *The Journal of Cell Biology, 178*, 71–84.
Karpinar, D. P., Balija, M. B., Kugler, S., Opazo, F., Rezaei-Ghaleh, N., Wender, N., et al. (2009). Pre-fibrillar alpha-synuclein variants with impaired beta-structure increase neurotoxicity in Parkinson's disease models. *The EMBO Journal, 28*, 3256–3268.
Kim, K., & Yim, J. (2013). Glutathione S-transferase omega suppresses the defective phenotypes caused by PINK1 loss-of-function in Drosophila. *Biochemical and Biophysical Research Communications, 437*, 615–619.
Kinghorn, K. J., Castillo-Quan, J. I., Bartolome, F., Angelova, P. R., Li, L., Pope, S., et al. (2015). Loss of PLA2G6 leads to elevated mitochondrial lipid peroxidation and mitochondrial dysfunction. *Brain, 138*, 1801–1816.
Klein, P., Muller-Rischart, A. K., Motori, E., Schonbauer, C., Schnorrer, F., Winklhofer, K. F., et al. (2014). Ret rescues mitochondrial morphology and muscle degeneration of Drosophila Pink1 mutants. *The EMBO Journal, 33*, 341–355.
Kumar, A., Greggio, E., Beilina, A., Kaganovich, A., Chan, D., Taymans, J. M., et al. (2010). The Parkinson's disease associated LRRK2 exhibits weaker in vitro phosphorylation of 4E-BP compared to autophosphorylation. *PloS One, 5*, e8730.
Lai, Y. C., Kondapalli, C., Lehneck, R., Procter, J. B., Dill, B. D., Woodroof, H. I., et al. (2015). Phosphoproteomic screening identifies Rab GTPases as novel downstream targets of PINK1. *The EMBO Journal, 34*, 2840–2861.
Laplante, M., & Sabatini, D. M. (2012). mTOR signaling in growth control and disease. *Cell, 149*, 274–293.
Lavara-Culebras, E., Munoz-Soriano, V., Gomez-Pastor, R., Matallana, E., & Paricio, N. (2010). Effects of pharmacological agents on the lifespan phenotype of Drosophila DJ-1beta mutants. *Gene, 462*, 26–33.

Law, B. M., Spain, V. A., Leinster, V. H., Chia, R., Beilina, A., Cho, H. J., et al. (2014). A direct interaction between leucine-rich repeat kinase 2 and specific beta-tubulin isoforms regulates tubulin acetylation. *The Journal of Biological Chemistry*, *289*, 895–908.

Lawal, H. O., Chang, H. Y., Terrell, A. N., Brooks, E. S., Pulido, D., Simon, A. F., et al. (2010). The Drosophila vesicular monoamine transporter reduces pesticide-induced loss of dopaminergic neurons. *Neurobiology of Disease*, *40*, 102–112.

Lazarou, M., Thorburn, D. R., Ryan, M. T., & McKenzie, M. (2009). Assembly of mitochondrial complex I and defects in disease. *Biochimica et Biophysica Acta*, *1793*, 78–88.

Lee, S. B., Kim, W., Lee, S., & Chung, J. (2007). Loss of LRRK2/PARK8 induces degeneration of dopaminergic neurons in Drosophila. *Biochemical and Biophysical Research Communications*, *358*, 534–539.

Lee, F. K., Wong, A. K., Lee, Y. W., Wan, O. W., Chan, H. Y., & Chung, K. K. (2009). The role of ubiquitin linkages on alpha-synuclein induced-toxicity in a Drosophila model of Parkinson's disease. *Journal of Neurochemistry*, *110*, 208–219.

Linhart, R., Wong, S. A., Cao, J., Tran, M., Huynh, A., Ardrey, C., et al. (2014). Vacuolar protein sorting 35 (Vps35) rescues locomotor deficits and shortened lifespan in Drosophila expressing a Parkinson's disease mutant of leucine-rich repeat kinase 2 (LRRK2). *Molecular Neurodegeneration*, *9*, 23.

Liu, W., Acin-Perez, R., Geghman, K. D., Manfredi, G., Lu, B., & Li, C. (2011). Pink1 regulates the oxidative phosphorylation machinery via mitochondrial fission. *Proceedings of the National Academy of Sciences of the United States of America*, *108*, 12920–12924.

Liu, S., & Lu, B. (2010). Reduction of protein translation and activation of autophagy protect against PINK1 pathogenesis in Drosophila melanogaster. *PLoS Genetics*, *6*, e1001237.

Liu, S., Sawada, T., Lee, S., Yu, W., Silverio, G., Alapatt, P., et al. (2012). Parkinson's disease-associated kinase PINK1 regulates Miro protein level and axonal transport of mitochondria. *PLoS Genetics*, *8*, e1002537.

MacLeod, D. A., Rhinn, H., Kuwahara, T., Zolin, A., Di Paolo, G., McCabe, B. D., et al. (2013). RAB7L1 interacts with LRRK2 to modify intraneuronal protein sorting and Parkinson's disease risk. *Neuron*, *77*, 425–439.

Malhotra, A., Edelman-Novemsky, I., Xu, Y., Plesken, H., Ma, J., Schlame, M., et al. (2009). Role of calcium-independent phospholipase A2 in the pathogenesis of Barth syndrome. *Proceedings of the National Academy of Sciences of the United States of America*, *106*, 2337–2341.

Malik, B. R., Godena, V. K., & Whitworth, A. J. (2015). VPS35 pathogenic mutations confer no dominant toxicity but partial loss of function in Drosophila and genetically interact with parkin. *Human Molecular Genetics*, *24*, 6106–6117.

Maor, G., Cabasso, O., Krivoruk, O., Rodriguez, J., Steller, H., Segal, D., et al. (2016). The contribution of mutant GBA to the development of Parkinson disease in Drosophila. *Human Molecular Genetics*. pii: ddw129. [Epub ahead of print].

Maor, G., Rencus-Lazar, S., Filocamo, M., Steller, H., Segal, D., & Horowitz, M. (2013). Unfolded protein response in Gaucher disease: From human to Drosophila. *Orphanet Journal of Rare Diseases*, *8*, 140.

Martin, I., Kim, J. W., Lee, B. D., Kang, H. C., Xu, J. C., Jia, H., et al. (2014). Ribosomal protein s15 phosphorylation mediates LRRK2 neurodegeneration in Parkinson's disease. *Cell*, *157*, 472–485.

Matta, S., Van Kolen, K., da Cunha, R., van den Bogaart, G., Mandemakers, W., Miskiewicz, K., et al. (2012). LRRK2 controls an EndoA phosphorylation cycle in synaptic endocytosis. *Neuron*, *75*, 1008–1021.

McGough, I. J., Steinberg, F., Jia, D., Barbuti, P. A., McMillan, K. J., Heesom, K. J., et al. (2014). Retromer binding to FAM21 and the WASH complex is perturbed by the Parkinson disease-linked VPS35(D620N) mutation. *Current Biology*, *24*, 1670–1676.

McLelland, G. L., Soubannier, V., Chen, C. X., McBride, H. M., & Fon, E. A. (2014). Parkin and PINK1 function in a vesicular trafficking pathway regulating mitochondrial quality control. *The EMBO Journal*, *33*, 282–295.

Meulener, M., Whitworth, A. J., Armstrong-Gold, C. E., Rizzu, P., Heutink, P., Wes, P. D., et al. (2005). Drosophila DJ-1 mutants are selectively sensitive to environmental toxins associated with Parkinson's disease. *Current Biology*, *15*, 1572–1577.

Meulener, M. C., Xu, K., Thompson, L., Ischiropoulos, H., & Bonini, N. M. (2006). Mutational analysis of DJ-1 in Drosophila implicates functional inactivation by oxidative damage and aging. *Proceedings of the National Academy of Sciences of the United States of America*, *103*, 12517–12522.

Miura, E., Hasegawa, T., Konno, M., Suzuki, M., Sugeno, N., Fujikake, N., et al. (2014). VPS35 dysfunction impairs lysosomal degradation of alpha-synuclein and exacerbates neurotoxicity in a Drosophila model of Parkinson's disease. *Neurobiology of Disease*, *71*, 1–13.

Morais, V. A., Haddad, D., Craessaerts, K., De Bock, P. J., Swerts, J., Vilain, S., et al. (2014). PINK1 loss-of-function mutations affect mitochondrial complex I activity via NdufA10 ubiquinone uncoupling. *Science (New York, N.Y.)*, *344*, 203–207.

Morais, V. A., Verstreken, P., Roethig, A., Smet, J., Snellinx, A., Vanbrabant, M., et al. (2009). Parkinson's disease mutations in PINK1 result in decreased complex I activity and deficient synaptic function. *EMBO Molecular Medicine*, *1*, 99–111.

Munoz-Soriano, V., & Paricio, N. (2011). Drosophila models of Parkinson's disease: Discovering relevant pathways and novel therapeutic strategies. *Parkinson's Disease*, *2011*, 520640.

Narendra, D. P., Jin, S. M., Tanaka, A., Suen, D. F., Gautier, C. A., Shen, J., et al. (2010). PINK1 is selectively stabilized on impaired mitochondria to activate Parkin. *PLoS Biology*, *8*, e1000298.

Narendra, D., Tanaka, A., Suen, D. F., & Youle, R. J. (2008). Parkin is recruited selectively to impaired mitochondria and promotes their autophagy. *The Journal of Cell Biology*, *183*, 795–803.

Navarro, J. A., Hessner, S., Yenisetti, S. C., Bayersdorfer, F., Zhang, L., Voigt, A., et al. (2014). Analysis of dopaminergic neuronal dysfunction in genetic and toxin-induced models of Parkinson's disease in Drosophila. *Journal of Neurochemistry*, *131*, 369–382.

O'Connell, M. A., & Hayes, J. D. (2015). The Keap1/Nrf2 pathway in health and disease: From the bench to the clinic. *Biochemical Society Transactions*, *43*, 687–689.

Park, J., Kim, S. Y., Cha, G. H., Lee, S. B., Kim, S., & Chung, J. (2005). Drosophila DJ-1 mutants show oxidative stress-sensitive locomotive dysfunction. *Gene*, *361*, 113–119.

Park, J., Lee, S. B., Lee, S., Kim, Y., Song, S., Kim, S., et al. (2006). Mitochondrial dysfunction in Drosophila PINK1 mutants is complemented by parkin. *Nature*, *441*, 1157–1161.

Park, S. S., Schulz, E. M., & Lee, D. (2007). Disruption of dopamine homeostasis underlies selective neurodegeneration mediated by alpha-synuclein. *The European Journal of Neuroscience*, *26*, 3104–3112.

Paterna, J. C., Leng, A., Weber, E., Feldon, J., & Bueler, H. (2007). DJ-1 and Parkin modulate dopamine-dependent behavior and inhibit MPTP-induced nigral dopamine neuron loss in mice. *Molecular Therapy*, *15*, 698–704.

Pickrell, A. M., & Youle, R. J. (2015). The roles of PINK1, parkin, and mitochondrial fidelity in Parkinson's disease. *Neuron*, *85*, 257–273.

Pogson, J. H., Ivatt, R. M., Sanchez-Martinez, A., Tufi, R., Wilson, E., Mortiboys, H., et al. (2014). The complex I subunit NDUFA10 selectively rescues Drosophila pink1 mutants through a mechanism independent of mitophagy. *PLoS Genetics*, *10*, e1004815.

Poole, A. C., Thomas, R. E., Andrews, L. A., McBride, H. M., Whitworth, A. J., & Pallanck, L. J. (2008). The PINK1/Parkin pathway regulates mitochondrial morphology. *Proceedings of the National Academy of Sciences of the United States of America, 105*, 1638–1643.
Riemensperger, T., Isabel, G., Coulom, H., Neuser, K., Seugnet, L., Kume, K., et al. (2011). Behavioral consequences of dopamine deficiency in the Drosophila central nervous system. *Proceedings of the National Academy of Sciences of the United States of America, 108*, 834–839.
Rosen, K. M., Veereshwarayya, V., Moussa, C. E., Fu, Q., Goldberg, M. S., Schlossmacher, M. G., et al. (2006). Parkin protects against mitochondrial toxins and beta-amyloid accumulation in skeletal muscle cells. *The Journal of Biological Chemistry, 281*, 12809–12816.
Sang, T. K., Chang, H. Y., Lawless, G. M., Ratnaparkhi, A., Mee, L., Ackerson, L. C., et al. (2007). A Drosophila model of mutant human parkin-induced toxicity demonstrates selective loss of dopaminergic neurons and dependence on cellular dopamine. *The Journal of Neuroscience, 27*, 981–992.
Sayre, L. M., Perry, G., & Smith, M. A. (2008). Oxidative stress and neurotoxicity. *Chemical Research in Toxicology, 21*, 172–188.
Schwarz, T. L. (2013). Mitochondrial trafficking in neurons. *Cold Spring Harbor Perspectives in Biology, 5*(6), a011304.
Scialo, F., Sriram, A., Fernandez-Ayala, D., Gubina, N., Lohmus, M., Nelson, G., et al. (2016). Mitochondrial ROS produced via reverse electron transport extend animal lifespan. *Cell Metabolism, 23*, 725–734.
Seaman, M. N. (2012). The retromer complex—Endosomal protein recycling and beyond. *Journal of Cell Science, 125*, 4693–4702.
Steger, M., Tonelli, F., Ito, G., Davies, P., Trost, M., Vetter, M., et al. (2016). Phosphoproteomics reveals that Parkinson's disease kinase LRRK2 regulates a subset of Rab GTPases. *eLife, 5*, e12813.
Strausfeld, N. J., & Hirth, F. (2013). Deep homology of arthropod central complex and vertebrate basal ganglia. *Science (New York, N.Y.), 340*, 157–161.
Sugiura, A., McLelland, G. L., Fon, E. A., & McBride, H. M. (2014). A new pathway for mitochondrial quality control: Mitochondrial-derived vesicles. *The EMBO Journal, 33*, 2142–2156.
Suzuki, M., Fujikake, N., Takeuchi, T., Kohyama-Koganeya, A., Nakajima, K., Hirabayashi, Y., et al. (2015). Glucocerebrosidase deficiency accelerates the accumulation of proteinase K-resistant alpha-synuclein and aggravates neurodegeneration in a Drosophila model of Parkinson's disease. *Human Molecular Genetics, 24*, 6675–6686.
Suzuki, T., Shimoda, M., Ito, K., Hanai, S., Aizawa, H., Kato, T., et al. (2013). Expression of human Gaucher disease gene GBA generates neurodevelopmental defects and ER stress in Drosophila eye. *PloS One, 8*, e69147.
Tain, L. S., Mortiboys, H., Tao, R. N., Ziviani, E., Bandmann, O., & Whitworth, A. J. (2009). Rapamycin activation of 4E-BP prevents parkinsonian dopaminergic neuron loss. *Nature Neuroscience, 12*, 1129–1135.
Tanaka, A., Cleland, M. M., Xu, S., Narendra, D. P., Suen, D. F., Karbowski, M., et al. (2010). Proteasome and p97 mediate mitophagy and degradation of mitofusins induced by Parkin. *The Journal of Cell Biology, 191*, 1367–1380.
Tang, F. L., Liu, W., Hu, J. X., Erion, J. R., Ye, J., Mei, L., et al. (2015). VPS35 deficiency or mutation causes dopaminergic neuronal loss by impairing mitochondrial fusion and function. *Cell Reports, 12*, 1631–1643.
Trancikova, A., Mamais, A., Webber, P. J., Stafa, K., Tsika, E., Glauser, L., et al. (2012). Phosphorylation of 4E-BP1 in the mammalian brain is not altered by LRRK2 expression or pathogenic mutations. *PloS One, 7*, e47784.

Trinh, K., Andrews, L., Krause, J., Hanak, T., Lee, D., Gelb, M., et al. (2010). Decaffeinated coffee and nicotine-free tobacco provide neuroprotection in Drosophila models of Parkinson's disease through an NRF2-dependent mechanism. *The Journal of Neuroscience*, *30*, 5525–5532.

Tsai, P. I., Course, M. M., Lovas, J. R., Hsieh, C. H., Babic, M., Zinsmaier, K. E., et al. (2014). PINK1-mediated phosphorylation of Miro inhibits synaptic growth and protects dopaminergic neurons in Drosophila. *Scientific Reports*, *4*, 6962.

Tsika, E., Glauser, L., Moser, R., Fiser, A., Daniel, G., Sheerin, U. M., et al. (2014). Parkinson's disease-linked mutations in VPS35 induce dopaminergic neurodegeneration. *Human Molecular Genetics*, *23*, 4621–4638.

Tufi, R., Gandhi, S., de Castro, I. P., Lehmann, S., Angelova, P. R., Dinsdale, D., et al. (2014). Enhancing nucleotide metabolism protects against mitochondrial dysfunction and neurodegeneration in a PINK1 model of Parkinson's disease. *Nature Cell Biology*, *16*, 157–166.

Twig, G., Elorza, A., Molina, A. J., Mohamed, H., Wikstrom, J. D., Walzer, G., et al. (2008). Fission and selective fusion govern mitochondrial segregation and elimination by autophagy. *The EMBO Journal*, *27*, 433–446.

Verstreken, P., Koh, T. W., Schulze, K. L., Zhai, R. G., Hiesinger, P. R., Zhou, Y., et al. (2003). Synaptojanin is recruited by endophilin to promote synaptic vesicle uncoating. *Neuron*, *40*, 733–748.

Vilain, S., Esposito, G., Haddad, D., Schaap, O., Dobreva, M. P., Vos, M., et al. (2012). The yeast complex I equivalent NADH dehydrogenase rescues pink1 mutants. *PLoS Genetics*, *8*, e1002456.

Vincow, E. S., Merrihew, G., Thomas, R. E., Shulman, N. J., Beyer, R. P., MacCoss, M. J., et al. (2013). The PINK1-Parkin pathway promotes both mitophagy and selective respiratory chain turnover in vivo. *Proceedings of the National Academy of Sciences of the United States of America*, *110*, 6400–6405.

Vos, M., Esposito, G., Edirisinghe, J. N., Vilain, S., Haddad, D. M., Slabbaert, J. R., et al. (2012). Vitamin K2 is a mitochondrial electron carrier that rescues pink1 deficiency. *Science (New York, N.Y.)*, *336*, 1306–1310.

Wang, B., Liu, Q., Shan, H., Xia, C., & Liu, Z. (2015). Nrf2 inducer and cncC overexpression attenuates neurodegeneration due to alpha-synuclein in Drosophila. *Biochemistry and Cell Biology*, *93*, 351–358.

Wang, W., Wang, X., Fujioka, H., Hoppel, C., Whone, A. L., Caldwell, M. A., et al. (2016). Parkinson's disease-associated mutant VPS35 causes mitochondrial dysfunction by recycling DLP1 complexes. *Nature Medicine*, *22*, 54–63.

Wang, X., Winter, D., Ashrafi, G., Schlehe, J., Wong, Y. L., Selkoe, D., et al. (2011). PINK1 and Parkin target Miro for phosphorylation and degradation to arrest mitochondrial motility. *Cell*, *147*, 893–906.

Weihofen, A., Thomas, K. J., Ostaszewski, B. L., Cookson, M. R., & Selkoe, D. J. (2009). Pink1 forms a multiprotein complex with Miro and Milton, linking Pink1 function to mitochondrial trafficking. *Biochemistry*, *48*, 2045–2052.

Whitworth, A. J. (2011). Drosophila models of Parkinson's disease. *Advances in Genetics*, *73*, 1–50.

Whitworth, A. J., Theodore, D. A., Greene, J. C., Benes, H., Wes, P. D., & Pallanck, L. J. (2005). Increased glutathione S-transferase activity rescues dopaminergic neuron loss in a Drosophila model of Parkinson's disease. *Proceedings of the National Academy of Sciences of the United States of America*, *102*, 8024–8029.

Winklhofer, K. F. (2014). Parkin and mitochondrial quality control: Toward assembling the puzzle. *Trends in Cell Biology*, *24*, 332–341.

Yang, Y., Ouyang, Y., Yang, L., Beal, M. F., McQuibban, A., Vogel, H., et al. (2008). Pink1 regulates mitochondrial dynamics through interaction with the fission/fusion machinery.

Proceedings of the National Academy of Sciences of the United States of America, 105, 7070–7075.

Yin, G., Lopes da Fonseca, T., Eisbach, S. E., Anduaga, A. M., Breda, C., Orcellet, M. L., et al. (2014). alpha-Synuclein interacts with the switch region of Rab8a in a Ser129 phosphorylation-dependent manner. *Neurobiology of Disease, 70*, 149–161.

Yun, J., Puri, R., Yang, H., Lizzio, M. A., Wu, C., Sheng, Z. H., et al. (2014). MUL1 acts in parallel to the PINK1/parkin pathway in regulating mitofusin and compensates for loss of PINK1/parkin. *eLife, 3*, e01958.

Zhang, L., Karsten, P., Hamm, S., Pogson, J. H., Muller-Rischart, A. K., Exner, N., et al. (2013). TRAP1 rescues PINK1 loss-of-function phenotypes. *Human Molecular Genetics, 22*, 2829–2841.

Ziviani, E., Tao, R. N., & Whitworth, A. J. (2010). Drosophila parkin requires PINK1 for mitochondrial translocation and ubiquitinates mitofusin. *Proceedings of the National Academy of Sciences of the United States of America, 107*, 5018–5023.

CHAPTER SIX

Neurotoxicity Pathways in *Drosophila* Models of the Polyglutamine Disorders

M. Krench, J.T. Littleton[1]
The Picower Institute for Learning and Memory, Massachusetts Institute of Technology, Cambridge, MA, United States
[1]Corresponding author: e-mail address: troy@mit.edu

Contents

Abstract

Although polyglutamine expansion diseases are the most common genetically inherited neurodegenerative disorders, the key pathogenic mechanisms that lead to neuronal cell death are unclear. The expansion of a polyglutamine tract in specific proteins is the defining molecular insult, leading to cell-type and region-specific neuronal death. Intraneuronal aggregates of the affected protein can be found in the nucleus and/or cytoplasm and are a hallmark of these disorders. Whether and how aggregation leads to pathology, however, is under debate. In this chapter, we will review some of the key observations using *Drosophila* models of polyglutamine disorders that have highlighted a host of potential contributing pathologies, including defects in transcription, autophagy, and mitochondrial biology. We will also examine how genetic screening approaches have been used in *Drosophila* to provide insights into potential therapeutic approaches for polyglutamine disorders.

1. INTRODUCTION

The trinucleotide repeat disorders arise from genetically inherited expansions of unstable repetitive elements within specific loci. While some

Current Topics in Developmental Biology, Volume 121
ISSN 0070-2153
http://dx.doi.org/10.1016/bs.ctdb.2016.07.006

variability in repeat length is observed in normal alleles, pathology results when repeat length exceeds a specific threshold. Disease-causing expansions can occur in both coding and noncoding regions of genes depending upon the disorder (Orr & Zoghbi, 2007). The polyglutamine (polyQ) class of trinucleotide repeat disorders occur as a result of a CAG tract expansion in the coding region, leading to a protein product with an extended polyQ tract that forms intracellular aggregates. The nine well-known polyQ diseases include Huntington disease (HD), spinobulbar muscular atrophy (SBMA), dentatorubropallidoluysian atrophy (DRPLA), and spinocerebellar ataxia (SCA) types 1, 2, 3, 6, 7, and 17 (Orr & Zoghbi, 2007). Together, the polyQ diseases comprise the most common class of inherited neurodegenerative disorders.

A defining feature of the polyQ disorders is that different neuronal populations are uniquely vulnerable in each disease, despite the fact that the pathogenic proteins are widely expressed throughout the nervous system. For instance, D2-expressing GABAergic medium spiny neurons of the striatum are the first population to degenerate in HD (Zuccato, Valenza, & Cattaneo, 2010). Similarly, basal ganglia degeneration is also a hallmark of DRPLA, with neuronal loss occurring in the globus pallidus, the red nuclei of brainstem and dentate nuclei of cerebellum (Yamada, Sato, Tsuji, & Takahashi, 2008). Cerebellar degeneration is the primary neuropathology of the SCA diseases (Yamada et al., 2008). In contrast to the other polyQ diseases, SBMA affects lower motor neurons in the brainstem and spinal cord (Suzuki, Kastuno, Banno, & Sobue, 2009). How these cell-type-specific pathologies arise is a fundamental question in the field.

The specific protein disrupted in each of the polyQ disorders is unique. It is generally accepted that the disorders result in a dominant gain-of-function phenotype caused by the expanded repeat. However, there is also evidence that loss of function from disruption of the normal properties of the protein may contribute to aspects of toxicity (Cortes et al., 2014; Lam et al., 2006; Orr, 2012; Schulte & Littleton, 2011). Each of these disorders is highlighted by the presence of aggregated polyQ proteins within affected neurons. The proteins affected in the polyQ disorders are normally present in a variety of cellular compartments, from the plasma membrane to the cytosol to the nucleus. In the case of HD, the repeat expansion occurs within the first coding exon of huntingtin (HTT), a large HEAT repeat-containing cytosolic protein of unknown biological function (The Huntington's Disease Collaborative Research Group, 1993). The SCA family of disorders result from expansions in an unrelated subset of proteins. The nuclear transcriptional and splicing regulator ataxin-1 is disrupted in SCA1 (Orr et al., 1993). The cytoplasmic ataxin-2 protein is altered in SCA2 (Imbert et al., 1996; Pulst et al., 1996; Sanpei et al., 1996), while the nuclear ataxin-3 protein has an expanded repeat in SCA3 (Kawaguchi et al., 1994). SCA6 results from polyQ expansion in the plasma membrane-resident alpha 1 subunit of the P/Q-type calcium channel (Zhuchenko et al., 1997), a key regulator of neurotransmitter release. The nuclear and cytoplasmic ataxin-7 protein is part of an acetyltransferase complex and is mutated in SCA7 (David et al., 1997). SCA17 results from a polyQ expansion in the nuclear RNA polymerase component TATA-binding protein (Nakamura et al., 2001). SBMA results from expansion in the nuclear androgen receptor (La Spada, Wilson, Lubahn, Harding, & Fischbeck, 1991), while DRPLA results from expansion in atrophin-1 (Koide et al., 1994; Nagafuchi et al., 1994). Many of the polyQ proteins have also been shown to undergo proteolytic cleavage, which can lead to changes in their normal subcellular localization with nuclear accumulation a common feature. Beyond cell-type specific pathology, another key question in the polyQ field is determining whether the aggregated version of the protein is the pathogenic form. Similarly, determining how the disease-relevant form of the protein generates toxicity in a specific cellular compartment has also been challenging, with evidence suggesting both nuclear and cytoplasmic dysfunction in some cases.

This review highlights how the *Drosophila* model has contributed to the characterization of several toxic pathways implicated in the polyQ disorders. *Drosophila* is particularly amenable to dissecting pathogenic mechanisms in this class of neurodegenerative diseases (Perrimon, Bonini, & Dhillon, 2016). Since causative mutations have been characterized for the polyQ disorders, one can generate transgenic *Drosophila* expressing the human gene of interest or characterize *Drosophila* homologs to study loss-of-function phenotypes. The rapid generation time and ease of genetic manipulation make it possible to conduct large-scale forward genetic screens to identify disease modifiers (Ugur, Chen, & Bellen, 2016). There are also a wide variety of tools available to record neuronal activity in vivo, including physiology and live imaging (Harris & Littleton, 2015). A wide variety of *Drosophila* stocks can be used to examine the effects of loss-of-function or overexpression genetic interactions (Venken & Bellen, 2014). Where null and hypomorph mutants are not readily accessible, genetically encoded RNAi lines are available to knockdown nearly every gene in *Drosophila* (Mohr, Smith, Shamu, & Neumüller, 2014). The advent of CRISPR gene editing is making it increasingly easy to generate mutants or tag endogenous loci with fluorescent markers (Gratz, Harrison, Wildonger, & O'Connor-Giles, 2015). *Drosophila* also has the GAL4-inducible system that allows modulation of gene expression with temporal and spatial specificity (Elliott & Brand, 2008; Venken & Bellen, 2012). In cases where global expression of a transgene would prove lethal, one can also confine expression to non-essential cell populations such as the eye. A popular approach in the field takes advantage of suppressor and enhancer screens for morphological phenotypes induced by polyQ expression in the *Drosophila* eye.

The polyQ field is a large one with extensive data, suggesting that a host of cellular pathways are likely to be altered in dying neurons (Orr & Zoghbi, 2007; Weber, Sowa, Binder, & Hubener, 2014). One can find evidence for dysfunction of almost any molecular pathway of interest, creating a difficult arena for parsing out the most pathogenic insults created by polyQ expansion. To cope with this large and expansive literature, we have restricted our coverage of the polyQ field into four subsections. First, we survey evidence from the mammalian and *Drosophila* fields that transcriptional dysregulation in the polyQ models is likely to be a key factor in cell toxicity. We next highlight some key findings from modifier screens performed in *Drosophila* polyQ models that highlight several cellular pathways that are relevant for neurodegeneration in fly neurons. We also discuss issues surrounding mitochondrial function and mobility that have been linked to toxicity. Finally, we examine how *Drosophila* mutants and RNAi lines have been used to characterize the role of autophagy in the polyQ disorders. Although these pathways are likely to be only a subset of the toxicity mechanisms at play, they highlight how *Drosophila* can be used to model polyQ pathogenesis (Fig. 1).

2. TRANSCRIPTIONAL AND NUCLEAR DYSFUNCTION

Transcriptional aberrations were noted in early studies of polyQ patient brain tissue and supported by findings in mouse and *Drosophila* disease models. Transcriptional dysregulation has been studied most extensively in the context of HD (Sugars & Rubinsztein, 2003). Given that many of the pathogenic polyQ proteins accumulate in the nucleus, either as aggregates or as more soluble forms, there has been a heavy focus on whether and how these aggregates might alter nuclear biology. A common finding has been the ability of such nuclear aggregates to sequester various transcription factors or nuclear proteins, setting the stage for a loss-of-function phenotype in transcriptional regulation.

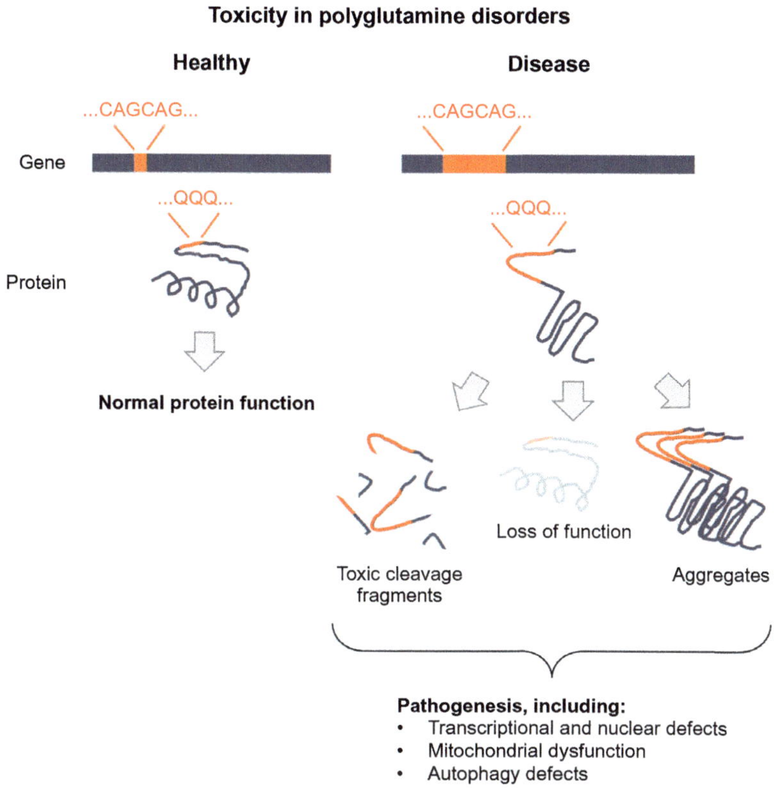

Fig. 1 Toxicity pathways in the polyQ disorders. CAG repeat expansions beyond a critical length (typically >40) result in an expanded polyQ tract within the affected protein and the formation of intracellular aggregates. Numerous downstream molecular pathways have been implicated in subsequent pathology, including defects in transcription, autophagy, and mitochondrial function. The expanded polyQ tract may also disrupt the normal function of the affected protein, leading to loss-of-function pathology as well. In addition, proteins containing the expanded polyQ tract may form novel protein–protein interactions that lead to aberrant signaling. Affected proteins can also undergo cleavage that may lead to altered cellular localization and enhance aggregate formation. It is still unclear which form of the polyQ protein (monomers, small aggregates, or larger inclusions) is the most toxic to neurons, and why only selected neuronal populations are highly sensitive to the expanded polyQ protein.

Microarray analysis has provided a powerful approach to measure gene expression changes in HD models. The first microarrays from R6/2 mouse striatum showed changes in less than 2% of 6000 genes analyzed. Most genes showed decreased expression, but a handful of genes, mostly related to inflammation, were increased. Genes central to neuronal signaling (neurotransmitter receptors, neurotransmitters, and neuropeptides), retinoid signaling, and calcium homeostasis were among those showing decreased expression (Luthi-Carter et al., 2000). In 2002, a consortium published a series of microarray experiments from various HD and polyQ disease models (Orr, 2002). Decreases in transcription were more commonly observed, with changes seen in genes associated with neurotransmission, intracellular signaling, calcium homeostasis, and transcriptional processes (Luthi-Carter, Hanson, et al., 2002; Luthi-Carter, Strand, et al., 2002). Interestingly, initial results indicated that expressing a short N-terminal fragment of HTT resulted in more dramatic changes than those observed in animals expressing longer fragments or full-length HTT transgenes. Short N-terminal fragments are expressed in R6/2 (67 amino acids) and N171-82Q (171 amino acids) models and were compared to longer fragments found in HD46 and HD100 mice (964 amino acids), and YAC72 mice (full-length, 3144 amino acids) (Chan et al., 2002; Pouladi, Morton, & Hayden, 2013). While R6/2 and N171-82Q mice have mRNA profiles similar to those observed in HD patients, the HD46, HD100, and YAC72 mice profiles were more subtle and qualitatively different (Chan et al., 2002). However, more recent analysis indicates that the longer full-length HD models have some similarities in transcriptional dysregulation to N-terminal fragment models and HD patients if older animals (1½ to 2 years old) are analyzed (Kuhn et al., 2007). As such, transcriptional dysregulation is commonly observed in all HD models, suggesting that a complex genomic response is occurring in HTT-expressing cells.

Several mechanisms may account for transcriptional changes in HD. One model proposes a physical interaction between expanded polyQ proteins and certain transcription factors. For example, pathogenic HTT and other mutant polyQ proteins have been shown to interact with and sequester CREB-binding protein (CBP), impairing regulation of CBP transcriptional targets (McCampbell et al., 2000; Nucifora et al., 2001; Steffan et al., 2000). The mechanism of transcription factor sequestration and transcriptional interference fits with the observation that nuclear localization of the polyQ protein is central to toxicity in many polyQ diseases, including HD. However, aberrant gene regulation must result from mechanisms other than just transcription factor sequestration, given that cellular dysfunction in polyQ disease models can occur before the appearance of nuclear inclusions, and some studies demonstrate a neuroprotective role for aggregates. Besides physical sequestration of transcription factors, HTT has been suggested to physically bind to DNA to directly repress transcription (Benn et al., 2008; Zhai, Jeong, Cui, Krainc, & Tjian, 2005). It has also been proposed that alterations in mRNA levels may be the result of physical changes to the nucleus (invagination of nuclear membrane) and nuclear pores (nuclear pore density) (Cha, 2000; Davies et al., 1997). Lastly, transcriptional abnormalities may be the result of chromatin modifications in polyQ diseases (Mohan, Abmayr, & Workman, 2014). HTT has been shown to interact with methyl-CpG-binding protein 2 (MeCP2), and expression of mutant HTT decreases CBP-dependent acetylation and increases methylation of H3K9 (McFarland et al., 2014; Nucifora et al., 2001; Ryu et al., 2006).

Early in vitro studies revealed that immunoprecipitation of HTT can pull down CBP, a cofactor in CREB-mediated transcriptional activation. Interactions with CBP were enhanced when the HTT fragment was expanded to include the polyproline domain, a proline-rich stretch of residues adjacent to the polyQ stretch near the N-terminus of HTT (Steffan et al., 2001). Additional in vitro experiments demonstrated that HTT also binds other acetyltransferases, including P/CAF. Furthermore, HTT binding impairs transcriptional activators' acetyltransferase activity, as measured by acetylation of histone H4 peptides (Steffan et al., 2001). In PC12 cell cultures transfected with a fragment of HTT-Q103 plus the polyproline region, H3 and H4 acetylation levels were restored after treatment with histone deacetylase (HDAC) inhibitors (Steffan et al., 2001).

In vivo experiments in *Drosophila* have supported a role for HDAC biology in polyQ pathology. Expression of a pathogenic HTT exon 1 fragment in *Drosophila* eyes caused progressive rhabdomere degeneration, along with ~70% lethality and reduced lifespan (Steffan et al., 2001). Rearing larvae on medium that contained HDAC inhibitors ameliorated degeneration in a concentration-dependent manner. HDAC inhibitors also increased viability and extended lifespan. Similar results were obtained with the expression of an isolated polyQ tract (Q48) (Steffan et al., 2001). HDAC inhibitors have also proven effective in models of SCA7 and SCA3 (Jung & Bonini, 2007; Latouche et al., 2007), suggesting a potential convergence on HDAC pathology.

Drosophila mutants have been used to further examine the role of histone acetylation state in polyQ diseases, specifically CBP-regulated acetylation. Using live-cell imaging, coexpression of GFP-tagged CBP and expanded HTT or ataxin-3 in cell culture resulted in CBP sequestration and subsequent immobilization after its incorporation into polyQ-containing aggregates (Chai, Shao, Miller, Williams, & Paulson, 2002). This adds to a body of evidence, suggesting that polyQ proteins impair transcription via CBP loss of function (Chai et al., 2002; McCampbell et al., 2000; Nucifora et al.,

2001; Steffan et al., 2000). Indeed, overexpression of *dCBP*, which encodes the *Drosophila* CBP ortholog, fully rescued Q127-induced eye neurodegeneration in *Drosophila* models (Marek et al., 2000). In addition to improvements in eye morphology, a visual challenge demonstrated that eye function was also restored by upregulation of *dCBP* (Taylor et al., 2003). How does dCBP overexpression rescue polyQ toxicity? Analysis of *Drosophila* homogenates showed that acetylation of histones H3 and H4 was reduced in flies expressing Q127 (Taylor et al., 2003). Acetylation levels were restored by the overexpression of dCBP. Given that reduced acetylation would promote a chromatin state that is less accessible to transcription factors, it was hypothesized that changes in H3 and H4 acetylation would be accompanied by altered gene expression.

Expanding upon this model, a set of *Drosophila* mutants and shRNA lines have been used to test whether loss of function or overexpression of *Drosophila* HDACs can modify HTT-induced pathology (Fernandez-Funez et al., 2000). *Drosophila* have 10 HDACs: 5 are considered classic zinc-dependent HDACs (Rpd3, HDAC3, HDAC4, HDAC6, and HDAC11) and the other 5 are NAD-1-dependent deacetylases (Sir2, Sirt2, Sirt4, Sirt6, and Sirt7). A combination of *Drosophila* mutants and shRNA silencing revealed that reduced levels of Rpd3, but not the other zinc-dependent HDACs, significantly improved eye degeneration and survival following pan-neuronal expression of an HTT-Q93 exon 1 fragment. Reducing levels of Sir2 (the *Drosophila* homolog of mammalian SIRT1) also improved eye degeneration and survival. Furthermore, the modifier effects were additive (Pallos et al., 2008), indicating that the two types of HDACs may act in parallel pathways to moderate transcription. The findings are also interesting in light of the fact that *Rpd3* and *Sir2* were identified in an SCA1 suppressor screen, suggesting that some toxic pathways may be shared among different polyQ diseases (Fernandez-Funez et al., 2000).

Modification of acetylation levels by genetic intervention in HD model flies has also been performed. *Drosophila Sin3A* encodes a corepressor protein that plays an essential role in HDAC complexes. Partial loss of function of Sin3A reduces histone deacetylation and also reduces HTT-induced eye degeneration and increases viability (Steffan et al., 2001). *Sin3A* was also identified in a forward genetics screen for modifiers of phenotypes in an SCA1 *Drosophila* model (Fernandez-Funez et al., 2000).

Forward genetic screening in *Drosophila* indicates that pathogenic polyQ proteins physically entering the nucleus may be detrimental (Doumanis, Wada, Kino, Moore, & Nukina, 2009). Using a combination of in vitro and in vivo model systems expressing an N-terminal fragment of expanded HTT, Doumanis et al. identified 21 high-confidence modifiers of aggregation and toxicity. In addition to genes with functions in general transcription and RNA processing, a set of hits were identified with critical roles in nuclear pore transport. In vivo knockdown of *CG4738*, the *Drosophila* homolog of human *NUP160,* suppressed aggregation and pathology in the brains of *Drosophila* expressing HTT-Q93, HTT-Q152, or nuclear-targeted HTT-Q48-NLS (Doumanis et al., 2009). Doumanis et al. hypothesized that this rescue may be due to reduced formation of HTT inclusions in the nucleus, which would match the evidence that nuclear localization is central to toxicity in several polyQ diseases, including SCA1, SCA7, DRPLA, and SBMA (Walsh, Storey, Stefani, Kelly, & Turnbull, 2005). Early observations from HD patient brain tissue showed that nonexpanded HTT is consistently localized to the cytoplasm (DiFiglia et al., 1995). However, pathogenic mutant HTT forms dense inclusions in the nuclei, as well as aggregates in the cytoplasm (DiFiglia et al., 1997). The appearance of nuclear inclusions coincides with disease onset in human patient tissue and mouse models of the disease (Davies et al., 1997; DiFiglia et al., 1997).

Experimental manipulations to prevent N-terminal HTT from entering the nucleus further supported a role for nuclear toxicity in HD. Cell culture experiments revealed that truncating the pathogenic HTT protein enhances its localization to the nucleus and exacerbates toxicity (Cooper et al., 1998; Hackam et al., 1998; Lunkes & Mandel, 1998; Martindale et al., 1998). Toxicity can also be suppressed by preventing cleavage and nuclear translocation of the truncated N-terminal fragment (Gafni et al., 2004). In another set of experiments, a nuclear localization signal (NLS) or nuclear export signal (NES) was added to truncated mutant HTT. The NES tag rescued toxicity, as measured by cell loss, while the NLS exacerbated the toxic effects of HTT (Peters et al., 1999). Although nuclear localization of mutant HTT is detrimental to neurons, both nuclear and cytoplasmic events likely contribute to pathogenesis (Benn et al., 2005; Lee, Yoshihara, & Littleton, 2004).

In addition to HD models, transcriptional dysregulation has also been analyzed in the context of other polyQ disorders, including SCA1. Overexpression of the *Drosophila* ataxin-1 homolog, dAtx-1, induces phenotypes similar to those observed from overexpression of an expanded human ataxin-1, including patterning defects and bristle loss in the wing margin (Tsuda et al., 2005). The bristle phenotype suggested that Sens, a zinc-finger transcription factor required for sensory organ development, may have abnormal activity following overexpression. Both dAtx-1 and human

Atx-1 physically interact with Sens. In vivo, dAtx-1 and Atx-1-Q82 decrease levels of endogenous Sens protein, and overexpression of Sens can suppress the loss-of-bristle phenotype caused by expression of *Drosophila* or human ataxin-1. The vertebrate homolog of Sens is Gfi-1. In mice, coimmunoprecipitation assays revealed that Atx-1 interacts with Gfi-1 in vivo. In SCA1 transgenic mice, which overexpress human Atx-1 in Purkinje cells, there was a marked decrease in Gfi-1 protein levels prior to Purkinje cell death. Indeed, human ataxin-1 enhanced proteosomal degradation of Gfi-1, and loss of Gfi-1 leads to Purkinje cell degeneration (Tsuda et al., 2005).

Ataxin-1 has also been shown to interact with the transcriptional repressor Capicua. Regulation of this interaction through mutation of a specific residue (S776A) in ataxin-1 suppresses SCA1 neuropathology (Lam et al., 2006), implicating another route for ataxin-1 to drive transcriptional alterations. Genetic approaches in both *Drosophila* and mouse SCA1 models revealed that alterations in the RAS–MAKP pathway could reduce SCA1 pathology (Park et al., 2013). The authors demonstrated that specific MAPK kinases phosphorylate residue S776 in ataxin-1 to regulate the stability of the protein. Inhibitors of the MAPK pathway reduced ataxin-1 levels, leading to less toxic protein in the nucleus, and providing a potential therapeutic approach for targeting SCA1 toxicity.

3. GENETIC AND PHARMACOLOGICAL SCREENS FOR SUPPRESSORS OF PolyQ PATHOLOGY

There are a variety of tools available to screen for modifiers of polyQ toxicity in *Drosophila*, including RNAi, transposable elements, chemical mutagenesis, and small-molecule assays. Kazemi-Esfarjani and Benzer (2000) performed one of the first screens searching for modifiers of polyQ toxicity in *Drosophila*. Using P-element mutagenesis, 7000 mutant strains were tested for their role in dominantly modifying the eye degeneration phenotypes observed in *Drosophila* following expression of a simple polyQ tract (Q127) using the *GMR-GAL4* eye driver. The screen identified 30 suppressors and 29 enhancers, including suppressor hits in *dHJD1*, the homolog of human *HSP40/HDJ1*, and *dTPR2*, the homolog of human *TPR2*. Both insertions increased transcription of the J (Hsp40) class of cochaperones that stimulate activity of Hsp70. Kazemi-Esfarjani and Benzer generated transgenic flies overexpressing *dHDJ1* or *dTPR2* and confirmed that upregulation of either gene was protective against eye degeneration caused by Q127 expression. This screen set the stage for future studies that explored the mechanisms by which cellular chaperones suppress toxicity in polyQ disorders and other neurodegenerative diseases (for reviews, see Muchowski & Wacker, 2005; Opal & Zoghbi, 2002; Paul & Mahanta, 2014).

In addition to in vivo screening in *Drosophila*, primary neurons can also be generated for in vitro screening assays of RNAi suppressors or candidate small-molecule therapeutics. Using *Drosophila* engineered to express a membrane-bound GFP (CD8-GFP) and a 12-exon fragment of HTT (HTT-Q138), primary neuronal cultures have been shown to accumulate cytoplasmic aggregates and result in altered neuronal morphology, including dystrophic neurites (Schulte, Sepp, Wu, Hong, & Littleton, 2011). These cultures were screened with RNAi libraries to identify suppressors of either aggregation or defects in neurite morphology. The screen revealed a novel target, lkb1, a kinase with roles in mTOR signaling and autophagy, as a potential suppressor of polyQ pathology in this neuronal population. Heterozygous loss of *lkb1* was sufficient to partially suppress lethality in adult HD model flies, suggesting that lkb1 may be linked to pathology in this model. Cultured neurons expressing pathogenic HTT were also screened using a collection of 2600 small molecules enriched for FDA-approved drugs. Sixty-two novel aggregation inhibitors, of which eight also improved the morphology of dystrophic neurites, were identified in this screen (Schulte et al., 2011). The topoisomerase inhibitors camptothecin or 10-hydroxycamptothecin extended the lifespan of HTT-expressing animals, indicating that the compounds are neuroprotective in vivo (Schulte et al., 2011).

In another screen designed to identify new FDA-approved modifiers of toxicity, Jimenz-Sanchez et al. used mammalian cell culture and *Drosophila* to search for targets that suppressed HTT-induced degeneration. Like the screens performed by Doumanis et al. and Schulte et al., Jimenez-Sanchez et al. began with an RNAi screen in vitro for suppressors of HTT-Q138 toxicity. Of the 257 RNAi screened, a subset with *Drosophila* homologs were selected for validation in vivo using neurodegeneration observed in the *Drosophila* eye following expression of a polyQ tract (Jimenez-Sanchez et al., 2015). In mammalian cell lines and primary neurons, siRNA against the gene encoding QPCT modified HTT-induced toxicity. RNAi against the *Drosophila* homologs, *Glutaminyl cyclase* (*QC*) and *iso Glutaminyl cyclase* (*isoQC*), partially rescued photoreceptor loss induced by expression of Q48 or an exon 1 fragment of HTT-Q120. Three lead compounds designed to inhibit QPCT activity demonstrated dose-dependent decreases in HTT aggregation and cellular toxicity in vitro. QPCT localizes to the endoplasmic reticulum and exhibits glutaminyl cyclase activity. It is thought to exert neuroprotective effects through increasing levels of the chaperone alpha-B-crystallin, a small heat-shock protein.

In vivo screens to search for modifiers of HTT aggregation and toxicity have also been performed in *Drosophila* expressing a larger 12-exon fragment of the human *HTT* gene with 138 CAG repeats (HTT-Q138). This longer fragment, which is tagged with mRFP to allow visualization of aggregation in live animals, maintains many key binding sites for HTT-interacting proteins, along with critical residues for posttranslational modification and caspase cleavage. Weiss, Kimura, Lee, and Littleton (2012) used a deficiency collection kit covering 80% of the *Drosophila* genome to screen for suppressors of HTT toxicity in a haploinsufficiency screen where 50% loss of the deleted gene product uncovered by the deficiency would modify toxicity. These experiments identified four gene regions that reduced HTT-Q138 aggregation when deleted. The deficiency collection was also used to find new suppressors of lethality induced by neuronal expression of HTT-Q138 following expression with the *elav-GAL4* pan-neuronal driver. On its own, *elav>HTT-Q138* is pharate lethal. In 11 deficiency backgrounds, viability improved 20–30%, and in another line, lethality was almost fully suppressed. These findings provided promising targets for future HD research (Weiss et al., 2012).

The connection between aggregation and toxicity is a perplexing relationship. On one hand, *Drosophila* screens performed by Doumanis et al., Desai et al., 2006, Schulte et al. and Weiss et al. all found correlation between aggregates and toxicity. In the in vivo screen by Weiss et al. the four deficiency lines identified in the screen for suppressors of aggregation were each identified in the lethality screen as well. However, four other lines that suppressed lethality did not have any observable changes in HTT-Q138 aggregate load (Weiss et al., 2012). Similarly, Schulte et al. found that only a subset of aggregation inhibitors were effective at suppressing cellular toxicity (Schulte et al., 2011). In the case of the screen performed by Doumanis et al., the top suppressor of aggregation in vitro (*CG1109*) strongly suppressed HTT-induced toxicity in vivo (Doumanis et al., 2009). More support for the connection between aggregation and toxicity comes from additional small-molecule screens that found strong inhibitors of aggregation suppressed neurodegeneration in *Drosophila* and mouse models of HD (Chopra et al., 2007; Zhang et al., 2005). However, other studies suggest that aggregates are not directly related to toxicity. For example, two in vitro studies show that HTT-induced toxicity is most strongly correlated with soluble HTT, not aggregates, and that aggregates may play a protective role (Arrasate, Mitra, Schweitzer, Segal, & Finkbeiner, 2004; Lajoie & Snapp, 2010). It remains unclear if aggregates are toxic, beneficial, or merely a by-product in polyQ diseases (Krench & Littleton, 2013).

4. MITOCHONDRIAL DYSFUNCTION

Many studies have demonstrated a role for mitochondrial dysfunction in HD pathogenesis (Damiano, Galvan, Déglon, & Brouillet, 2010). Research has shown that HTT can bind directly to mitochondria, impairing metabolic function and motility (Orr et al., 2008; Panov et al., 2002; Trushina et al., 2004). HTT also causes energy defects and neurodegeneration via its interference with CREB-Taf4-dependent transcription of PGC-1α, a regulator of mitochondrial biogenesis (Cui et al., 2006). Several models describe other mechanisms by which expanded HTT may affect mitochondria, including impairing trafficking or reducing their ability to buffer cytosolic calcium. One theory hypothesizes increased production of reactive oxygen species (ROS) and oxidative damage in HD. The vast majority of ROS in cells are produced by mitochondria (Balaban, Nemoto, & Finkel, 2005). In support of this theory, samples from postmortem tissue and blood samples from HD patients show increased oxidative damage and reduced mitochondrial function (Browne, Ferrante, & Beal, 1999; Chen et al., 2007). Further support comes from studies in yeast, which shows that expression of polyQ proteins leads to mitochondrial transport chain impairment and significantly higher levels of ROS (Solans, Zambrano, Rodríguez, & Barrientos, 2006). In vitro studies also show increased ROS generation induced by mutant HTT (Fukui & Moraes, 2007; Lim et al., 2008; Solans et al., 2006).

Defects in metabolism may be exacerbated by mislocalized mitochondria in HD. Mitochondria are generally distributed throughout the cell in response to local energy demands. However, there is ample evidence from *Drosophila* demonstrating that expression of pathogenic HTT impairs axonal transport (Lee et al., 2004). In a *Drosophila* HD model expressing an exon 1 fragment of HTT, neuronal expression of HTT-Q93 caused organelle accumulations along axons (Gunawardena et al., 2003). Live imaging of fluorescent transport proteins in flies expressing HTT-Q93 showed that trafficking of organelles and proteins along axons was impaired by mutant HTT (Sinadinos et al., 2009).

Animal models and postmortem brain tissue from HD patients show mitochondrial components and trafficking motors are also sequestered in mutant HTT aggregates. In vitro experiments suggest that aggregates impair the movement of mitochondria down the axon (Chang, Rintoul, Pandipati, & Reynolds, 2006). Mutant HTT has also been shown to disrupt the association of microtubule-based transport of mitochondria. As a consequence of reduced mitochondrial motility, distal axon terminals are starved of ATP (Orr et al., 2008). By impairing mitochondrial energy production and distribution, mutant HTT results in lower levels of cellular ATP, which may contribute to increased vulnerability (Zuccato et al., 2010). In addition to the toxic gain-of-function effects of mutant HTT, loss of function of normal HTT may also contribute to aberrant mitochondrial localization. Studies in both *Drosophila* and mammalian neurons demonstrate that reducing levels of endogenous HTT results in defective mitochondrial trafficking (Trushina et al., 2004).

5. AUTOPHAGY DEFECTS

Another emerging pathway that has been linked to cellular dysfunction in polyQ diseases is the bulk cellular degradation process of autophagy. In macroautophagy (hereafter referred to as autophagy), cytoplasmic cargo is sequestered and enveloped by a double-membrane vesicle termed an autophagosome. The autophagosome fuses with the lysosome, creating an autolysosome, where the membrane and contents of the autophagosome are degraded by lysosomal enzymes. As new proteins and organelles are synthesized, autophagic processes degrade old or damaged organelles, along with misfolded or aggregated proteins (Levine & Kroemer, 2008). Dysfunctions in autophagy have been implicated in a variety of diseases, ranging from neurodegenerative diseases to metabolic disorders (Cuervo, 2011; Levine & Kroemer, 2008). Some disease states may induce an "autophagic traffic jam," where autophagy induction is highly active, but there are not enough lysosomes to receive all of the cargo for degradation (Cuervo, 2011). Neurons with polyQ aggregates rely heavily on autophagy to clear these cellular obstructions. In doing so, expanded HTT may overwhelm the autophagy system, impairing routine housekeeping activities (Tooze & Schiavo, 2008). Furthermore, it has been suggested that the other main degradative pathway, the ubiquitin–proteasome system, is also impaired in HD (Tydlacka, Wang, Wang, Li, & Li, 2008; Wang et al., 2008), placing an additional burden on autophagy. The increased levels of autophagosomes found in HD brains may represent hyperactive autophagy that is attempting to degrade damaged components of the cell or intracellular aggregates (Levine & Kroemer, 2008). Another link to autophagy is the recent observation that one normal function for HTT is to scaffold autophagy cargo receptors and autophagosome components to initiate the autophagy process (Rui et al., 2015). PolyQ expansion within HTT could potentially disrupt this process, putting additional stress on the autophagic pathway.

In support of a neuroprotective theory of autophagy, increasing the process via inhibition of mammalian TOR (mTOR) is neuroprotective in *Drosophila* and mouse HD models. In *Drosophila* expressing a fragment of HTT-Q120, rhabdomere degeneration was reduced following treatment with rapamycin, which inhibits TOR function (Ravikumar et al., 2004). mTOR inhibition was also assayed in mammalian models. The rapamycin derivative CCI-779 enhanced clearance of expanded HTT in HD cell culture models and in the Ross/Borchelt HD mouse model (Huang & Houghton, 2001). Treatment with CCI-779 also reduced striatal aggregates and improved behavior and motor function (Ravikumar et al., 2004). These findings in *Drosophila* and mouse models of HD suggest a neuroprotective benefit when activating autophagy pathways.

Experiments in *Drosophila* also demonstrate that mutant HTT can be cleared when autophagy is stimulated through mTOR-independent pathways. Several autophagy-stimulating compounds were validated in vivo in *Drosophila* expressing a fragment of HTT-Q120 in the eye: verapamil (L-type calcium channel antagonist), clonidine (binds to alpha-2 adrenergic receptors and type I imidazoline receptors to activate G_i signaling pathways), and valproic acid (reduces inositol and IP_3 levels). Treatment with all three compounds suppressed rhabdomere degeneration (Williams et al., 2008). In addition to pharmacological efforts to increase autophagy, Pandey et al. used RNAi knockdown of *Drosophila* autophagy genes to study the interplay between autophagy and the UPS system in an SBMA model (AR-Q52). Knockdown of the autophagy genes *atg6* or *atg12* exacerbated eye degeneration in this SBMA model, indicating that the autophagy pathway was critical to clearance (Pandey et al., 2007).

Multiple studies have highlighted an interplay between autophagy and the UPS system (Pandey et al., 2007; Saitoh et al., 2015). The p62/sequestosome 1 protein (p62) allows for select ubiquitinated proteins to be targeted for autophagic degradation, bridging together the two main cellular degradation pathways. Previous *Drosophila* screens for modifiers of SCA3 toxicity have shown that truncated MJD-Q78 can undergo either autophagic degradation or degradation via the UPS system (Bilen & Bonini, 2007). In agreement with that finding, Saitoh et al. found that knockdown of either autophagy (*Atg12* or *alfy*) or proteasome (*Prosβ2*) genes exacerbated the rough eye phenotype induced by expression of truncated MJD-Q78. RNAi knockdown of p62 also enhanced the MJD-Q78 rough eye and increased aggregation of cytoplasmic MJD-Q78. These data support the model that p62 triggers the clearance of polyQ proteins via autophagic degradation (Saitoh et al., 2015).

In a *Drosophila* DRPLA model, transcriptional profiling offered new perspective into how autophagy is involved in pathogenic pathways triggered by expanded Atrophin. *Atrophin* (*Atro*), the *Drosophila* ortholog to human *Atrophin-1*, contains two polyQ stretches. DRPLA models have been generated by expanding the polyQ region near either the N-terminus (Atro-75QN) or the C-terminus (Atro-66QC). Atro-75QN contains a polyQ stretch in the middle of a polyproline domain, similar to where the polyQ stretch is found in human Atrophin-1 (Nisoli et al., 2010). Overexpression of expanded (and to a lesser extent wild-type) *Atro* in the eye caused a down-regulation of *ft*, a tumor suppressor gene that encodes a cadherin and regulates planar polarity through interactions with *Atro*. Loss-of-function *ft* mutants exacerbate Atro-75QN-induced retinal degeneration. Introducing additional mutants supported the idea that the Hippo tumor suppressor pathway was implicated in *ft* neurodegeneration, and the mechanism of toxicity is linked to defective autophagy (Napoletano et al., 2011).

Nisoli et al. also examined the role of autophagy in *Drosophila* DRPLA models. In contrast to HD and SBMA models, enhancing autophagy did not suppress toxicity in the DRPLA model. Photoreceptors overexpressing polyQ atrophins show an increase in autophagosomes and autophagic markers. How autophagy contributed to toxicity was assayed by using the autophagy mutant, $atg1^{\Delta 3D}$, a putative null allele of *atg1*. Blocking the induction of autophagy with the $atg1^{\Delta 3D}$ mutant enhanced retinal neurodegeneration by Atro-Q75N. Similarly, RNAi knockdown of *atg5* also enhanced Atro-Q75N neurodegeneration. Further study indicated that expanded atrophins block clearance at the lysosomal level after fusion between autophagosomes and lysosomes. These data indicated that increasing autophagy is insufficient to rescue neurodegeneration because downstream lysosomal functions remain impaired in the DRPLA model (Nisoli et al., 2010). Thus, while defective autophagy appears to be a common dysfunction among many polyQ diseases, enhancing autophagy may fail to rescue toxicity if defects exist downstream of the early autophagy steps.

6. CONCLUSION

The rapid generation time and the ease of genetic manipulations have made *Drosophila* an ideal model system to screen for modifiers of toxicity in the polyQ disorders. By using the *Drosophila* eye, suppressor or enhancers can be identified in a nonlethal cell population. In addition to screening for novel modifiers, *Drosophila* provides a complimentary model organism for testing hypotheses generated from in vitro experiments or candidates derived from other disease models. Research in *Drosophila* has provided a window into neurotoxic pathways in the polyQ disorders. As described in this review, toxicity in polyQ disorders may stem from a variety of mechanisms, including transcriptional and nuclear dysregulation, mitochondrial dysfunction, autophagy, and more. A number of additional mechanisms are also likely to contribute to the pathogenesis of these diseases. For example, it has been shown that overexpression of molecular chaperones can suppress aggregation and toxicity in polyQ disease models (reviewed in Muchowski & Wacker, 2005; Sakahira, Breuer, Hayer-Hartl, & Hartl, 2002; Xu, Tito, Rui, & Zhang, 2015). As the field of polyQ research progresses, it will be important to determine whether similar pathogenic mechanisms that occur downstream of polyQ protein expression in *Drosophila* are found in human HD tissue as well.

REFERENCES

Arrasate, M., Mitra, S., Schweitzer, E. S., Segal, M. R., & Finkbeiner, S. (2004). Inclusion body formation reduces levels of mutant huntingtin and the risk of neuronal death. *Nature*, *431*, 805–810.

Balaban, R. S., Nemoto, S., & Finkel, T. (2005). Mitochondria, oxidants, and aging. *Cell*, *120*, 483–495.

Benn, C. L., Landles, C., Li, H., Strand, A. D., Woodman, B., Sathasivam, K., et al. (2005). Contribution of nuclear and extranuclear polyQ to neurological phenotypes in mouse models of Huntington's disease. *Human Molecular Genetics*, *14*, 3065–3078.

Benn, C. L., Sun, T., Sadri-Vakili, G., McFarland, K. N., DiRocco, D. P., Yohrling, G. J., et al. (2008). Huntingtin modulates transcription, occupies gene promoters in vivo, and binds directly to DNA in a polyglutamine-dependent manner. *The Journal of Neuroscience*, *28*, 10720–10733.

Bilen, J., & Bonini, N. M. (2007). Genome-wide screen for modifiers of ataxin-3 neurodegeneration in *Drosophila*. *PLoS Genetics*, *3*, 1950–1964.

Browne, S. E., Ferrante, R. J., & Beal, M. F. (1999). Oxidative stress in Huntington's disease. *Brain Pathology*, *9*, 147–163.

Cha, J. H. J. (2000). Transcriptional dysregulation in Huntington's disease. *Trends in Neurosciences*, *23*, 387–392.

Chai, Y., Shao, J., Miller, V. M., Williams, A., & Paulson, H. L. (2002). Live-cell imaging reveals divergent intracellular dynamics of polyglutamine disease proteins and supports a sequestration model of pathogenesis. *Proceedings of the National Academy of Sciences of the United States of America, 99*, 9310–9315.

Chan, E. Y. W., Luthi-Carter, R., Strand, A., Solano, S. M., Hanson, S. A., DeJohn, M. M., et al. (2002). Increased huntingtin protein length reduces the number of polyglutamine-induced gene expression changes in mouse models of Huntington's disease. *Human Molecular Genetics, 11*, 1939–1951.

Chang, D. T. W., Rintoul, G. L., Pandipati, S., & Reynolds, I. J. (2006). Mutant huntingtin aggregates impair mitochondrial movement and trafficking in cortical neurons. *Neurobiology of Disease, 22*, 388–400.

Chen, C. M., Wu, Y. R., Cheng, M. L., Liu, J. L., Lee, Y. M., Lee, P. W., et al. (2007). Increased oxidative damage and mitochondrial abnormalities in the peripheral blood of Huntington's disease patients. *Biochemical and Biophysical Research Communications, 359*, 335–340.

Chopra, V., Fox, J. H., Lieberman, G., Dorsey, K., Matson, W., Waldmeier, P., et al. (2007). A small-molecule therapeutic lead for Huntington's disease: Preclinical pharmacology and efficacy of C2-8 in the R6/2 transgenic mouse. *Proceedings of the National Academy of Sciences of the United States of America, 104*, 16685–16689.

Cooper, J. K., Schilling, G., Peters, M. F., Herring, W. J., Sharp, A. H., Kaminsky, Z., et al. (1998). Truncated N-terminal fragments of huntingtin with expanded glutamine repeats form nuclear and cytoplasmic aggregates in cell culture. *Human Molecular Genetics, 7*, 783–790.

Cortes, C. J., Miranda, H. C., Frankowski, H., Batlevi, Y., Young, J. E., Le, A., et al. (2014). Polyglutamine-expanded androgen receptor interferes with TFEB to elicit autophagy defects in SBMA. *Nature Neuroscience, 17*, 1180–1189.

Cuervo, A. M. (2011). Cell biology. Autophagy's top chef. *Science, 332*, 1392–1393.

Cui, L., Jeong, H., Borovecki, F., Parkhurst, C. N., Tanese, N., & Krainc, D. (2006). Transcriptional repression of PGC-1α by mutant Huntingtin leads to mitochondrial dysfunction and neurodegeneration. *Cell, 127*, 59–69.

Damiano, M., Galvan, L., Déglon, N., & Brouillet, E. (2010). Mitochondria in Huntington's disease. *Biochimica et Biophysica Acta—Molecular Basis of Disease, 1802*, 52–61.

David, G., Abbas, N., Stevanin, G., Dürr, A., Yvert, G., Cancel, G., et al. (1997). Cloning of the SCA7 gene reveals a highly unstable CAG repeat expansion. *Nature Genetics, 17*, 65–70.

Davies, S. W., Turmaine, M., Cozens, B. A., DiFiglia, M., Sharp, A. H., Ross, C. A., et al. (1997). Formation of neuronal intranuclear inclusions underlies the neurological dysfunction in mice transgenic for the HD mutation. *Cell, 90*, 537–548.

Desai, U. A., Pallos, J., Ma, A. A., Stockwell, B. R., Thompson, L. M., Marsh, J. L., et al. (2006). Biologically active molecules that reduce polyglutamine aggregation and toxicity. *Human Molecular Genetics, 15*, 2114–2124.

DiFiglia, M., Sapp, E., Chase, K. O., Davies, S. W., Bates, G. P., Vonsattel, J. P., et al. (1997). Aggregation of huntingtin in neuronal intranuclear inclusions and dystrophic neurites in brain. *Science, 277*, 1990–1993.

DiFiglia, M., Sapp, E., Chase, K., Schwarz, C., Meloni, A., Young, C., et al. (1995). Huntingtin is a cytoplasmic protein associated with vesicles in human and rat brain neurons. *Neuron, 14*, 1075–1081.

Doumanis, J., Wada, K., Kino, Y., Moore, A. W., & Nukina, N. (2009). RNAi screening in *Drosophila* cells identifies new modifiers of mutant huntingtin aggregation. *PLoS One, 4*, e7275.

Elliott, D. A., & Brand, A. H. (2008). The GAL4 system: A versatile system for the expression of genes. *Methods in Molecular Biology, 420*, 79–95.

Fernandez-Funez, P., Nino-Rosales, M. L., de Gouyon, B., She, W. C., Luchak, J. M., Martinez, P., et al. (2000). Identification of genes that modify ataxin-1-induced neurodegeneration. *Nature, 408*, 101–106.

Fukui, H., & Moraes, C. T. (2007). Extended polyglutamine repeats trigger a feedback loop involving the mitochondrial complex III, the proteasome and huntingtin aggregates. *Human Molecular Genetics*, *16*, 783–797.

Gafni, J., Hermel, E., Young, J. E., Wellington, C. L., Hayden, M. R., & Ellerby, L. M. (2004). Inhibition of calpain cleavage of Huntingtin reduces toxicity: Accumulation of calpain/caspase fragments in the nucleus. *The Journal of Biological Chemistry*, *279*, 20211–20220.

Gratz, S. J., Harrison, M. M., Wildonger, J., & O'Connor-Giles, K. M. (2015). Precise genome editing of *Drosophila* with CRISPR RNA- guided cas9. *Methods in Molecular Biology*, *1311*, 335–348.

Gunawardena, S., Her, L. S., Brusch, R. G., Laymon, R. a., Niesman, I. R., Gordesky-Gold, B., et al. (2003). Disruption of axonal transport by loss of huntingtin or expression of pathogenic polyQ proteins in *Drosophila*. *Neuron*, *40*, 25–40.

Hackam, A. S., Singaraja, R., Wellington, C. L., Metzler, M., McCutcheon, K., Zhang, T., et al. (1998). The influence of huntingtin protein size on nuclear localization and cellular toxicity. *The Journal of Cell Biology*, *141*, 1097–1105.

Harris, K. P., & Littleton, J. T. (2015). Transmission, development, and plasticity of synapses. *Genetics*, *201*, 345–375.

Huang, S., & Houghton, P. J. (2001). Mechanisms of resistance to rapamycins. *Drug Resistance Updates*, *4*, 378–391.

Imbert, G., Saudou, F., Yvert, G., Devys, D., Trottier, Y., Garnier, J. M., et al. (1996). Cloning of the gene for spinocerebellar ataxia 2 reveals a locus with high sensitivity to expanded CAG/glutamine repeats. *Nature Genetics*, *14*, 285–291.

Jimenez-Sanchez, M., Lam, W., Hannus, M., Sönnichsen, B., Imarisio, S., Fleming, A., et al. (2015). siRNA screen identifies QPCT as a druggable target for Huntington's disease. *Nature Chemical Biology*, *11*, 347–354.

Jung, J., & Bonini, N. (2007). CREB-binding protein modulates repeat instability in a *Drosophila* model for polyQ disease. *Science*, *315*, 1857–1859.

Kawaguchi, Y., Okamoto, T., Taniwaki, M., Aizawa, M., Inoue, M., Katayama, S., et al. (1994). CAG expansions in a novel gene for Machado-Joseph Disease at chromosome 14q32.1. *Nature Genetics*, *8*, 221–228.

Kazemi-Esfarjani, P., & Benzer, S. (2000). Genetic suppression of polyglutamine toxicity in *Drosophila*. *Science*, *287*, 1837–1840.

Koide, R., Ikeuchi, T., Onodera, O., Tanaka, H., Igarashi, S., Endo, K., et al. (1994). Unstable expansion of CAG repeat in hereditary dentatorubral-pallidoluysian atrophy (DRPLA). *Nature Genetics*, *6*, 9–13.

Krench, M., & Littleton, J. T. (2013). Modeling Huntington disease in *Drosophila*: Insights into axonal transport defects and modifiers of toxicity. *Fly (Austin)*, 7, 229–236.

Kuhn, A., Goldstein, D. R., Hodges, A., Strand, A. D., Sengstag, T., Kooperberg, C., et al. (2007). Mutant huntingtin's effects on striatal gene expression in mice recapitulate changes observed in human Huntington's disease brain and do not differ with mutant huntingtin length or wild-type huntingtin dosage. *Human Molecular Genetics*, *16*, 1845–1861.

Lajoie, P., & Snapp, E. L. (2010). Formation and toxicity of soluble polyglutamine oligomers in living cells. *PLoS One*, *5*, e15245.

Lam, Y. C., Bowman, A. B., Jafar-Nejad, P., Lim, J., Richman, R., Fryer, J. D., et al. (2006). ATAXIN-1 interacts with the repressor Capicua in its native complex to cause SCA1 neuropathology. *Cell*, *127*, 1335–1347.

La Spada, A. R., Wilson, E. M., Lubahn, D. B., Harding, A. E., & Fischbeck, K. H. (1991). Androgen receptor gene mutations in X-linked spinal and bulbar muscular atrophy. *Nature*, *352*, 77–79.

Latouche, M., Lasbleiz, C., Martin, E., Monnier, V., Debeir, T., Mouatt-Prigent, A., et al. (2007). A conditional pan-neuronal *Drosophila* model of spinocerebellar ataxia 7 with a reversible adult phenotype suitable for identifying modifier genes. *The Journal of Neuroscience, 27*, 2483–2492.

Lee, W. C., Yoshihara, M., & Littleton, J. T. (2004). Cytoplasmic aggregates trap polyglutamine-containing proteins and block axonal transport in a *Drosophila* model of Huntington's disease. *Proceedings of the National Academy of Sciences of the United States of America, 101*, 3224–3229.

Levine, B., & Kroemer, G. (2008). Autophagy in the pathogenesis of disease. *Cell, 132*, 27–42.

Lim, D., Fedrizzi, L., Tartari, M., Zuccato, C., Cattaneo, E., Brini, M., et al. (2008). Calcium homeostasis and mitochondrial dysfunction in striatal neurons of Huntington disease. *The Journal of Biological Chemistry, 283*, 5780–5789.

Lunkes, A., & Mandel, J. L. (1998). A cellular model that recapitulates major pathogenic steps of Huntington's disease. *Human Molecular Genetics, 7*, 1355–1361.

Luthi-Carter, R., Hanson, S. A., Strand, A. D., Bergstrom, D. A., Chun, W., Peters, N. L., et al. (2002). Dysregulation of gene expression in the R6/2 model of polyglutamine disease: Parallel changes in muscle and brain. *Human Molecular Genetics, 11*, 1911–1926.

Luthi-Carter, R., Strand, A. D., Hanson, S. A., Kooperberg, C., Schilling, G., La Spada, A. R., et al. (2002). Polyglutamine and transcription: Gene expression changes shared by DRPLA and Huntington's disease mouse models reveal context-independent effects. *Human Molecular Genetics, 11*, 1927–1937.

Luthi-Carter, R., Strand, A., Peters, N. L., Solano, S. M., Hollingsworth, Z. R., Menon, A. S., et al. (2000). Decreased expression of striatal signaling genes in a mouse model of Huntington's disease. *Human Molecular Genetics, 9*, 1259–1271.

Marek, K. W., Ng, N., Fetter, R., Smolik, S., Goodman, C. S., & Davis, G. W. (2000). A genetic analysis of synaptic development: Pre- and postsynaptic dCBP control transmitter release at the *Drosophila* NMJ. *Neuron, 25*, 537–547.

Martindale, D., Hackam, A., Wieczorek, A., Ellerby, L., Wellington, C., McCutcheon, K., et al. (1998). Length of huntingtin and its polyglutamine tract influences localization and frequency of intracellular aggregates. *Nature Genetics, 18*, 150–154.

McCampbell, A., Taylor, J. P., Taye, A. A., Robitschek, J., Li, M., Walcott, J., et al. (2000). CREB-binding protein sequestration by expanded polyglutamine. *Human Molecular Genetics, 9*, 2197–2202.

McFarland, K. N., Huizenga, M. N., Darnell, S. B., Sangrey, G. R., Berezovska, O., Cha, J. H. J., et al. (2014). MeCP2: A novel huntingtin interactor. *Human Molecular Genetics, 23*, 1036–1044.

Mohan, R. D., Abmayr, S. M., & Workman, J. L. (2014). The expanding role for chromatin and transcription in polyglutamine disease. *Current Opinion in Genetics & Development, 26C*, 96–104.

Mohr, S. E., Smith, J. A., Shamu, C. E., & Neumüller, R. (2014). RNAi screening comes of age: Improved techniques and complementary approaches. *Nature Reviews. Molecular Cell Biology, 15*, 591–600.

Muchowski, P. J., & Wacker, J. L. (2005). Modulation of neurodegeneration by molecular chaperones. *Nature Reviews. Neuroscience, 6*, 11–22.

Nagafuchi, S., Yanagisawa, H., Sato, K., Shirayama, T., Ohsaki, E., Bundo, M., et al. (1994). Dentatorubral and pallidoluysian atrophy expansion of an unstable CAG trinucleotide on chromosome 12p. *Nature Genetics, 6*, 14–18.

Nakamura, K., Jeong, S. Y., Uchihara, T., Anno, M., Nagashima, K., Nagashima, T., et al. (2001). SCA17, a novel autosomal dominant cerebellar ataxia caused by an expanded polyglutamine in TATA-binding protein. *Human Molecular Genetics, 10*, 1441–1448.

Napoletano, F., Occhi, S., Calamita, P., Volpi, V., Blanc, E., Charroux, B., et al. (2011). Polyglutamine Atrophin provokes neurodegeneration in *Drosophila* by repressing *fat*. *The EMBO Journal, 30*, 945–958.

Nisoli, I., Chauvin, J. P., Napoletano, F., Calamita, P., Zanin, V., Fanto, M., et al. (2010). Neurodegeneration by polyglutamine Atrophin is not rescued by induction of autophagy. *Cell Death and Differentiation*, *17*, 1577–1587.

Nucifora, F. C., Sasaki, M., Peters, M. F., Huang, H., Cooper, J. K., Yamada, M., et al. (2001). Interference by huntingtin and atrophin-1 with CPB-mediated transcription leading to cellular toxicity. *Science*, *291*, 2423–2428.

Opal, P., & Zoghbi, H. Y. (2002). The role of chaperones in polyglutamine disease. *Trends in Molecular Medicine*, *8*, 232–236.

Orr, H. T. (2002). Microarrays and polyglutamine disorders: Reports from the hereditary disease array group. *Human Molecular Genetics*, *11*, 1909–1910.

Orr, H. T. (2012). Polyglutamine neurodegeneration: Expanded glutamines enhance native functions. *Current Opinion in Genetics & Development*, *22*, 251–255.

Orr, H. T., Chung, M. Y., Banfi, S., Kwiatkowski, T. J., Servadio, A., Beaudet, A. L., et al. (1993). Expansion of an unstable trinucleotide CAG repeat in spinocerebellar ataxia type 1. *Nature Genetics*, *4*, 221–226.

Orr, A. L., Li, S., Wang, C.-E., Li, H., Wang, J., Rong, J., et al. (2008). N-terminal mutant huntingtin associates with mitochondria and impairs mitochondrial trafficking. *The Journal of Neuroscience*, *28*, 2783–2792.

Orr, H. T., & Zoghbi, H. Y. (2007). Trinucleotide repeat disorders. *Annual Review of Neuroscience*, *30*, 575–621.

Pallos, J., Bodai, L., Lukacsovich, T., Purcell, J. M., Steffan, J. S., Thompson, L. M., et al. (2008). Inhibition of specific HDACs and sirtuins suppresses pathogenesis in a *Drosophila* model of Huntington's disease. *Human Molecular Genetics*, *17*, 3767–3775.

Pandey, U. B., Nie, Z., Batlevi, Y., McCray, B. A., Ritson, G. P., Nedelsky, N. B., et al. (2007). HDAC6 rescues neurodegeneration and provides an essential link between autophagy and the UPS. *Nature*, *447*, 859–863.

Panov, A. V., Gutekunst, C.-A., Leavitt, B. R., Hayden, M. R., Burke, J. R., Strittmatter, W. J., et al. (2002). Early mitochondrial calcium defects in Huntington's disease are a direct effect of polyglutamines. *Nature Neuroscience*, *5*, 731–736.

Park, J., Al-Ramahi, I., Tan, Q., Mollema, N., Diaz-Garcia, J. R., Gallego-Flores, T., et al. (2013). RAS-MAPK-MSK1 pathway modulates ataxin 1 protein levels and toxicity in SCA1. *Nature*, *498*, 325–331.

Paul, S., & Mahanta, S. (2014). Association of heat-shock proteins in various neurodegenerative disorders: Is it a master key to open the therapeutic door? *Molecular and Cellular Biochemistry*, *386*, 45–61.

Perrimon, N., Bonini, N. M., & Dhillon, P. (2016). Fruit flies on the front line: The translational impact of *Drosophila. Disease Models & Mechanisms*, *9*, 229–231.

Peters, M. F., Nucifora, F. C., Kushi, J., Seaman, H. C., Cooper, J. K., Herring, W. J., et al. (1999). Nuclear targeting of mutant huntingtin increases toxicity. *Molecular and Cellular Neurosciences*, *14*, 121–128.

Pouladi, M. A., Morton, A. J., & Hayden, M. R. (2013). Choosing an animal model for the study of Huntington's disease. *Nature Reviews. Neuroscience*, *14*, 708–721.

Pulst, S. M., Nechiporuk, A., Nechiporuk, T., Gispert, S., Chen, X. N., Lopes-Cendes, I., et al. (1996). Moderate expansion of a normally biallelic trinucleotide repeat in spinocerebellar ataxia type 2. *Nature Genetics*, *14*, 269–276.

Ravikumar, B., Vacher, C., Berger, Z., Davies, J. E., Luo, S., Oroz, L. G., et al. (2004). Inhibition of mTOR induces autophagy and reduces toxicity of polyglutamine expansions in fly and mouse models of Huntington disease. *Nature Genetics*, *36*, 585–595.

Rui, Y.-N., Xu, Z., Patel, B., Chen, Z., Chen, D., Tito, A., et al. (2015). Huntingtin functions as a scaffold for selective macroautophagy. *Nature Cell Biology*, *17*, 262–275.

Ryu, H., Lee, J., Hagerty, S. W., Soh, B. Y., McAlpin, S. E., Cormier, K. A., et al. (2006). ESET/SETDB1 gene expression and histone H3 (K9) trimethylation in Huntington's disease. *Proceedings of the National Academy of Sciences of the United States of America, 103*, 19176–19181.

Saitoh, Y., Fujikake, N., Okamoto, Y., Popiel, H. A., Hatanaka, Y., Ueyama, M., et al. (2015). P62 plays a protective role in the autophagic degradation of polyglutamine protein oligomers in polyglutamine disease model flies. *The Journal of Biological Chemistry, 290*, 1442–1453.

Sakahira, H., Breuer, P., Hayer-Hartl, M. K., & Hartl, F. U. (2002). Molecular chaperones as modulators of polyglutamine protein aggregation and toxicity. *Proceedings of the National Academy of Sciences of the United States of America, 99*, 16412–16418.

Sanpei, K., Takano, H., Igarashi, S., Sato, T., Oyake, M., Sasaki, H., et al. (1996). Identification of the spinocerebellar ataxia type 2 gene using a direct identification of repeat expansion and cloning technique, DIRECT. *Nature Genetics, 14*, 277–284.

Schulte, J., & Littleton, J. T. (2011). The biological function of the Huntingtin protein and its relevance to Huntington's disease pathology. *Current Trends in Neurology, 5*, 65–78.

Schulte, J., Sepp, K. J., Wu, C., Hong, P., & Littleton, J. T. (2011). High-content chemical and RNAi screens for suppressors of neurotoxicity in a Huntington's disease model. *PLoS One, 6*, e23841.

Sinadinos, C., Burbidge-King, T., Soh, D., Thompson, L. M., Marsh, J. L., Wyttenbach, A., et al. (2009). Live axonal transport disruption by mutant huntingtin fragments in *Drosophila* motor neuron axons. *Neurobiology of Disease, 34*, 389–395.

Solans, A., Zambrano, A., Rodríguez, M., & Barrientos, A. (2006). Cytotoxicity of a mutant huntingtin fragment in yeast involves early alterations in mitochondrial OXPHOS complexes II and III. *Human Molecular Genetics, 15*, 3063–3081.

Steffan, J. S., Bodai, L., Pallos, J., Poelman, M., McCampbell, A., Apostol, B. L., et al. (2001). Histone deacetylase inhibitors arrest polyglutamine-dependent neurodegeneration in *Drosophila*. *Nature, 413*, 739–743.

Steffan, J. S., Kazantsev, A., Spasic-Boskovic, O., Greenwald, M., Zhu, Y. Z., Gohler, H., et al. (2000). The Huntington's disease protein interacts with p53 and CREB-binding protein and represses transcription. *Proceedings of the National Academy of Sciences of the United States of America, 97*, 6763–6768.

Sugars, K. L., & Rubinsztein, D. C. (2003). Transcriptional abnormalities in Huntington disease. *Trends in Genetics, 19*, 233–238.

Suzuki, K., Kastuno, M., Banno, H., & Sobue, G. (2009). Pathogenesis-targeting therapeutics for spinal and bulbar muscular atrophy (SBMA): Symposium: Clinicopathological aspects of neuromuscular disorders—A new horizon. *Neuropathology, 29*, 509–516.

Taylor, J. P., Taye, A. A., Campbell, C., Kazemi-Esfarjani, P., Fischbeck, K. H., & Min, K.-T. (2003). Aberrant histone acetylation, altered transcription, and retinal degeneration in a *Drosophila* model of polyglutamine disease are rescued by CREB-binding protein. *Genes & Development, 17*, 1463–1468.

The Huntington's Disease Collaborative Research Group. (1993). A novel gene containing a trinucleotide repeat that is expanded and unstable on Huntington's disease chromosomes. *Cell, 72*, 971–983.

Tooze, S. a., & Schiavo, G. (2008). Liaisons dangereuses: Autophagy, neuronal survival and neurodegeneration. *Current Opinion in Neurobiology, 18*, 504–515.

Trushina, E., Dyer, R. B., Ii, J. D. B., Eide, L., Tran, D. D., Vrieze, B. T., et al. (2004). Mutant Huntingtin impairs axonal trafficking in mammalian neurons in vivo and in vitro. *Molecular and Cellular Biology, 24*, 8195–8209.

Tsuda, H., Jafar-Nejad, H., Patel, A. J., Sun, Y., Chen, H.-K., Rose, M. F., et al. (2005). The AXH domain of Ataxin-1 mediates neurodegeneration through its interaction with Gfi-1/senseless proteins. *Cell, 122*, 633–644.

Tydlacka, S., Wang, C.-E., Wang, X., Li, S., & Li, X.-J. (2008). Differential activities of the ubiquitin-proteasome system in neurons versus glia may account for the preferential accumulation of misfolded proteins in neurons. *The Journal of Neuroscience, 28*, 13285–13295.

Ugur, B., Chen, K., & Bellen, H. J. (2016). *Drosophila* tools and assays for the study of human diseases. *Disease Models & Mechanisms, 9*, 235–244.

Venken, K. J. T., & Bellen, H. J. (2012). Genome-wide manipulations of Drosophila melanogaster with transposons, Flp recombinase, and ϕC31 integrase. *Methods in Molecular Biology, 859*, 203–228.

Venken, K. J. T., & Bellen, H. J. (2014). Chemical mutagens, transposons, and transgenes to interrogate gene function in *Drosophila melanogaster. Methods, 68*, 15–28.

Walsh, R., Storey, E., Stefani, D., Kelly, L., & Turnbull, V. (2005). The roles of proteolysis and nuclear localisation in the toxicity of the polyglutamine diseases. A review. *Neurotoxicity Research*, 7, 43–57.

Wang, J., Wang, C.-E., Orr, A., Tydlacka, S., Li, S.-H., & Li, X.-J. (2008). Impaired ubiquitin-proteasome system activity in the synapses of Huntington's disease mice. *The Journal of Cell Biology, 180*, 1177–1189.

Weber, J. J., Sowa, A. S., Binder, T., & Hubener, J. (2014). From pathways to targets: Understanding the mechanisms behind polyglutamine disease. *BioMed Research International, 2014*, 701758.

Weiss, K. R., Kimura, Y., Lee, W. C. M., & Littleton, J. T. (2012). Huntingtin aggregation kinetics and their pathological role in a *Drosophila* Huntington's disease model. *Genetics, 190*, 581–600.

Williams, A., Sarkar, S., Cuddon, P., Ttofi, E. K., Saiki, S., Siddiqi, F. H., et al. (2008). Novel targets for Huntington's disease in an mTOR-independent autophagy pathway. *Nature Chemical Biology, 4*, 295–305.

Xu, Z., Tito, A., Rui, Y.-N., & Zhang, S. (2015). Studying polyglutamine diseases in *Drosophila. Experimental Neurology, 274*, 25–41.

Yamada, M., Sato, T., Tsuji, S., & Takahashi, H. (2008). CAG repeat disorder models and human neuropathology: Similarities and differences. *Acta Neuropathologica, 115*, 71–86.

Zhai, W., Jeong, H., Cui, L., Krainc, D., & Tjian, R. (2005). In vitro analysis of huntingtin-mediated transcriptional repression reveals multiple transcription factor targets. *Cell, 123*, 1241–1253.

Zhang, X., Smith, D. L., Meriin, A. B., Engemann, S., Russel, D. E., Roark, M., et al. (2005). A potent small molecule inhibits polyglutamine aggregation in Huntington's disease neurons and suppresses neurodegeneration in vivo. *Proceedings of the National Academy of Sciences of the United States of America, 102*, 892–897.

Zhuchenko, O., Bailey, J., Bonnen, P., Ashizawa, T., Stockton, D. W., Amos, C., et al. (1997). Autosomal dominant cerebellar ataxia (SCA6) associated with small polyglutamine expansions in the alpha 1A-voltage-dependent calcium channel. *Nature Genetics, 15*, 62–69.

Zuccato, C., Valenza, M., & Cattaneo, E. (2010). Molecular mechanisms and potential therapeutical targets in Huntington's disease. *Physiological Reviews, 90*, 905–981.

CHAPTER SEVEN

AxGxE: Using Flies to Interrogate the Complex Etiology of Neurodegenerative Disease

C. Burke, K. Trinh, V. Nadar, S. Sanyal[1]
Neurology Research, Biogen, Cambridge, MA United States
[1]Corresponding author: e-mail address: subhabrata.sanyal@biogen.com

Contents

Abstract

Progressive and late-onset neurological disorders such as Parkinson's disease and Alzheimer's disease affect up to 50 million people globally—a number postulated to double every 20 years in a continually aging population. While predisposing allelic variants in several genes clearly confer risk, individual age and specific environmental influences are equally important discriminators of disease onset age and progression. However, none of these factors can independently predict disease with significant precision. Therefore, we must actively develop models that accommodate contributions from all factors, potentially resulting in an A × G × E (age–gene–environment)

Current Topics in Developmental Biology, Volume 121
ISSN 0070-2153
http://dx.doi.org/10.1016/bs.ctdb.2016.07.007

metric that reflects individual cumulative risk and reliably forecasts disease outcomes. This effort can only be enabled by a deep quantitative understanding of the contribution of these factors to neurodegenerative disease, both individually and in combination. This is also an important consideration because neuronal loss typically precedes clinical presentation and disease-modifying therapies are contingent on early diagnosis that is likely to be informed by an accurate estimation of individual risk. Although epidemiological studies continue to make strong advances in these areas with the advent of powerful "omics"-based approaches, systematic phenotypic modeling of AxGxE interactions is currently more feasible in model organisms such as *Drosophila melanogaster* where all three parameters can be manipulated with manageable experimental burden. Here, we outline the advantages of using fruit flies for investigating these complex interactions and highlight potential approaches that might help synthesize existing information from diverse fields into a cogent description of age-dependent, environmental, and genetic risk factors in the pathophysiology of neurological disorders.

1. AGE, ENVIRONMENTAL INSULT AND GENETIC RISK IN HUMAN NEURODEGENERATIVE DISEASE

Cell biological mechanisms underlying healthy aging are likely to impact age-dependent diseases in concert with specific genetic backgrounds and cumulative exposure to various environmental factors (Fig. 1). Diseases of old age are typically characterized by late onset and their progressive nature; i.e., the disease and its symptoms worsen over time. For many such disorders, disease-modifying therapies—interventions that will slow, stop, or reverse the course of the disease—do not exist. Barriers to the development of such therapeutics are rooted in a current lack of early diagnostic capability and a poor mechanistic understanding of disease pathophysiology. This includes the contributing factors of underlying genetics and environmental exposure, combined with an incomplete understanding of healthy aging. In this review, we describe our current understanding of these factors as they apply to human neurodegenerative disease (with specific examples from Parkinson's disease (PD)), followed by a summary of contributions in these fields made by *Drosophila* research. A synergistic and sequential target identification pipeline, which includes fly research and combines it with network biology-driven hypothesis testing in human cell lines and rodent models, is also proposed.

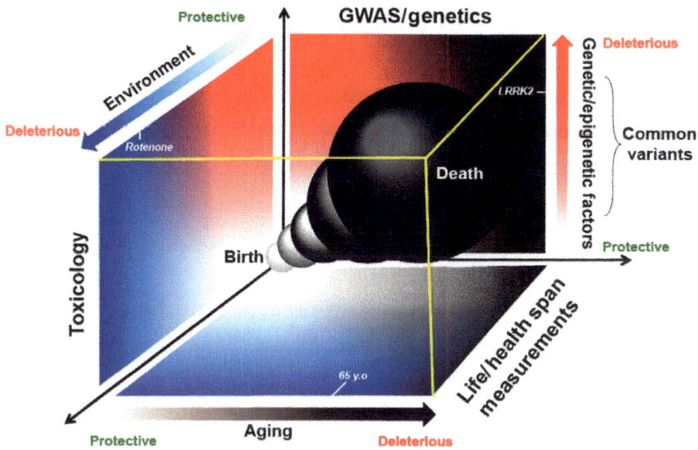

Fig. 1 Schematic representation of how genetic background (*white* to *red*), exposure to environmental factors (*white* to *blue*), and age (*white* to *black*) cooperate in the incidence of neurodegenerative disease. Each factor is represented as capable of being protective or deleterious. Thus, allelic variants can confer protection or predispose to disease. Similarly, environmental factor may protect or enhance risk. Aging appears along a continuum that gets progressively worse, i.e., the risk of disease increases with chronological age. The probability of disease incidence is depicted as the color and diameter of spheres in the 3D space bound by the three axes. The origin denotes a situation at birth of an individual who has the largest complement of protective allelic variants and who has only been exposed to environmental factors that are protective. All humans are postulated to fall within a certain coordinate in this 3D space. Ideally, this coordinate can be represented as a "cumulative risk score" that is both quantitative and has predictive value. While this is clearly a very difficult goal, research in simpler systems such as *Drosophila* might help to drive a more quantitative understanding of how these three factors influence disease pathways. Expanding epidemiological and "omics" studies in patient populations need to be cross-correlated with these findings to finally achieve the goal of measuring and assigning cumulative disease risk. This knowledge might also help determine avenues for personalized intervention in the future.

2. CHALLENGES IN INTERROGATING CONTRIBUTIONS FROM AGING, GENETIC RISK FACTORS AND ENVIRONMENTAL INSULT TO HUMAN NEURODEGENERATION

2.1 Age

Aging is the single largest risk factor for neurodegenerative diseases in humans. Thus, diseases such as PD and Alzheimer's disease (AD) are typically late-onset diseases of old age. For both AD and PD, early-onset heritable forms of the disease do exist, but these include no more than 5% of all cases. As a result, the vast majority of cases are sporadic or idiopathic, where a clear-cut genetic cause is difficult to identify. Although age is clearly a defining feature, how age impacts disease onset and rate of progression remains mysterious.

Many theories of aging have suggested that aging comprises a slow accumulation of deleterious cellular changes that are persistent and irreversible. This idea can perhaps be most readily appreciated in the context of protein damage and turnover. Proteins within cells are constantly under attack from various agents such as free radicals (e.g., oxidation, nitrosylation) that can lead to the irreversible modification of active sites and loss of activity (Dauer & Przedborski, 2003; Dias, Junn, & Mouradian, 2013; Nakamura et al., 2013; Okamoto & Lipton, 2015; Radak, Zhao, Goto, & Koltai, 2011; Ryan et al., 2013). Robust cellular defense mechanisms (such as autophagy) usually degrade these proteins within discrete intracellular compartments (such as the autophagosome or lysosome) and recycle constituents or refold misfolded proteins through the activity of dedicated chaperones (Boya, Reggiori, & Codogno, 2013; Mizushima, Levine, Cuervo, & Klionsky, 2008). However, current understanding suggests that these mechanisms of protein degradation or recovery might be impaired or blocked altogether in many pathological situations and are themselves subject to age-dependent regulation (Dehay et al., 2013; Harris & Rubinsztein, 2012; Martinez-Lopez, Athonvarangkul, & Singh, 2015; Menzies, Fleming, & Rubinsztein, 2015; Tanaka & Matsuda, 2014; Wang & Mao, 2014; Wong & Holzbaur, 2015). Under these conditions, cells accumulate damaged proteins, which can interfere in the function of normal proteins through the gain of new deleterious activities, or through the formation of toxic intracellular aggregates, ultimately leading to cellular pathology and/or death. Cumulative damage can also be observed in cellular organelles such as mitochondria. Mitochondria sustain damage to proteins as well as mitochondrial DNA through oxidative damage that then triggers a program of repair, removal, and new organellar biogenesis (Bratic & Larsson, 2013; Muftuoglu, Mori, & de Souza-Pinto, 2014; Osiewacz & Bernhardt, 2013). For instance, mitophagy, the process of autophagic removal of entire mitochondria, is thought to be impaired in diseases such as PD (Geisler et al., 2010; Hasson et al., 2013; Ryan, Hoek, Fon, & Wade-Martins, 2015; Shaltouki et al., 2015; Vincow et al., 2013). In sum, aging-related changes can impair the ability of a cell to repair both macromolecules such as proteins and DNA and organelles such as mitochondria. Once this balance shifts toward a net accumulation of damage and a threshold is crossed, pathophysiological changes are set in motion, eventually leading to cell dysfunction and death. Significantly, this process of age-related damage suggests the presence of a preventive and protective therapeutic window during which interventions can be designed to slow, stop, or reverse this trend (Fig. 2). By contrast, a sudden insult leading to pathology is expected to follow a different time course. For instance, age-dependent changes in individual cells might occur suddenly (e.g., in cancer cells), but the probability of their occurrence increases with age. One possibility is that somatic spontaneous mutations in neurons might lead to a sudden change that renders it selectively vulnerable.

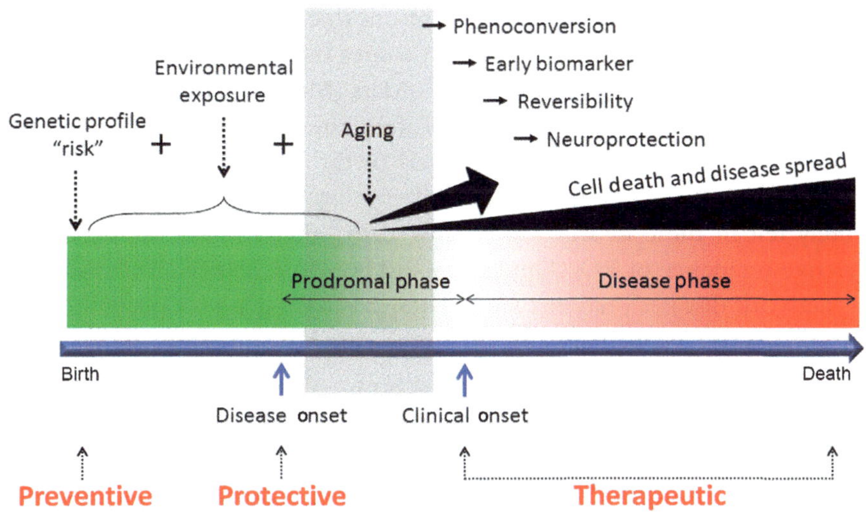

Fig. 2 Timeline for a late-onset neurodegenerative disorder such as Parkinson's disease. The start of disease usually precedes clinical onset (appearance of symptoms), thus defining a prodromal phase in these disorders that may or may not be detectable based on symptoms that are usually difficult to uniquely associate with disease. Disease onset is dictated by genetic risk, environmental exposure, and age and can be defined as the beginning of cellular pathology that will eventually lead to cell death and disease spread. The precise time of disease onset varies from person to person based on their cumulative risk and will, therefore, dictate the age at clinical onset of disease. The prodromal phase (shown in *gray*) is an attractive time period for therapeutic intervention since there is a window for stopping or reversing the disease through neuroprotective therapies. Identification of this time point requires a greater understanding of genetic risk, environmental exposure, and aging in disease. This can be done through longitudinal studies of at-risk humans who go on to develop disease ("pheno-conversion"), as well as the development of early biomarkers of disease. Our ability to design disease-modifying therapies relies critically on these factors.

Another manner in which age might impact neuronal health is through a change in the properties of protein networks (Guruharsha et al., 2011; Huttlin et al., 2015; Rolland et al., 2014). These networks offer a view of biology that incorporates the "output" of a protein complex rather than single proteins working in isolation (for example, the mitogen-activated protein kinase pathway (Trempolec, Dave-Coll, & Nebreda, 2013)). Decades of intense research have shown bewildering complexity that surrounds such pathways with evidence for elaborate cross talk, multiple substrates, and homeostatic regulation. It is reasonable to speculate that some of these networks might be subject to age-dependent decay (Kogan, Molodtsov, Menshikov, Shmookler Reis, & Fedichev, 2015), such that altering network "node" protein expression or activity could lead to loss of network homeostatic regulation, negatively affecting output. Such an idea is also consistent with recent demonstrations that specific protein networks are more highly represented in certain tissues than others correlating well with the tissue specificity of several disorders (Menche et al., 2015). Thus, although Parkin is a widely expressed protein, the network in which it functions might be more highly represented in the brain and, specifically, in dopaminergic neurons. As such, an aberration in this network is more acutely felt in these neurons as compared to other tissues, leading to the selective incidence of PD. Thinking about complex neurological disorders as "network diseases" rather than as classical single-gene Mendelian disorders offers several advantages including the possibility of integrating genetic risk factors and age-dependent decay (Barabasi, Gulbahce, & Loscalzo, 2011).

Whether the effect of age is slow or sudden, it is extremely difficult to explore it mechanistically in humans. Though we understand the signs and symptoms of aging, and that pathological aging differs from healthy aging—the ultimate question of what aging is requires models that are more amenable and in which natural variation can be kept to a minimum. As we outline later, *Drosophila* offers unique advantages in this regard and has already proved to be a model system that has provided unprecedented insights into fundamental mechanisms underlying aging.

2.2 Genetics

Our ability to connect genetic variants with disease has progressed rapidly in the era of next-generation sequencing technologies. In the context of neurodegenerative disorders such as PD, mutations in single genes, such as *α-synuclein* and *parkin*, have been causally linked with early-onset heritable forms of the disease. While these observations inform our ability to understand the pathophysiological basis for PD, they do not explain the vast majority of late-onset PD cases where the patient seemingly does not carry a mutation, or the reason why, in some cases (such as carriers for mutations in *lrrk2*), disease onset is late. Since the advent of more powerful statistical genetics methods, genome-wide association studies (GWAS) have been applied to numerous human disorders including PD and AD. These studies have highlighted several "risk factors"—putative allelic variants of specific genes. However, GWAS identified that risk factors need more experiments and further analysis to be concretely linked to a given disease. First, GWAS identify the relative risk (odds ratio) of getting a disease when someone carries a region of DNA identified through linkage disequilibrium of single nucleotide polymorphisms (SNPs). In rare cases, the statistical peak of a given SNP correlates very strongly to a single genetic locus and might be bolstered by the concomitant presence of a known expression variant at that locus (e.g., eQTL). More frequently, however, a disease SNP identifies a region of DNA that contains several genes, each of which has the likelihood of being linked to a specific disease. Furthermore, the SNP may also identify a region of DNA that contains *cis*-regulatory elements such as enhancers and regions participating in long-range interactions. These elements may ultimately control the expression of a downstream gene in the best-case scenario or a gene much further away that is itself "unlinked" to the disease (Smemo et al., 2014). Simpler model organisms such as *Drosophila*, that are genetically streamlined (i.e., contain fewer homologs/isoforms per gene), have excellent tools for genetic analysis, share many of the genes found in humans, and offer the possibility of analyzing the role of each gene putatively suggested by GWAS in a given disease. Finally, most GWAS have led to the identification of common allelic variants that confer small risk (compare this to rare alleles that confer large risk, e.g., mutations that are completely penetrant and assure disease presence). This means that these so-called disease alleles are widely represented in the population at a percentage much higher than a given disease percentage. What determines whether a person carrying a certain complement of risk alleles will develop a disease? Is it simply a matter of cumulative burden (the more risk factors you have, the more likely you are to develop the disease)? Or is it a consequence of interactions between each individual's risk complement together with their environmental exposure and age? Modeling these possibilities is an exercise in correlation when working with humans but in models such as flies, researchers are able to perform controlled experiments to gain better understanding of complex interactions.

2.3 Environment

Environmental influences in disease have been appreciated for a long time, and yet this remains a catch-all for any extra-genetic factor that contributes to a disease. As a result, environmental factors can vary from dietary habits to infection to toxin exposure. In the context of neurodegenerative disorders such as PD, exposure to toxins that target mitochondria deserves a special mention (Sherer, Betarbet, & Greenamyre, 2001). Through some exceptional investigational work, it was discovered in the early 1980s that a group of patients that had reported to the clinic with substance abuse and overdose were showing classical Parkinsonian symptoms. It was then deduced that the cause of this sudden onset of PD-like behavior was contamination of a prescribed drug (Meperidine) with small amounts of 1-methyl-4-phenyl-1,2,3,6-tetrahydropyridine (MPTP) (Langston, Ballard, Tetrud, & Irwin, 1983). Over the next several years, the cellular impact of MPTP exposure (and that of its active derivative MPP+) was clarified. MPP+ predominantly inhibits mitochondrial Complex I, resulting in the production of abnormal amounts of reactive oxygen species (ROS), eventual mitochondrial inactivity, and cell death. MPP+ seems to preferentially affect DA neurons since it is taken up by these neurons through the DA transporter (DAT). Similar to MPP+, several other toxins such as Rotenone, Paraquat, and 6-hydroxydopamine (6-OHDA) can cause selective death of DA neurons in the substantia nigra pars compacta, leading to PD-like pathology and phenotypes in mammalian models (Betarbet et al., 2000; Blesa & Przedborski, 2014; Martinez & Greenamyre, 2012). Not surprisingly, these toxins have found wide use in the laboratory for the generation of animal models of PD but also continue to be present in the environment in the form of pesticides and herbicides (e.g., Paraquat and Rotenone) (Caudle, Guillot, Lazo, & Miller, 2012). Thus, their effect on the population and their potential to be major contributors to PD prevalence is appreciable.

Several studies have suggested that DA neurons are preferentially sensitive to high ROS (Wang & Michaelis, 2010). This may be due to the DA biosynthetic process, the high demand on mitochondrial respiration (leading to a generally high level of ROS in these cells), or some other idiosyncrasy of these neurons such as their prolific branching. On the whole, however, increased and chronic exposure to ROS leads to oxidative damage to both

DNA and proteins, leading to cellular signaling events that might may lead to age-related cellular damage, as outlined earlier. In fact, a parsimonious explanation for such damage is that they accelerate or cooperate with normal aging process in a cell. If this were correct, cellular markers of aging are likely to be observed in cells exposed to high levels of ROS. While this is generally true in that protein and DNA damage, changes in mitochondrial bioenergetics, and autophagy are indeed observed, more precise and unique identifiers of aging are required to definitively test this hypothesis.

2.4 Noncell Autonomous Interactions

Although one often tacitly assumes a cell-autonomous basis for cellular pathology in neurodegenerative diseases, the situation is undoubtedly more complex. For instance, various epidemiological studies have shown a clear connection between Type II diabetes and PD/AD (Aviles-Olmos, Limousin, Lees, & Foltynie, 2013). These observations support the idea of an interaction between aging, environment, and genes in these diseases on one hand (since insulin signaling is a prominent and conserved regulator of lifespan in various species) and on the other underscore the potential for tissue interaction (noncell autonomous interactions). Simply put, the health or demise of a neuron in the brain will rely not only on intrinsic signaling pathways within the cell but will also be affected either by factors from adjoining cells, such as glia, or by molecules from other tissues that can cross the blood–brain barrier. Finding the locus of noncell autonomous interactions is crucial not just for understanding the mechanism of pathology but also for developing drugs that might target a tissue that, while not the primary site of the disease, "colludes" with the target tissue in a manner that predisposes or helps in the progression of disease. Simpler systems such as *Drosophila*, with a history of research in noncell autonomous signaling pathways, offer a powerful platform to rigorously test this idea in the context of disease.

2.5 Clinical Considerations

The interaction of gene, environment, and aging in neurodegeneration has important implications for clinical research. Although these diseases are late onset, it is widely believed that most, if not all, have a significant prodromal phase (Fig. 2). This phase, predicted to last several years, is usually clinically cryptic in that various symptoms and endo-phenotypes, either do not lead to significant reduction in quality of life to merit a visit to the clinic or are so nonspecific in nature that they cannot be correlated to an impending neurological disorder with accuracy. For example, it has been known for some time that PD patients often develop olfactory dysfunction, REM behavior sleep disorder, and chronic constipation long before they visit the clinic with typical early motor symptoms of PD (Edwards, Pfeiffer, Quigley, Hofman, & Balluff, 1991; Kranick & Duda, 2008; Postuma, Gagnon, & Montplaisir, 2013; Schenck, Bundlie, & Mahowald, 1996; Tan, Salgado, & Fahn, 1996; Uchiyama et al., 1995). However, none of these early indications in and of themselves are diagnostic of PD. One solution might be to discover early biomarkers of disease that are derived either from studies of aging or environmental insult—studies that are more amenable in model systems such as *Drosophila*. In sum, the objective is to define disease onset not as the clinical manifestation of disease, but as the *onset of cellular pathology* as defined by a combination of pathological endpoints that reflect both changes in physiology and cell biology.

3. *DROSOPHILA* AS A VERSATILE MODEL FOR INVESTIGATING AGING, GENETICS AND ENVIRONMENTAL FACTORS INVOLVED IN NEURODEGENERATION

3.1 Aging

Invertebrate model systems such as *Drosophila* and *Caenorhabditis elegans* are paradigms of choice for the investigation of mechanisms underlying aging (Kenyon, 2011; Partridge, 2011). Due to the ease of maintaining large numbers of animals under controlled conditions, the presence of several isogenic control strains, and a relatively short laboratory lifespan of around 10 weeks, flies have been used extensively to study the genetic basis for physiological aging. The general approach to understanding lifespan regulation in the fly has been to search for genetic perturbations that extend median lifespan beyond control values. Since many deleterious mutations might reduce fitness and shorten lifespan, this approach provides a filter through which genes that selectively affect lifespan can be identified. Among the many genes that have been identified as lifespan regulators, some, such as *methuselah* (*mth*), are probably fly specific (the GPCR encoded by *mth* does not have any human orthologs) (Lin, Seroude, & Benzer, 1998). However, others such as components of the insulin-like growth factor-1/insulin-like signaling pathway (IIS) or the target of rapamycin (TOR) signaling cascade have emerged as key regulators of lifespan across species (Clancy et al., 2001; Harrison et al., 2009; Holzenberger et al., 2003; Hwangbo, Gershman, Tu,

Palmer, & Tatar, 2004; Johnson, Rabinovitch, & Kaeberlein, 2013; Kapahi et al., 2004; Kenyon, Chang, Gensch, Rudner, & Tabtiang, 1993; Lin, Hsin, Libina, & Kenyon, 2001; McElwee, Bubb, & Thomas, 2003; Murphy et al., 2003; Tu, Epstein, & Tatar, 2002; Vellai et al., 2003). Furthermore, mechanistic connections between the IIS pathway and caloric restriction—a widely accepted method of extending lifespan—highlight the importance of this pathway in aging and lifespan regulation in mammals (Giannakou, Goss, & Partridge, 2008; Min, Yamamoto, Buch, Pankratz, & Tatar, 2008).

The IIS pathway reached prominence in the aging field through initial demonstrations in *C. elegans* and then in *Drosophila*, that lowering or knocking down signaling through this pathway conferred striking lifespan extension (Clancy et al., 2001; Kenyon et al., 1993; Lin et al., 2001; Tu et al., 2002). Research in several laboratories went on to show that IIS signaling through the insulin receptor inhibits activity of the transcription factor FOXO (Forkhead Box O family of transcription factors; dFoxo in flies and Daf-16 in worms) (Hwangbo et al., 2004). Thus, inhibition of the insulin receptor (or the receptor substrate Chico in flies) results in lifespan extension in a FOXO-dependent manner. Similarly, elevating FOXO levels can extend lifespan consistent with the idea that the IIS pathway controls aging through the transcriptional activity of FOXO. Further studies in worms went on to elaborate the precise pathway downstream of Insulin receptors that leads to FOXO inhibition as involving the PI3/Akt signaling cascade (Lin et al., 2001). Mechanisms underlying IIS signaling-dependent regulation of aging are more complex, however, and extend beyond the regulation of FOXO. For instance, IIS signaling can impact both proliferation and glucose uptake through related but parallel pathways. Once the role of the IIS pathway in lifespan regulation had been established in invertebrate models, mammalian experiments corroborated the conservation of this pathway in mammalian aging (Harrison et al., 2009; Holzenberger et al., 2003).

FOXO has emerged as the prime regulator of aging and has stimulated intense research into the identification of its target genes. Several studies in flies and other species have led to the identification of many FOXO target genes through both mRNA expression profiling in mutants and ChIP-Seq-based experiments to directly identify DNA-binding regions for FOXO (Alic et al., 2011; Lee, Kennedy, Tolonen, & Ruvkun, 2003; McElwee et al., 2003; Oh et al., 2006). These experiments have revealed, perhaps not surprisingly, that the effect of FOXO on lifespan extension is nonautonomous and involves communication among tissues. In flies,

FOXO expression in the fat body (a tissue analogous to the mammalian liver) extends lifespan as does its activity in the brain and musculature (Alic et al., 2011, 2014; Bai, Kang, Hernandez, & Tatar, 2013; Demontis & Perrimon, 2010; Flatt et al., 2008; Giannakou et al., 2008; Hwangbo et al., 2004). Thus, it seems reasonable that FOXO-dependent gene expression in these tissues regulates aging. For example, experiments have shown that FOXO expression in the adult musculature can also extend lifespan and confer protection against proaging agents such as increased ROS insult (Demontis & Perrimon, 2010). Insulin-like peptides (dilps), particularly dilp2, are secreted by neurosecretory cells in the fly brain and act at a distance (potentially in the gut or the fat body). *C. elegans* experiments show that signaling from neurons is required for lifespan extension, and conversely, FOXO activity in neurons is cell-autonomously required for axonal regeneration and neuronal aging (Byrne et al., 2014). Consistent with differential effects of FOXO on gene expression, recent experiments in the worm using cell sorted neurons show that the complement of genes regulated by FOXO in neurons is different from that in other tissues (Kaletsky et al., 2016). A similar situation is likely to be the case in flies since FOXO perturbation in various tissues can impact lifespan and aging phenotypes.

A primary means by which FOXO impacts aging is believed to be through the regulation of autophagy, mediated by inhibition of TOR (dTOR in flies) as part of the TOR complex TORC1 (Hansen et al., 2008; Partridge, Alic, Bjedov, & Piper, 2011). This model is in line with the general idea that increased autophagy is beneficial and is antiaging. However, emerging evidence suggests that FOXO acts in concert with other transcription factors such as HNF4A and might function in parallel to analogous signaling modules such as the recently described REPTOR/REPTOR-BP complex that functions downstream of TORC1 (Ganjam, Dimova, Unterman, & Kietzmann, 2009; Goudeau et al., 2011; Hirota et al., 2008; Tiebe et al., 2015). Cooperation between HNF4A and FOXO is interesting because the regulation is complex and context specific. In some instances, binding of both HNF4A and FOXO is required to initiate gene expression, while, in other cases, they function antagonistically. In sum, combinatorial control of gene expression by FOXO and HNF4A in concert with the activity of other key transcription factors such as REPTOR is likely to regulate aging-related mechanisms in a tissue-specific manner (Calnan & Brunet, 2008).

Connections between cellular pathways that regulate aging (such as IIS) and neurodegeneration are beginning to be explored. In an elegant set of experiments, Knight et al. have shown that perturbing IIS signaling in both worms and flies confers protection against the toxic effects of mutant α-synuclein expression (although this protection is independent of the transcriptional activity of FOXO) (Knight et al., 2014). Though both flies and worms lack any endogenous synuclein, these experiments uncover fundamental biology that can be extended to both cell lines and rodents. These experiments provide strong rationale to explore such interactions further in the background of additional mutations in neurodegeneration-associated genes to examine the generality of the observed neuroprotection (Fatima et al., 2014). Similarly, cell nonautonomous interactions can now be explored using the unique tools available to *Drosophila* researchers. The IIS signaling pathway is likely to be an important player in AD and PD in particular due to the strong epidemiological connections between these diseases and Type II diabetes, and the observation from meta-analysis that peripheral blood gene expression changes in HNF4A and PTBP1 (a protein that regulates insulin gene expression) correlate with PD progression (Santiago & Potashkin, 2014, 2015). Together, these data strongly support the need to investigate IIS signaling in the pathophysiology of neurodegenerative diseases. Further support for this possibility comes from human postmortem patient brain methylome studies which show a high enrichment of differentially methylated regions in the vicinity of genes involved in insulin signaling (Sanchez-Mut et al., 2016). *Drosophila* can serve as an excellent model system to explore this question further.

The IIS pathway appears to be a regulator not just of lifespan but also of "healthspan." As a result, in addition to living longer, flies that are deficient in the insulin signaling pathway show an improvement in other age-related phenotypes such as sleep fragmentation (Katewa et al., 2016; Metaxakis et al., 2014; Yamazaki et al., 2012). Sleep reflects a composite output of the nervous system and is significantly disturbed in neurodegenerative diseases such as PD (reviewed in this issue). Thus, not only is sleep more disturbed in these patients as compared to healthy controls, consolidated, or sound, sleep can confer a clear therapeutic benefit. Consistent with this clinical observation, several recent experiments in *Drosophila* have shown that sleep fragments with age and with exposure to ROS-generating toxins such as Paraquat (Koh, Evans, Hendricks, & Sehgal, 2006). More strikingly, knocking down the IIS pathway improves sleep such that older flies show much better consolidation as compared to similarly aged wild-type controls (Metaxakis et al., 2014). Whether this translates into a rescue of sleep phenotypes in models of neurodegeneration in the fly remains to be determined.

3.2 Genetics

Drosophila has been a powerhouse for the genetic analysis of almost every aspect of biology including disease modeling. Two approaches have been widely used to try and understand the genetic contributions in neurodegenerative diseases using *Drosophila* as a model—traditional forward genetic screens based upon unbiased searches for genes that cause or modify a phenotype of interest, and reverse genetics—in which candidate disease genes are perturbed in flies to test their ability to mimic phenotypes and cell biological outcomes of interest.

Forward genetics is facilitated by two important observations: (a) large numbers of flies can be screened in the laboratory and (b) heritable single-gene variants (mutants) that produce discernible phenotypes can be identified in the fly. Therefore, two principle themes in designing contemporary genetic screens in *Drosophila* involve a method to *randomly* perturb genes throughout the genome at a high frequency and, a well-defined, observable phenotype that maximizes signal to noise; i.e., it offers the ability to detect small changes without being overwhelmed with false positives arising due to normal variation in a given phenotype. A particularly interesting use of transposons is the generation of a large number of independent insertions that allow inducible expression of short-hairpin RNA molecules targeted against specific genes (Ni et al., 2009). In principle, these lines permit RNA interference (RNAi)-mediated gene-specific knockdown in tissues of choice (with the GAL4–UAS technique described later (Brand & Perrimon, 1993)).

Reverse genetics entails a hypothesis-driven search for gene function where the experimenter has circumstantial reason to believe that a gene of interest is involved in a biological process or pathology. The RNAi library described in the previous section offers the easiest way to knock down a given gene in a spatiotemporal context that is relevant to the biological question at hand. However, unambiguous assignment of function to a given gene comes from analyzing genetic mutations. Two important criteria for the success of a reverse genetic approach are the ability to *direct* mutagenesis to the gene of interest and a measurable and relevant phenotype. Transposon-based mutagenesis is a method of choice for reverse genetics since some transposons such as the P-element can be imprecisely excised from their insertion site by adding a source of transposase to the germline (Spradling & Rubin, 1982). Genetic deletions can also be created in a targeted manner by using site-specific homologous recombination or, more recently, the CRISPR–Cas9 system (Bassett, Tibbit, Ponting, & Liu,

2013; Gratz et al., 2013; Yu et al., 2013). Although homologous recombination with an externally introduced piece of DNA occurs very rarely in *Drosophila*, recombination-mediated DNA exchange can be stimulated by exogenously supplied recombinase enzymes and the introduction of specific DNA sequences that are recognized by these enzymes. CRISPR–Cas9-based methods rely on the ability of the Cas9 nuclease to create double-strand breaks in a region of DNA that is specified through a small RNA sequence (called guide RNA or gRNA) designed to be homologous to a gene of interest. Collectively, these tools allow experimental access to virtually every gene in the *Drosophila* genome. A combination of forward and reverse genetic methodologies makes flies an excellent system for interrogating the function of any gene in a given biological process or pathophysiology.

Conceptually, genes involved in neurodegenerative disorders such as AD and PD can be divided into three broad categories. We propose that the first class of genes (Class I) are those in which human loss- or gain-of-function mutations leads to a very high probability of disease onset, identified through linkage analysis in pedigrees. The homologs of these genes have often been knocked out in flies and have resulted in phenotypes that have richly informed our understanding of disease pathology (see other reviews in this series for detailed examples). Class II genes, on the other hand, are genetic risk factors identified through GWAS. Finally, Class III genes are the large collection of genes that have been identified through either single experiments or large-scale forward genetic screens using mutagenesis strategies outlined earlier. One way to reconcile these three groups of genes would be to consider complex neurodegenerative diseases as resulting from the perturbation of protein interaction networks—the same networks that might be compromised during aging. Thus, network modeling approaches might result in the identification of discrete disease modules consisting of pertinent Class I genes whose activity is modified by allelic variants in Class II genes. These pathways are likely to be further modified by Class III genes that can only be identified through the types of large-scale screens that are the traditional strength of *Drosophila*. Functional connections between these various classes of genes can be elaborated using classical epistasis analysis in the fly where the relative position of one gene/protein is assigned in a pathway with respect to the known function of another gene/protein. Experiments of this type are rapid and convenient in this model system since a large number of publicly curated reagents that increase or decrease the function of any gene of interest, already exist.

3.3 Environmental Exposure

To systematically test the role of exposure to various toxins such as xenobiotics, one must be able to control exposure of the model organism to a given agent. Flies can be exposed to a variety of toxins through food quite conveniently (Vrailas-Mortimer et al., 2011; Vrailas-Mortimer, Gomez, Dowse, & Sanyal, 2012). Depending on the precision required, the amount of food ingested can be monitored either in a coarse manner or with great degree of precision using methods such as the CAFÉ assay (Ja et al., 2007). Another advantage of using the fly for such experiments is the ability to expose the nervous system to toxins that have to transition across a more primitive glia-based blood–brain barrier (though this imposes some limits on the translational potential of these findings). Cellular responses to ROS insult have also been monitored through the development of new reagents such as ro-GFP (short-term response) as well as reporter constructs based on GSTD1 transcription in response to oxidative stress (ARE::GFP) (Liu, Celotto, Romero, Wipf, & Palladino, 2012; Sykiotis & Bohmann, 2008). Some pitfalls of using feeding to introduce ROS are uneven exposure based on varying dietary intake and systemic exposure to ROS that might confound analysis of tissue-specific responses. These concerns can be partially mitigated either by monitoring and grouping flies based on how much toxin they ingest or by standardizing exposure through more precise feeding. Additionally, dietary delivery of ROS-generating agents is also likely to directly impact the gut and, through interaction with gut-resident microbes, influence various aspects of physiology. A more directed method of delivering ROS insult to specific tissues involves genetically encoded ROS-generating agents such as KillerRed—a variant of GFP that produces ROS upon stimulation with green light (Bulina et al., 2006). KillerRed has been used widely in *C. elegans* for cell ablation experiments and in mammalian cell lines and we have recently made these reagents in the fly based on the GAL4–UAS system (Kobayashi et al., 2013). Initial testing shows that KillerRed expression can be controlled predictably with the GAL4 system and our hope is that we can target and calibrate ROS generation by controlling expression, strength, and duration of exposure to green light.

Interactions of ROS with genetic mutants have often been tested, for example, to examine the sensitivity of *parkin* mutants to oxidative stress (Trinh et al., 2008; Whitworth et al., 2005). In general, readouts for these experiments are lifespan and motor behavior, which, though reasonable and measurable, do not offer insight into precise cellular changes. We are currently investing heavily in profiling experiments (both gene expression and quantitative proteomics) to identify genes that are regulated by a combination of genetic background, aging, and toxin exposure. We believe that intersectional experiments such as these are likely to lead to the identification of genes that are uniquely regulated by a combination of these factors. Of course, the power of the *Drosophila* system is that once identified, causal relationships and interactions between single or multiple genes and pathways in aging, toxin response, and neurodegeneration can be readily tested. Additionally, both deleterious and protective responses identified through such studies can be compared to gene expression changes seen in patient samples to verify relevance. Components of the IIS signaling pathway are prime candidates and expression changes in HNF4A and PTBP1 during PD progression in patients offers hope that this approach will bear fruit.

3.4 Noncell Autonomous Interactions

Flies are an excellent system for testing complex noncell autonomous interactions between signaling pathways. Spatiotemporally regulated expression of genes of interest and RNAi with the GAL4–UAS system has been a workhorse ever since its implementation some 25 years ago. Over the years, the GAL4–UAS system has been modified and made more versatile by the addition of temporal control (GeneSwitch system; GAL80 repressor system) and by the generation of several thousand GAL4 driver lines, many of which have very narrow range of expression with some expression in only a handful of cells (Jenett et al., 2012; Osterwalder, Yoon, White, & Keshishian, 2001). In order to perturb two cell types or tissues independently of one another, an additional method called the LexA-AOP binary system has been developed—similar in principle to the GAL4–UAS system (Lai & Lee, 2006). In this method, targeted expression of the transcription factor LexA drives expression of a gene of interest placed downstream of the LexA operator sequence (AOP). As a result, one can easily design an experiment combining GAL4–UAS with LexA-AOP to perturb two tissues in a controlled manner at once. Although this composite system has mostly been applied to developmental biology and lineage tracing, it can equally be used to decipher noncell autonomous interactions in the context of aging and neurodegeneration.

3.5 Selective Advantages and Drawbacks of *Drosophila*

The fly affords many advantages over other model systems as is readily apparent from the series of reviews in this issue (see Fig. 3 for SWOT analysis).

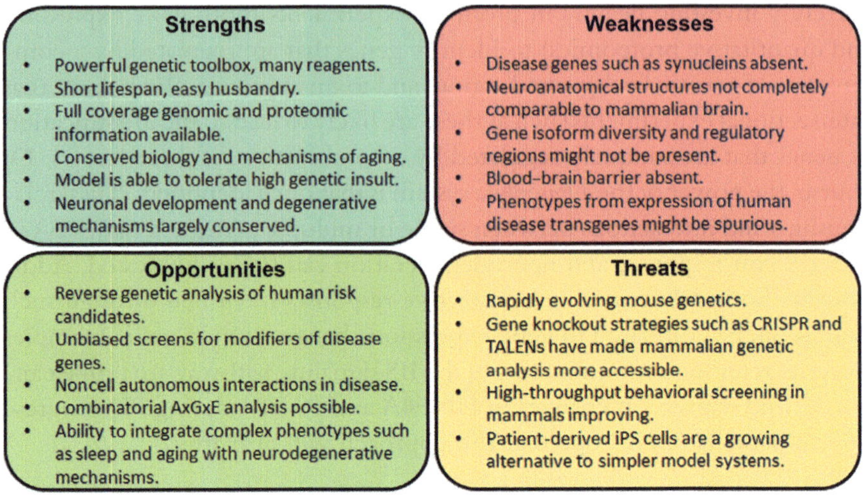

Fig. 3 An SWOT analysis of the *Drosophila* model system in neurodegeneration research. Traditional strengths of flies are coupled with emerging technologies to result in specific opportunities to addressing the question of gene, environment, and age interactions. Several potential weaknesses are also mentioned that must be kept in mind by the investigator especially when translating fly findings to clinical research. Finally, mammalian models are improving rapidly due to newer techniques of gene targeting such as CRISPR, but flies retain some significant advantages such as low cost, short generation time, and unparalleled tools for genetic analysis.

The fly community has one of the largest preexisting collections of reagents. This includes not just the various stock centers in Bloomington, Vienna, and Tokyo but also the large number of laboratories that use the fruit fly as a model. When combined with the general rapidity of fly genetics and a short lifespan, this leads to quick hypothesis testing. This feature is particularly attractive when interrogating a large number of genes (e.g., human risk factor candidates) for their potential involvement in the pathophysiology of a given disease.

Another advantage to using the fly is the ability of this animal to tolerate a large amount of "genetic insult." For example, the fly eye has been extensively used to study neurodegeneration through either the expression of mutant proteins (fly and human) selectively in the eye, or the knockout (or knockdown) of specific genes. As a result, the eye serves as an in vivo test tube in which to carry out genetic analysis. Excellent genetic tools allow the experimenter to perturb multiple genes in a controlled manner and position candidates within molecular pathways with ease using tests for epistasis.

Experiments of this type remain relatively cumbersome and time consuming in mammalian models.

In spite of all these remarkable advantages, there remain some limitations and constraints of the fly model. The fly genome is highly orthologous to vertebrates, but the absence of evolutionarily driven isoforms found in humans also precludes a nuanced analysis of these genes. In addition, several disease-causing genes are completely absent in flies, for example, *α-synuclein*, *mecp2*, or *c9orf72*. Under these conditions, investigators have often expressed human disease-related mutant forms of these proteins in flies and tested for pathology. For instance, expression of multiple nucleotide repeats in the *c9orf72* gene that are correlated with amyotrophic lateral sclerosis in the fly eye leads to progressive cell death and degeneration (Xu et al., 2013). Assuming that such phenotypes reflect gain-of-function pathology that closely approximates pathological mechanisms in human neurons, these models can be used for further hypothesis testing and screening. However, whether the pathology is indeed similar or not needs deeper investigation on a case-by-case basis. Given that human disease proteins have been expressed in as evolutionarily distant organisms as yeast to model protein dysfunction, and that studies have generally shown a conservation of protein-binding partners and regulatory interactions in such a naïve environment, such approaches are likely to remain relevant and powerful (Tardiff et al., 2013).

In the context of neurodegeneration in particular, an additional point to consider is phenotypic endpoint. For instance, many studies have used impaired motor function in flies (flying, negative geotaxis, etc.) as a cognate for motor impairments in patients with diseases such as PD. While it is true that impaired motor activity in flies is likely to report on aberrant nervous system function, it remains unclear whether the pathology is truly analogous to that in patients. Often flies that are generally unwell are poor movers and it is not clear in which tissue the locomotor defect resides. It is important in all these cases to identify the cellular and molecular basis for a behavioral phenotype so that it can be most parsimoniously related to a specific human defect.

4. THE RELEVANCE OF *DROSOPHILA* IN A DRUG DISCOVERY PIPELINE

Drosophila is an exceptional system for analyzing gene function and for examining the role for various cellular pathways in health and disease. However, owing to limitations mentioned earlier and a greater need for findings in the fly to be translated to mammalian systems—the general perception in clinical research is that "patients make the best models." Certainly, this view is not without merit, but human studies are likely to offer experimental limitations that might limit a deep understanding of pathophysiological mechanisms. Moreover, in spite of genome-wide studies of large populations, the discovery of additional genes and, therefore, drug targets cannot be carried out in humans, and for that matter, not very easily in mammalian models. It is here that systems such as *Drosophila* excel, although recent progress in gene targeting methods have significantly enhanced the speed with which genes can be knocked out and manipulated in mammalian models. As a final note, another area in which the fly has a clear advantage is the possibility of combining genetics, the effect of age and environmental exposure in the same animal. One way to better connect fly biomedical research with the clinical field is to use human cell lines or, more preferably, patient-derived iPSC neurons as an intermediate model for hypothesis testing. This approach could be followed by experiments in a relevant rodent or nonhuman primate model(s) that are best suited for a specific disease (Fig. 4). Aiding this

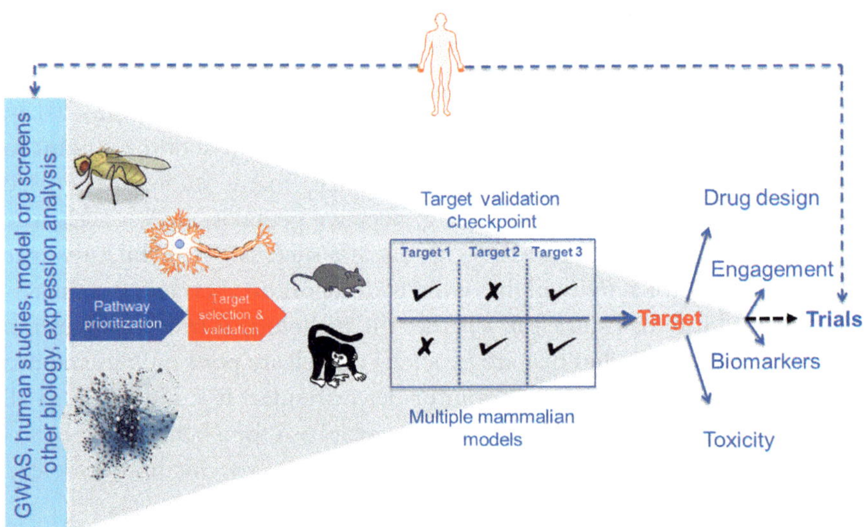

Fig. 4 A target/drug discovery pipeline that integrates *Drosophila* research as a discovery engine. Both forward genetic screens and reverse genetic analysis of putative disease-related genes from humans can rapidly generate biological understanding of disease pathways in *Drosophila*. Network biology and cell-based assays (mammalian cell lines or patient-derived iPS cells) can then be used to filter/confirm these findings before more targeted hypothesis testing in relevant rodent or nonhuman primate models of disease. Several disease models can be used to increase confidence in a given target, and if validated, this target is then prioritized for drug development, biomarker identification, and ultimately becomes the focus of clinical trials.

sequential strategy can be network biology that helps identify particular disease modules that operate within specific tissues. Finally, this pipeline has the potential to inform better synthetic models of disease in flies that go beyond the single gene—single disease model and more capably capture the complexity of human neurodegenerative diseases.

REFERENCES

Alic, N., Andrews, T. D., Giannakou, M. E., Papatheodorou, I., Slack, C., Hoddinott, M. P., ... Partridge, L. (2011). Genome-wide dFOXO targets and topology of the transcriptomic response to stress and insulin signalling. *Molecular Systems Biology*, 7, 502. http://dx.doi.org/10.1038/msb.2011.36.

Alic, N., Tullet, J. M., Niccoli, T., Broughton, S., Hoddinott, M. P., Slack, C., ... Partridge, L. (2014). Cell-nonautonomous effects of dFOXO/DAF-16 in aging. *Cell Reports*, *6*(4), 608–616. http://dx.doi.org/10.1016/j.celrep.2014.01.015.

Aviles-Olmos, I., Limousin, P., Lees, A., & Foltynie, T. (2013). Parkinson's disease, insulin resistance and novel agents of neuroprotection. *Brain*, *136*(Pt 2), 374–384. http://dx.doi.org/10.1093/brain/aws009.

Bai, H., Kang, P., Hernandez, A. M., & Tatar, M. (2013). Activin signaling targeted by insulin/dFOXO regulates aging and muscle proteostasis in Drosophila. *PLoS Genetics*, *9*(11), e1003941. http://dx.doi.org/10.1371/journal.pgen.1003941.

Barabasi, A. L., Gulbahce, N., & Loscalzo, J. (2011). Network medicine: A network-based approach to human disease. *Nature Reviews. Genetics*, *12*(1), 56–68. http://dx.doi.org/10.1038/nrg2918.

Bassett, A. R., Tibbit, C., Ponting, C. P., & Liu, J. L. (2013). Highly efficient targeted mutagenesis of Drosophila with the CRISPR/Cas9 system. *Cell Reports*, *4*(1), 220–228. http://dx.doi.org/10.1016/j.celrep.2013.06.020.

Betarbet, R., Sherer, T. B., MacKenzie, G., Garcia-Osuna, M., Panov, A. V., & Greenamyre, J. T. (2000). Chronic systemic pesticide exposure reproduces features of Parkinson's disease. *Nature Neuroscience*, *3*(12), 1301–1306. http://dx.doi.org/10.1038/81834.

Blesa, J., & Przedborski, S. (2014). Parkinson's disease: Animal models and dopaminergic cell vulnerability. *Frontiers in Neuroanatomy*, *8*, 155. http://dx.doi.org/10.3389/fnana.2014.00155.

Boya, P., Reggiori, F., & Codogno, P. (2013). Emerging regulation and functions of autophagy. *Nature Cell Biology*, *15*(7), 713–720. http://dx.doi.org/10.1038/ncb2788.

Brand, A. H., & Perrimon, N. (1993). Targeted gene expression as a means of altering cell fates and generating dominant phenotypes. *Development*, *118*(2), 401–415.

Bratic, A., & Larsson, N. G. (2013). The role of mitochondria in aging. *Journal of Clinical Investigation*, *123*(3), 951–957. http://dx.doi.org/10.1172/JCI64125.

Bulina, M. E., Chudakov, D. M., Britanova, O. V., Yanushevich, Y. G., Staroverov, D. B., Chepurnykh, T. V., ... Lukyanov, K. A. (2006). A genetically encoded photosensitizer. *Nature Biotechnology*, *24*(1), 95–99. http://dx.doi.org/10.1038/nbt1175.

Byrne, A. B., Walradt, T., Gardner, K. E., Hubbert, A., Reinke, V., & Hammarlund, M. (2014). Insulin/IGF1 signaling inhibits age-dependent axon regeneration. *Neuron*, *81*(3), 561–573. http://dx.doi.org/10.1016/j.neuron.2013.11.019.

Calnan, D. R., & Brunet, A. (2008). The FoxO code. *Oncogene*, *27*(16), 2276–2288. http://dx.doi.org/10.1038/onc.2008.21.

Caudle, W. M., Guillot, T. S., Lazo, C. R., & Miller, G. W. (2012). Industrial toxicants and Parkinson's disease. *Neurotoxicology*, *33*(2), 178–188. http://dx.doi.org/10.1016/j.neuro.2012.01.010.

Clancy, D. J., Gems, D., Harshman, L. G., Oldham, S., Stocker, H., Hafen, E., … Partridge, L. (2001). Extension of life-span by loss of CHICO, a Drosophila insulin receptor substrate protein. *Science*, *292*(5514), 104–106. http://dx.doi.org/10.1126/science.1057991.

Dauer, W., & Przedborski, S. (2003). Parkinson's disease: Mechanisms and models. *Neuron*, *39*(6), 889–909.

Dehay, B., Martinez-Vicente, M., Caldwell, G. A., Caldwell, K. A., Yue, Z., Cookson, M. R., … Bezard, E. (2013). Lysosomal impairment in Parkinson's disease. *Movement Disorders*, *28*(6), 725–732. http://dx.doi.org/10.1002/mds.25462.

Demontis, F., & Perrimon, N. (2010). FOXO/4E-BP signaling in Drosophila muscles regulates organism-wide proteostasis during aging. *Cell*, *143*(5), 813–825. http://dx.doi.org/10.1016/j.cell.2010.10.007.

Dias, V., Junn, E., & Mouradian, M. M. (2013). The role of oxidative stress in Parkinson's disease. *Journal of Parkinson's Disease*, *3*(4), 461–491. http://dx.doi.org/10.3233/JPD-130230.

Edwards, L. L., Pfeiffer, R. F., Quigley, E. M., Hofman, R., & Balluff, M. (1991). Gastrointestinal symptoms in Parkinson's disease. *Movement Disorders*, *6*(2), 151–156. http://dx.doi.org/10.1002/mds.870060211.

Fatima, S., Haque, R., Jadiya, P., Shamsuzzama, Kumar, L., & Nazir, A. (2014). Ida-1, the Caenorhabditis elegans orthologue of mammalian diabetes autoantigen IA-2, potentially acts as a common modulator between Parkinson's disease and diabetes: Role of Daf-2/Daf-16 insulin like signalling pathway. *PLoS One*, *9*(12), e113986. http://dx.doi.org/10.1371/journal.pone.0113986.

Flatt, T., Min, K. J., D'Alterio, C., Villa-Cuesta, E., Cumbers, J., Lehmann, R., … Tatar, M. (2008). Drosophila germ-line modulation of insulin signaling and lifespan. *Proceedings of the National Academy of Sciences of the United States of America*, *105*(17), 6368–6373. http://dx.doi.org/10.1073/pnas.0709128105.

Ganjam, G. K., Dimova, E. Y., Unterman, T. G., & Kietzmann, T. (2009). FoxO1 and HNF-4 are involved in regulation of hepatic glucokinase gene expression by resveratrol. *Journal of Biological Chemistry*, *284*(45), 30783–30797. http://dx.doi.org/10.1074/jbc.M109.045260.

Geisler, S., Holmstrom, K. M., Treis, A., Skujat, D., Weber, S. S., Fiesel, F. C., … Springer, W. (2010). The PINK1/Parkin-mediated mitophagy is compromised by PD-associated mutations. *Autophagy*, *6*(7), 871–878.

Giannakou, M. E., Goss, M., & Partridge, L. (2008). Role of dFOXO in lifespan extension by dietary restriction in Drosophila melanogaster: Not required, but its activity modulates the response. *Aging Cell*, 7(2), 187–198. http://dx.doi.org/10.1111/j.1474-9726.2007.00362.x.

Goudeau, J., Bellemin, S., Toselli-Mollereau, E., Shamalnasab, M., Chen, Y., & Aguilaniu, H. (2011). Fatty acid desaturation links germ cell loss to longevity through NHR-80/HNF4 in C. elegans. *PLoS Biology*, *9*(3), e1000599. http://dx.doi.org/10.1371/journal.pbio.1000599.

Gratz, S. J., Cummings, A. M., Nguyen, J. N., Hamm, D. C., Donohue, L. K., Harrison, M. M., … O'Connor-Giles, K. M. (2013). Genome engineering of Drosophila with the CRISPR RNA-guided Cas9 nuclease. *Genetics*, *194*(4), 1029–1035. http://dx.doi.org/10.1534/genetics.113.152710.

Guruharsha, K. G., Rual, J. F., Zhai, B., Mintseris, J., Vaidya, P., Vaidya, N., … Artavanis-Tsakonas, S. (2011). A protein complex network of Drosophila melanogaster. *Cell*, *147*(3), 690–703. http://dx.doi.org/10.1016/j.cell.2011.08.047.

Hansen, M., Chandra, A., Mitic, L. L., Onken, B., Driscoll, M., & Kenyon, C. (2008). A role for autophagy in the extension of lifespan by dietary restriction in C. elegans. *PLoS Genetics*, *4*(2), e24. http://dx.doi.org/10.1371/journal.pgen.0040024.

Harris, H., & Rubinsztein, D. C. (2012). Control of autophagy as a therapy for neurodegenerative disease. *Nature Reviews. Neurology*, *8*(2), 108–117. http://dx.doi.org/10.1038/nrneurol.2011.200.

Harrison, D. E., Strong, R., Sharp, Z. D., Nelson, J. F., Astle, C. M., Flurkey, K., … Miller, R. A. (2009). Rapamycin fed late in life extends lifespan in genetically heterogeneous mice. *Nature*, *460*(7253), 392–395. http://dx.doi.org/10.1038/nature08221.

Hasson, S. A., Kane, L. A., Yamano, K., Huang, C. H., Sliter, D. A., Buehler, E., … Youle, R. J. (2013). High-content genome-wide RNAi screens identify regulators of parkin upstream of mitophagy. *Nature*, *504*(7479), 291–295. http://dx.doi.org/10.1038/nature12748.

Hirota, K., Sakamaki, J., Ishida, J., Shimamoto, Y., Nishihara, S., Kodama, N., … Fukamizu, A. (2008). A combination of HNF-4 and Foxo1 is required for reciprocal transcriptional regulation of glucokinase and glucose-6-phosphatase genes in response to fasting and feeding. *Journal of Biological Chemistry*, *283*(47), 32432–32441. http://dx.doi.org/10.1074/jbc.M806179200.

Holzenberger, M., Dupont, J., Ducos, B., Leneuve, P., Geloen, A., Even, P. C., … Le Bouc, Y. (2003). IGF-1 receptor regulates lifespan and resistance to oxidative stress in mice. *Nature*, *421*(6919), 182–187. http://dx.doi.org/10.1038/nature01298.

Huttlin, E. L., Ting, L., Bruckner, R. J., Gebreab, F., Gygi, M. P., Szpyt, J., … Gygi, S. P. (2015). The BioPlex network: A systematic exploration of the human interactome. *Cell*, *162*(2), 425–440. http://dx.doi.org/10.1016/j.cell.2015.06.043.

Hwangbo, D. S., Gershman, B., Tu, M. P., Palmer, M., & Tatar, M. (2004). Drosophila dFOXO controls lifespan and regulates insulin signalling in brain and fat body. *Nature*, *429*(6991), 562–566. http://dx.doi.org/10.1038/nature02549.

Ja, W. W., Carvalho, G. B., Mak, E. M., de la Rosa, N. N., Fang, A. Y., Liong, J. C., … Benzer, S. (2007). Prandiology of Drosophila and the CAFE assay. *Proceedings of the National Academy of Sciences of the United States of America*, *104*(20), 8253–8256. http://dx.doi.org/10.1073/pnas.0702726104.

Jenett, A., Rubin, G. M., Ngo, T. T., Shepherd, D., Murphy, C., Dionne, H., … Zugates, C. T. (2012). A GAL4-driver line resource for Drosophila neurobiology. *Cell Reports*, *2*(4), 991–1001. http://dx.doi.org/10.1016/j.celrep.2012.09.011.

Johnson, S. C., Rabinovitch, P. S., & Kaeberlein, M. (2013). mTOR is a key modulator of ageing and age-related disease. *Nature*, *493*(7432), 338–345. http://dx.doi.org/10.1038/nature11861.

Kaletsky, R., Lakhina, V., Arey, R., Williams, A., Landis, J., Ashraf, J., & Murphy, C. T. (2016). The C. elegans adult neuronal IIS/FOXO transcriptome reveals adult phenotype regulators. *Nature*, *529*(7584), 92–96. http://dx.doi.org/10.1038/nature16483.

Kapahi, P., Zid, B. M., Harper, T., Koslover, D., Sapin, V., & Benzer, S. (2004). Regulation of lifespan in Drosophila by modulation of genes in the TOR signaling pathway. *Current Biology*, *14*(10), 885–890. http://dx.doi.org/10.1016/j.cub.2004.03.059.

Katewa, S. D., Akagi, K., Bose, N., Rakshit, K., Camarella, T., Zheng, X., … Kapahi, P. (2016). Peripheral circadian clocks mediate dietary restriction-dependent changes in lifespan and fat metabolism in Drosophila. *Cell Metabolism*, *23*(1), 143–154. http://dx.doi.org/10.1016/j.cmet.2015.10.014.

Kenyon, C. (2011). The first long-lived mutants: Discovery of the insulin/IGF-1 pathway for ageing. *Philosophical Transactions of the Royal Society of London. Series B, Biological Sciences*, *366*(1561), 9–16. http://dx.doi.org/10.1098/rstb.2010.0276.

Kenyon, C., Chang, J., Gensch, E., Rudner, A., & Tabtiang, R. (1993). A C. elegans mutant that lives twice as long as wild type. *Nature*, *366*(6454), 461–464. http://dx.doi.org/10.1038/366461a0.

Knight, A. L., Yan, X., Hamamichi, S., Ajjuri, R. R., Mazzulli, J. R., Zhang, M. W., … Caldwell, G. A. (2014). The glycolytic enzyme, GPI, is a functionally conserved modifier of dopaminergic neurodegeneration in Parkinson's models. *Cell Metabolism*, *20*(1), 145–157. http://dx.doi.org/10.1016/j.cmet.2014.04.017.

Kobayashi, J., Shidara, H., Morisawa, Y., Kawakami, M., Tanahashi, Y., Hotta, K., & Oka, K. (2013). A method for selective ablation of neurons in C. elegans using the phototoxic fluorescent protein, KillerRed. *Neuroscience Letters*, *548*, 261–264. http://dx.doi.org/10.1016/j.neulet.2013.05.053.

Kogan, V., Molodtsov, I., Menshikov, L. I., Shmookler Reis, R. J., & Fedichev, P. (2015). Stability analysis of a model gene network links aging, stress resistance, and negligible senescence. *Science Reports*, *5*, 13589. http://dx.doi.org/10.1038/srep13589.

Koh, K., Evans, J. M., Hendricks, J. C., & Sehgal, A. (2006). A Drosophila model for age-associated changes in sleep:wake cycles. *Proceedings of the National Academy of Sciences of the United States of America*, *103*(37), 13843–13847. http://dx.doi.org/10.1073/pnas.0605903103.

Kranick, S. M., & Duda, J. E. (2008). Olfactory dysfunction in Parkinson's disease. *Neurosignals*, *16*(1), 35–40. http://dx.doi.org/10.1159/000109757.

Lai, S. L., & Lee, T. (2006). Genetic mosaic with dual binary transcriptional systems in Drosophila. *Nature Neuroscience*, *9*(5), 703–709. http://dx.doi.org/10.1038/nn1681.

Langston, J. W., Ballard, P., Tetrud, J. W., & Irwin, I. (1983). Chronic Parkinsonism in humans due to a product of meperidine-analog synthesis. *Science*, *219*(4587), 979–980.

Lee, S. S., Kennedy, S., Tolonen, A. C., & Ruvkun, G. (2003). DAF-16 target genes that control C. elegans life-span and metabolism. *Science*, *300*(5619), 644–647. http://dx.doi.org/10.1126/science.1083614.

Lin, K., Hsin, H., Libina, N., & Kenyon, C. (2001). Regulation of the Caenorhabditis elegans longevity protein DAF-16 by insulin/IGF-1 and germline signaling. *Nature Genetics*, *28*(2), 139–145. http://dx.doi.org/10.1038/88850.

Lin, Y. J., Seroude, L., & Benzer, S. (1998). Extended life-span and stress resistance in the Drosophila mutant methuselah. *Science*, *282*(5390), 943–946.

Liu, Z., Celotto, A. M., Romero, G., Wipf, P., & Palladino, M. J. (2012). Genetically encoded redox sensor identifies the role of ROS in degenerative and mitochondrial disease pathogenesis. *Neurobiology of Disease*, *45*(1), 362–368. http://dx.doi.org/10.1016/j.nbd.2011.08.022.

Martinez, T. N., & Greenamyre, J. T. (2012). Toxin models of mitochondrial dysfunction in Parkinson's disease. *Antioxidants & Redox Signaling*, *16*(9), 920–934. http://dx.doi.org/10.1089/ars.2011.4033.

Martinez-Lopez, N., Athonvarangkul, D., & Singh, R. (2015). Autophagy and aging. *Advances in Experimental Medicine and Biology*, *847*, 73–87. http://dx.doi.org/10.1007/978-1-4939-2404-2_3.

McElwee, J., Bubb, K., & Thomas, J. H. (2003). Transcriptional outputs of the Caenorhabditis elegans forkhead protein DAF-16. *Aging Cell*, *2*(2), 111–121.

Menche, J., Sharma, A., Kitsak, M., Ghiassian, S. D., Vidal, M., Loscalzo, J., & Barabasi, A. L. (2015). Disease networks. Uncovering disease-disease relationships through the incomplete interactome. *Science*, *347*(6224), 1257601. http://dx.doi.org/10.1126/science.1257601.

Menzies, F. M., Fleming, A., & Rubinsztein, D. C. (2015). Compromised autophagy and neurodegenerative diseases. *Nature Reviews. Neuroscience*, *16*(6), 345–357. http://dx.doi.org/10.1038/nrn3961.

Metaxakis, A., Tain, L. S., Gronke, S., Hendrich, O., Hinze, Y., Birras, U., & Partridge, L. (2014). Lowered insulin signalling ameliorates age-related sleep fragmentation in Drosophila. *PLoS Biology*, *12*(4), e1001824. http://dx.doi.org/10.1371/journal.pbio.1001824.

Min, K. J., Yamamoto, R., Buch, S., Pankratz, M., & Tatar, M. (2008). Drosophila lifespan control by dietary restriction independent of insulin-like signaling. *Aging Cell*, 7(2), 199–206. http://dx.doi.org/10.1111/j.1474-9726.2008.00373.x.

Mizushima, N., Levine, B., Cuervo, A. M., & Klionsky, D. J. (2008). Autophagy fights disease through cellular self-digestion. *Nature*, *451*(7182), 1069–1075. http://dx.doi.org/10.1038/nature06639.

Muftuoglu, M., Mori, M. P., & de Souza-Pinto, N. C. (2014). Formation and repair of oxidative damage in the mitochondrial DNA. *Mitochondrion*, *17*, 164–181. http://dx.doi.org/10.1016/j.mito.2014.03.007.

Murphy, C. T., McCarroll, S. A., Bargmann, C. I., Fraser, A., Kamath, R. S., Ahringer, J., ... Kenyon, C. (2003). Genes that act downstream of DAF-16 to influence the lifespan of Caenorhabditis elegans. *Nature*, *424*(6946), 277–283. http://dx.doi.org/10.1038/nature01789.

Nakamura, T., Tu, S., Akhtar, M. W., Sunico, C. R., Okamoto, S., & Lipton, S. A. (2013). Aberrant protein s-nitrosylation in neurodegenerative diseases. *Neuron*, *78*(4), 596–614. http://dx.doi.org/10.1016/j.neuron.2013.05.005.

Ni, J. Q., Liu, L. P., Binari, R., Hardy, R., Shim, H. S., Cavallaro, A., ... Perrimon, N. (2009). A Drosophila resource of transgenic RNAi lines for neurogenetics. *Genetics*, *182*(4), 1089–1100. http://dx.doi.org/10.1534/genetics.109.103630.

Oh, S. W., Mukhopadhyay, A., Dixit, B. L., Raha, T., Green, M. R., & Tissenbaum, H. A. (2006). Identification of direct DAF-16 targets controlling longevity, metabolism and diapause by chromatin immunoprecipitation. *Nature Genetics*, *38*(2), 251–257. http://dx.doi.org/10.1038/ng1723.

Okamoto, S., & Lipton, S. A. (2015). S-Nitrosylation in neurogenesis and neuronal development. *Biochimica et Biophysica Acta*, *1850*(8), 1588–1593. http://dx.doi.org/10.1016/j.bbagen.2014.12.013.

Osiewacz, H. D., & Bernhardt, D. (2013). Mitochondrial quality control: Impact on aging and life span—A mini-review. *Gerontology*, *59*(5), 413–420. http://dx.doi.org/10.1159/000348662.

Osterwalder, T., Yoon, K. S., White, B. H., & Keshishian, H. (2001). A conditional tissue-specific transgene expression system using inducible GAL4. *Proceedings of the National Academy of Sciences of the United States of America*, *98*(22), 12596–12601. http://dx.doi.org/10.1073/pnas.221303298.

Partridge, L. (2011). Some highlights of research on aging with invertebrates, 2010. *Aging Cell*, *10*(1), 5–9. http://dx.doi.org/10.1111/j.1474-9726.2010.00649.x.

Partridge, L., Alic, N., Bjedov, I., & Piper, M. D. (2011). Ageing in Drosophila: The role of the insulin/Igf and TOR signalling network. *Experimental Gerontology*, *46*(5), 376–381. http://dx.doi.org/10.1016/j.exger.2010.09.003.

Postuma, R. B., Gagnon, J. F., & Montplaisir, J. (2013). Rapid eye movement sleep behavior disorder as a biomarker for neurodegeneration: The past 10 years. *Sleep Medicine*, *14*(8), 763–767. http://dx.doi.org/10.1016/j.sleep.2012.09.001.

Radak, Z., Zhao, Z., Goto, S., & Koltai, E. (2011). Age-associated neurodegeneration and oxidative damage to lipids, proteins and DNA. *Molecular Aspects of Medicine*, *32*(4–6), 305–315. http://dx.doi.org/10.1016/j.mam.2011.10.010.

Rolland, T., Tasan, M., Charloteaux, B., Pevzner, S. J., Zhong, Q., Sahni, N., ... Vidal, M. (2014). A proteome-scale map of the human interactome network. *Cell*, *159*(5), 1212–1226. http://dx.doi.org/10.1016/j.cell.2014.10.050.

Ryan, S. D., Dolatabadi, N., Chan, S. F., Zhang, X., Akhtar, M. W., Parker, J., ... Lipton, S. A. (2013). Isogenic human iPSC Parkinson's model shows nitrosative stress-induced dysfunction in MEF2-PGC1alpha transcription. *Cell*, *155*(6), 1351–1364. http://dx.doi.org/10.1016/j.cell.2013.11.009.

Ryan, B. J., Hoek, S., Fon, E. A., & Wade-Martins, R. (2015). Mitochondrial dysfunction and mitophagy in Parkinson's: From familial to sporadic disease. *Trends in Biochemical Sciences*, *40*(4), 200–210. http://dx.doi.org/10.1016/j.tibs.2015.02.003.

Sanchez-Mut, J. V., Heyn, H., Vidal, E., Moran, S., Sayols, S., Delgado-Morales, R., ... Esteller, M. (2016). Human DNA methylomes of neurodegenerative diseases show common epigenomic patterns. *Translational Psychiatry*, *6*, e718. http://dx.doi.org/10.1038/tp.2015.214.

Santiago, J. A., & Potashkin, J. A. (2014). System-based approaches to decode the molecular links in Parkinson's disease and diabetes. *Neurobiology of Disease*, *72*(Pt A), 84–91. http://dx.doi.org/10.1016/j.nbd.2014.03.019.

Santiago, J. A., & Potashkin, J. A. (2015). Network-based metaanalysis identifies HNF4A and PTBP1 as longitudinally dynamic biomarkers for Parkinson's disease. *Proceedings of the National Academy of Sciences of the United States of America*, *112*(7), 2257–2262. http://dx.doi.org/10.1073/pnas.1423573112.

Schenck, C. H., Bundlie, S. R., & Mahowald, M. W. (1996). Delayed emergence of a parkinsonian disorder in 38% of 29 older men initially diagnosed with idiopathic rapid eye movement sleep behaviour disorder. *Neurology*, *46*(2), 388–393.

Shaltouki, A., Sivapatham, R., Pei, Y., Gerencser, A. A., Momcilovic, O., Rao, M. S., & Zeng, X. (2015). Mitochondrial alterations by PARKIN in dopaminergic neurons using PARK2 patient-specific and PARK2 knockout isogenic iPSC lines. *Stem Cell Reports*, *4*(5), 847–859. http://dx.doi.org/10.1016/j.stemcr.2015.02.019.

Sherer, T. B., Betarbet, R., & Greenamyre, J. T. (2001). Pathogenesis of Parkinson's disease. *Current Opinion in Investigational Drugs*, *2*(5), 657–662.

Smemo, S., Tena, J. J., Kim, K. H., Gamazon, E. R., Sakabe, N. J., Gomez-Marin, C., ... Nobrega, M. A. (2014). Obesity-associated variants within FTO form long-range functional connections with IRX3. *Nature*, *507*(7492), 371–375. http://dx.doi.org/10.1038/nature13138.

Spradling, A. C., & Rubin, G. M. (1982). Transposition of cloned P elements into Drosophila germ line chromosomes. *Science*, *218*(4570), 341–347.

Sykiotis, G. P., & Bohmann, D. (2008). Keap1/Nrf2 signaling regulates oxidative stress tolerance and lifespan in Drosophila. *Developmental Cell*, *14*(1), 76–85. http://dx.doi.org/10.1016/j.devcel.2007.12.002.

Tan, A., Salgado, M., & Fahn, S. (1996). Rapid eye movement sleep behavior disorder preceding Parkinson's disease with therapeutic response to levodopa. *Movement Disorders*, *11*(2), 214–216. http://dx.doi.org/10.1002/mds.870110216.

Tanaka, K., & Matsuda, N. (2014). Proteostasis and neurodegeneration: The roles of proteasomal degradation and autophagy. *Biochimica et Biophysica Acta*, *1843*(1), 197–204. http://dx.doi.org/10.1016/j.bbamcr.2013.03.012.

Tardiff, D. F., Jui, N. T., Khurana, V., Tambe, M. A., Thompson, M. L., Chung, C. Y., ... Lindquist, S. (2013). Yeast reveal a "druggable" Rsp5/Nedd4 network that ameliorates alpha-synuclein toxicity in neurons. *Science*, *342*(6161), 979–983. http://dx.doi.org/10.1126/science.1245321.

Tiebe, M., Lutz, M., De La Garza, A., Buechling, T., Boutros, M., & Teleman, A. A. (2015). REPTOR and REPTOR-BP regulate organismal metabolism and transcription downstream of TORC1. *Developmental Cell*, *33*(3), 272–284. http://dx.doi.org/10.1016/j.devcel.2015.03.013.

Trempolec, N., Dave-Coll, N., & Nebreda, A. R. (2013). SnapShot: p38 MAPK signaling. *Cell*, *152*(3), 656–656.e1. http://dx.doi.org/10.1016/j.cell.2013.01.029.

Trinh, K., Moore, K., Wes, P. D., Muchowski, P. J., Dey, J., Andrews, L., & Pallanck, L. J. (2008). Induction of the phase II detoxification pathway suppresses neuron loss in Drosophila models of Parkinson's disease. *Journal of Neuroscience: The Official Journal of the Society for Neuroscience*, *28*(2), 465–472. http://dx.doi.org/10.1523/JNEUROSCI.4778-07.2008.

Tu, M. P., Epstein, D., & Tatar, M. (2002). The demography of slow aging in male and female Drosophila mutant for the insulin-receptor substrate homologue chico. *Aging Cell*, *1*(1), 75–80.

Uchiyama, M., Isse, K., Tanaka, K., Yokota, N., Hamamoto, M., Aida, S., ... Okawa, M. (1995). Incidental Lewy body disease in a patient with REM sleep behavior disorder. *Neurology, 45*(4), 709–712.

Vellai, T., Takacs-Vellai, K., Zhang, Y., Kovacs, A. L., Orosz, L., & Muller, F. (2003). Genetics: Influence of TOR kinase on lifespan in C. elegans. *Nature, 426*(6967), 620. http://dx.doi.org/10.1038/426620a.

Vincow, E. S., Merrihew, G., Thomas, R. E., Shulman, N. J., Beyer, R. P., MacCoss, M. J., & Pallanck, L. J. (2013). The PINK1-Parkin pathway promotes both mitophagy and selective respiratory chain turnover in vivo. *Proceedings of the National Academy of Sciences of the United States of America, 110*(16), 6400–6405. http://dx.doi.org/10.1073/pnas.1221132110.

Vrailas-Mortimer, A., del Rivero, T., Mukherjee, S., Nag, S., Gaitanidis, A., Kadas, D., ... Sanyal, S. (2011). A muscle-specific p38 MAPK/Mef2/MnSOD pathway regulates stress, motor function, and life span in Drosophila. *Developmental Cell, 21*(4), 783–795. http://dx.doi.org/10.1016/j.devcel.2011.09.002.

Vrailas-Mortimer, A., Gomez, R., Dowse, H., & Sanyal, S. (2012). A survey of the protective effects of some commercially available antioxidant supplements in genetically and chemically induced models of oxidative stress in Drosophila melanogaster. *Experimental Gerontology, 47*(9), 712–722. http://dx.doi.org/10.1016/j.exger.2012.06.016.

Wang, G., & Mao, Z. (2014). Chaperone-mediated autophagy: Roles in neurodegeneration. *Translational Neurodegeneration, 3*, 20. http://dx.doi.org/10.1186/2047-9158-3-20.

Wang, X., & Michaelis, E. K. (2010). Selective neuronal vulnerability to oxidative stress in the brain. *Frontiers in Aging Neuroscience, 2*, 12. http://dx.doi.org/10.3389/fnagi.2010.00012.

Whitworth, A. J., Theodore, D. A., Greene, J. C., Benes, H., Wes, P. D., & Pallanck, L. J. (2005). Increased glutathione S-transferase activity rescues dopaminergic neuron loss in a Drosophila model of Parkinson's disease. *Proceedings of the National Academy of Sciences of the United States of America, 102*(22), 8024–8029. http://dx.doi.org/10.1073/pnas.0501078102.

Wong, Y. C., & Holzbaur, E. L. (2015). Autophagosome dynamics in neurodegeneration at a glance. *Journal of Cell Science, 128*(7), 1259–1267. http://dx.doi.org/10.1242/jcs.161216.

Xu, Z., Poidevin, M., Li, X., Li, Y., Shu, L., Nelson, D. L., ... Jin, P. (2013). Expanded GGGGCC repeat RNA associated with amyotrophic lateral sclerosis and frontotemporal dementia causes neurodegeneration. *Proceedings of the National Academy of Sciences of the United States of America, 110*(19), 7778–7783. http://dx.doi.org/10.1073/pnas.1219643110.

Yamazaki, M., Tomita, J., Takahama, K., Ueno, T., Mitsuyoshi, M., Sakamoto, E., ... Kume, K. (2012). High calorie diet augments age-associated sleep impairment in Drosophila. *Biochemical and Biophysical Research Communications, 417*(2), 812–816. http://dx.doi.org/10.1016/j.bbrc.2011.12.041.

Yu, Z., Ren, M., Wang, Z., Zhang, B., Rong, Y. S., Jiao, R., & Gao, G. (2013). Highly efficient genome modifications mediated by CRISPR/Cas9 in Drosophila. *Genetics, 195*(1), 289–291. http://dx.doi.org/10.1534/genetics.113.153825.

CHAPTER EIGHT

Unraveling the Neurobiology of Sleep and Sleep Disorders Using *Drosophila*

L. Chakravarti, E.H. Moscato, M.S. Kayser[1]
Perelman School of Medicine at the University of Pennsylvania, Philadelphia, PA, United States
[1]Corresponding author: e-mail address: kayser@mail.med.upenn.edu

Contents

Abstract

Sleep disorders in humans are increasingly appreciated to be not only widespread but also detrimental to multiple facets of physical and mental health. Recent work has begun to shed light on the mechanistic basis of sleep disorders like insomnia, restless legs syndrome, narcolepsy, and a host of others, but a more detailed genetic and molecular understanding of how sleep goes awry is lacking. Over the past 15 years, studies in *Drosophila* have yielded new insights into basic questions regarding sleep function and regulation. More recently, powerful genetic approaches in the fly have been applied toward studying primary human sleep disorders and other disease states associated with dysregulated sleep. In this review, we discuss the contribution of *Drosophila* to the landscape of sleep biology, examining not only fundamental advances in sleep neurobiology but also how flies have begun to inform pathological sleep states in humans.

Current Topics in Developmental Biology, Volume 121
ISSN 0070-2153
http://dx.doi.org/10.1016/bs.ctdb.2016.07.010

1. WHAT IS SLEEP, AND WHAT CONTROLS SLEEP?

Sleep, a reversible state of quiescence controlled by homeostatic and circadian factors, is a highly conserved behavior across most species (Campbell & Tobler, 1984). Despite the ubiquitous need for sleep, we lack a complete understanding of how sleep is regulated and what purpose it serves. Over the past 15 years, work in a genetic model organism, *Drosophila melanogaster*, has provided numerous insights into these questions. Here, we review factors and brain regions found to control sleep–wake states and some proposed functions of sleep revealed from studies in *Drosophila*. We then discuss the use of *Drosophila* as a model system for studying primary sleep disorders and the role of sleep disturbances in other neuropsychiatric conditions.

1.1 Defining Sleep in *Drosophila*: Behavioral and Electrophysiological Correlates

Sleep can be defined by behavioral and electrophysiological criteria. The behavioral definition of sleep includes: (1) prolonged behavioral quiescence, (2) assumption of a stereotypical posture, (3) increased arousal threshold, and (4) state reversibility upon stimulation (Campbell & Tobler, 1984). In the late 1930s, electroencephalogram (EEG) studies in humans described electrical signatures of sleep (Davis, Davis, Loomis, Harvey, & Hobart, 1937; Dynes, 1947). Sleep stages were further defined, with EEG correlates of wake, rapid eye movement (REM), and stage I–IV nonrapid eye movement (NREM) sleep (Williams, 1971). Due to the focus on the EEG as a method to study sleep, animal models were mainly limited to mammals and birds, organisms with cortical structures that produce EEG correlates (Sehgal & Mignot, 2011). However, the characterization of behaviorally defined sleep-like states in hundreds of vertebrate and invertebrate species suggests that sleep is an evolutionarily conserved behavior (Campbell & Tobler, 1984). Simpler organisms can thus be used to dissect mechanisms pertaining to sleep, and are attractive due to the relative ease of completing large, unbiased screens, and the myriad forward and reverse genetics approaches.

Seminal studies in the 2000s used behavioral criteria to define sleep-like states in *Drosophila* and later in *Caenorhabditis elegans* and *Danio rerio* (zebrafish) (Hendricks et al., 2000; Raizen et al., 2008; Shaw, Cirelli, Greenspan, & Tononi, 2000; Zhdanova, Wang, Leclair, & Danilova, 2001). In addition, electrophysiological and imaging correlates of sleep–wake states show decreased neuronal activity concurrent with behaviorally defined sleep in *Drosophila*. Monitoring of local field potentials (LFPs) in the *Drosophila* brain during locomotor activity and quiescence shows that prolonged rest correlates with decreased spike-like activity and diminished LFP power across all frequencies (Nitz, van Swinderen, Tononi, & Greenspan, 2002). More recently, in vivo calcium imaging showed decreased intracellular neuronal calcium levels during sleep and larger calcium transients in response to stimuli during wake vs sleep states (Bushey, Tononi, & Cirelli, 2015). Electrophysiological and other studies also provide evidence for different sleep stages in *Drosophila*, indicating that sleep in *Drosophila* is not a homogenous rest state (van Alphen, Yap, Kirszenblat, Kottler, & van Swinderen, 2013).

Given that both behavioral and electrophysiological features of sleep in *Drosophila* are reminiscent of mammalian sleep, it stands to reason that many underlying molecular mechanisms will also be conserved from *Drosophila* to humans. Indeed, factors controlling sleep–wake states borne out of genetic approaches in *Drosophila* have already emerged as having analogous functions in humans. Thirty years after the circadian gene *period* (PER) was discovered in *Drosophila*, genetic analyses of humans with familial advanced sleep phase syndrome, delayed sleep phase syndrome, and extreme diurnal preference revealed various mutations in human PER homologs (Kay & Wager-Smith, 2000; Konopka & Benzer, 1971). Additionally, forward genetic screens in flies have implicated potassium channels in the regulation of sleep (discussed in detail later), and autoantibodies against potassium channels are associated with the human disease Morvan's syndrome, which often manifests with severe insomnia (Josephs et al., 2004). Lastly, drugs such as caffeine, amphetamine, and modafinil have similar effects on sleep and wake in flies and humans, demonstrating the molecular conservation of sleep regulatory mechanisms (Cirelli, 2009).

Genes discovered through human studies appear to have similar functions in flies as well: a study of human short sleepers uncovered a mutation in *Dec2*; introducing this mutation into *Drosophila* produced a similarly short-sleeping phenotype (He et al., 2009). In addition, human genome-wide association studies for sleep duration identified a variant in a gene encoding an ATP-sensitive potassium channel subunit, ABCC9, that accounts for genetic differences in sleep time, and disruption of the conserved *Drosophila* homolog leads to a significant loss of sleep (Allebrandt et al., 2013). These findings link *Drosophila* and human sleep and circadian rhythms at the functional and molecular levels, validating the use of *Drosophila* for modeling sleep and sleep disorders.

1.2 Factors Regulating Sleep–Wake States in *Drosophila*

1.2.1 Ion Channels

The finding that *Drosophila* sleep, along with use of the high-throughput *Drosophila* activity monitoring (DAM) system to monitor rest–activity states, set the stage for forward genetic sleep screens, heralding discovery of new sleep-regulating genes (Hendricks et al., 2000; Shaw et al., 2000). Many of the genes uncovered by unbiased screens encode ion channels and channel regulators (Cirelli, 2009). A mutation in the voltage-dependent potassium channel, *Shaker*, reduces sleep to one-third of wild-type amounts (Cirelli et al., 2005). Loss-of-function mutations in a modulatory subunit of *Shaker*, *Hyperkinetic* (*Hk*), produce a similar short-sleeping phenotype with correlated memory deficits, suggesting sleep amount may not be reflective of sleep need (Bushey, Huber, Tononi, & Cirelli, 2007). A separate screen uncovered *SLEEPLESS* (*sss*), a GPI-anchored membrane protein that regulates protein levels and activity of the Shaker channel. Disruption of SSS function dramatically reduces sleep and impairs sleep homeostasis (Koh et al., 2008). In addition to its regulation of Shaker, SSS also antagonizes nAChRs, reducing excitatory synaptic transmission with a large effect on sleep. Pharmacological antagonism or RNAi specific to nAChR subunits can rescue sleep deficits of *sss* mutants (Wu, Robinson, & Joiner, 2014). SSS thus interacts with multiple ion channels to regulate sleep.

The nAChR was implicated in a more recent screen for short-sleeping mutants that identified *Redeye* (*rye*), a gene encoding a nAChR α subunit (Shi, Yue, Kuryatov, Lindstrom, & Sehgal, 2014). *rye* mutants are short sleepers and RYE protein levels cycle independent of circadian rhythms and increase with increased waking time and after sleep deprivation, suggestive of a molecular marker of sleep drive. Overall, these studies along with the association of potassium channels in a human disease with insomnia indicate a robust and cross-species role for ion channels and their regulators in modulating sleep amount (Abou-Zeid, Boursoulian, Metzer, & Gundogdu, 2012).

1.2.2 Neurotransmitter Systems

Sleep- or wake-promoting roles of several neurotransmitter systems are conserved between *Drosophila* and mammals (for a detailed review, see Nall & Sehgal, 2014). Dopamine, octopamine (an analog of mammalian norepinephrine), and histamine are wake-promoting in *Drosophila*, while GABA and serotonin promote sleep; these effects are similar to those seen in mammals with the exception of serotonin, which has a less clear role in mammalian sleep (Sehgal & Mignot, 2011).

Hyperactive *fumin* flies and another short-sleeping line contain mutations in the dopamine transporter (DAT), which is needed for dopamine reuptake from the synaptic cleft (Kume, Kume, Park, Hirsh, & Jackson, 2005; Wu, Koh, Yue, Joiner, & Sehgal, 2008). Loss of dopamine receptor (dDopR) function or inhibitors of dopamine synthesis promote sleep, while drugs that increase dopaminergic tone promote arousal in *Drosophila* (Andretic, van Swinderen, & Greenspan, 2005; Lebestky et al., 2009). Specific brain areas are involved in mediating the effects of dopamine in *Drosophila*. Dopamine receptor D1 (DA1) mutant flies do not respond to wake-promoting effects of caffeine, a deficit that is rescued by expression of DA1 in the *Drosophila* mushroom body (MB), which is involved in learning and memory as well as regulation of sleep (Andretic, Kim, Jones, Han, & Greenspan, 2008; Joiner, Crocker, White, & Sehgal, 2006; Pitman, McGill, Keegan, & Allada, 2006). Recent work identified MB-projecting dopaminergic neurons in the Paired Anterior Medial (PAM) cluster as necessary for the wake-promoting effect of caffeine (Nall et al., 2016). Further, dopaminergic inputs to the MB specifically activate wake-promoting mushroom body output neurons (MBONs) (Sitaraman, Aso, Rubin, & Nitabach, 2015) and activation of DA1 in the MB specifically rescues sleep deprivation-related learning impairments (Seugnet, Suzuki, Vine, Gottschalk, & Shaw, 2008).

Dopamine also acts in the dorsal fan-shaped body (dFSB), a sleep-promoting area (Donlea, Thimgan, Suzuki, Gottschalk, & Shaw, 2011). Sensitivity to dopamine in the dFSB mediates its wake-promoting effect, as the short-sleeping phenotype of DAT mutants depends on DA1 expression in the dFSB (Liu, Liu, Kodama, Driscoll, & Wu, 2012; Ueno et al., 2012). Activation of specific dopaminergic projections to the dFSB drives wakefulness in a DA1-dependent manner, possibly through suppression of normal sleep-promoting dFSB activity (Liu et al., 2012; Ueno et al., 2012).

Like norepinephrine in mammals, octopamine promotes wake in *Drosophila*: reduced levels of octopamine lead to increases in sleep amount and arousal threshold, and excitation of specific octopaminergic cells in the anterior superior medial cluster induces wakefulness (Crocker & Sehgal, 2008; Nall & Sehgal, 2014). Wake-promoting effects of octopamine are mediated through PKA activity in the pars intercerebralis (PI), a *Drosophila* analog of the mammalian hypothalamus (Crocker, Shahidullah,

Levitan, & Sehgal, 2010). Octopamine signaling may regulate the balance between metabolic demand and sleep need, as PI cells signal to insulin-producing cells, which regulate production of triglycerides (Nall & Sehgal, 2014). Norepinephrine also regulates various behaviors besides sleep in humans, such as aggression (Yanowitch & Coccaro, 2011). Interestingly, recent work in *Drosophila* showed that sleep deprivation suppresses normal male aggressive behavior in an octopamine-dependent manner; evidence suggests that the effect of sleep deprivation on aggression might be conserved in humans (Cote, McCormick, Geniole, Renn, & MacAulay, 2013; Kayser, Mainwaring, Yue, & Sehgal, 2015).

Other neurotransmitters appear to have conserved roles from flies to humans as well. The wake-promoting effect of histamine, while long appreciated in mammals, was more recently described in *Drosophila* (Oh, Jang, Sonn, & Choe, 2013), and GABA promotes sleep in both mammals and flies (Agosto et al., 2008; Gmeiner et al., 2013). Finally, while serotonin may have different effects on REM vs NREM sleep in mammals, serotonin acts in *Drosophila* on d5HT1A receptors in the MB to promote sleep (Yuan, Joiner, & Sehgal, 2006). Going forward, perhaps work in flies can help determine differential roles of serotonergic signaling within other specific sleep- and wake-promoting circuits.

Intriguingly, recent work using flies has provided new insights into the broad question of whether all wake states are the same, and conversely, whether different types of sleep deprivation are equal. While independent activation of cholinergic, dopaminergic, and octopaminergic neuronal populations can each suppress nighttime sleep, cholinergic-mediated sleep deprivation leads to significantly more rebound sleep than does dopaminergic-driven sleep loss (Seidner et al., 2015). Further, octopamine-induced deprivation actually prevents rebound sleep, coupled with an inability to recover from sleep deprivation-related learning impairments (Seidner et al., 2015). These findings may help us understand how sleep homeostasis can be induced by various interventions, such as stress or pharmacologic manipulations, based on unique profiles of wake-promoting neurotransmitters/circuits that are engaged. This work also suggests that there may be distinct consequences of sleep deprivation depending on the mechanism of sleep loss.

1.2.3 Intracellular Signaling Molecules

Intracellular signaling pathways regulate sleep–wake states in a conserved manner. The cAMP-protein kinase A (PKA)–CREB intracellular signaling pathway drives wakefulness in both flies and mammals (Graves et al., 2003;

Hendricks et al., 2001). Mutations in the adenylate cyclase *rutabaga* result in decreased cAMP and increased sleep, and disruption of *dunce* (phosphodiesterase 4) increases cAMP and decreases sleep (Hendricks et al., 2001). PKA is also necessary for the wake-promoting effects of octopamine (Crocker & Sehgal, 2008). In general, the wake-promoting role of this pathway is consistent with evidence that sleep is accompanied by reductions in neuronal activity and the idea that cAMP concentrations may be higher in active neurons during wake and lower during periods of less neuronal activity in sleep (Flavell & Greenberg, 2008; Nitz et al., 2002). As learning and memory are intimately tied to sleep in adulthood across species, cAMP/PKA signaling may link the physiological state of sleep with some of its functions (Rasch & Born, 2013).

Recently, protein degradation pathways have been implicated in controlling sleep. A forward genetic screen for wake-promoting factors uncovered *insomniac*, which encodes a BTB-domain adaptor protein for the E3-ubiquitin ligase Cul3 (Stavropoulos & Young, 2011). Reductions in both Insomniac and Cul3 decrease the duration and consolidation of sleep and impair sleep homeostasis. Pharmacologic and genetic evidence suggests Insomniac and Cul3 may function downstream of dopamine synthesis, degrading dopamine receptors or effectors to decrease arousal and allow sleep (Pfeiffenberger & Allada, 2012). Interestingly, Cul3 and a different BTB protein, BTBD9, have been implicated in the human disease restless legs syndrome (RLS; see later).

Molecules involved in several other intracellular functions also play roles in regulating sleep. Reduction of the histone acetyltransferase Elp3 during neural development results in hyperactivity and sleep loss in adult flies (Singh, Lorbeck, Zervos, Zimmerman, & Elefant, 2010). The conserved fragile X mental retardation protein, dFMRP, negatively regulates sleep, as loss of function leads to increases in sleep time, while overexpression decreases sleep time (Bushey, Tononi, & Cirelli, 2009). Heat-shock proteins BiP and Hsp83 (Naidoo, Casiano, Cater, Zimmerman, & Pack, 2007; Shaw, Tononi, Greenspan, & Robinson, 2002) and regulators of energy metabolism such as *Drosophila* insulin-like peptides and triglyceride storage genes also influence sleep (Cong et al., 2015; Thimgan, Suzuki, Seugnet, Gottschalk, & Shaw, 2010). Finally, cell cycle regulators Cyclin A and TARANIS act together to promote sleep, suggesting a repurposing of cell cycle machinery within arousal centers to control sleep in adulthood (Afonso et al., 2015; Rogulja & Young, 2012). In sum, the diverse repertoire, high redundancy, and evolutionary conservation of sleep-regulating genes suggests that sleep is intimately tied to modulation of both intracellular and extracellular processes, positioning *Drosophila* as a strategic model for dissecting these mechanisms.

1.2.4 Circadian Outputs

The classical two-process model of sleep involves interaction between sleep homeostasis, termed Process S, and a circadian process, or Process C (Borbély, 1982). How do the outputs of the circadian process control sleep, and have clock genes been implicated in regulation of sleep and its function? A *Drosophila* screen identified *wide awake* (*wake*), a cycling gene that links circadian clock output to sleep onset at dusk in flies and rodents (Liu et al., 2014). WAKE expressed in clock neurons peaks in expression at dusk and acts downstream of Clock and Cycle to upregulate $GABA_A$ receptors, which drives sleep. Additionally, pigment-dispersing factor (PDF) secreted from small ventral lateral neurons (LNvs), part of the circadian clock, acts on a subset of dorsal circadian neurons (DN1s) to suppress sleep in anticipation of dawn; DN1s respond to PDF by increasing secretion of wake-promoting diuretic hormone 31 (DH31), the fly homolog of calcitonin gene-related peptide (Kunst et al., 2014). These studies thus provide genetic and molecular insights into how the circadian arm of the two-process model exerts effects on sleep.

1.3 Brain Regions Involved in Regulating Sleep

Studies in *Drosophila* have identified specific brain regions that can be independently activated to control arousal states. Despite anatomical differences, the neural logic underpinning how specific *Drosophila* brain regions interact to control sleep and wakefulness can prove immensely useful for understanding similar processes in humans. The MBs, a learning and memory center in insects, can promote both sleep and wake depending on which neuronal subpopulations are activated (Joiner et al., 2006; Pitman et al., 2006). The MB plays a key role in relaying sensory information, and may act like the mammalian thalamus to gate sensory perception during sleep (Stopfer, 2014). Wake-promoting MB microcircuits activate glutamatergic MBONs, while the sleep-promoting circuit activates cholinergic MBONs and becomes more active after sleep deprivation, forming a component of the sleep homeostat (Sitaraman, Aso, Jin, et al., 2015). GABAergic MBONs also promote sleep (Aso, Sitaraman, et al., 2014). The MB is known to be involved in encoding associative memory, and in particular, drives experience-based attraction or repulsion to learned stimuli (Aso, Hattori, et al., 2014). Intriguingly, MBONs whose activity drives attraction were found to be sleep-promoting; separate MBONs driving repulsion promote wake (Aso, Sitaraman, et al., 2014). This may point to an adaptive relationship between processing of noxious stimuli and increased arousal.

The *Drosophila* dFSB also acts as a sleep-promoting center, as specific activation of dFSB neurons induces sleep while inhibition promotes wake (Donlea et al., 2011; Liu et al., 2012). Not only does the dFSB appear to have a role in driving baseline sleep, but recent work suggests that projections to the dFSB act as the output arm of the sleep homeostat (Donlea, Pimentel, & Miesenböck, 2014). As a sleep-promoting center, the dFSB may integrate signals from both components in the two-process model of sleep, as sleep-promoting inputs to the dFSB are differently effective at controlling sleep depending on the time of day (Cavanaugh, Vigderman, Dean, Garbe, & Sehgal, 2016). Mammalian analogs of the MB and dFSB are not immediately apparent. However, the regulation of a sleep-promoting center, the dFSB, by wake-promoting dopaminergic inputs can invoke a similar mechanism to that proposed in mammals of a "flip–flop switch" (Liu et al., 2012; Lu, Sherman, Devor, & Saper, 2006).

2. MODELING SLEEP IN HEALTH AND DISEASE USING *DROSOPHILA*

While significant progress has been made in identifying factors that control sleep and wake, we are in more nascent stages of understanding the function of sleep and its role in disease. The discovery of many sleep-regulating molecules and brain regions in *Drosophila* make it an ideal organism for studying sleep disturbances that may have perturbations in these regulatory pathways.

2.1 Sleep and Brain Development

Most species have the highest levels of sleep in early life (Jouvet-Mounier, Astic, & Lacote, 1970; Kayser, Yue, & Sehgal, 2014; Roffwarg, Muzio, & Dement, 1966; Shaw et al., 2000). Despite this highly conserved behavior, little is known about the underlying neural circuitry mediating ontogenetic sleep changes. Recent work in *Drosophila* provided new insights into this circuitry, as well as the role of early life sleep in brain development (Kayser et al., 2014; Seugnet, Suzuki, Donlea, Gottschalk, & Shaw, 2011). Compared to mature flies, young flies have increased total sleep times, higher arousal thresholds, and increased resistance to sleep deprivation

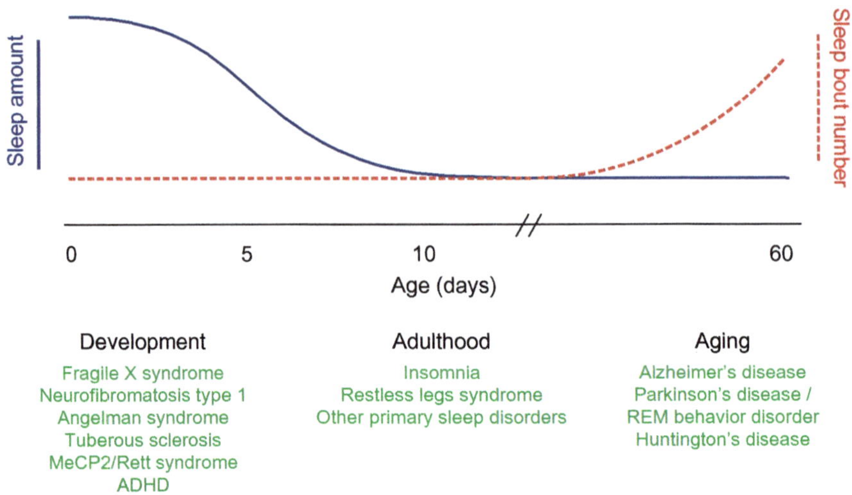

Fig. 1 Sleep in *Drosophila* recapitulates sleep changes in humans throughout the lifespan, with more, deep sleep in early life (*blue solid line*), and less consolidated sleep with aging (increased bout number, *red dashed line*; also reduced bout length, not shown). The conservation of sleep ontogenetic changes presents the opportunity to investigate sleep disorders/dysregulation associated with various diseases occurring throughout life.

(Fig. 1). In early life, hypoactivity of dopaminergic inputs to the dFSB allows increased dFSB activity, promoting sleep (Kayser et al., 2014). Since dopamine has a conserved wake-promoting role, it is possible that decreased dopaminergic signaling underlies ontogenetic changes in sleep in other species as well, including humans.

In humans, sleep disruption in infancy and childhood can lead to long-lasting cognitive impairment as well as behavioral disorders (Ednick et al., 2009; O'Brien, 2009). Likewise in *Drosophila*, sleep deprivation of juvenile flies leads to persistent deficits in short-term memory (STM) and response inhibition (Seugnet, Suzuki, Donlea, et al., 2011). How does sleep loss in early life contribute to cognitive and behavioral deficits? Certain brain regions in *Drosophila*, such as the olfactory system, undergo significant changes in young flies (Devaud, Acebes, Ramaswami, & Ferrús, 2003). Sleep deprivation specifically of young flies impedes growth of an olfactory glomerulus, VA1v, known to play a role in courtship behaviors (Dweck et al., 2015). This structural change is associated with deficits in courtship later in adulthood. The VA1v glomerulus normally undergoes rapid growth in the first few days posteclosion, and the reduction in volume seen with sleep loss is likely due to impaired addition of synapses (Kayser et al.,

2014). Sleep during development thus promotes structural brain maturation required for an innate behavior. Interestingly, glomeruli not undergoing significant growth during this period are unaffected by sleep deprivation, suggesting that rapidly growing regions of brain are most susceptible to sleep loss (Kayser et al., 2014). Thus, sleep disturbances during different developmental periods could lead to variable structural deficits, possibly with unique cognitive and behavioral outcomes. The clear delineation of circuitry controlling early life sleep in *Drosophila* makes it uniquely suited for studying cellular and genetic factors coupling early life sleep and brain development.

2.2 Functions of Sleep in the Adult Brain

At the behavioral level, it is clear that learning, memory, attention, and mood all depend on adequate sleep (Diekelmann & Born, 2010; Durmer & Dinges, 2005; Kirszenblat & van Swinderen, 2015). It has long been suggested that the behavioral quiescence and associated reduction or synchronization of neural activity during sleep serve a restorative purpose at the cellular level, which may be effected through several processes, including metabolite clearance, neuronal plasticity and synaptic homeostasis, and macromolecule biosynthesis (Cirelli, 2013; Mackiewicz et al., 2007; Xie et al., 2013).

Studies in *Drosophila* have contributed in particular to our understanding of how sleep impacts learning, memory, and behavior. We know from decades of evidence that sleep facilitates learning and memory (Rasch & Born, 2013). Precise control of sleep circuitry in the fly has allowed for a refined examination of sleep/memory interactions. Activation of the dFSB to induce sleep posttraining promotes long-term memory (LTM) in a courtship conditioning protocol that would not otherwise produce LTM, indicating a causal role for sleep in memory consolidation (Donlea et al., 2011). By contrast, sleep deprivation decreases memory formation in multiple memory paradigms (Li, Yu, & Guo, 2009; Seidner et al., 2015; Seugnet et al., 2008). Studies of short-sleeping *Drosophila* mutants also show defects in learning and memory: *Shaker* and *Hyperkinetic* mutants have impaired STM, and memory formation is compromised in short-sleeping *fumin* mutants (Bushey et al., 2007; Zhang, Yin, Lu, & Guo, 2008). Notably, studies of *Drosophila* mutants originally discovered for deficits in learning and memory also reveal sleep disruptions, potentially providing clues to molecular links between sleep, learning, and memory (Hendricks et al., 2001). Genetically or pharmacologically driving sleep has even been suggested to rescue memory defects in *dunce* and *rutabaga* mutants (Dissel et al., 2015).

Work in *Drosophila* has further revealed specific circuitry and even single neurons that link sleep to learning and memory. The *Drosophila* MBs, long implicated in learning and memory, may serve analogous functions to the mammalian hippocampus in memory encoding (Cervantes-Sandoval, Martin-Peña, Berry, & Davis, 2013). Genetic manipulations of the MB influence sensitivity to sleep deprivation-related learning deficits (Li et al., 2009; Seugnet et al., 2008; Seugnet, Suzuki, Donlea, et al., 2011; Seugnet, Suzuki, Merlin, et al., 2011). Recent work identified specific neuronal populations that link sleep to memory: dorsal paired medial (DPM) neurons, which innervate the MB to promote olfactory memory consolidation, release inhibitory GABA, and 5HT onto wake-promoting MB α'/β' neurons to drive sleep (Haynes, Christmann, & Griffith, 2015). The *Drosophila* learning and memory mutant *amnesiac* (*amn*) is deficient in a neuropeptide expressed in DPM neurons and exhibits fragmented sleep, reduced sleep latency, and impaired sleep homeostasis, indicating that DPM-derived AMN peptide may also act on the MB to promote sleep (Liu, Guo, Lu, & Guo, 2008).

Another intriguing theory emerging from work in the fly is that decreased sensory processing during sleep prevents interference-based forgetting, allowing for memory formation (Mednick, Cai, Shuman, Anagnostaras, & Wixted, 2011). A subset of dopaminergic neurons in the *Drosophila* brain innervate the MB to promote forgetting of olfactory memories, and sleep facilitates memory retention by decreasing activity of this circuit (Berry, Cervantes-Sandoval, Chakraborty, & Davis, 2015). The remarkable ability to distil aspects of the sleep–learning–memory relationship to interactions between a few neurons and brain structures highlights the usefulness of *Drosophila* for understanding key mechanisms that may be conserved in mammals.

In mammals as well as in flies, sleep appears to have a key role in modulating synaptic strength (Frank, 2015b). The synaptic homeostasis hypothesis (SHY) posits that synapses are potentiated in a specific, experience-dependent manner during wake and proportionally downscaled during sleep to increase signal-to-noise ratio and maintain plasticity in the following waking period (Tononi & Cirelli, 2003). Studies in *Drosophila* provided much of the initial evidence for SHY, including structural evidence from multiple brain regions suggesting that synaptic weights increase after normal waking or sleep deprivation and decrease after sleep (Bushey, Tononi, & Cirelli, 2011). Work in mammals suggests the relationship of sleep to synaptic homeostasis may be more complicated, as sleep loss impairs long-term potentiation of synaptic strength in some but not all brain regions, and different studies provide electrophysiological and molecular evidence for both synaptic strengthening and weakening during sleep (Frank, 2015a). Thus, although sleep seems to have a role in synaptic plasticity, the exact nature of this relationship remains to be determined. Combining existing tools of electrophysiology with behavioral experiments and structural analysis in *Drosophila* may yield new insights into how sleep influences the structure and function of synaptic connections to promote learning and memory.

2.3 *Drosophila* Studies of Primary Sleep Disorders

Since regulators of sleep–wake states are well described in *Drosophila*, it has potential to be a powerful organism for modeling primary sleep disorders. Insomnia is characterized by persistent difficulty initiating or maintaining sleep despite adequate sleep opportunity, along with associated daytime impairment (Roth, 2007). Short-sleeping *Drosophila* mutants discovered through forward genetic screens might be particularly useful for understanding the molecular underpinnings of insomnia. Decreased sleep in *Shaker*, *sss*, and *Hk* mutants mainly derives from decreased sleep bout length, indicating that flies initiate but cannot maintain sleep (Bushey et al., 2007; Cirelli et al., 2005; Koh et al., 2008). It is unlikely that these short sleepers simply require less sleep, since all three mutations reduce lifespan and *Shaker* and *Hk* mutants have memory deficits. *Insomniac* mutants also have decreased sleep bout lengths and show an increased number of sleep episodes, indicating that they repeatedly initiate sleep to meet a sleep need, but cannot stay asleep (Stavropoulos & Young, 2011). *fumin* flies, with a mutation in the DAT, have significant reductions in sleep time as well as decreased arousal threshold and memory problems, again suggesting mutants are not able to fulfill their sleep need, as with human insomnia (Kume et al., 2005; Zhang et al., 2008).

Analysis of different GABA signaling pathways may allow for a mechanistic distinction between sleep onset and sleep maintenance insomnia. *Rdl* (ionotropic $GABA_A$ receptor) mutants have increased sleep latency, as in sleep onset insomnia, whereas mutations in the metabotropic $GABA_B$ receptor reduce sleep bout duration primarily in the second half of the night, consistent with sleep maintenance insomnia (Agosto et al., 2008; Gmeiner et al., 2013). Both effects are mediated through PDF-expressing lLNvs, invoking an interesting model where faster temporal dynamics in ionotropic signaling control the switch from an awake to sleeping state, while slower, more sustained metabotropic signaling maintains a preestablished sleep state (Gmeiner et al., 2013; Parisky et al., 2008). It may be possible to combine recently discovered molecular or structural markers of sleep drive with well-established assays of learning and memory to distinguish decreased sleep need from decreased sleep ability (Liu, Liu, Tabuchi, & Wu, 2016; Shi et al., 2014).

In addition to short-sleeping mutants, selection for features of insomnia over several generations produced another *Drosophila* model, insomnia-like (ins-l). Ins-l flies sleep only 60 min a day and have trouble initiating and maintaining sleep with related impairment in a learning assay (Seugnet, Suzuki, et al., 2009). Ins-l flies have decreased lifespan and increased levels of dopamine, triglycerides, cholesterol, and free fatty acids, consistent with evidence that sleep loss increases risk of obesity (Knutson & Van Cauter, 2008), and further emphasizing the tight link between sleep and metabolism (Yurgel, Masek, DiAngelo, & Keene, 2015). In sum, *Drosophila* short-sleeping mutants provide an avenue toward a deeper understanding of the cellular/molecular basis of insomnia, perhaps pointing to novel therapeutic interventions.

Work on another primary sleep disorder, RLS, has demonstrated the power of reverse genetic approaches in *Drosophila*. RLS symptoms include an urge to move one's legs that worsens at night, fragmented sleep, and periodic limb movements during sleep; RLS has also been associated with reduced brain iron and dopamine levels (Earley, 2003; Freeman & Rye, 2013). GWAS studies revealed RLS-associated polymorphisms in BTBD9, an adaptor (like Insomniac) for the E3 ubiquitin ligase Cul3, which had been independently implicated in controlling sleep in *Drosophila* (Pfeiffenberger & Allada, 2012; Stavropoulos & Young, 2011). Loss of the *Drosophila* BTBD9 homolog, CG1826, results in fragmented sleep, increased motor activity, and decreased brain dopamine. Further analysis suggested a mechanism whereby Cul3/BTBD9 normally degrades iron-regulatory proteins to maintain normal ferritin levels and iron stores (Freeman et al., 2012). Studies in this model may help reveal other Cul3/BTBD9 complex interactors and substrates that could serve as therapeutic targets in RLS. Disorders other than RLS can be studied in *Drosophila* using a similar reverse genetics approach based on human GWAS, as discussed earlier with examination of the ABCC9 homolog in flies (Allebrandt et al., 2013). *Drosophila* models thus represent a largely untapped resource for studying primary sleep disorders.

2.4 The Role of Sleep in Neurodevelopmental Disorders

2.4.1 Single Gene Disorders and Autism Spectrum Disorder

The relationship of early sleep to structural brain development and behavior in *Drosophila* can be harnessed to deepen our understanding of pathological changes in neurodevelopmental disorders. These disorders manifest before adulthood with deficits in emotion, cognition, behavior, and motor skills (Hu, Chahrour, & Walsh, 2014). Sleep disturbances are prevalent among children with neurodevelopmental disorders and it is likely that impaired sleep–wake function both arises from and contributes to abnormalities in nervous system development (Phillips & Appleton, 2004; Robinson-Shelton & Malow, 2015). Fly models have been immensely useful for studying neurodevelopmental disorders caused by single gene disruption and more complex genetic lesions, as well as the involvement of sleep in these conditions.

Fragile X syndrome (FXS), a leading cause of inherited intellectual disability, most commonly results from trinucleotide repeat expansions in the Fragile X mental retardation 1 (*FMR1*) gene that lead to loss of Fragile X mental retardation protein (FMRP) (Bagni, Tassone, Neri, & Hagerman, 2012). Flies with loss of function of the *Drosophila* homolog, dFMRP, exhibit deficits in learning, memory, and social interaction similar to those seen in human patients; mutant flies also exhibit increased sleep duration (Bushey et al., 2009; Dockendorff et al., 2002; McBride et al., 2005). At the structural level, loss of dFMRP leads to overgrown dendritic trees and enlarged synaptic boutons, with defects in pruning (Bushey et al., 2009; Pan, Zhang, Woodruff, & Broadie, 2004; Tessier & Broadie, 2008). dFMRP overexpression has opposite effects on both synaptic structure and sleep time, but both loss and gain of dFMRP function mutants exhibit abnormal sleep homeostasis, with no sleep rebound (Bushey et al., 2009). Interestingly, sleep deprivation increases dFMRP level, suggesting a reciprocal interaction between sleep and dFMRP (Bushey et al., 2009). As dFMRP is highly expressed during the period of increased sleep time immediately following eclosion, the physiological state of sleep and dFMRP may act in concert to regulate synapse formation in a developmentally timed manner (Bushey et al., 2009; Kayser et al., 2014). These studies in *Drosophila* suggest that correcting early sleep abnormalities in FXS could benefit brain development to improve neurocognitive function.

Drosophila models of other neurodevelopmental disorders have likewise provided new insight into potential pathology at the intersection of sleep and development. The neurodevelopmental disorder neurofibromatosis type 1

(NF1) is marked by benign peripheral nerve tumors, attention and learning deficits, visual problems, and sometimes seizures (Jett & Friedman, 2010). Mutants for the *Drosophila* homolog, dNf1, have learning and growth defects resembling human disease and abnormal circadian rhythms (The et al., 1997; Williams, Su, Bernards, Field, & Sehgal, 2001). It was recently shown that loss of dNf1 also leads to a short-sleeping phenotype (Bai & Sehgal, 2015). dNf1 limits synaptic growth at the neuromuscular junction (NMJ) and interacts with the receptor tyrosine kinase Anaplastic lymphoma kinase, another regulator of synaptogenesis (Bai & Sehgal, 2015; Tsai et al., 2012). As both dFMRP and dNf1 mutants exhibit altered sleep and synaptic structure, continued investigation of FMRP and Nf1 may provide insight into both physiological and pathological aspects of the sleep–synapse relationship.

Several other single gene neurodevelopmental disorders that involve learning disabilities, autism-like features, and motor abnormalities, including MeCP2-spectrum disorders, tuberous sclerosis, and Angelman syndrome, also present with sleep disturbances in humans (Clayton-Smith & Laan, 2003; Islam & Roach, 2015; Wong, Leonard, Jacoby, Ellaway, & Downs, 2015). Overexpression of MeCP2 in *Drosophila* astrocytes, octopaminergic, and dopaminergic neurons can each independently reduce total sleep amount, with sleep loss at different times of day depending on cell type (Hess-Homeier, Fan, Gupta, Chiang, & Certel, 2014). Additionally, MeCP2 overexpression in *Drosophila* motorneurons leads to alterations in dendritic structure and consequent motor defects, suggesting links between sleep and synapse regulation (Vonhoff, Williams, Ryglewski, & Duch, 2012). Likewise, mutations in the mTOR regulators Tsc1 and Tsc2 (giga) in a *Drosophila* model of tuberous sclerosis (TS) are associated with synaptic overgrowth at the NMJ (Natarajan, Trivedi-Vyas, & Wairkar, 2013; Pan, Dong, Zhang, & Gao, 2004). TSC interactors Akt and TOR determine the length of the circadian period in *Drosophila*, indicating that circadian dysfunction may also contribute to sleep problems in TS (Zheng & Sehgal, 2010). Finally, disruption of the *Drosophila* homolog of Angelman syndrome (AS) gene UBE3A, dUbe3a, which is enriched in the learning and memory center, the MB, causes impaired locomotion, LTM, and circadian rhythm abnormalities (Wu, Bolduc, et al., 2008). As with MeCP2 manipulation, loss of dUbe3a function also affects dendrite morphogenesis, with decreased dendritic branching of sensory neurons (Lu et al., 2009). Decreased dUbe3a leads to reduced brain dopamine levels, while increased dUbe3a increases brain dopamine, pointing to a potential biochemical explanation for altered regulation of sleep–wake state in AS

(Ferdousy et al., 2011). Further analysis of sleep disturbances in these existing *Drosophila* models of neurodevelopmental disease will likely reveal interesting ways in which sleep modulates symptomology and how disease-relevant mutations interfere with sleep.

Autism spectrum disorder (ASD) encompasses conditions ranging in severity but characterized by impaired social interaction and communication, and repetitive behaviors, also associated with abnormalities in brain connectivity (American Psychiatric Association, 2013). Aside from *Drosophila* models of single gene disorders with autism-like features, many other ASD susceptibility genes, such as NRXN1 (neurexin), NLGN1 (neuroligin), and CUL3 (Codina-Solà et al., 2015; Glessner et al., 2009), have been studied in *Drosophila*. Loss of NRXN1 in *Drosophila* leads to fragmented sleep and loss of neuroligin-4 disrupts sleep through impaired GABA signaling (Li et al., 2013). As discussed earlier, the ubiquitin ligase Cul3 promotes sleep, likely via reduction of dopaminergic signaling, so it is possible that the association of Cul3 variants with ASD results from alterations in sleep architecture that increase ASD susceptibility (Pfeiffenberger & Allada, 2012; Stavropoulos & Young, 2011). In addition, several ASD-associated genes are involved in activity-dependent synaptic plasticity, and the activity-dependent cAMP–PKA pathway is known to regulate sleep (Bourgeron, 2015; Hendricks et al., 2001). Together with previously discussed sleep effects in single gene disorders, these studies in *Drosophila* point to a strong link between ASD pathogenesis and sleep.

2.4.2 Attention Deficit Hyperactivity Disorder

Another common neurodevelopmental disorder, attention deficit hyperactivity disorder (ADHD), emerges from complex combinations of genetic variants and environmental factors (Li, Chang, Zhang, Gao, & Wang, 2014; Thapar & Cooper, 2016). *Drosophila* models have identified measures of complex processes such as attention and others with striking parallels to human behavior. Children with ADHD tend to have sleep disturbances, and primary sleep disorders are often misdiagnosed as ADHD due to significant effects of sleep loss on attention (Chervin, Dillon, Bassetti, Ganoczy, & Pituch, 1997; Cortese, Faraone, Konofal, & Lecendreux, 2009). The *Drosophila* learning and memory mutant *radish* displays increased visual distractibility during a maze task and an inability to maintain attention at both the behavioral and electrophysiological level, offering a compelling fly model of ADHD in which sleep disturbance can be evaluated (van Swinderen & Brembs, 2010).

One interesting theory relating attention to sleep, corroborated by electrophysiological evidence in *Drosophila*, stems from the fact that both processes require stimuli to be ignored: attention represents an increase in activity of neurons relevant to the attended stimulus and decrease in activity of other populations, as though attending neurons are "awake," while nonattending neurons are "asleep" (Kirszenblat & van Swinderen, 2015). Chronic disruption of neurotransmitter signaling could interfere with this functional differentiation in activity state, leading to symptoms of inattention. Dopamine deficiency has been extensively documented in patients with ADHD, along with variants in the dopamine transporter DAT and dopamine receptors DRD4/5 (Tarver, Daley, & Sayal, 2014). *Drosophila* mutants for the dopamine receptor DA1 exhibit prolonged hyperactivity after a startle-inducing stimulus and have increased sleep times (Lebestky et al., 2009). Pan-neuronal knockdown of DAT or another ADHD-associated gene, latrophilin, produces hyperactivity and decreased sleep in *Drosophila* that can be rescued by treatment with the dopamine reuptake inhibitor, methylphenidate, a common ADHD medication (van der Voet, Harich, Franke, & Schenck, 2016). Interestingly, sleep depriving wild-type juvenile flies leads to a persistent increase in DA1 expression but decreased sensitivity to wake-promoting effects of dopamine, suggesting that early life sleep loss might permanently interfere with dopaminergic signaling and perhaps contribute to ADHD-like symptoms (Seugnet, Suzuki, Donlea, et al., 2011). Since pharmacologic agents that influence dopaminergic signaling have similar effects on sleep and attention in flies and humans, fly models may be useful for investigating long-term effects of these drugs on multiple behavioral measures (Andretic et al., 2008; Nall et al., 2016).

2.5 Sleep and Psychiatric Illness

The development of *Drosophila* models of psychiatric illnesses such as schizophrenia and bipolar disorder is limited by the difficulty of recapitulating behavioral symptoms in flies and, until more recent genome-wide studies, dearth of clear genetic lesions. However, patients with psychiatric illnesses often have sleep impairments, raising the possibility that sleep is an ideal output for modeling these disorders in *Drosophila* (Winokur, 2015). Interestingly, many of the aforementioned genes that connect sleep to synaptogenesis in *Drosophila* models of neurodevelopmental disorders have also been implicated in psychiatric diseases. For example, genes associated with schizophrenia include UBE3A, NRXN2 (neurexin 2), and genes whose mRNAs are targets of FMRP (Aberg et al., 2013; Fromer et al., 2014). Analyses of de novo copy number variants in schizophrenia have consistently revealed postsynaptic density proteins known to regulate synaptic plasticity, and studies of *Drosophila* homologs may uncover relationships to sleep (Kirov et al., 2012).

Indeed, fly models based on manipulation of other candidate genes from human studies are beginning to provide insights. Both bipolar disorder and schizophrenia share the major susceptibility gene disrupted-in-schizophrenia-1 (DISC1) (Hennah et al., 2009). Transgenic expression of a disease-associated form of human DISC1 in the *Drosophila* MB impairs sleep homeostasis and regulates CREB-dependent transcription, suggesting a role in activity-dependent synaptic plasticity (Sawamura et al., 2008). In addition, the *Drosophila* learning and memory mutant *dunce* has a mutation in a DISC1-interacting gene, PDE4B, which reduces total sleep time and has been independently associated with schizophrenia (Fatemi et al., 2008; Hendricks et al., 2001). A forward genetic screen for *Drosophila* mutants with impaired synaptic neurotransmission revealed *dysbindin*, another schizophrenia, and bipolar disorder-associated gene (Breen et al., 2006; Dickman & Davis, 2009). Loss of dysbindin in *Drosophila* glial cells decreases expression of a dopamine-inactivating enzyme, increasing dopamine levels and locomotor activity (Shao et al., 2011). Finally, using sleep as a quantifiable behavioral output, reverse genetic approaches can be extended to high-throughput assays with thousands of susceptibility genes, deepening our understanding of mechanisms underlying psychiatric diseases (Fig. 2).

2.6 Sleep in Normal Aging and Neurodegenerative Disease

Sleep behaviors follow a robust and reproducible trajectory throughout the life span, with the highest amounts of sleep occurring early in development and the lowest in the elderly (Huber & Tononi, 2009; Roffwarg et al., 1966). Similar to increased sleep in young life, this pattern of sleep decay with age is evident across species (Shaw et al., 2000; Van Gool & Mirmiran, 1983). In humans, aging is associated with decreased sleep, impaired sleep consolidation, and reduced sleep efficiency (Cajochen, Münch, Knoblauch, Blatter, & Wirz-Justice, 2006). In *Drosophila*, aging is most frequently accompanied by increased sleep fragmentation and changes in sleep timing (Bushey, Hughes, Tononi, & Cirelli, 2010; Koh, Evans, Hendricks, & Sehgal, 2006; Shaw et al., 2000) (Fig. 1). Raising flies at a lower ambient temperature prolongs lifespan and slows this breakdown

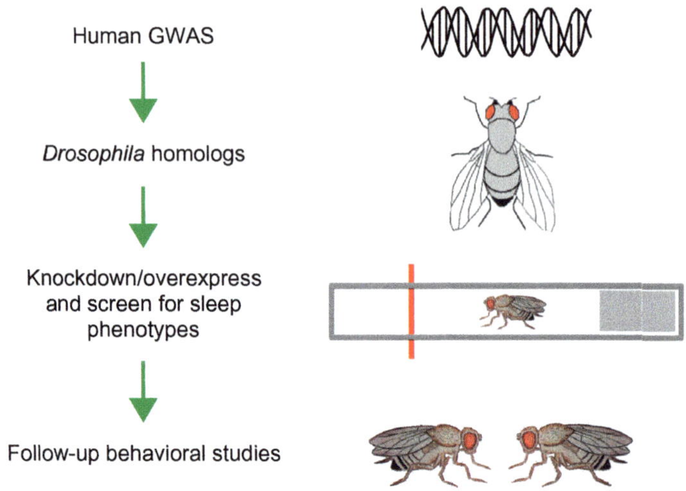

Fig. 2 Reverse genetic strategy for using *Drosophila* to discover mechanisms of human sleep disorders. Human GWAS studies identify candidate genes underlying primary sleep disorders or diseases that manifest with dysregulated sleep, such as autism. *Drosophila* homologues of candidate genes are mutated or knocked down with RNA interference, and sleep is screened using *Drosophila* activity monitors. Follow-up studies can be designed to determine genetic mechanisms of sleep dysregulation and to monitor other complex behaviors, such as learning and memory, aggression, and courtship.

in sleep structure, while an oxidative stress-inducing reagent accelerates this process, linking physiological changes associated with aging to changes in sleep architecture (Koh et al., 2006).

The role of sleep disruption in neurodegenerative diseases has recently become well appreciated. *Drosophila* models of Alzheimer's, Huntington's, and Parkinson's disease (PD) have enhanced our understanding of how sleep abnormalities signify or contribute to neuronal dysfunction and death (Mattis & Sehgal, 2016; McGurk et al., 2015). Alzheimer's disease is characterized by progressive impairment in memory and cognition; buildup of amyloid-β (Aβ) peptide plaques in the brain is thought to be a major cause (Wimo, Jönsson, Bond, Prince, & Winblad, 2013). Evidence from human studies and *Drosophila* models points to a reciprocal interaction between Aβ buildup and sleep disturbance (Lim, Gerstner, & Holtzman, 2014). Recent work in a *Drosophila* model showed that Aβ accumulation results in decreased and fragmented sleep, and sleep loss increases Aβ burden (Tabuchi et al., 2015). Sleep deprivation and Aβ accumulation both independently increase intrinsic neuronal excitability, and sleep loss exacerbates

Aβ-induced excitability, suggesting a pathological cycle between sleep disturbance and Aβ buildup. Treatment with a pharmacologic agent that decreases neuronal excitability prolongs lifespan, suggesting that sleep-related changes in excitability may be relevant to disease pathogenesis and serve as a therapeutic target (Tabuchi et al., 2015). Abnormalities in circadian control of sleep–wake states may also contribute to Alzheimer's pathogenesis: disruption of the interaction between the histone acetyltransferase Tip60 and amyloid precursor protein leads to retraction of sLNV synaptic arbors, decreased PDF signaling, and dysregulated sleep–wake cycles (Pirooznia, Chiu, Chan, Zimmerman, & Elefant, 2012). Treatment with a sleep-promoting compound has been demonstrated to rescue certain age-related LTM deficits in a different *Drosophila* model of AD, supporting the idea that sleep is a potential therapeutic target in AD (Dissel et al., 2015).

In PD, primary sleep disorders often precede neurodegenerative symptom onset: up to 80% of patients diagnosed with REM sleep behavior disorder go on to develop PD (Fyfe, 2016; Schulte & Winkelmann, 2011). Symptoms of PD are thought to result from aggregation of the presynaptic protein α-synuclein (αS) (Breydo, Wu, & Uversky, 2012; Jankovic, 2008). Expression of αS in serotonergic and dopaminergic neurons in *Drosophila* leads to PD-like motor symptoms as well as altered day and night sleep times, sleep episode length, and disrupted light anticipation at dawn. Genetic inhibition of these neuronal populations recapitulates sleep–wake abnormalities, suggesting that neuronal dysfunction rather than cell death is responsible for early sleep disturbances (Gajula Balija, Griesinger, Herzig, Zweckstetter, & Jäckle, 2011). αS flies at an intermediate stage in pathology (16 days old) are most susceptible to sleep deprivation-induced defects in STM, and blocking D1 receptors rescues these defects (Seugnet, Galvin, Suzuki, Gottschalk, & Shaw, 2009). Since sleep deprivation alters dopamine receptor expression and dopaminergic dysfunction features prominently in PD pathology, correction of early primary sleep disorders may reduce symptom severity (Seugnet, Suzuki, Donlea, et al., 2011).

Drosophila models of Huntington's disease have also provided mechanistic links between disease pathology and sleep. Transgenic expression of mutant Huntingtin and knockdown of endogenous dHtt in *Drosophila* both lead to sleep fragmentation (Gonzales & Yin, 2010). Problems with sleep homeostasis, initiation, maintenance, and nighttime hyperactivity emerge in early adulthood in flies, consistent with the idea of sleep disturbance as prodromal in HD (Gonzales, Tanenhaus, Zhang, Chaffee, & Yin, 2016; Roos, 2010). Interestingly, genetic reduction of PKA signaling corrects many of these defects and extends lifespan; since increased PKA signaling generally results from increased neuronal activity, there could be a similar relationship between sleep and neuronal activity in HD to that seen in the *Drosophila* model of Alzheimer's (Gonzales et al., 2016; Tabuchi et al., 2015). As evidenced by work in *Drosophila*, correction of sleep disturbances might have significant therapeutic potential for several neurodegenerative disorders. Additionally, recognizing specific prodromal changes in sleep–wake regulation and increasing monitoring for neurodegenerative symptoms can allow for earlier intervention with existing treatments.

3. CONCLUSIONS

Work in *Drosophila* has in many ways shaped modern behavioral genetics, with arguably no greater impact than in sleep and circadian biology. Unbiased genetic screens and precise microcircuit mapping have provided a new depth to understanding sleep at its most rudimentary level. While a singular genetic/molecular basis for sleep has not yet emerged (in contrast to the circadian clock), sleep research in *Drosophila* has implicated a novel, ontologically related group of genes (ion channels) and catalyzed major advances in elucidating key functions of sleep in synaptic homeostasis, brain development, and learning. More recently, flies have been leveraged to inform disordered sleep in humans, ranging from reverse genetic approaches modeling RLS to basic mechanistic insights regarding sleep and Alzheimer's. Going forward, the study of sleep in *Drosophila* is poised to promote breakthroughs in key areas like interactions between sleep and development in health and disease, the role of sleep in synaptic function, and a molecular/neural basis for insomnia. All the while, flies will surely continue to contribute to influential theories of why we sleep in the first place.

ACKNOWLEDGMENTS

We thank Alex Keene for thoughtful input on the manuscript. This work was supported by NIH grants T32HL07713 (E.H.M.), T32GM007170 (L.C.), and K08NS090461 (M.S.K.), and the Burroughs Wellcome Fund Career Award for Medical Scientists (M.S.K.).

REFERENCES

Aberg, K. A., Liu, Y., Bukszár, J., McClay, J. L., Khachane, A. N., Andreassen, O. A., … van den Oord, E. J. (2013). A comprehensive family-based replication study of schizophrenia genes. *JAMA Psychiatry*, *70*(6), 573–581. http://doi.org/10.1001/jamapsychiatry.2013.288.

Abou-Zeid, E., Boursoulian, L. J., Metzer, W. S., & Gundogdu, B. (2012). Morvan syndrome: A case report and review of the literature. *Journal of Clinical Neuromuscular Disease*, *13*(4), 214–227. http://doi.org/10.1097/CND.0b013e31822b1977.

Afonso, D. J. S., Liu, D., Machado, D. R., Pan, H., Jepson, J. E. C., Rogulja, D., & Koh, K. (2015). TARANIS functions with Cyclin A and Cdk1 in a novel arousal center to control sleep in Drosophila. *Current Biology: CB*, *25*(13), 1717–1726. http://doi.org/10.1016/j.cub.2015.05.037.

Agosto, J., Choi, J. C., Parisky, K. M., Stilwell, G., Rosbash, M., & Griffith, L. C. (2008). Modulation of GABAA receptor desensitization uncouples sleep onset and maintenance in Drosophila. *Nature Neuroscience*, *11*(3), 354–359. http://doi.org/10.1038/nn2046.

Allebrandt, K. V., Amin, N., Müller-Myhsok, B., Esko, T., Teder-Laving, M., Azevedo, R. V. D. M., … Roenneberg, T. (2013). A K(ATP) channel gene effect on sleep duration: From genome-wide association studies to function in Drosophila. *Molecular Psychiatry*, *18*(1), 122–132. http://doi.org/10.1038/mp.2011.142.

American Psychiatric Association. (2013). Diagnostic and statistical manual of mental disorders (5th ed.). *American Journal of Psychiatry*, 50–59. http://doi.org/10.1176/appi.books.9780890425596.744053.

Andretic, R., Kim, Y.-C., Jones, F. S., Han, K.-A., & Greenspan, R. J. (2008). Drosophila D1 dopamine receptor mediates caffeine-induced arousal. *Proceedings of the National Academy of Sciences of the United States of America*, *105*(51), 20392–20397. http://doi.org/10.1073/pnas.0806776105.

Andretic, R., van Swinderen, B., & Greenspan, R. J. (2005). Dopaminergic modulation of arousal in Drosophila. *Current Biology: CB*, *15*(13), 1165–1175. http://doi.org/10.1016/j.cub.2005.05.025.

Aso, Y., Hattori, D., Yu, Y., Johnston, R. M., Iyer, N. A., Ngo, T.-T. B., … Rubin, G. M. (2014). The neuronal architecture of the mushroom body provides a logic for associative learning. *eLife*, *3*, e04577. http://doi.org/10.7554/eLife.04577.

Aso, Y., Sitaraman, D., Ichinose, T., Kaun, K. R., Vogt, K., Belliart-Guérin, G., … Rubin, G. M. (2014). Mushroom body output neurons encode valence and guide memory-based action selection in Drosophila. *eLife*, *3*, e04580. http://doi.org/10.7554/eLife.04580.

Bagni, C., Tassone, F., Neri, G., & Hagerman, R. (2012). Fragile X syndrome: Causes, diagnosis, mechanisms, and therapeutics. *The Journal of Clinical Investigation*, *122*(12), 4314–4322. http://doi.org/10.1172/JCI63141.

Bai, L., & Sehgal, A. (2015). Anaplastic lymphoma kinase acts in the Drosophila mushroom body to negatively regulate sleep. *PLoS Genetics*, *11*(11), e1005611. http://doi.org/10.1371/journal.pgen.1005611.

Berry, J. A., Cervantes-Sandoval, I., Chakraborty, M., & Davis, R. L. (2015). Sleep facilitates memory by blocking dopamine neuron-mediated forgetting. *Cell*, *161*(7), 1656–1667. http://doi.org/10.1016/j.cell.2015.05.027.

Borbély, A. A. (1982). A two process model of sleep regulation. *Human Neurobiology*, *1*(3), 195–204. Retrieved from, http://www.ncbi.nlm.nih.gov/pubmed/7185792.

Bourgeron, T. (2015). From the genetic architecture to synaptic plasticity in autism spectrum disorder. *Nature Reviews. Neuroscience*, *16*(9), 551–563. http://doi.org/10.1038/nrn3992.

Breen, G., Prata, D., Osborne, S., Munro, J., Sinclair, M., Li, T., … Collier, D. (2006). Association of the dysbindin gene with bipolar affective disorder. *American Journal of Psychiatry*, *163*(9), 1636–1638. http://doi.org/10.1176/ajp.2006.163.9.1636.

Breydo, L., Wu, J. W., & Uversky, V. N. (2012). α-Synuclein misfolding and Parkinson's disease. *Biochimica et Biophysica Acta (BBA)—Molecular Basis of Disease*, *1822*(2), 261–285. http://doi.org/10.1016/j.bbadis.2011.10.002.

Bushey, D., Huber, R., Tononi, G., & Cirelli, C. (2007). Drosophila hyperkinetic mutants have reduced sleep and impaired memory. *The Journal of Neuroscience: The Official Journal of the Society for Neuroscience*, *27*(20), 5384–5393. http://doi.org/10.1523/JNEUROSCI.0108-07.2007.

Bushey, D., Hughes, K. A., Tononi, G., & Cirelli, C. (2010). Sleep, aging, and lifespan in Drosophila. *BMC Neuroscience*, *11*, 56. http://doi.org/10.1186/1471-2202-11-56.

Bushey, D., Tononi, G., & Cirelli, C. (2009). The Drosophila fragile X mental retardation gene regulates sleep need. *The Journal of Neuroscience: The Official Journal of the Society for Neuroscience*, *29*(7), 1948–1961. http://doi.org/10.1523/JNEUROSCI.4830-08.2009.

Bushey, D., Tononi, G., & Cirelli, C. (2011). Sleep and synaptic homeostasis: Structural evidence in Drosophila. *Science (New York, N.Y.)*, *332*(6037), 1576–1581. http://doi.org/10.1126/science.1202839.

Bushey, D., Tononi, G., & Cirelli, C. (2015). Sleep- and wake-dependent changes in neuronal activity and reactivity demonstrated in fly neurons using in vivo calcium imaging. *Proceedings of the National Academy of Sciences of the United States of America*, *112*(15), 4785–4790. http://doi.org/10.1073/pnas.1419603112.

Cajochen, C., Münch, M., Knoblauch, V., Blatter, K., & Wirz-Justice, A. (2006). Age-related changes in the circadian and homeostatic regulation of human sleep. *Chronobiology International*, *23*(1–2), 461–474. http://doi.org/10.1080/07420520500545813.

Campbell, S. S., & Tobler, I. (1984). Animal sleep: A review of sleep duration across phylogeny. *Neuroscience and Biobehavioral Reviews*, *8*(3), 269–300. Retrieved from, http://www.ncbi.nlm.nih.gov/pubmed/6504414.

Cavanaugh, D. J., Vigderman, A. S., Dean, T., Garbe, D. S., & Sehgal, A. (2016). The Drosophila circadian clock gates sleep through time-of-day dependent modulation of sleep-promoting neurons. *Sleep*, *39*(2), 345–356. http://dx.doi.org/10.5665/sleep.5442. Retrieved from, http://www.ncbi.nlm.nih.gov/pubmed/26350473.

Cervantes-Sandoval, I., Martin-Peña, A., Berry, J. A., & Davis, R. L. (2013). System-like consolidation of olfactory memories in Drosophila. *The Journal of Neuroscience: The Official Journal of the Society for Neuroscience*, *33*(23), 9846–9854. http://doi.org/10.1523/JNEUROSCI.0451-13.2013.

Chervin, R. D., Dillon, J. E., Bassetti, C., Ganoczy, D. A., & Pituch, K. J. (1997). Symptoms of sleep disorders, inattention, and hyperactivity in children. *Sleep*, *20*(12), 1185–1192. Retrieved from, http://www.ncbi.nlm.nih.gov/pubmed/9493930.

Cirelli, C. (2009). The genetic and molecular regulation of sleep: From fruit flies to humans. *Nature Reviews. Neuroscience*, *10*(8), 549–560. http://doi.org/10.1038/nrn2683.

Cirelli, C. (2013). Sleep and synaptic changes. *Current Opinion in Neurobiology*, *23*(5), 841–846. http://doi.org/10.1016/j.conb.2013.04.001.

Cirelli, C., Bushey, D., Hill, S., Huber, R., Kreber, R., Ganetzky, B., & Tononi, G. (2005). Reduced sleep in Drosophila Shaker mutants. *Nature*, *434*(7037), 1087–1092. http://doi.org/10.1038/nature03486.

Clayton-Smith, J., & Laan, L. (2003). Angelman syndrome: A review of the clinical and genetic aspects. *Journal of Medical Genetics*, *40*(2), 87–95. Retrieved from, http://www.pubmedcentral.nih.gov/articlerender.fcgi?artid=1735357&tool=pmcentrez&rendertype
=abstract.

Codina-Solà, M., Rodríguez-Santiago, B., Homs, A., Santoyo, J., Rigau, M., Aznar-Laín, G., … Cuscó, I. (2015). Integrated analysis of whole-exome sequencing and transcriptome profiling in males with autism spectrum disorders. *Molecular Autism*, *6*, 21. http://doi.org/10.1186/s13229-015-0017-0.

Cong, X., Wang, H., Liu, Z., He, C., An, C., & Zhao, Z. (2015). Regulation of sleep by insulin-like peptide system in Drosophila melanogaster. *Sleep*, *38*(7), 1075–1083. http://doi.org/10.5665/sleep.4816.

Cortese, S., Faraone, S. V., Konofal, E., & Lecendreux, M. (2009). Sleep in children with attention-deficit/hyperactivity disorder: Meta-analysis of subjective and objective studies. *Journal of the American Academy of Child and Adolescent Psychiatry*, *48*(9), 894–908. http://doi.org/10.1097/CHI.0b013e3181ac09c9.

Cote, K. A., McCormick, C. M., Geniole, S. N., Renn, R. P., & MacAulay, S. D. (2013). Sleep deprivation lowers reactive aggression and testosterone in men. *Biological Psychology, 92*(2), 249–256. http://doi.org/10.1016/j.biopsycho.2012.09.011.

Crocker, A., & Sehgal, A. (2008). Octopamine regulates sleep in drosophila through protein kinase A-dependent mechanisms. *The Journal of Neuroscience: The Official Journal of the Society for Neuroscience, 28*(38), 9377–9385. http://doi.org/10.1523/JNEUROSCI.3072-08a.2008.

Crocker, A., Shahidullah, M., Levitan, I. B., & Sehgal, A. (2010). Identification of a neural circuit that underlies the effects of octopamine on sleep:wake behavior. *Neuron, 65*(5), 670–681. http://doi.org/10.1016/j.neuron.2010.01.032.

Davis, H., Davis, P. A., Loomis, A. L., Harvey, E. N., & Hobart, G. (1937). Changes in human brain potentials during the onset of sleep. *Science (New York, N.Y.), 86*(2237), 448–450. http://doi.org/10.1126/science.86.2237.448.

Devaud, J.-M., Acebes, A., Ramaswami, M., & Ferrús, A. (2003). Structural and functional changes in the olfactory pathway of adult Drosophila take place at a critical age. *Journal of Neurobiology, 56*(1), 13–23. http://doi.org/10.1002/neu.10215.

Dickman, D. K., & Davis, G. W. (2009). The schizophrenia susceptibility gene dysbindin controls synaptic homeostasis. *Science (New York, N.Y.), 326*(5956), 1127–1130. http://doi.org/10.1126/science.1179685.

Diekelmann, S., & Born, J. (2010). The memory function of sleep. *Nature Reviews. Neuroscience, 11*(2), 114–126. http://doi.org/10.1038/nrn2762.

Dissel, S., Angadi, V., Kirszenblat, L., Suzuki, Y., Donlea, J., Klose, M., … Shaw, P. J. (2015). Sleep restores behavioral plasticity to Drosophila mutants. *Current Biology: CB, 25*(10), 1270–1281. http://doi.org/10.1016/j.cub.2015.03.027.

Dockendorff, T. C., Su, H. S., McBride, S. M. J., Yang, Z., Choi, C. H., Siwicki, K. K., … Jongens, T. A. (2002). Drosophila lacking dfmr1 activity show defects in circadian output and fail to maintain courtship interest. *Neuron, 34*(6), 973–984. Retrieved from, http://www.ncbi.nlm.nih.gov/pubmed/12086644.

Donlea, J. M., Pimentel, D., & Miesenböck, G. (2014). Neuronal machinery of sleep homeostasis in Drosophila. *Neuron, 81*(4), 860–872. http://doi.org/10.1016/j.neuron.2013.12.013.

Donlea, J. M., Thimgan, M. S., Suzuki, Y., Gottschalk, L., & Shaw, P. J. (2011). Inducing sleep by remote control facilitates memory consolidation in Drosophila. *Science (New York, N.Y.), 332*(6037), 1571–1576. http://doi.org/10.1126/science.1202249.

Durmer, J. S., & Dinges, D. F. (2005). Neurocognitive consequences of sleep deprivation. *Seminars in Neurology, 25*(1), 117–129. http://doi.org/10.1055/s-2005-867080.

Dweck, H. K. M., Ebrahim, S. A. M., Thoma, M., Mohamed, A. A. M., Keesey, I. W., Trona, F., … Hansson, B. S. (2015). Pheromones mediating copulation and attraction in Drosophila. *Proceedings of the National Academy of Sciences of the United States of America, 112*(21), E2829–E2835. http://doi.org/10.1073/pnas.1504527112.

Dynes, J. B. (1947). Objective method for distinguishing sleep from the hypnotic trance. *Archives of Neurology and Psychiatry, 57*(1), 84–93. Retrieved from, http://www.ncbi.nlm.nih.gov/pubmed/20284361.

Earley, C. J. (2003). Clinical practice. Restless legs syndrome. *The New England Journal of Medicine, 348*(21), 2103–2109. http://doi.org/10.1056/NEJMcp021288.

Ednick, M., Cohen, A. P., McPhail, G. L., Beebe, D., Simakajornboon, N., & Amin, R. S. (2009). A review of the effects of sleep during the first year of life on cognitive, psychomotor, and temperament development. *Sleep, 32*(11), 1449–1458. Retrieved from, http://www.pubmedcentral.nih.gov/articlerender.fcgi?artid=2768951&tool=pmcentrez&rendertype=abstract.

Fatemi, S. H., King, D. P., Reutiman, T. J., Folsom, T. D., Laurence, J. A., Lee, S., … Menniti, F. S. (2008). PDE4B polymorphisms and decreased PDE4B expression are associated with schizophrenia. *Schizophrenia Research, 101*(1–3), 36–49. http://doi.org/10.1016/j.schres.2008.01.029.

Ferdousy, F., Bodeen, W., Summers, K., Doherty, O., Wright, O., Elsisi, N., … Reiter, L. T. (2011). Drosophila Ube3a regulates monoamine synthesis by increasing GTP cyclohydrolase I activity via a non-ubiquitin ligase mechanism. *Neurobiology of Disease, 41*(3), 669–677. http://doi.org/10.1016/j.nbd.2010.12.001.

Flavell, S. W., & Greenberg, M. E. (2008). Signaling mechanisms linking neuronal activity to gene expression and plasticity of the nervous system. *Annual Review of Neuroscience, 31*, 563–590. http://doi.org/10.1146/annurev.neuro.31.060407.125631.

Frank, M. G. (2015a). Sleep and synaptic plasticity in the developing and adult brain. *Current Topics in Behavioral Neurosciences, 25*, 123–149. http://doi.org/10.1007/7854_2014_305.

Frank, M. G. (2015b). *Sleep, neuronal plasticity and brain function. In P. Meerlo, R. M. Benca, & T. Abel (Eds.), Current topics in behavioral neurosciences:* Vol. 25. Berlin, Heidelberg: Springer. http://doi.org/10.1007/978-3-662-46878-4.

Freeman, A., Pranski, E., Miller, R. D., Radmard, S., Bernhard, D., Jinnah, H. A., … Sanyal, S. (2012). Sleep fragmentation and motor restlessness in a Drosophila model of Restless Legs Syndrome. *Current Biology: CB, 22*(12), 1142–1148. http://doi.org/10.1016/j.cub.2012.04.027.

Freeman, A. A. H., & Rye, D. B. (2013). The molecular basis of restless legs syndrome. *Current Opinion in Neurobiology, 23*(5), 895–900. http://doi.org/10.1016/j.conb.2013.07.001.

Fromer, M., Pocklington, A. J., Kavanagh, D. H., Williams, H. J., Dwyer, S., Gormley, P., … O'Donovan, M. C. (2014). De novo mutations in schizophrenia implicate synaptic networks. *Nature, 506*(7487), 179–184. http://doi.org/10.1038/nature12929.

Fyfe, I. (2016). Parkinson disease: Sleep disorder deficits suggest signature for early Parkinson disease. *Nature Reviews Neurology, 12*(1), 3. http://doi.org/10.1038/nrneurol.2015.232.

Gajula Balija, M. B., Griesinger, C., Herzig, A., Zweckstetter, M., & Jäckle, H. (2011). Pre-fibrillar α-synuclein mutants cause Parkinson's disease-like non-motor symptoms in Drosophila. *PloS One, 6*(9), e24701. http://doi.org/10.1371/journal.pone.0024701.

Glessner, J. T., Wang, K., Cai, G., Korvatska, O., Kim, C. E., Wood, S., … Hakonarson, H. (2009). Autism genome-wide copy number variation reveals ubiquitin and neuronal genes. *Nature, 459*(7246), 569–573. http://doi.org/10.1038/nature07953.

Gmeiner, F., Kołodziejczyk, A., Yoshii, T., Rieger, D., Nässel, D. R., & Helfrich-Förster, C. (2013). GABA(B) receptors play an essential role in maintaining sleep during the second half of the night in Drosophila melanogaster. *The Journal of Experimental Biology, 216*(Pt 20), 3837–3843. http://doi.org/10.1242/jeb.085563.

Gonzales, E. D., Tanenhaus, A. K., Zhang, J., Chaffee, R. P., & Yin, J. C. P. (2016). Early onset sleep defects in Drosophila models of Huntington's Disease reflect alterations of PKA/CREB signaling. *Human Molecular Genetics, 25*(5), 837–852. Epub 2015 Nov 2. http://doi.org/10.1093/hmg/ddv482.

Gonzales, E., & Yin, J. (2010). Drosophila models of Huntington's disease exhibit sleep abnormalities. *PLOS Currents Huntington Disease.* http://dx.doi.org/10.1371/currents.RRN1185.

Graves, L. A., Hellman, K., Veasey, S., Blendy, J. A., Pack, A. I., & Abel, T. (2003). Genetic evidence for a role of CREB in sustained cortical arousal. *Journal of Neurophysiology, 90*(2), 1152–1159. http://doi.org/10.1152/jn.00882.2002.

Haynes, P. R., Christmann, B. L., & Griffith, L. C. (2015). A single pair of neurons links sleep to memory consolidation in Drosophila melanogaster. *eLife, 4*. http://doi.org/10.7554/eLife.03868.

He, Y., Jones, C. R., Fujiki, N., Xu, Y., Guo, B., Holder, J. L., … Fu, Y.-H. (2009). The transcriptional repressor DEC2 regulates sleep length in mammals. *Science (New York, N.Y.), 325*(5942), 866–870. http://doi.org/10.1126/science.1174443.

Hendricks, J. C., Finn, S. M., Panckeri, K. A., Chavkin, J., Williams, J. A., Sehgal, A., & Pack, A. I. (2000). Rest in Drosophila is a sleep-like state. *Neuron*, *25*(1), 129–138. Retrieved from, http://www.ncbi.nlm.nih.gov/pubmed/10707978.

Hendricks, J. C., Williams, J. A., Panckeri, K., Kirk, D., Tello, M., Yin, J. C., & Sehgal, A. (2001). A non-circadian role for cAMP signaling and CREB activity in Drosophila rest homeostasis. *Nature Neuroscience*, *4*(11), 1108–1115. http://doi.org/10.1038/nn743.

Hennah, W., Thomson, P., McQuillin, A., Bass, N., Loukola, A., Anjorin, A., … Porteous, D. (2009). DISC1 association, heterogeneity and interplay in schizophrenia and bipolar disorder. *Molecular Psychiatry*, *14*(9), 865–873. http://doi.org/10.1038/mp.2008.22.

Hess-Homeier, D. L., Fan, C.-Y., Gupta, T., Chiang, A.-S., & Certel, S. J. (2014). Astrocyte-specific regulation of hMeCP2 expression in Drosophila. *Biology Open*, *3*(11), 1011–1019. http://doi.org/10.1242/bio.20149092.

Hu, W. F., Chahrour, M. H., & Walsh, C. A. (2014). The diverse genetic landscape of neurodevelopmental disorders. *Annual Review of Genomics and Human Genetics*, *15*, 195–213. http://doi.org/10.1146/annurev-genom-090413-025600.

Huber, R., & Tononi, G. (2009). Sleep and waking across the lifespan. In G. G. Berntson & J. T. Cacioppo (Eds.), *Handbook of neuroscience for the behavioral sciences. Vol. 1.* (pp. 461–481). Hoboken, NJ: John Wiley & Sons.

Islam, M. P., & Roach, E. S. (2015). Tuberous sclerosis complex. *Handbook of Clinical Neurology*, *132*, 97–109. http://doi.org/10.1016/B978-0-444-62702-5.00006-8.

Jankovic, J. (2008). Parkinson's disease: Clinical features and diagnosis. *Journal of Neurology, Neurosurgery, and Psychiatry*, *79*(4), 368–376. http://doi.org/10.1136/jnnp.2007.131045.

Jett, K., & Friedman, J. M. (2010). Clinical and genetic aspects of neurofibromatosis 1. *Genetics in Medicine: Official Journal of the American College of Medical Genetics*, *12*(1), 1–11. http://doi.org/10.1097/GIM.0b013e3181bf15e3.

Joiner, W. J., Crocker, A., White, B. H., & Sehgal, A. (2006). Sleep in Drosophila is regulated by adult mushroom bodies. *Nature*, *441*(7094), 757–760. http://doi.org/10.1038/nature04811.

Josephs, K. A., Silber, M. H., Fealey, R. D., Nippoldt, T. B., Auger, R. G., & Vernino, S. (2004). Neurophysiologic studies in Morvan syndrome. *Journal of Clinical Neurophysiology: Official Publication of the American Electroencephalographic Society*, *21*(6), 440–445. Retrieved from, http://www.ncbi.nlm.nih.gov/pubmed/15622131.

Jouvet-Mounier, D., Astic, L., & Lacote, D. (1970). Ontogenesis of the states of sleep in rat, cat, and guinea pig during the first postnatal month. *Developmental Psychobiology*, *2*(4), 216–239. http://doi.org/10.1002/dev.420020407.

Kay, S. A., & Wager-Smith, K. (2000). Circadian rhythm genetics: From flies to mice to humans. *Nature Genetics*, *26*(1), 23–27. http://doi.org/10.1038/79134.

Kayser, M. S., Mainwaring, B., Yue, Z., & Sehgal, A. (2015). Sleep deprivation suppresses aggression in Drosophila. *eLife*, *4*, e07643. http://doi.org/10.7554/eLife.07643.

Kayser, M. S., Yue, Z., & Sehgal, A. (2014). A critical period of sleep for development of courtship circuitry and behavior in Drosophila. *Science (New York, N.Y.)*, *344*(6181), 269–274. http://doi.org/10.1126/science.1250553.

Kirov, G., Pocklington, A. J., Holmans, P., Ivanov, D., Ikeda, M., Ruderfer, D., … Owen, M. J. (2012). De novo CNV analysis implicates specific abnormalities of postsynaptic signalling complexes in the pathogenesis of schizophrenia. *Molecular Psychiatry*, *17*(2), 142–153. http://doi.org/10.1038/mp.2011.154.

Kirszenblat, L., & van Swinderen, B. (2015). The Yin and Yang of sleep and attention. *Trends in Neurosciences*, *38*(12), 776–786. http://doi.org/10.1016/j.tins.2015.10.001.

Knutson, K. L., & Van Cauter, E. (2008). Associations between sleep loss and increased risk of obesity and diabetes. *Annals of the New York Academy of Sciences*, *1129*, 287–304. http://doi.org/10.1196/annals.1417.033.

Koh, K., Evans, J. M., Hendricks, J. C., & Sehgal, A. (2006). A Drosophila model for age-associated changes in sleep:wake cycles. *Proceedings of the National Academy of Sciences of the United States of America*, *103*(37), 13843–13847. http://doi.org/10.1073/pnas.0605903103.

Koh, K., Joiner, W. J., Wu, M. N., Yue, Z., Smith, C. J., & Sehgal, A. (2008). Identification of SLEEPLESS, a sleep-promoting factor. *Science (New York, N.Y.)*, *321*(5887), 372–376. http://doi.org/10.1126/science.1155942.

Konopka, R. J., & Benzer, S. (1971). Clock mutants of Drosophila melanogaster. *Proceedings of the National Academy of Sciences of the United States of America*, *68*(9), 2112–2116. Retrieved from, http://www.pubmedcentral.nih.gov/articlerender.fcgi?artid=389363&tool=pmcentrez&rendertype=abstract.

Kume, K., Kume, S., Park, S. K., Hirsh, J., & Jackson, F. R. (2005). Dopamine is a regulator of arousal in the fruit fly. *The Journal of Neuroscience: The Official Journal of the Society for Neuroscience*, *25*(32), 7377–7384. http://doi.org/10.1523/JNEUROSCI.2048-05.2005.

Kunst, M., Hughes, M. E., Raccuglia, D., Felix, M., Li, M., Barnett, G., … Nitabach, M. N. (2014). Calcitonin gene-related peptide neurons mediate sleep-specific circadian output in Drosophila. *Current Biology: CB*, *24*(22), 2652–2664. http://doi.org/10.1016/j.cub.2014.09.077.

Lebestky, T., Chang, J.-S. C., Dankert, H., Zelnik, L., Kim, Y.-C., Han, K.-A., … Anderson, D. J. (2009). Two different forms of arousal in Drosophila are oppositely regulated by the dopamine D1 receptor ortholog DopR via distinct neural circuits. *Neuron*, *64*(4), 522–536. http://doi.org/10.1016/j.neuron.2009.09.031.

Li, Z., Chang, S.-H., Zhang, L.-Y., Gao, L., & Wang, J. (2014). Molecular genetic studies of ADHD and its candidate genes: A review. *Psychiatry Research*, *219*(1), 10–24. http://doi.org/10.1016/j.psychres.2014.05.005.

Li, X., Yu, F., & Guo, A. (2009). Sleep deprivation specifically impairs short-term olfactory memory in Drosophila. *Sleep*, *32*(11), 1417–1424. Retrieved from, http://www.pubmedcentral.nih.gov/articlerender.fcgi?artid=2768948&tool=pmcentrez&rendertype=abstract.

Li, Y., Zhou, Z., Zhang, X., Tong, H., Li, P., Zhang, Z. C., … Han, J. (2013). Drosophila neuroligin 4 regulates sleep through modulating GABA transmission. *The Journal of Neuroscience: The Official Journal of the Society for Neuroscience*, *33*(39), 15545–15554. http://doi.org/10.1523/JNEUROSCI.0819-13.2013.

Lim, M. M., Gerstner, J. R., & Holtzman, D. M. (2014). The sleep-wake cycle and Alzheimer's disease: What do we know? *Neurodegenerative Disease Management*, *4*(5), 351–362. http://doi.org/10.2217/nmt.14.33.

Liu, W., Guo, F., Lu, B., & Guo, A. (2008). amnesiac regulates sleep onset and maintenance in Drosophila melanogaster. *Biochemical and Biophysical Research Communications*, *372*(4), 798–803. http://doi.org/10.1016/j.bbrc.2008.05.119.

Liu, S., Lamaze, A., Liu, Q., Tabuchi, M., Yang, Y., Fowler, M., … Wu, M. N. (2014). WIDE AWAKE mediates the circadian timing of sleep onset. *Neuron*, *82*(1), 151–166. http://doi.org/10.1016/j.neuron.2014.01.040.

Liu, Q., Liu, S., Kodama, L., Driscoll, M. R., & Wu, M. N. (2012). Two dopaminergic neurons signal to the dorsal fan-shaped body to promote wakefulness in Drosophila. *Current Biology: CB*, *22*(22), 2114–2123. http://doi.org/10.1016/j.cub.2012.09.008.

Liu, S., Liu, Q., Tabuchi, M., & Wu, M. N. (2016). Sleep drive is encoded by neural plastic changes in a dedicated circuit. *Cell*, *165*(6), 1347–1360. Epub 2016 May 19. http://doi.org/10.1016/j.cell.2016.04.013.

Lu, J., Sherman, D., Devor, M., & Saper, C. B. (2006). A putative flip-flop switch for control of REM sleep. *Nature*, *441*(7093), 589–594. http://doi.org/10.1038/nature04767.

Lu, Y., Wang, F., Li, Y., Ferris, J., Lee, J.-A., & Gao, F.-B. (2009). The Drosophila homologue of the Angelman syndrome ubiquitin ligase regulates the formation of terminal dendritic branches. *Human Molecular Genetics*, *18*(3), 454–462. http://doi.org/10.1093/hmg/ddn373.

Mackiewicz, M., Shockley, K. R., Romer, M. A., Galante, R. J., Zimmerman, J. E., Naidoo, N., ... Pack, A. I. (2007). Macromolecule biosynthesis: A key function of sleep. *Physiological Genomics*, *31*(3), 441–457. http://doi.org/10.1152/physiolgenomics.00275.2006.

Mattis, J., & Sehgal, A. (2016). Circadian rhythms, sleep, and disorders of aging. *Trends in Endocrinology & Metabolism*, *27*(4), 192–203. http://doi.org/10.1016/j.tem.2016.02.003.

McBride, S. M. J., Choi, C. H., Wang, Y., Liebelt, D., Braunstein, E., Ferreiro, D., ... Jongens, T. A. (2005). Pharmacological rescue of synaptic plasticity, courtship behavior, and mushroom body defects in a Drosophila model of fragile X syndrome. *Neuron*, *45*(5), 753–764. http://doi.org/10.1016/j.neuron.2005.01.038.

McGurk, L., Berson, A., Bonini, N. M., Aarli, J. A., Dua, T., Janca, A., ... Pletnikova, O. (2015). Drosophila as an in vivo model for human neurodegenerative disease. *Genetics*, *201*(2), 377–402. http://doi.org/10.1534/genetics.115.179457.

Mednick, S. C., Cai, D. J., Shuman, T., Anagnostaras, S., & Wixted, J. T. (2011). An opportunistic theory of cellular and systems consolidation. *Trends in Neurosciences*, *34*(10), 504–514. http://doi.org/10.1016/j.tins.2011.06.003.

Naidoo, N., Casiano, V., Cater, J., Zimmerman, J., & Pack, A. I. (2007). A role for the molecular chaperone protein BiP/GRP78 in Drosophila sleep homeostasis. *Sleep*, *30*(5), 557–565. Retrieved from, http://www.ncbi.nlm.nih.gov/pubmed/17552370.

Nall, A., & Sehgal, A. (2014). Monoamines and sleep in Drosophila. *Behavioral Neuroscience*, *128*(3), 264–272. http://doi.org/10.1037/a0036209.

Nall, A. H., Shakhmantsir, I., Cichewicz, K., Birman, S., Hirsh, J., & Sehgal, A. (2016). Caffeine promotes wakefulness via dopamine signaling in Drosophila. *Scientific Reports*, *6*, 20938. http://doi.org/10.1038/srep20938.

Natarajan, R., Trivedi-Vyas, D., & Wairkar, Y. P. (2013). Tuberous sclerosis complex regulates Drosophila neuromuscular junction growth via the TORC2/Akt pathway. *Human Molecular Genetics*, *22*(10), 2010–2023. http://doi.org/10.1093/hmg/ddt053.

Nitz, D. A., van Swinderen, B., Tononi, G., & Greenspan, R. J. (2002). Electrophysiological correlates of rest and activity in Drosophila melanogaster. *Current Biology*, *12*(22), 1934–1940. http://doi.org/10.1016/S0960-9822(02)01300-3.

O'Brien, L. M. (2009). The neurocognitive effects of sleep disruption in children and adolescents. *Child and Adolescent Psychiatric Clinics of North America*, *18*(4), 813–823. http://doi.org/10.1016/j.chc.2009.04.008.

Oh, Y., Jang, D., Sonn, J. Y., & Choe, J. (2013). Histamine-HisCl1 receptor axis regulates wake-promoting signals in Drosophila melanogaster. *PloS One*, *8*(7), e68269. http://doi.org/10.1371/journal.pone.0068269.

Pan, D., Dong, J., Zhang, Y., & Gao, X. (2004). Tuberous sclerosis complex: From Drosophila to human disease. *Trends in Cell Biology*, *14*(2), 78–85. http://doi.org/10.1016/j.tcb.2003.12.006.

Pan, L., Zhang, Y. Q., Woodruff, E., & Broadie, K. (2004). The Drosophila fragile X gene negatively regulates neuronal elaboration and synaptic differentiation. *Current Biology: CB*, *14*(20), 1863–1870. http://doi.org/10.1016/j.cub.2004.09.085.

Parisky, K. M., Agosto, J., Pulver, S. R., Shang, Y., Kuklin, E., Hodge, J. J. L., ... Griffith, L. C. (2008). PDF cells are a GABA-responsive wake-promoting component of the Drosophila sleep circuit. *Neuron*, *60*(4), 672–682. http://doi.org/10.1016/j.neuron.2008.10.042.

Pfeiffenberger, C., & Allada, R. (2012). Cul3 and the BTB adaptor insomniac are key regulators of sleep homeostasis and a dopamine arousal pathway in Drosophila. *PLoS Genetics*, *8*(10), e1003003. http://doi.org/10.1371/journal.pgen.1003003.
Phillips, L., & Appleton, R. E. (2004). Systematic review of melatonin treatment in children with neurodevelopmental disabilities and sleep impairment. *Developmental Medicine and Child Neurology*, *46*(11), 771–775. Retrieved from, http://www.ncbi.nlm.nih.gov/pubmed/15540639.
Pirooznia, S. K., Chiu, K., Chan, M. T., Zimmerman, J. E., & Elefant, F. (2012). Epigenetic regulation of axonal growth of Drosophila pacemaker cells by histone acetyltransferase tip60 controls sleep. *Genetics*, *192*(4), 1327–1345. http://doi.org/10.1534/genetics.112.144667.
Pitman, J. L., McGill, J. J., Keegan, K. P., & Allada, R. (2006). A dynamic role for the mushroom bodies in promoting sleep in Drosophila. *Nature*, *441*(7094), 753–756. http://doi.org/10.1038/nature04739.
Raizen, D. M., Zimmerman, J. E., Maycock, M. H., Ta, U. D., You, Y., Sundaram, M. V., & Pack, A. I. (2008). Lethargus is a Caenorhabditis elegans sleep-like state. *Nature*, *451*(7178), 569–572. http://doi.org/10.1038/nature06535.
Rasch, B., & Born, J. (2013). About sleep's role in memory. *Physiological Reviews*, *93*(2), 681–766. http://doi.org/10.1152/physrev.00032.2012.
Robinson-Shelton, A., & Malow, B. A. (2015). Sleep disturbances in neurodevelopmental disorders. *Current Psychiatry Reports*, *18*(1), 6. http://doi.org/10.1007/s11920-015-0638-1.
Roffwarg, H. P., Muzio, J. N., & Dement, W. C. (1966). Ontogenetic development of the human sleep-dream cycle. *Science*, *152*(3722), 604–619. http://doi.org/10.1126/science.152.3722.604.
Rogulja, D., & Young, M. W. (2012). Control of sleep by cyclin A and its regulator. *Science (New York, N.Y.)*, *335*(6076), 1617–1621. http://doi.org/10.1126/science.1212476.
Roos, R. A. C. (2010). Huntington's disease: A clinical review. *Orphanet Journal of Rare Diseases*, *5*, 40. http://doi.org/10.1186/1750-1172-5-40.
Roth, T. (2007). Insomnia: Definition, prevalence, etiology, and consequences. *Journal of Clinical Sleep Medicine: JCSM: Official Publication of the American Academy of Sleep Medicine*, *3*(5 Suppl.), S7–S10. Retrieved from, http://www.pubmedcentral.nih.gov/articlerender.fcgi?artid=1978319&tool=pmcentrez&rendertype=abstract.
Sawamura, N., Ando, T., Maruyama, Y., Fujimuro, M., Mochizuki, H., Honjo, K., … Sawa, A. (2008). Nuclear DISC1 regulates CRE-mediated gene transcription and sleep homeostasis in the fruit fly. *Molecular Psychiatry*, *13*(12), 1138–1148. 1069, http://doi.org/10.1038/mp.2008.101.
Schulte, E. C., & Winkelmann, J. (2011). When Parkinson's disease patients go to sleep: Specific sleep disturbances related to Parkinson's disease. *Journal of Neurology*, *258*(Suppl. 2), S328–S335. http://doi.org/10.1007/s00415-011-5933-0.
Sehgal, A., & Mignot, E. (2011). Genetics of sleep and sleep disorders. *Cell*, *146*(2), 194–207. http://doi.org/10.1016/j.cell.2011.07.004.
Seidner, G., Robinson, J. E., Wu, M., Worden, K., Masek, P., Roberts, S. W., … Joiner, W. J. (2015). Identification of neurons with a privileged role in sleep homeostasis in Drosophila melanogaster. *Current Biology*, *25*(22), 2928–2938. http://doi.org/10.1016/j.cub.2015.10.006.
Seugnet, L., Galvin, J. E., Suzuki, Y., Gottschalk, L., & Shaw, P. J. (2009). Persistent short-term memory defects following sleep deprivation in a drosophila model of Parkinson disease. *Sleep*, *32*(8), 984–992. Retrieved from, http://www.pubmedcentral.nih.gov/articlerender.fcgi?artid=2717205&tool=pmcentrez&rendertype=abstract.
Seugnet, L., Suzuki, Y., Donlea, J. M., Gottschalk, L., & Shaw, P. J. (2011a). Sleep deprivation during early-adult development results in long-lasting learning deficits in adult Drosophila. *Sleep*, *34*(2), 137–146. Retrieved from, http://www.pubmedcentral.nih.gov/articlerender.fcgi?artid=3022932&tool=pmcentrez&rendertype=abstract.

Seugnet, L., Suzuki, Y., Merlin, G., Gottschalk, L., Duntley, S. P., & Shaw, P. J. (2011b). Notch signaling modulates sleep homeostasis and learning after sleep deprivation in Drosophila. *Current Biology: CB*, *21*(10), 835–840. http://doi.org/10.1016/j.cub.2011.04.001.

Seugnet, L., Suzuki, Y., Thimgan, M., Donlea, J., Gimbel, S. I., Gottschalk, L., … Shaw, P. J. (2009). Identifying sleep regulatory genes using a Drosophila model of insomnia. *The Journal of Neuroscience: The Official Journal of the Society for Neuroscience*, *29*(22), 7148–7157. http://doi.org/10.1523/JNEUROSCI.5629-08.2009.

Seugnet, L., Suzuki, Y., Vine, L., Gottschalk, L., & Shaw, P. J. (2008). D1 receptor activation in the mushroom bodies rescues sleep-loss-induced learning impairments in Drosophila. *Current Biology: CB*, *18*(15), 1110–1117. http://doi.org/10.1016/j.cub.2008.07.028.

Shao, L., Shuai, Y., Wang, J., Feng, S., Lu, B., Li, Z., … Zhong, Y. (2011). Schizophrenia susceptibility gene dysbindin regulates glutamatergic and dopaminergic functions via distinctive mechanisms in Drosophila. *Proceedings of the National Academy of Sciences of the United States of America*, *108*(46), 18831–18836. http://doi.org/10.1073/pnas.1114569108.

Shaw, P. J., Cirelli, C., Greenspan, R. J., & Tononi, G. (2000). Correlates of sleep and waking in Drosophila melanogaster. *Science (New York, N.Y.)*, *287*(5459), 1834–1837. Retrieved from, http://www.ncbi.nlm.nih.gov/pubmed/10710313.

Shaw, P. J., Tononi, G., Greenspan, R. J., & Robinson, D. F. (2002). Stress response genes protect against lethal effects of sleep deprivation in Drosophila. *Nature*, *417*(6886), 287–291. http://doi.org/10.1038/417287a.

Shi, M., Yue, Z., Kuryatov, A., Lindstrom, J. M., & Sehgal, A. (2014). Identification of Redeye, a new sleep-regulating protein whose expression is modulated by sleep amount. *eLife*, *3*, e01473. http://doi.org/10.7554/eLife.01473.

Singh, N., Lorbeck, M. T., Zervos, A., Zimmerman, J., & Elefant, F. (2010). The histone acetyltransferase Elp3 plays in active role in the control of synaptic bouton expansion and sleep in Drosophila. *Journal of Neurochemistry*, *115*(2), 493–504. http://doi.org/10.1111/j.1471-4159.2010.06892.x.

Sitaraman, D., Aso, Y., Jin, X., Chen, N., Felix, M., Rubin, G. M., & Nitabach, M. N. (2015). Propagation of homeostatic sleep signals by segregated synaptic microcircuits of the Drosophila mushroom body. *Current Biology: CB*, *25*(22), 2915–2927. http://doi.org/10.1016/j.cub.2015.09.017.

Sitaraman, D., Aso, Y., Rubin, G. M., & Nitabach, M. N. (2015). Control of sleep by dopaminergic inputs to the Drosophila mushroom body. *Frontiers in Neural Circuits*, *9*, 73. http://doi.org/10.3389/fncir.2015.00073.

Stavropoulos, N., & Young, M. W. (2011). insomniac and Cullin-3 regulate sleep and wakefulness in Drosophila. *Neuron*, *72*(6), 964–976. http://doi.org/10.1016/j.neuron.2011.12.003.

Stopfer, M. (2014). Central processing in the mushroom bodies. *Current Opinion in Insect Science*, *6*, 99–103. http://doi.org/10.1016/j.cois.2014.10.009.

Tabuchi, M., Lone, S. R., Liu, S., Liu, Q., Zhang, J., Spira, A. P., & Wu, M. N. (2015). Sleep interacts with aβ to modulate intrinsic neuronal excitability. *Current Biology: CB*, *25*(6), 702–712. http://doi.org/10.1016/j.cub.2015.01.016.

Tarver, J., Daley, D., & Sayal, K. (2014). Attention-deficit hyperactivity disorder (ADHD): An updated review of the essential facts. *Child: Care, Health and Development*, *40*(6), 762–774. http://doi.org/10.1111/cch.12139.

Tessier, C. R., & Broadie, K. (2008). Drosophila fragile X mental retardation protein developmentally regulates activity-dependent axon pruning. *Development (Cambridge, England)*, *135*(8), 1547–1557. http://doi.org/10.1242/dev.015867.

Thapar, A., & Cooper, M. (2016). Attention deficit hyperactivity disorder. *Lancet (London, England)*, *387*(10024), 1240–1250. http://doi.org/10.1016/S0140-6736(15)00238-X.

The, I., Hannigan, G. E., Cowley, G. S., Reginald, S., Zhong, Y., Gusella, J. F., … Bernards, A. (1997). Rescue of a Drosophila NF1 mutant phenotype by protein kinase A. *Science (New York, N.Y.)*, *276*(5313), 791–794. Retrieved from, http://www.ncbi.nlm.nih.gov/pubmed/9115203.

Thimgan, M. S., Suzuki, Y., Seugnet, L., Gottschalk, L., & Shaw, P. J. (2010). The perilipin homologue, lipid storage droplet 2, regulates sleep homeostasis and prevents learning impairments following sleep loss. *PLoS Biology*, *8*(8). http://doi.org/10.1371/journal.pbio.1000466.

Tononi, G., & Cirelli, C. (2003). Sleep and synaptic homeostasis: A hypothesis. *Brain Research Bulletin*, *62*(2), 143–150. Retrieved from, http://www.ncbi.nlm.nih.gov/pubmed/14638388.

Tsai, P.-I., Wang, M., Kao, H.-H., Cheng, Y.-J., Walker, J. A., Chen, R.-H., & Chien, C.-T. (2012). Neurofibromin mediates FAK signaling in confining synapse growth at Drosophila neuromuscular junctions. *The Journal of Neuroscience: The Official Journal of the Society for Neuroscience*, *32*(47), 16971–16981. http://doi.org/10.1523/JNEUROSCI.1756-12.2012.

Ueno, T., Tomita, J., Tanimoto, H., Endo, K., Ito, K., Kume, S., & Kume, K. (2012). Identification of a dopamine pathway that regulates sleep and arousal in Drosophila. *Nature Neuroscience*, *15*(11), 1516–1523. http://doi.org/10.1038/nn.3238.

van Alphen, B., Yap, M. H. W., Kirszenblat, L., Kottler, B., & van Swinderen, B. (2013). A dynamic deep sleep stage in Drosophila. *The Journal of Neuroscience: The Official Journal of the Society for Neuroscience*, *33*(16), 6917–6927. http://doi.org/10.1523/JNEUROSCI.0061-13.2013.

van der Voet, M., Harich, B., Franke, B., & Schenck, A. (2016). ADHD-associated dopamine transporter, latrophilin and neurofibromin share a dopamine-related locomotor signature in Drosophila. *Molecular Psychiatry*, *21*(4), 565–573. http://doi.org/10.1038/mp.2015.55.

Van Gool, W. A., & Mirmiran, M. (1983). Age-related changes in the sleep pattern of male adult rats. *Brain Research*, *279*(1-2), 394–398. [pii] ET-1983/11/21, http://doi.org/0006-8993(83)90217-2.

van Swinderen, B., & Brembs, B. (2010). Attention-like deficit and hyperactivity in a Drosophila memory mutant. *The Journal of Neuroscience: The Official Journal of the Society for Neuroscience*, *30*(3), 1003–1014. http://doi.org/10.1523/JNEUROSCI.4516-09.2010.

Vonhoff, F., Williams, A., Ryglewski, S., & Duch, C. (2012). Drosophila as a model for MECP2 gain of function in neurons. *PloS One*, 7(2), e31835. http://doi.org/10.1371/journal.pone.0031835.

Williams, H. L. (1971). The new biology of sleep. *Journal of Psychiatric Research*, *8*(3), 445–478. Retrieved from, http://www.ncbi.nlm.nih.gov/pubmed/4331539.

Williams, J. A., Su, H. S., Bernards, A., Field, J., & Sehgal, A. (2001). A circadian output in Drosophila mediated by neurofibromatosis-1 and Ras/MAPK. *Science (New York, N.Y.)*, *293*(5538), 2251–2256. http://doi.org/10.1126/science.1063097.

Wimo, A., Jönsson, L., Bond, J., Prince, M., & Winblad, B. (2013). The worldwide economic impact of dementia 2010. *Alzheimer's & Dementia: The Journal of the Alzheimer's Association*, *9*(1). http://doi.org/10.1016/j.jalz.2012.11.006. 1–11.e3.

Winokur, A. (2015). The relationship between sleep disturbances and psychiatric disorders: Introduction and overview. *The Psychiatric Clinics of North America*, *38*(4), 603–614. http://doi.org/10.1016/j.psc.2015.07.001.

Wong, K., Leonard, H., Jacoby, P., Ellaway, C., & Downs, J. (2015). The trajectories of sleep disturbances in Rett syndrome. *Journal of Sleep Research*, *24*(2), 223–233. http://doi.org/10.1111/jsr.12240.

Wu, Y., Bolduc, F. V., Bell, K., Tully, T., Fang, Y., Sehgal, A., & Fischer, J. A. (2008). A Drosophila model for Angelman syndrome. *Proceedings of the National Academy of*

Sciences of the United States of America, *105*(34), 12399–12404. http://doi.org/10.1073/pnas.0805291105.

Wu, M. N., Koh, K., Yue, Z., Joiner, W. J., & Sehgal, A. (2008). A genetic screen for sleep and circadian mutants reveals mechanisms underlying regulation of sleep in Drosophila. *Sleep*, *31*(4), 465 472. Retrieved from, http://www.pubmedcentral.nih.gov/articlerender.fcgi?artid=2279757&tool=pmcentrez&rendertype=abstract.

Wu, M., Robinson, J. E., & Joiner, W. J. (2014). SLEEPLESS is a bifunctional regulator of excitability and cholinergic synaptic transmission. *Current Biology: CB*, *24*(6), 621–629. http://doi.org/10.1016/j.cub.2014.02.026.

Xie, L., Kang, H., Xu, Q., Chen, M. J., Liao, Y., Thiyagarajan, M., … Nedergaard, M. (2013). Sleep drives metabolite clearance from the adult brain. *Science (New York, N.Y.)*, *342*(6156), 373–377. http://doi.org/10.1126/science.1241224.

Yanowitch, R., & Coccaro, E. F. (2011). The neurochemistry of human aggression. *Advances in Genetics*, *75*, 151–169. http://doi.org/10.1016/B978-0-12-380858-5.00005-8.

Yuan, Q., Joiner, W. J., & Sehgal, A. (2006). A sleep-promoting role for the Drosophila serotonin receptor 1A. *Current Biology: CB*, *16*(11), 1051–1062. http://doi.org/10.1016/j.cub.2006.04.032.

Yurgel, M. E., Masek, P., DiAngelo, J., & Keene, A. C. (2015). Genetic dissection of sleep-metabolism interactions in the fruit fly. *Journal of Comparative Physiology A, Neuroethology, Sensory, Neural, and Behavioral Physiology*, *201*(9), 869–877. http://doi.org/10.1007/s00359-014-0936-9.

Zhang, S., Yin, Y., Lu, H., & Guo, A. (2008). Increased dopaminergic signaling impairs aversive olfactory memory retention in Drosophila. *Biochemical and Biophysical Research Communications*, *370*(1), 82–86. http://doi.org/10.1016/j.bbrc.2008.03.015.

Zhdanova, I. V., Wang, S. Y., Leclair, O. U., & Danilova, N. P. (2001). Melatonin promotes sleep-like state in zebrafish. *Brain Research*, *903*(1–2), 263–268. Retrieved from, http://www.ncbi.nlm.nih.gov/pubmed/11382414.

Zheng, X., & Sehgal, A. (2010). AKT and TOR signaling set the pace of the circadian pacemaker. *Current Biology: CB*, *20*(13), 1203–1208. http://doi.org/10.1016/j.cub.2010.05.027.

CHAPTER NINE

Modeling Human Cancers in *Drosophila*

M. Sonoshita*,†, R.L. Cagan*,1
*Icahn School of Medicine at Mount Sinai, New York, NY, United States
†Kyoto University Graduate School of Medicine, Kyoto, Japan
[1]Corresponding author: e-mail address: ross.cagan@mssm.edu

Contents

Abstract

Cancer is a complex disease that affects multiple organs. Whole-body animal models provide important insights into oncology that can lead to clinical impact. Here, we review novel concepts that *Drosophila* studies have established for cancer biology, drug discovery, and patient therapy. Genetic studies using *Drosophila* have explored the roles of oncogenes and tumor-suppressor genes that when dysregulated promote cancer formation, making *Drosophila* a useful model to study multiple aspects of transformation. Not limited to mechanism analyses, *Drosophila* has recently been showing its value in facilitating drug development. Flies offer rapid, efficient platforms by which novel classes of drugs can be identified as candidate anticancer leads. Further, we discuss the use of *Drosophila* as a platform to develop therapies for individual patients by modeling the tumor's genetic complexity. *Drosophila* provides both a classical and a novel tool to identify new therapeutics, complementing other more traditional cancer tools.

Current Topics in Developmental Biology, Volume 121
ISSN 0070-2153
http://dx.doi.org/10.1016/bs.ctdb.2016.07.008

1. INTRODUCTION

The fruit fly *Drosophila* is a holometabolous insect that develops through three stages: embryo, larva, and pupa. The generation time is rapid, 11–12 days at 25°C, permitting the rapid building and expansion of new strains for a variety of assays. *Drosophila* researchers benefit from a century of genetic tool building, allowing for a detailed dissection of the functions of genes in development and, relevant to this essay, in disease. Further, gene misexpression and gene knockdown systems have evolved to permit gene manipulation with spatial and temporal precision. Until recently, the *Drosophila* field has been perhaps best known for its work in delineating developmental processes. In doing so, it has made fundamental contributions in the discovery and exploration of most major signaling pathways including Wnt, Hippo, Hedgehog, Notch, and Dpp (Ugur, Chen, & Bellen, 2016).

Recently, an increasing number of *Drosophila* biologists have begun to focus on studies of disease. Cancer is an especially strong fit for *Drosophila*: it is a predominantly genetic disease that remains one of the leading causes of mortality worldwide (Siegel, Miller, & Jemal, 2016). Cancer therapeutics remains a key unmet need despite considerable efforts to develop effective therapeutics over many decades. Cell culture and mouse studies have provided important contributions to our understanding of cancer and have identified numerous useful therapeutics. However, low success rates in clinical trials, high toxicity, and rapidly emergent resistance emphasize that cancer therapeutics would benefit from additional approaches that complement current efforts. *Drosophila* has proven a useful model organism for exploring cancer mechanisms, drug discovery, and personalized medicine, as discussed later.

2. FLY CANCER MODELS

2.1 Neoplasia

2.1.1 Cell Polarity

The idea of using *Drosophila* in cancer research came from early recognition of mutant flies that exhibited massive overgrowth phenotypes. Mutations in the gene *lethal giant larvae* (*lgl*)—first identified in the 1930s—represent the first recessive alleles that directed neoplasia in *Drosophila* (Froldi et al., 2008). Mutant larvae exhibited abnormal overproliferation of tissues including the brain, imaginal discs, and hematopoietic organs (Gateff, 1978).

The neoplasms grew quickly in a cell-autonomous manner; in addition, transformed tissue invaded into neighboring regions, one aspect of metastasis. When isolated and transplanted into wild-type flies, these neoplasms formed large tumors that eventually killed the host. Bryant et al. reported *lethal(2)giant discs* (*l(2)gd*) mutant flies, which arose spontaneously in their stocks, as one of the first models of a tumor suppressor (Bryant & Schuewer, 1971). Genotypically *l(2)gd* larvae demonstrated dramatic overgrowth of developing wing and leg tissues (imaginal discs). Interestingly, other alleles such as *discs large* (*dlg*) and *scribble* (*scrib*) also generated neoplastic tissues, and further analyses demonstrated compromised apical–basal polarity of epithelial cells in these mutants (Bilder, 2004; Froldi et al., 2008).

Later studies demonstrated that Dlg and Scrib proteins colocalized at the cell's septate junction (corresponding to the tight junction in vertebrates), partially overlapping with Lgl (Fig. 1A). Mutant cells displayed a loss in

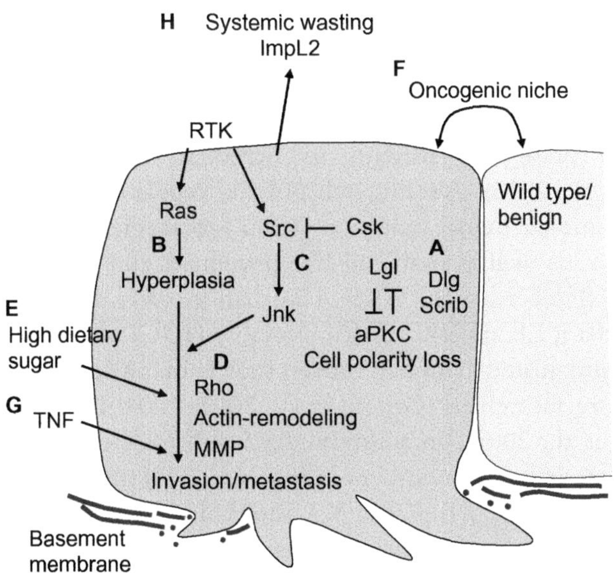

Fig. 1 A subset of key effectors and intercellular interactions in cancer progression identified in *Drosophila* studies. Junction proteins such as Lgl, Dlg, and Scrib inhibit tumorigenesis by maintaining cell polarity (A). *Ras* causes tissue hyperplasia (B), whereas *Src* and its effectors (C and D) and high dietary sugar (E) contribute to malignant progression. Cell–cell interactions between neoplastic (*pink*) and wild-type cells (*blue*) in tumor microenvironment also determine oncogenic potential of neoplastic cells within the oncogenic niche (F). Juxtaposed TNF promotes expansion of *dRas*^{G12V}; *scrib* tumors (G). By secreting ImpL2, neoplastic cells promote a systemic wasting syndrome similar to cachexia (H). See text for details.

apical/basal polarity that led to deregulated tissue growth (Bilder, Li, & Perrimon, 2000). Many of the neoplastic cells migrated to distant organs when transplanted to new hosts, causing mortality; this migration also phenocopies important aspects of metastasis (Froldi et al., 2008). Exploring *Drosophila* neuroblasts, another study demonstrated that atypical protein kinase C (aPKC) phosphorylated Lgl to promote its release from the plasma membrane. This cascade inhibited proper distribution of polarity determinant proteins such as Miranda (Betschinger, Mechtler, & Knoblich, 2003). In turn, loss of *lgl* led to ectopic cortical localization of aPKC and symmetric cell division; the result was a proliferation of neuroblasts (Lee, Robinson, & Doe, 2006).

In multicellular organisms, epithelial cells must retain correct cell polarity to maintain proper tissue integrity (Nelson, 2003). These fly studies highlight the significance of properly regulating cell polarity as a key defense against transformation; loss of apical–basal polarity is one of the key characteristics of malignant cells in humans (Hanahan & Weinberg, 2011). An example in mammals is the role of the adherence junction protein E-cadherin, which contributes to the maintenance of cell polarity: germline mutations in the *CDH1* gene—encoding E-cadherin—compromise cell polarity and predispose patients to hereditary diffuse gastric cancer (Carneiro et al., 2012). Altering cell polarity regulators can also promote epithelial-to-mesenchymal transition (EMT) to accelerate cancer invasion and metastasis, as well as stem cell-like properties and chemoresistance of cancer cells (Fischer et al., 2015; Hanahan & Weinberg, 2011; Ye & Weinberg, 2015; Zheng et al., 2015). Notably *HUGL-1*, a human ortholog of *lgl*, is downregulated in human cancers such as breast, lung, prostate, ovarian, and colorectal cancers (Grifoni et al., 2007, 2004; Kuphal et al., 2006; Schimanski et al., 2005). In addition, *Lgl1*-null mice exhibit severe brain dysplasia due to increased number of progenitor cells that fail to differentiate (Klezovitch, Fernandez, Tapscott, & Vasioukhin, 2004), suggesting parallel mechanisms for tumor suppression across species. These studies emphasize the importance of *Drosophila* models in exploring the mechanisms that direct epithelial-based tumors.

2.1.2 Inducible Expression of Cancer Genes

A key tool provided by *Drosophila* is its sophisticated use of targeted gene activation and inactivation. One way to drive transgene expression is to use inducible drivers such as heat-shock promoters: placing flies in a warm incubator activates the associated transgenes, allowing temporal control

(Greenspan, 2004). This method allows inducible expression; however, induction is uncontrolled, yielding expression throughout the body.

To avoid these concerns, Brand and Perrimon developed the binary *gal4–UAS* system (Brand & Perrimon, 1993). This induction system utilizes two components: (i) the yeast transcription activator gene *gal4* placed downstream of a promoter/enhancer, and (ii) GAL4 target sequences (*UAS*, short for *U*pstream *A*ctivation *S*equence) placed 5′ of the gene of interest. When a cell has both of these elements—typically achieved by standard genetic crossing—the promoter activates GAL4 expression, which in turn binds to a cell's UAS elements to induce transgene expression in a temporally and spatially precise manner (Brand, Manoukian, & Perrimon, 1994; Brand & Perrimon, 1993; Brumby & Richardson, 2005). More recently, other tools have added to the fly researcher's repertoire (Venken & Bellen, 2012). Together, these (often off-the-shelf) tools allow researchers to activate/inactivate multiple genes in a single tissue or even cell.

2.1.3 Ras

RAS genes are among the most frequently mutated across human cancers; tumors harboring activating *RAS* mutations are among the most difficult to treat (Stephen, Esposito, Bagni, & McCormick, 2014). To dissect the role of *RAS* transformation in vivo, Karim and Rubin expressed the oncogenic *Drosophila RAS* isoform $dRas1^{G12V}$ in developing imaginal discs, leading to tissue hyperplasia (Karim & Rubin, 1998). They also observed tissue-wide cell death even for cells that were apart from $dRas1^{G12V}$-expressing cells, a form of compensatory apoptosis. The canonical MAPK signaling pathway was necessary for these phenotypes, as loss-of-function mutations in *raf*, *mek*, and *mapk* dominantly suppressed these phenotypes. Using the FLP–FRT system (Golic & Lindquist, 1989), to generate small homozygous clones, Richardson and colleagues observed similar transformation phenotypes (Brumby & Richardson, 2003, 2005). Thus, studies such as these helped established the key roles of RAS–MAPK signaling pathway in regulating proper tissue growth and in promoting transformation (Fig. 1B).

In another study, the Ras pathway was artificially activated specifically in the developing eye epithelium by reducing activity of Ksr (Huang & Rubin, 2000), an interesting Ras effector identified as a genetic modifier of RAS activity in flies and worms (Kornfeld, Hom, & Horvitz, 1995; Sundaram & Han, 1995; Therrien et al., 1995). Huang and Rubin then used the overexpression "EP" system (Rørth, 1996; Rørth et al., 1998) to screen genes for the ability to alter the resulting hyperplasia. Their "genetic modifier screen" successfully identified four enhancers and eight suppressors of the hyperplastic phenotype. For example, they identified *Lk6*, a kinase downstream of MAPK that has proven a significant tumor suppressor (Huang & Rubin, 2000; Proud, 2007). These studies emphasize the power of loss-of-function and gain-of-function screens to identify oncogenes and tumor suppressors, providing an opportunity to explore these factors with single-cell resolution and to place them into cancer networks in situ.

2.2 Fly Models for Invasion and Metastasis

2.2.1 Ras-Based Models

When tumors are limited to the primary sites, several treatment options can be available, such as surgical resection, chemotherapy, and radiotherapy. Once tumors metastasize, however, mortality increases significantly. Metastasis consists of multiple steps: growth of the primary tumor, invasion into surrounding stroma, intravasation into blood and lymphatic circulations, extravasation to the secondary organs, and growth of secondary tumors (Fidler, 2003; Steeg, 2006).

Metastasis is considered to occur when multiple mutations accumulate (Kinzler & Vogelstein, 1996). For example, in the colon, tumorigenesis starts with mutations in *APC* (*Adenomatous Polyposis Coli*) gene. However, *APC* alteration by itself is not sufficient to cause malignancy; mutations in additional genes such as members of the RAS family are required for cancer progression (Morris et al., 2008). Using the $dRas1^{G12V}$ "benign" tumor model, Pagliarini and Xu found that reducing activity of cell polarity genes *scrib*, *lgl*, *dlg*, *bazooka* (*baz*), *stardust* (*sdt*), or *cdc42* provoked $dRas1^{G12V}$ cells to progress towards dissemination and secondary growth (Pagliarini & Xu, 2003). A variety of phenotypic similarities were observed between $dRas^{G12V}$; *scrib* and human tumors; for example, the basement membrane (BM) was degraded and cells invaded into neighboring tissues, a behavior regulated by E-cadherin (Pagliarini & Xu, 2003). These results indicate a key role of maintaining proper cell polarity in preventing metastatic progression of benign cells with oncogenic *RAS* isoforms.

2.2.2 Src-Based Models

In a variety of human cancers including melanoma, breast, and colorectal cancers, SRC family kinases (SFKs) are activated by various cues such as growth factors and cell–cell contact (Yeatman, 2004). SFKs are linked to malignant progression of human cancers, and in particular, their activity is frequently associated with metastatic potential. As such they serve as attractive therapeutic targets, but their precise roles in cancer progression remain to be clarified.

SFKs contain a negative regulatory C-terminal domain. C-terminal SRC kinase (CSK) phosphorylates a key regulatory tyrosine residue, causing conformational change of the domain to inactivate kinase activity. Indeed, *Csk*-null mice exhibit early embryonic lethality due to hyperactivation of SFKs (Imamoto & Soriano, 1993). *Drosophila Csk* (*dCsk*) also antagonizes *Drosophila Src* (*dSrc*), and reducing *dCsk* activity led to increased Src activity and increased cell proliferation (Read, Bach, & Cagan, 2004; Stewart, Li, Hung, & Xu, 2003) (Fig. 1C).

Interestingly, our knockdown experiments for *dCsk* using different *gal4* drivers caused distinct phenotypes. *dCsk* knockdown in a whole tissue increased the size of the tissue due to overproliferation of affected cells. The *patched* (*ptc*) promoter directed expression to a stripe of a few rows of cells along the anterior/posterior (A/P) boundary in wing discs (Speicher, Thomas, Hinz, & Knust, 1994). In contrast to the whole-tissue knockdown, *ptc–gal4*-driven *dCsk* knockdown promoted apoptosis of affected cells in wing discs, which was similar to *dCsk* clones (Vidal, Larson, & Cagan, 2006). Noticeably, mutant cells near the A/P boundary had dropped out of the epithelial monolayer and migrated basally to posterior compartment away from the *ptc* region. Genetic screening for modifiers of *dCsk*-induced migration phenotype identified *Drosophila* orthologs of E-cadherin, Jnk, and Mmp1, as well as actin-remodeling genes such as Rho1. This indicates that SFKs coordinate invasion by multiple signaling pathways, altering the transforming cells' cellular network and promoting EMT (Fig. 1D) (Rudrapatna, Bangi, & Cagan, 2014; Vidal et al., 2006).

2.2.3 Ras/Src Models and the Importance of Diet

Simultaneous activation of RAS and SRC occurs in a broad palette of human cancers including breast, colorectal, and pancreatic (Ishizawar & Parsons, 2004). By itself, the oncogenic isoform $dRas^{G12V}$ led to low-level outgrowth as discussed earlier. Introducing a *dCsk*-null allele led to massive overgrowth and invasion into the brain (Vidal, Warner, Read, & Cagan, 2007). These results indicate that coactivation of RAS and SRC can enhance tumor progression (Fig. 1D). Although $dRas1^{G12V}$; $dCsk^{null}$ tumor cells show invasive phenotypes, they did not disseminate to promote secondary tumors (Hirabayashi, Baranski, & Cagan, 2013; Vidal et al., 2007), indicating that they require additional biological factors to more fully exhibit metastasis-like behavior. One factor may be dietary sugar.

Accumulating evidence indicates that diabetic patients show increased risk of specific types of cancer (Giovannucci et al., 2010). Does metabolic dysfunction affect carcinogenesis?

Feeding flies high dietary sugar (HDS) led to insulin resistance, hyperglycemia, hyperinsulinemia, and accumulation of fat, mimicking key aspects of type 2 diabetes (Musselman et al., 2011; Na et al., 2013). Feeding flies HDS converted localized $dRas1^{G12V}$; $dCsk^{null}$ tumors into aggressively "metastasizing" tumors that displayed increased growth, invasion, and insertion into secondary sites around the fly (Fig. 1E) (Hirabayashi et al., 2013). Interestingly, $dRas1^{G12V}$; $dCsk^{null}$ cells retained insulin sensitivity, while surrounding tissues developed insulin resistance. Salt-inducible kinase (Sik) acted as a downstream mediator of Ras/Src signaling to inhibit Hippo pathway activity by phosphorylating Salvador; this led to the dissociation of the Hippo complex and activation of the transcription factor Yorkie (Yki) to induce the Wnt ortholog Wingless (Wg). Induced Wg in turn upregulated insulin receptor (InR) in $dRas1^{G12V}$; $dCsk^{null}$ cells in Tcf-dependent manner (Hirabayashi et al., 2013; Hirabayashi & Cagan, 2015). This complex Ras/Src/glucose–Hippo–Wnt–insulin pathway appears to form a feed-forward network that drives up InR expression in the tumor, allowing it to absorb glucose in a body that is otherwise insulin resistant.

These studies uncover the impact of diet on cancer development and raise several relevant points of therapeutic intervention such as RAS/MAPK, SRC, SIK, WNT, and InR against which inhibitors are already under development and available. Drugs including the alpha-glucosidase inhibitor acarbose, an inhibitor of the canonical Wnt signaling pathway pyrvinium, the SIK inhibitor HG-9-91-01, and the multi-kinase inhibitor (KI) AD81 all showed the efficacy in flies that is predicted by the pathway model (Anderson, 2005; Clark et al., 2012; Dar, Das, Shokat, & Cagan, 2012; Hirabayashi et al., 2013; Hirabayashi & Cagan, 2015; Thorne et al., 2010).

The Hippo pathway has been implicated in cancer development because it controls cell proliferation, cell fate, and organ size (Halder & Johnson, 2011; Hariharan, 2015; Harvey & Tapon, 2007). Perhaps surprisingly, Hippo signaling also plays a central role in high sugar-induced cancer progression. From a therapeutic standpoint Hippo pathway is an attractive target, as YAP1 was shown to promote resistance to inhibitors against RAF and MEK (Lin et al., 2015).

2.2.4 Brain Tumor Models

Gliomas are the most common tumors of the central nervous system. Especially glioblastoma (GBM) is rapidly fatal, with median survival of patients being less than 1 year (Stupp et al., 2005). In most cases GBM is hard to cure despite surgery, intensive chemotherapy, and radiotherapy. To establish effective therapeutics, significant effort has been focused on determining the mechanisms of GBM formation.

The most frequent genetic alterations include activation of EGFR (Epidermal Growth Factor Receptor) and PI3K (PhosphatidylInositol-3 Kinase) signaling pathways (Maher et al., 2001). To test the effects of these abnormalities in GBM development, Read et al. expressed activated isoforms of *dEgfr* and *dp110* transgenes specifically in the glia. Coactivation of these two pathways led to dramatic invasive overgrowth of glial cells, resulting in lethality at late larval stage (Read, Cavenee, Furnari, & Thomas, 2009). Activation of either pathway alone showed milder or no effects, indicating that their concurrent activation was necessary for GBM formation. The authors also found that the neoplastic transformation required multiple pathways dysregulated in human GBM, including cyclins–Cdks and RB-E2F, suggesting new therapeutic strategies for slowing GBM progression.

2.2.5 Medullary Thyroid Cancer Models

By some measures, thyroid cancer has been the most rapidly increasing cancer type (Siegel et al., 2016); the incidence of thyroid cancer may exceed colorectal cancer as the fourth leading cancer diagnosis by 2030 (Rahib et al., 2014). Medullary thyroid cancer (MTC) arises from transformation of the parafollicular C cells in the thyroid and represents 3–5% of thyroid cancers (Hadoux, Furio, Tuttle, & Schlumberger, 2016). Activating mutations in the protooncogene *RET* (*REarranged during Transfection*) represent the dominant cause of MTC: most hereditary MTC and 40–50% of sporadic MTC cases also have *RET* mutations (Cerrato, De Falco, & Santoro, 2009; Elisei et al., 2008; Hadoux et al., 2016). MTC is typically slow growing; if resected early patients' prognosis is good. However, patients with metastatic disease show a significant rate of mortality. Metastasized MTC in distant organs such as liver is often refractory to cytotoxic chemotherapy, and new models were needed to explore mechanism and identify candidate therapies.

In MEN2 (Multiple Endocrine Neoplasia 2) patients, two major classes of point mutations have been reported: MEN2A patients typically have mutations in an extracellular cysteine that promotes ligand-independent homodimerization of RET; MEN2B patients have an M918T mutation that leads to uncontrolled activity of the intracellular kinase cleft (Cerrato et al., 2009). Expressing these mutant Ret isoforms in the fly eye led to aberrant growth and development similar to $dRas1^{G12V}$ (Read et al., 2005). To determine signaling pathways required for *RET*-dependent MTC development, a dominant genetic modifier screen identified 140 genes required to regulate oncogenic Ret signaling in the eye. They defined signaling pathways including Ras, Src, Jnk, and PI3K.

The dominant modifier screen also identified *Drosophila Sin3a* (*dSin3a*) as a strong suppressor downstream of $dRet^{MEN2}$. Subsequent work showed that *dSin3a*, as a cofactor of HDAC, regulated a "module of genes" pivotal for invasion (Das, Sangodkar, Negre, Narla, & Cagan, 2013). *SIN3A* expression is similarly lowered in a broad cross section of human tumors and was required to reduce receptor tyrosine kinase signaling and cell migration in cultured human cancer cells (Das et al., 2013). This demonstrates how genetic modifier screens can open new windows into the mechanisms that drive tumors and the network "checkpoints" that resist tumor progression.

2.3 Microenvironment

A key contribution of *Drosophila* to understanding the biology of tumor progression is work on the local interactions between transformed and normal cells within an epithelium. Accumulating evidence in *Drosophila* studies indicates that genotype differences between neighboring epithelial cells in a tumor affect tumor growth. Flies mutant for *Minute*—encoding for ribosomal proteins—show a characteristic developmental delay. In genotypically *Minute* mosaic wings, *Minute* clones showed slower dividing time than wild-type cells and were eventually outcompeted and removed from the epithelium (Morata & Ripoll, 1975). Curiously the final size of the wing was not affected, suggesting that this phenomenon, termed "cell competition," ensures proper tissue size. Cell competition was also observed in $dRas1^{G12V}$ flies (Prober & Edgar, 2000), and the involvement of cell competition in the formation of the oncogenic niche has drawn much attention (Enomoto, Vaughen, & Igaki, 2015) (Fig. 1F). *Ras* activation directed upregulation of *Drosophila Myc*, inducing *cyclin E* and increasing the rate of cell growth (Prober & Edgar, 2000). Notably, clones containing higher levels of *dMyc*-induced apoptotic removal of neighboring cells with lower *dMyc* (de la Cova, Abril, Bellosta, Gallant, & Johnston, 2004), further suggesting a role for cell competition in the development of human cancers (Enomoto et al., 2015; Moreno, 2008).

Another aspect of tumorigenesis shaped by tumor microenvironment came from studies on tumor necrosis factor (TNF), a pleiotropic cytokine. It regulates various biological reactions such as infections, inflammation, and tissue homeostasis (Igaki & Miura, 2014; Kalliolias & Ivashkiv, 2016). The *Drosophila* TNF ortholog Eiger activates downstream Jnk signaling as in mammals (Igaki et al., 2002; Kalliolias & Ivashkiv, 2016). When expressed in the developing eye epithelium, Eiger triggers caspase-independent cell death, resulting in a "reduced-eye" phenotype. It is known that clones of cells mutant for tumor suppressors such as *scrib* show loss of polarity and are eventually eliminated by cell death. In the absence of Eiger, however, Jnk activation does not take place to eliminate *scrib* cells; the result is tumorous growth (Igaki, Pastor-Pareja, Aonuma, Miura, & Xu, 2009). These results suggest a tumor *suppressor* role for Eiger as a key environmental regulator. This is consistent with reports on its tumor suppressor function in mammals: TNF-induced necrotic death of human cancer cells xenografted in mice (Balkwill et al., 1986), after which the cytokine was named. These experiments position TNF-dependent killing of cancer cells as an evolutionarily conserved antitumor system.

Remarkably, TNF also has a tumor *promoting* role as demonstrated in mice (Moore et al., 1999). In flies, Eiger promoted the growth of tumors in the genetic context of $dRas^{G12V}$; *scrib* (Cordero et al., 2010) (Fig. 1G). Flies have circulating blood cells, hemocytes, that share functional characteristics with mammalian blood cells and constitute the fly immune system (Hartenstein, 2006). In mammals, immune cells such as macrophages accumulate in tumors (Lavin, Mortha, Rahman, & Merad, 2015). Similarly, in flies, hemocytes are found attached to $dRas^{G12V}$; *scrib* tumors where the BM has been disrupted, but not in $dRas^{G12V}$ benign tumors where the BM remains intact (Pastor-Pareja, Wu, & Xu, 2008). Curiously, these tumor-associated hemocytes produced Eiger and stimulated the expansion of $dRas^{G12V}$; *scrib* tumors (Cordero et al., 2010), suggesting the pivotal role of *RAS* in converting the effects of TNF signaling in the course of tumorigenesis from anti- to protumor.

Cancers frequently exhibit mitochondrial dysfunction (Brandon, Baldi, & Wallace, 2006; Modica-Napolitano, Kulawiec, & Singh, 2007), but its effects on progression were unclear. Ohsawa et al. found that $dRas^{G12V}$ clones harboring mitochondrial dysfunction generated reactive oxygen species, which led to a chain of signaling responses: (i) Jnk was activated; (ii) Jnk inhibited Hippo signaling to activate the transcription factor Yki, a fly ortholog of human YAP1; (iii) Yki-induced unpaired (Upd), a JAK/STAT-activating cytokine related to interleukin (IL)-6; and (iv) secreted Upd inactivated Jnk signaling in neighboring benign $dRas^{G12V}$ cells, causing their proliferation and invasion (Ohsawa et al.,

2012). These results highlight how the tumor microenvironment can act as a potent oncogenic niche that enhances tumor emergence and progression (Fig. 1F).

2.4 Cachexia

Cachexia is a multifactorial wasting syndrome associated with chronic disorders including cancers. It is a type of energy balance disorder in which an imbalance emerges between decreased energy intake and its increased consumption (Aoyagi, Terracina, Raza, Matsubara, & Takabe, 2015; Fearon, Arends, & Baracos, 2013). Cachectic cancer patients suffer from significant weight loss primarily due to loss of skeletal muscle and fat in the body. It occurs in 40% to more than 80% of cancer patients depending on cancer type, and it accounts for about 20% of cancer deaths (Argilés, Busquets, Stemmler, & López-Soriano, 2014). At present, there is no cure for patients suffering from cachexia nor a biomarker to identify patients at high risk of developing it. How cachexia emerges in cancer patients has not been thoroughly determined yet, but recent reports using *Drosophila* as models have given intriguing insights on its mechanisms.

When derepressed in Hippo signaling, *Drosophila* and mammalian Yki/YAP1-induced cell proliferation (Halder & Johnson, 2011). Kwon et al. reported induction and secretion of ImpL2—an IGF (Insulin-like Growth Factor)-binding protein (IGFBP)—from Yki-dependent hyperplastic intestinal cells (Kwon et al., 2015). They provided evidence that secreted ImpL2 is a key mediator of the wasting phenotype observed in distant organs such as muscle, fat body, and ovaries (Fig. 1H). Because the mammalian ortholog IGFBP is known to antagonize insulin/IGF signaling (Baxter, 2014), this study suggests that proper control of insulin/IGF signaling is required to prevent wasting symptoms.

In a separate study using a tumor transplantation model in flies, Figueroa-Clarevega and Bilder found that malignant but not benign tumors led to wasting phenotypes in distant organs such as muscles and adipose tissue (Figueroa-Clarevega & Bilder, 2015). ImpL2 was highly upregulated in malignant tumors but not benign tumors; artificially overexpressed *ImpL2* was sufficient to induce tissue wasting, while *ImpL2* knockdown in tumor cells reduced the severity of the wasting phenotype. Of note, hemolymph sugar levels were higher in flies with resident tumors, suggesting the emergence of insulin resistance; this mirrors the insulin resistant that frequently emerges in cancer patients. These studies again indicate that impaired insulin/IGF signaling contributes to cachexia development. It will be intriguing to ascertain if IGFBP can be a useful therapeutic target for this devastating syndrome.

3. DRUG DISCOVERY

3.1 Flies as a Therapeutics Screening Platform

Drosophila is perhaps the simplest commonly used model organism with epithelial structure, organs, and signaling pathways that are closely conserved to mammals. As a model for solid tumors, they have important advantages including cost, speed, and powerful genetic tools. Flies also have plenty of disadvantages, including differences in the details of signaling pathways as well as likely differences in drug ADME (absorption, distribution, metabolism, and excretion) properties. Still, thousands of cancer trials are currently ongoing; why do we need a genetically accessible, whole-animal screening platform to find more hits? One answer is that whole-animal screening identifies a different class of hits from "tumor focused" screening (cell lines, organoids, etc.): to reduce tumor progression, positives can act on any number of targets and in multiple parts of the animal. For example, if altering whole-body metabolism is helpful, than these activities may be included in a hit. Multiple screens have been done in *Drosophila* S2 cells (Gladstone & Su, 2011b), but here we consider the fly as a whole-animal screening platform and highlight a particular class of compounds that may best fit its unique characteristics: polypharmacology.

Targeted drugs have a long history of use as a "chemical genetic" tool to inhibit specific targets (see Gladstone & Su, 2011b, for an excellent review on this topic). For example, Radimerski et al. (2002) used the chemical pathway inhibitor RAD001 to explore regulation of the endogenous *Drosophila* PI3K pathway, while Bhandari and Shashidhara validated the ability of indomethacin to attack its target APC when expressed in flies (Bhandari & Shashidhara, 2001), and Micchelli et al. (2003) validated a γ-secretase inhibitor as reducing Notch pathway activity. Indeed, companies such as Novartis have explored *Drosophila* as a tool for validating drug specificity in situ (Bangi, Garza, & Hild, 2011). Regarding cancer our own laboratory—working with the oncologist Samuel Wells and with AstraZeneca—used the *Drosophila* $dRet^{MEN2B}$ model to help validate the lead compound ZD6474 as a candidate therapeutic for MTC (Vidal, Wells, Ryan, & Cagan, 2005); now named vandetanib, the drug was approved in 2011 as the first targeted therapy for MTC.

There have been several pioneers in the use of flies as a screening platform to identify candidate cancer therapeutics including the Richardson and Perrimon laboratories; our laboratory has also developed a high-throughput Center for Personalized Cancer Therapeutics that relies on robotics-based screening of fly cancer models to develop personalized treatment recommendations. The basic approach is to move food plus drug into tubes or a 96-well plate (Chang et al., 2008), then develop a rapid screen for scoring efficacy such as decreased GFP-marked growth or rescue of animal viability (e.g., Chang et al., 2008; Markstein et al., 2014; Willoughby et al., 2013).

Using this approach, Willoughby et al. (2013) identified cancer-relevant drugs such as acivicin as effective in suppressing Ras-dependent transformation. Using an oncogenic isoform of the Ras pathway effector Raf, Markstein and Perrimon identified multiple drugs that targeted transformed stem cells in the *Drosophila* gut (Markstein et al., 2014). In a focused screen of 88 FDA-approved drugs, they identified 14 that reduced gut transformation. Leveraging the advantages of whole-animal screening, they demonstrated both efficacy but also potential activities in promoting proliferation of normal stem cells. The ability to explore the details of a drug's effects on the animals' tissues is an important opportunity provided by screening simple model systems.

While targeted therapies remain a "sweet spot" for flies, the more common approach to cancer therapeutics is DNA-damaging agents including DNA intercalators and radiation. Su and colleagues have pioneered the use of *Drosophila* for whole-animal screening of "radiosensitizers," drugs that optimize DNA damage-mediated cell killing (Gladstone & Su, 2011a). Using X-rays to create damage and lethality, they identify the protein elongation inhibitor drug bouvardin as a potent radiosensitizer, validating this drug in flies, human cell lines, and mouse xenografts (Gladstone et al., 2012; Stickel, Gomes, Frederick, Raben, & Su, 2015).

3.2 *Drosophila* and the Case for Polypharmacology

Despite significant progress in understanding the mechanisms of cancer development as described earlier, cancer is still difficult to treat especially if it has metastasized to distant organs. Current therapeutics relies largely on small-molecule therapies. However, conventional cytotoxic chemotherapeutics such as DNA-damaging agents and cytoskeleton disruptors act on not only cancers but also normal parts of the body, which frequently results in severe side effects with only limited efficacy. To address this problem, a "targeted therapy" approach has emerged over the past two decades: cleanly targeting known cancer "driver" proteins may prove to have a better therapeutic index. Protein kinases have attracted particular attention as rate-limiting regulators of various aspects during carcinogenesis, and the field of KI chemistry is rapidly maturing (Fleuren, Zhang, Wu, & Daly, 2016).

The first success came from the development of a potent KI targeting chronic myelogenous leukemia (CML): imatinib has proven strongly effective, shifting CML a fatal disease to a curable disorder (Apperley, 2015). This success established targeted therapy as a promising strategy, and KIs have proven successful in helping patients across a broad spectrum of tumors. But cleanly targeting KIs also have their limitations; for example, as a class KIs have proven somewhat more toxic than standard cytotoxic chemotherapeutics (Gharwan & Groninger, 2016). In one example, an approved multi-KI sorafenib causes severe side effects such as diarrhea, severe rash, pancreatic atrophy, and even death in treated patients (Gharwan & Groninger, 2016; Hescot, Vignaux, & Goldwasser, 2013; Lam et al., 2010). In addition, KIs are strongly susceptible to acquired drug resistance. These difficulties have contributed to cancer's overall poor success rate in clinical trials, the lowest among major diseases (Hay, Thomas, Craighead, Economides, & Rosenthal, 2014). Such low success rates of drug discovery have been problematic especially for rare cancer types, where risk can exceed reward for pharmaceutical companies.

Targeting multiple kinases can ameliorate some of these issues, but most screening platforms do not offer the ability to (i) rationally identify multiple targets that act in concert to (ii) alter the overall tumor network while (iii) accounting for whole-body complexity. Simply targeting multiple known cancer pathways without accounting for the latter can lead to (and has led to) unexpected toxicities. The ability to use *Drosophila* as a whole-animal screening platform offers a way forward to identifying new classes of lead compounds.

We have been working together with the Kevan Shokat's and Arvin Dar's chemistry laboratories to establish a rational, genetics-based method for developing drugs that emphasize "polypharmacology," inhibiting multiple targets. To generate a fly model of medullary thyroid carcinoma, oncogenic Ret ($\mathrm{dRet}^{\mathrm{MEN2B}}$) was expressed broadly to direct animal lethality (Fig. 2A), and we identified AD57 as a novel KI that rescued $dRet^{MEN2B}$ flies more efficiently than vandetanib. Our previous work identified the Ras, Src, and PI3K pathways as key mediators of Ret transformation (Fig. 2B) (Read et al., 2005). Reducing activity of the PI3K effector Tor ($tor^{-/+}$)—a physical

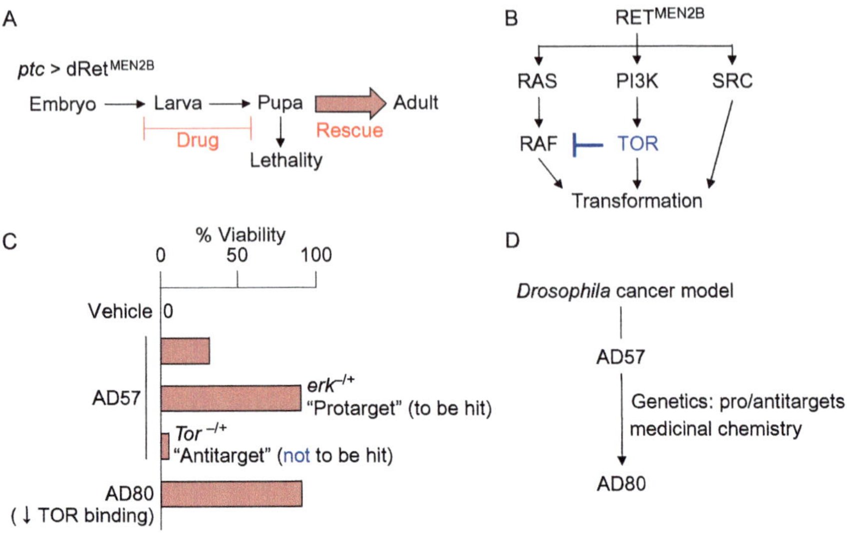

Fig. 2 *Drosophila* cancer model as a platform for drug development. (A) Viability assay using a *ptc>dRet*^{MEN2B} model for drug efficacy. The *patched* (*ptc*) promoter drives expression of oncogenic *Drosophila Ret* (*dRet*^{MEN2B}), causing complete lethality before adulthood. Feeding flies with food containing drugs can rescue lethality. (B) Major signaling pathways downstream of RET. TOR inhibits RAS pathway activity by interfering with RAF. (C) Stepwise development of novel anticancer drugs. A kinase inhibitor AD57 rescues *ptc>dRet*^{MEN2B} flies by inhibiting RAS, PI3K, and SRC pathways. Fly genetics identified Erk as a "prospective (pro-) target" that should be kept inhibited, and Tor as an "antitarget" to be kept uninhibited. AD80 and AD81, derivatives of AD57 with inhibition of RAS and SRC signaling but not TOR, show improved efficacy as compared with AD57. (D) *Drosophila* cancer models, combined with genetic approaches and medicinal chemistry, help achieve rational polypharmacology to accelerate development of novel anticancer drugs.

target of AD57—increased whole-animal toxicity of AD57, suggesting that Tor was an "antitarget" that should be removed from AD57's activities (Fig. 2C).

Through chemical modeling, our laboratories hypothesized that modifying the terminal phenyl ring could prevent AD57 from binding to Tor. The resulting compounds AD80 and AD81 showed reduced activity against Tor and, importantly, significantly improved efficacy in flies, human cell lines, and mouse xenografts (Fig. 2C and D) (Dar et al., 2012; Das et al., 2013). Together, studies from a variety of laboratories have shown the promise of *Drosophila* as an accessible whole-animal screening platform. Combining genetics as part of the compound screening process can further leverage the power of *Drosophila* to identify unique therapeutic spaces.

To leverage this promise further, we are using genetic modifier screens to identify new activities that can be "dialed in" or "dialed out" of currently approved drugs. This requires fly genetics, medicinal chemistry, and computational chemistry in an integrated approach to identify new "protargets" and "antitargets," then modeling to move into better chemical spaces (unpublished results). These approaches represent just one of many new and exciting paths that we anticipate will be taken by researchers who are taking advantage of the remarkable promise of *Drosophila* as a therapeutics discovery platform.

4. CONCLUSION

Drosophila has a long and proud history of solving complex problems with powerful genetics. In the developmental field, flies have proven an excellent discovery tool: genetic screens in particular identify new and surprising mechanisms, a "hypothesis-building" tool that is rapid and inexpensive. These same qualities make model systems such as *Drosophila* useful in translational work: surprising new mechanisms and therapeutics can be identified that address the complexity of diseases such as cancer. Similar to problems in development, the power of *Drosophila* lies in its ability to take a fresh, whole animal look at a disease that has joined heart disease as the major sources of mortality of Americans. Flies will not replace mammalian models, but similar to their role in development they provide a powerful and complementary tool.

ACKNOWLEDGMENTS

We thank members of the Cagan laboratory for important discussions. M.S. and R.C. were supported by National Institutes of Health grants U54OD020353, R01-CA170495, and R01-CA109730 and Department of Defense Grant W81XWH-15-1-0111.

REFERENCES

Anderson, D. C., Jr. (2005). Pharmacologic prevention or delay of type 2 diabetes mellitus. *The Annals of Pharmacotherapy*, *39*(1), 102–109.

Aoyagi, T., Terracina, K. P., Raza, A., Matsubara, H., & Takabe, K. (2015). Cancer cachexia, mechanism and treatment. *World Journal of Gastrointestinal Oncology*, 7(4), 17–29.

Apperley, J. F. (2015). Chronic myeloid leukaemia. *The Lancet*, *385*(9976), 1447–1459.

Argilés, J. M., Busquets, S., Stemmler, B., & López-Soriano, F. J. (2014). Cancer cachexia: Understanding the molecular basis. *Nature Reviews. Cancer*, *14*(11), 754–762.

Balkwill, F. R., Lee, A., Aldam, G., Moodie, E., Thomas, J. A., Tavernier, J., et al. (1986). Human tumor xenografts treated with recombinant human tumor necrosis factor alone or in combination with interferons. *Cancer Research*, *46*(8), 3990–3993.

Bangi, E., Garza, D., & Hild, M. (2011). In vivo analysis of compound activity and mechanism of action using epistasis in Drosophila. *Journal of Chemical Biology*, *4*, 55–68.

Baxter, R. C. (2014). IGF binding proteins in cancer: Mechanistic and clinical insights. *Nature Reviews. Cancer*, *14*(5), 329–341.

Betschinger, J., Mechtler, K., & Knoblich, J. A. (2003). The Par complex directs asymmetric cell division by phosphorylating the cytoskeletal protein Lgl. *Nature*, *422*(6929), 326–330.

Bhandari, P., & Shashidhara, L. S. (2001). Studies on human colon cancer gene APC by targeted expression in Drosophila. *Oncogene*, *20*(47), 6871–6880.

Bilder, D. (2004). Epithelial polarity and proliferation control: Links from the Drosophila neoplastic tumor suppressors. *Genes & Development*, *18*(16), 1909–1925.

Bilder, D., Li, M., & Perrimon, N. (2000). Cooperative regulation of cell polarity and growth by Drosophila tumor suppressors. *Science*, *289*(5476), 113–116.

Brand, A. H., Manoukian, A. S., & Perrimon, N. (1994). Ectopic expression in Drosophila. *Methods in Cell Biology*, *44*, 635–654.

Brand, A. H., & Perrimon, N. (1993). Targeted gene expression as a means of altering cell fates and generating dominant phenotypes. *Development*, *118*(2), 401–415.

Brandon, M., Baldi, P., & Wallace, D. C. (2006). Mitochondrial mutations in cancer. *Oncogene*, *25*(34), 4647–4662.

Brumby, A. M., & Richardson, H. E. (2003). Scribble mutants cooperate with oncogenic Ras or Notch to cause neoplastic overgrowth in Drosophila. *The EMBO Journal*, *22*(21), 5769–5779.

Brumby, A. M., & Richardson, H. E. (2005). Using Drosophila melanogaster to map human cancer pathways. *Nature Reviews. Cancer*, *5*(8), 626–639.

Bryant, P. J., & Schuewer, G. (1971). Giant and duplicated imaginal discs in a new lethal mutant of Drosophila melanogaster. *Developmental Biology*, *24*, 233–263.

Carneiro, P., Fernandes, M. S., Figueiredo, J., Caldeira, J., Carvalho, J., Pinheiro, H., et al. (2012). E-cadherin dysfunction in gastric cancer—Cellular consequences, clinical applications and open questions. *FEBS Letters*, *586*(18), 2981–2989.

Cerrato, A., De Falco, V., & Santoro, M. (2009). Molecular genetics of medullary thyroid carcinoma: The quest for novel therapeutic targets. *Journal of Molecular Endocrinology*, *43*(4), 143–155.

Chang, S., Bray, S. M., Li, Z., Zarnescu, D. C., He, C., Jin, P., et al. (2008). Identification of small molecules rescuing fragile X syndrome phenotypes in Drosophila. *Nature Chemical Biology*, *4*(4), 256–263.

Clark, K., MacKenzie, K. F., Petkevicius, K., Kristariyanto, Y., Zhang, J., Choi, H. G., et al. (2012). Phosphorylation of CRTC3 by the salt-inducible kinases controls the interconversion of classically activated and regulatory macrophages. *Proceedings of the National Academy of Sciences of the United States of America*, *109*(42), 16986–16991.

Cordero, J. B., Macagno, J. P., Stefanatos, R. K., Strathdee, K. E., Cagan, R. L., & Vidal, M. (2010). Oncogenic Ras diverts a host TNF tumor suppressor activity into tumor promoter. *Developmental Cell*, *18*(6), 999–1011.

Dar, A. C., Das, T. K., Shokat, K. M., & Cagan, R. L. (2012). Chemical genetic discovery of targets and anti-targets for cancer polypharmacology. *Nature*, *486*, 80–84.

Das, T. K., Sangodkar, J., Negre, N., Narla, G., & Cagan, R. L. (2013). Sin3a acts through a multi-gene module to regulate invasion in Drosophila and human tumors. *Oncogene*, *32*(26), 3184–3197.

de la Cova, C., Abril, M., Bellosta, P., Gallant, P., & Johnston, L. A. (2004). Drosophila Myc regulates organ size by inducing cell competition. *Cell*, *117*(1), 107–116.

Elisei, R., Cosci, B., Romei, C., Bottici, V., Renzini, G., Molinaro, E., et al. (2008). Prognostic significance of somatic RET oncogene mutations in sporadic medullary thyroid cancer: A 10-year follow-up study. *The Journal of Clinical Endocrinology and Metabolism*, *93*(3), 682–687.

Enomoto, M., Vaughen, J., & Igaki, T. (2015). Non-autonomous overgrowth by oncogenic niche cells: Cellular cooperation and competition in tumorigenesis. *Cancer Science*, *106*(12), 1651–1658.

Fearon, K., Arends, J., & Baracos, V. (2013). Understanding the mechanisms and treatment options in cancer cachexia. *Nature Reviews. Clinical Oncology*, *10*(2), 90–99.

Fidler, I. J. (2003). The pathogenesis of cancer metastasis: The 'seed and soil' hypothesis revisited. *Nature Reviews Cancer*, *3*(6), 453–458.

Figueroa-Clarevega, A., & Bilder, D. (2015). Malignant Drosophila tumors interrupt insulin signaling to induce cachexia-like wasting. *Developmental Cell*, *33*(1), 47–55.

Fischer, K. R., Durrans, A., Lee, S., Sheng, J., Li, F., Wong, S. T. C., et al. (2015). Epithelial-to-mesenchymal transition is not required for lung metastasis but contributes to chemoresistance. *Nature*, *527*(7579), 472–476.

Fleuren, E. D. G., Zhang, L., Wu, J., & Daly, R. J. (2016). The kinome 'at large' in cancer. *Nature Reviews. Cancer*, *16*(2), 83–98.

Froldi, F., Ziosi, M., Tomba, G., Parisi, F., Garoia, F., Pession, A., et al. (2008). Drosophila lethal giant larvae neoplastic mutant as a genetic tool for cancer modeling. *Current Genomics*, *9*(3), 147–154.

Gateff, E. (1978). Malignant neoplasms of genetic origin in Drosophila melanogaster. *Science*, *200*(4349), 1448–1459.

Gharwan, H., & Groninger, H. (2016). Kinase inhibitors and monoclonal antibodies in oncology: Clinical implications. *Nature Reviews. Clinical Oncology*, *13*(4), 209–227.

Giovannucci, E., Harlan, D. M., Archer, M. C., Bergenstal, R. M., Gapstur, S. M., Habel, L. A., et al. (2010). Diabetes and cancer: A consensus report. *CA: A Cancer Journal for Clinicians*, *60*(4), 207–221.

Gladstone, M., Frederick, B., Zheng, D., Edwards, A., Yoon, P., Stickel, S., et al. (2012). A translation inhibitor identified in a Drosophila screen enhances the effect of ionizing radiation and taxol in mammalian models of cancer. *Disease Models & Mechanisms*, *5*, 342–350.

Gladstone, M., & Su, T. T. (2011a). Screening for radiation sensitizers of Drosophila checkpoint mutants. *Methods in Molecular Biology*, *782*, 105–117.

Gladstone, M., & Su, T. T. (2011b). Chemical genetics and drug screening in Drosophila cancer models. *Journal of Genetics and Genomics = Yi Chuan Xue Bao*, *38*(10), 497–504.

Golic, K. G., & Lindquist, S. (1989). The FLP recombinase of yeast catalyzes site-specific recombination in the Drosophila genome. *Cell*, *59*(3), 499–509.

Greenspan, R. J. (2004). Analysis of mutations. In M. Regulski (Ed.), *Fly pushing: The theory and practice of Drosophila genetics* (pp. 115–117). New York: Cold Spring Harbor Laboratory Press.

Grifoni, D., Garoia, F., Bellosta, P., Parisi, F., De Biase, D., Collina, G., et al. (2007). aPKCzeta cortical loading is associated with Lgl cytoplasmic release and tumor growth in Drosophila and human epithelia. *Oncogene*, *26*(40), 5960–5965.

Grifoni, D., Garoia, F., Schimanski, C. C., Schmitz, G., Laurenti, E., Galle, P. R., et al. (2004). The human protein Hugl-1 substitutes for Drosophila lethal giant larvae tumour suppressor function in vivo. *Oncogene*, *23*(53), 8688–8694.

Hadoux, J., Furio, P., Tuttle, R. M., & Schlumberger, M. (2016). Management of advanced medullary thyroid cancer. *The Lancet Diabetes & Endocrinology*, *4*(1), 64–71.

Halder, G., & Johnson, R. L. (2011). Hippo signaling: Growth control and beyond. *Development*, *138*(1), 9–22.

Hanahan, D., & Weinberg, R. A. (2011). Hallmarks of cancer: The next generation. *Cell*, *144*(5), 646–674.

Hariharan, I. K. (2015). Organ size control: Lessons from Drosophila. *Developmental Cell*, *34*(3), 255–265.

Hartenstein, V. (2006). Blood cells and blood cell development in the animal kingdom. *Annual Review of Cell and Developmental Biology*, *22*, 677–712.

Harvey, K., & Tapon, N. (2007). The Salvador-Warts-Hippo pathway—An emerging tumour-suppressor network. *Nature Reviews. Cancer*, 7(3), 182–191.

Hay, M., Thomas, D. W., Craighead, J. L., Economides, C., & Rosenthal, J. (2014). Clinical development success rates for investigational drugs. *Nature Biotechnology*, *32*(1), 40–51.

Hescot, S., Vignaux, O., & Goldwasser, F. (2013). Pancreatic atrophy—A new late toxic effect of sorafenib. *The New England Journal of Medicine*, *369*(15), 1475–1476.

Hirabayashi, S., Baranski, T. J., & Cagan, R. L. (2013). Transformed Drosophila cells evade diet-mediated insulin resistance through wingless signaling. *Cell*, *154*(3), 664–675.

Hirabayashi, S., & Cagan, R. L. (2015). Salt-inducible kinases mediate nutrient-sensing to link dietary sugar and tumorigenesis in Drosophila. *eLife*, *4*, e08501.

Huang, A. M., & Rubin, G. M. (2000). A misexpression screen identifies genes that can modulate Ras1 pathway signaling in Drosophila melanogaster. *Genetics*, *156*(3), 1219–1230.

Igaki, T., Kanda, H., Yamamoto-Goto, Y., Kanuka, H., Kuranaga, E., Aigaki, T., et al. (2002). Eiger, a TNF superfamily ligand that triggers the Drosophila JNK pathway. *The EMBO Journal*, *21*(12), 3009–3018.

Igaki, T., & Miura, M. (2014). The Drosophila TNF ortholog Eiger: Emerging physiological roles and evolution of the TNF system. *Seminars in Immunology*, *26*(3), 267–274.

Igaki, T., Pastor-Pareja, J. C., Aonuma, H., Miura, M., & Xu, T. (2009). Intrinsic tumor suppression and epithelial maintenance by endocytic activation of Eiger/TNF signaling in Drosophila. *Developmental Cell*, *16*(3), 458–465.

Imamoto, A., & Soriano, P. (1993). Disruption of the Csk gene, encoding a negative regulator of Src family tyrosine kinases, leads to neural tube defects and embryonic lethality in mice. *Cell*, *73*(6), 1117–1124.

Ishizawar, R., & Parsons, S. J. (2004). C-Src and cooperating partners in human cancer. *Cancer Cell*, *6*(3), 209–214.

Kalliolias, G. D., & Ivashkiv, L. B. (2016). TNF biology, pathogenic mechanisms and emerging therapeutic strategies. *Nature Reviews. Rheumatology*, *12*(1), 49–62.

Karim, F. D., & Rubin, G. M. (1998). Ectopic expression of activated Ras1 induces hyperplastic growth and increased cell death in Drosophila imaginal tissues. *Development*, *125*(1), 1–9.

Kinzler, K. W., & Vogelstein, B. (1996). Lessons from hereditary colorectal cancer. *Cell*, *87*(2), 159–170.

Klezovitch, O., Fernandez, T. E., Tapscott, S. J., & Vasioukhin, V. (2004). Loss of cell polarity causes severe brain dysplasia in Lgl1 knockout mice. *Genes & Development*, *18*(5), 559–571.

Kornfeld, K., Hom, D. B., & Horvitz, H. R. (1995). The Ksr-1 gene encodes a novel protein kinase involved in Ras-mediated signaling in *C. elegans*. *Cell*, *83*(6), 903–913.

Kuphal, S., Wallner, S., Schimanski, C. C., Bataille, F., Hofer, P., Strand, S., et al. (2006). Expression of Hugl-1 is strongly reduced in malignant melanoma. *Oncogene*, *25*(1), 103–110.

Kwon, Y., Song, W., Droujinine, I. A., Hu, Y., Asara, J. M., & Perrimon, N. (2015). Systemic organ wasting induced by localized expression of the secreted insulin/IGF antagonist ImpL2. *Developmental Cell*, *33*(1), 36–46.

Lam, E. T., Ringel, M. D., Kloos, R. T., Prior, T. W., Knopp, M. V., Liang, J., et al. (2010). Phase II clinical trial of sorafenib in metastatic medullary thyroid cancer. *Journal of Clinical Oncology: Official Journal of the American Society of Clinical Oncology*, *28*(14), 2323–2330.

Lavin, Y., Mortha, A., Rahman, A., & Merad, M. (2015). Regulation of macrophage development and function in peripheral tissues. *Nature Reviews. Immunology*, *15*(12), 731–744.

Lee, C.-Y., Robinson, K. J., & Doe, C. Q. (2006). Lgl, pins and aPKC regulate neuroblast self-renewal versus differentiation. *Nature*, *439*(7076), 594–598.

Lin, L., Sabnis, A. J., Chan, E., Olivas, V., Cade, L., Pazarentzos, E., et al. (2015). The Hippo effector YAP promotes resistance to RAF- and MEK-targeted cancer therapies. *Nature Genetics*, *47*(3), 250–256.

Maher, E. A., Furnari, F. B., Bachoo, R. M., Rowitch, D. H., Louis, D. N., Cavenee, W. K., et al. (2001). Malignant glioma: Genetics and biology of a grave matter. *Genes & Development*, *15*(11), 1311–1333.

Markstein, M., Dettorre, S., Cho, J., Neumüller, R. A., Craig-Müller, S., & Perrimon, N. (2014). Systematic screen of chemotherapeutics in Drosophila stem cell tumors. *Proceedings of the National Academy of Sciences of the United States of America*, *111*(12), 4530–4535.

Micchelli, C. A., Esler, W. P., Kimberly, W. T., Jack, C., Berezovska, O., Kornilova, A., et al. (2003). Gamma-secretase/presenilin inhibitors for Alzheimer's disease phenocopy Notch mutations in Drosophila. *FASEB Journal: Official Publication of the Federation of American Societies for Experimental Biology*, *17*, 79–81.

Modica-Napolitano, J. S., Kulawiec, M., & Singh, K. K. (2007). Mitochondria and human cancer. *Current Molecular Medicine*, 7(1), 121–131.

Moore, R. J., Owens, D. M., Stamp, G., Arnott, C., Burke, F., East, N., et al. (1999). Mice deficient in tumor necrosis factor-alpha are resistant to skin carcinogenesis. *Nature Medicine*, *5*(7), 828–831.

Morata, G., & Ripoll, P. (1975). Minutes: Mutants of Drosophila autonomously affecting cell division rate. *Developmental Biology*, *42*(2), 211–221.

Moreno, E. (2008). Is cell competition relevant to cancer? *Nature Reviews. Cancer*, *8*(2), 141–147.

Morris, E. J., Ji, J.-Y., Yang, F., Di Stefano, L., Herr, A., Moon, N.-S., et al. (2008). E2F1 represses beta-catenin transcription and is antagonized by both pRB and CDK8. *Nature*, *455*(7212), 552–556.

Musselman, L. P., Fink, J. L., Narzinski, K., Ramachandran, P. V., Hathiramani, S. S., Cagan, R. L., et al. (2011). A high-sugar diet produces obesity and insulin resistance in wild-type Drosophila. *Disease Models & Mechanisms*, *4*(6), 842–849.

Na, J., Musselman, L. P., Pendse, J., Baranski, T. J., Bodmer, R., Ocorr, K., et al. (2013). A Drosophila model of high sugar diet-induced cardiomyopathy. *PLoS Genetics*, *9*(1), e1003175.

Nelson, W. J. (2003). Adaptation of core mechanisms to generate cell polarity. *Nature*, *422*(6933), 766–774.

Ohsawa, S., Sato, Y., Enomoto, M., Nakamura, M., Betsumiya, A., & Igaki, T. (2012). Mitochondrial defect drives non-autonomous tumour progression through Hippo signalling in Drosophila. *Nature*, *490*(7421), 547–551.

Pagliarini, R. A., & Xu, T. (2003). A genetic screen in Drosophila for metastatic behavior. *Science*, *302*(5648), 1227–1231.

Pastor-Pareja, J. C., Wu, M., & Xu, T. (2008). An innate immune response of blood cells to tumors and tissue damage in Drosophila. *Disease Models & Mechanisms*, *1*(2–3), 144–154. Discussion 153.

Prober, D. A., & Edgar, B. A. (2000). Ras1 promotes cellular growth in the Drosophila wing. *Cell*, *100*(4), 435–446.

Proud, C. G. (2007). Signalling to translation: How signal transduction pathways control the protein synthetic machinery. *Biochemical Journal*, *403*(2), 217–234.

Radimerski, T., Montagne, J., Rintelen, F., Stocker, H., van der Kaay, J., Downes, C. P., et al. (2002). dS6K-regulated cell growth Is dPKB/dPI(3)K-independent, but requires dPDK1. *Nature Cell Biology*, *4*(3), 251–255.

Rahib, L., Smith, B. D., Aizenberg, R., Rosenzweig, A. B., Fleshman, J. M., & Matrisian, L. M. (2014). Projecting cancer incidence and deaths to 2030: The unexpected burden of thyroid, liver, and pancreas cancers in the United States. *Cancer Research*, *74*(11), 2913–2921.

Read, R. D., Bach, E. A., & Cagan, R. L. (2004). Drosophila C-terminal Src kinase negatively regulates organ growth and cell proliferation through inhibition of the Src, Jun N-terminal kinase, and STAT pathways. *Molecular and Cellular Biology*, *24*(15), 6676–6689.

Read, R. D., Cavenee, W. K., Furnari, F. B., & Thomas, J. B. (2009). A Drosophila model for EGFR-Ras and PI3K-dependent human glioma. *PLoS Genetics*, *5*(2), e1000374.

Read, R. D., Goodfellow, P. J., Mardis, E. R., Novak, N., Armstrong, J. R., & Cagan, R. L. (2005). A Drosophila model of multiple endocrine neoplasia type 2. *Genetics*, *171*(3), 1057–1081.

Rørth, P. (1996). A modular misexpression screen in Drosophila detecting tissue-specific phenotypes. *Proceedings of the National Academy of Sciences of the United States of America*, *93*(22), 12418–12422.

Rørth, P., Szabo, K., Bailey, A., Laverty, T., Rehm, J., Rubin, G. M., et al. (1998). Systematic gain-of-function genetics in Drosophila. *Development*, *125*(6), 1049–1057.

Rudrapatna, V. A., Bangi, E., & Cagan, R. L. (2014). A Jnk-Rho-Actin remodeling positive feedback network directs Src-driven invasion. *Oncogene*, *33*(21), 2801–2806.

Schimanski, C. C., Schmitz, G., Kashyap, A., Bosserhoff, A. K., Bataille, F., Schäfer, S. C., et al. (2005). Reduced expression of Hugl-1, the human homologue of Drosophila tumour suppressor gene Lgl, contributes to progression of colorectal cancer. *Oncogene*, *24*(19), 3100–3109.

Siegel, R. L., Miller, K. D., & Jemal, A. (2016). Cancer statistics, 2016. *CA: A Cancer Journal for Clinicians*, *66*(1), 7–30.

Speicher, S. A., Thomas, U., Hinz, U., & Knust, E. (1994). The serrate locus of Drosophila and its role in morphogenesis of the wing imaginal discs: Control of cell proliferation. *Development*, *120*(3), 535–544.

Steeg, P. S. (2006). Tumor metastasis: Mechanistic insights and clinical challenges. *Nature Medicine*, *12*(8), 895–904.

Stephen, A. G., Esposito, D., Bagni, R. K., & McCormick, F. (2014). Dragging Ras back in the ring. *Cancer Cell*, *25*(3), 272–281.

Stewart, R. A., Li, D.-M., Hung, H., & Xu, T. (2003). A genetic screen for modifiers of the lats tumor suppressor gene identifies C-terminal Src kinase as a regulator of cell proliferation in Drosophila. *Oncogene*, *22*(41), 6436–6444.

Stickel, S. A., Gomes, N. P., Frederick, B., Raben, D., & Su, T. T. (2015). Bouvardin is a radiation modulator with a novel mechanism of action. *Radiation Research*, *184*(4), 392–403.

Stupp, R., Mason, W. P., van den Bent, M. J., Weller, M., Fisher, B., Taphoorn, M. J. B., et al. (2005). Radiotherapy plus concomitant and adjuvant temozolomide for glioblastoma. *The New England Journal of Medicine*, *352*(10), 987–996.

Sundaram, M., & Han, M. (1995). The *C. elegans* Ksr-1 gene encodes a novel Raf-related kinase involved in Ras-mediated signal transduction. *Cell*, *83*(6), 889–901.

Therrien, M., Chang, H. C., Solomon, N. M., Karim, F. D., Wassarman, D. A., & Rubin, G. M. (1995). KSR, a novel protein kinase required for RAS signal transduction. *Cell*, *83*(6), 879–888.

Thorne, C. A., Hanson, A. J., Schneider, J., Tahinci, E., Orton, D., Cselenyi, C. S., et al. (2010). Small-molecule inhibition of Wnt signaling through activation of casein kinase 1α. *Nature Chemical Biology*, *6*(11), 829–836.

Ugur, B., Chen, K., & Bellen, H. J. (2016). Drosophila tools and assays for the study of human diseases. *Disease Models & Mechanisms*, *9*(3), 235–244.

Venken, K. J. T., & Bellen, H. J. (2012). Genome-wide manipulations of Drosophila melanogaster with transposons, Flp recombinase, and ΦC31 integrase. *Methods in Molecular Biology*, *859*, 203–228.

Vidal, M., Larson, D. E., & Cagan, R. L. (2006). Csk-deficient boundary cells are eliminated from normal Drosophila epithelia by exclusion, migration, and apoptosis. *Developmental Cell, 10*(1), 33–44.

Vidal, M., Warner, S., Read, R., & Cagan, R. L. (2007). Differing Src signaling levels have distinct outcomes in Drosophila. *Cancer Research, 67*(21), 10278–10285.

Vidal, M., Wells, S., Ryan, A., & Cagan, R. (2005). ZD6474 suppresses oncogenic RET isoforms in a Drosophila model for type 2 multiple endocrine neoplasia syndromes and papillary thyroid carcinoma. *Cancer Research, 65*, 3538–3541.

Willoughby, L. F., Schlosser, T., Manning, S. A., Parisot, J. P., Street, I. P., Richardson, H. E., et al. (2013). An in vivo large-scale chemical screening platform using Drosophila for anti-cancer drug discovery. *Disease Models & Mechanisms, 6*, 521–529.

Ye, X., & Weinberg, R. A. (2015). Epithelial–mesenchymal plasticity: A central regulator of cancer progression. *Trends in Cell Biology, 25*(11), 675–686.

Yeatman, T. J. (2004). A renaissance for SRC. *Nature Reviews. Cancer, 4*(6), 470–480.

Zheng, X., Carstens, J. L., Kim, J., Scheible, M., Kaye, J., Sugimoto, H., et al. (2015). Epithelial-to-mesenchymal transition is dispensable for metastasis but induces chemoresistance in pancreatic cancer. *Nature, 527*(7579), 525–530.

CHAPTER TEN

Stem-Cell-Based Tumorigenesis in Adult *Drosophila*

S.X. Hou[1], S.R. Singh
The Basic Research Laboratory, National Cancer Institute at Frederick, National Institutes of Health, Frederick, MD, United States
[1]Corresponding author: e-mail address: hous@mail.nih.gov

Contents

Abstract

Recent studies suggest that a small subset of cells within a tumor, the so-called cancer stem cells (CSCs), are responsible for tumor propagation, relapse, and the eventual death of most cancer patients. CSCs may derive from a few tumor-initiating cells, which are either transformed normal stem cells or reprogrammed differentiated cells after acquiring initial cancer-causing mutations. CSCs and normal stem cells share some properties, but CSCs differ from normal stem cells in their tumorigenic ability. Notably, CSCs are usually resistant to chemo- and radiation therapies. Despite the apparent roles

Current Topics in Developmental Biology, Volume 121
ISSN 0070-2153
http://dx.doi.org/10.1016/bs.ctdb.2016.07.013

of CSCs in human cancers, the biology underlying their behaviors remains poorly understood. Over the past few years, studies in *Drosophila* have significantly contributed to this new frontier of cancer research. Here, we first review how stem-cell tumors are initiated and propagated in *Drosophila*, through niche appropriation in the posterior midgut and through stem-cell competition for niche occupancy in the testis. We then discuss the differences between normal and tumorigenic stem cells, revealed by studying Ras^{V12}-transformed stem-cell tumors in the *Drosophila* kidney. Finally, we review the biology behind therapy resistance, which has been elucidated through studies of stem-cell resistance and sensitivity to death inducers using female germline stem cells and intestinal stem cells of the posterior midgut. We expect that screens using adult *Drosophila* neoplastic stem-cell tumor models will be valuable for identifying novel and effective compounds for treating human cancers.

1. CANCER STEM CELLS

Therapeutic developments over the last several decades have enabled most cancer patients today to attain major clinical responses, including reduced side effects and an improved quality of life. However, most cancers still relapse and result in the eventual death of the patient. Although the biological and clinical importance of cancer stem cells (CSCs) are still under debate, accumulating data suggest that a small population of CSCs may be responsible for tumor metastasis, relapse, and the eventual death of most cancer patients (Malanchi et al., 2012). This hypothesis emerged from the initial finding that some solid tumors and leukemias harbor a small number of self-renewing cells that regenerate complete tumors when transplanted into mice (Al-Hajj, Wicha, Benito-Hernandez, Morrison, & Clarke, 2003; Lapidot et al., 1994; Ricci-Vitiani et al., 2007; Singh et al., 2004). In the brain, these CSCs usually express normal stem cells markers and are dependent on the niche microenvironments that regulate normal stem cells (Calabrese et al., 2007). Studies suggest that CSCs are a subset of cells in the tumor that assist tumor propagation (Kreso & Dick, 2014). CSCs are defined by two attributes, self-renewal and multipotency. However, the cells from which CSCs are derived, called tumor-initiating cells (TICs) or the cells-of-origin for cancer (COCs), appear to vary in different types of tumors.

An early study of colon cancer (Fearon & Vogelstein, 1990) showed that tumorigenesis occurs through a clonal evolution process, in which early adenomas lead to invasive carcinomas through a stepwise addition of mutations. However, in quickly proliferating tissues like the intestine or colon, this model of tumorigenesis suggests that only stem cells would acquire these mutations (Asfaha et al., 2015). In fact, in the mouse intestine and prostate, normal stem cells were proposed to be the COCs because they are susceptible to neoplastic transformation (Barker et al., 2009; Leong, Wang, Johnson, & Gao, 2008; Wang et al., 2009; Zhu et al., 2009). In contrast, slowly proliferating tissues such as the brain and liver, both differentiated cells and normal stem cells are cells-of-origin for cancer. In the adult mouse brain, an initial study found that loss of tumor suppressors caused malignant astrocytomas that originated from neural stem/progenitor cells, but not from the more differentiated cells (Alcantara et al., 2009). However, a later study revealed that gliomas could originate from astrocytes or even from mature neurons in the adult mouse brain (Friedmann-Morvinski et al., 2012). They proposed that most differentiated cells in the central nervous system can dedifferentiate to a stem-cell-like cell-of-origin for cancer after undergoing defined genetic alterations (Friedmann-Morvinski et al., 2012). In the liver, studies in numerous mouse models suggest that the cells-of-origin for liver cancer are mature hepatocytes that have undergone reprogramming and dedifferentiation in response to injury (reviewed in Marquardt, Andersen, & Thorgeirsson, 2015). However, a population of stem cells was recently identified adjacent to the central vein in the liver lobule suggested to be the cells-of-origin for liver cancer (Wang, Zhao, Fish, Logan, & Nusse, 2015). In chronic myelogenous leukemia (CML), hematopoietic stem cells (HSCs) are the cells-of-origin for acute forms of the disease, while granulocyte–macrophage progenitors that gain ability to self-renew appear to be the cells-of-origin in blast-crisis CML (Jamieson et al., 2004). In conclusion, both transformed normal stem cells and differentiated cells can serve as the cell-of-origin for cancer.

For most tumors, it takes years before a sizable propagating and heterogenetic tumor mass is established. How a TIC/COC becomes a CSC and the relationship between TICs and CSCs is not well understood. Notably, the CSCs are significantly enriched after conventional treatment in breast and pancreatic cancers, due to their resistance to radiation and chemotherapy (Creighton et al., 2009; Li et al., 2008; Mueller et al., 2009; Phillips, McBride, & Pajonk, 2006). However, the mechanisms for the CSCs' refractoriness to these therapies are still poorly understood, and no obvious molecular targets have been identified for rational drug design. Thus, to discover effective treatments, it is critical to clarify the differences between normal stem cells and CSCs and to understand the biology underlying therapy resistance. In this endeavor, the fruit fly *Drosophila melanogaster*, an established model organism with well-defined genetics, has already contributed valuable insights to the field of CSC biology.

In the last decade, studies using *Drosophila* have made significant contributions to cancer research, including the modeling of tumor metastasis, the demonstration of neoplastic transformation by the loss of cell polarity or loss of the asymmetric division of larval neuroblasts, and recent screens of chemical compounds that have led to Food and Drug Administration (FDA)-approved drugs and drugs in clinical trials. Since several outstanding reviews have already covered the general role of *Drosophila* in cancer research (Gonzalez, 2007; Pastor-Pareja & Xu, 2013; Patel & Edgar, 2014; Sonoshita & Cagan, 2017), this review will focus on stem-cell-based tumorigenesis in adult *Drosophila*.

2. STEM-CELL-BASED TUMORIGENESIS IN THE ADULT *DROSOPHILA* MIDGUT

2.1 Intestinal Stem Cells in the Adult *Drosophila* Posterior Midgut

The *Drosophila* posterior midgut is closely analogous to the mammalian small intestine and functions in digesting food (Hakim, Baldwin, & Smagghe, 2010). The midgut epithelium is maintained by self-renewing intestinal stem cells (ISCs) (Micchelli & Perrimon, 2006; Ohlstein & Spradling, 2006). ISCs reside near basement membrane (BM) and divide almost every day to produce either absorptive enterocytes (ECs) or secretory enteroendocrine (EE) cells. The *Drosophila* midgut is an attractive model system for studying adult stem-cell-mediated tissue homeostasis and regeneration.

ISCs asymmetrically divide to yield a new ISC and an immature daughter, an enteroblast (EB), that differentiates into an EC or EE cell (Micchelli & Perrimon, 2006; Ohlstein & Spradling, 2006). ISC self-renewal and differentiation are regulated by Notch (N) signaling (Micchelli & Perrimon, 2006; Ohlstein & Spradling, 2006, 2007). The Notch ligand Delta (Dl) is exclusively expressed in ISCs and induces N signaling in the neighboring EB cell (Ohlstein & Spradling, 2007) to promote their differentiation to an EC and restrict its differentiation to an EE cell (Micchelli & Perrimon, 2006; Ohlstein & Spradling, 2006, 2007). Later studies showed that an ISC can divide asymmetrically to produce one new ISC (self-renewal) and one EB (which differentiates into an EC) or a pre-enteroendocrine (pre-EE) cell (which ultimately matures into an EE cell) (Biteau & Jasper, 2014; Zeng & Hou, 2015; Zielke et al., 2014). Recently, this model was further revised (Guo & Ohlstein, 2015) as follows: a parent ISC divides asymmetrically to produce either a basal son ISC (sISC) and an apical EB or a basal

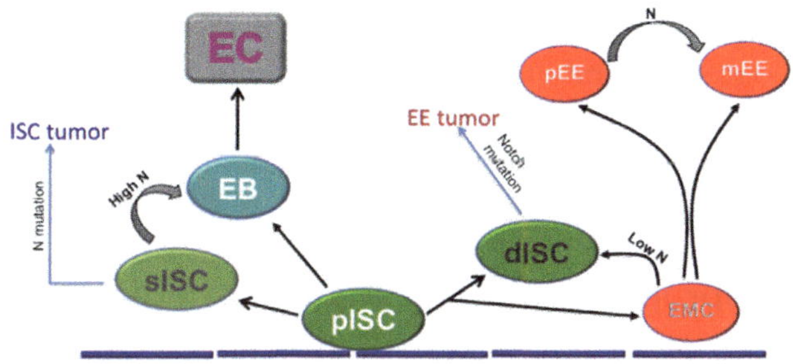

Fig. 1 Model of Notch activity in the *Drosophila* adult posterior midgut. Details are provided in the main text. *Modified after Guo, Z., & Ohlstein, B. (2015). Stem cell regulation. Bidirectional Notch signaling regulates Drosophila intestinal stem cell multipotency.* Science, 350, *aab0988.*

pre-EE (renamed an enteroendocrine mother cell, EMC) and an apical daughter ISC (dISC). The sISC expresses Dl and switches on high N signaling in the neighboring EB to promote EB-to-EC differentiation; the EMC expresses Dl and switches on low N signaling in the dISC to inhibit dISC-to-EMC differentiation. The EMC can differentiate into an immature EE, which can be converted into a mature EE through N signaling (Fig. 1). Therefore, the loss of N signaling results in sISC accumulation in the sISC-EB branch and premature EE cell formation in the dISC-EMC branch; these events are observed in N mutant stem-cell tumors (Biteau & Jasper, 2014; Guo & Ohlstein, 2015; Ohlstein & Spradling, 2006, 2007; Zeng & Hou, 2015). In addition to N signaling, other pathways/factors also regulate ISCs behavior (Kohlmaier et al., 2015; Kolahgar et al., 2015; Tian et al., 2015).

2.2 Spontaneous Somatic Mutations of *N* Lead to Neoplasias in Aged Flies

While somatic genetic mutation in cancers is well documented, it is not clear how such mutation is linked to stem-cell tumor. A recent study addressed this question (Siudeja et al., 2015).

In *Drosophila*, the *N* gene is on the X chromosome. Since male flies have only one copy of this gene, one somatic mutation can result in complete loss of the gene, which leads to stem-cell tumors in the posterior midgut. Using this convenient system, Siudeja et al. (2015) measured *N* mutations and stem-cell tumor formation. They found that very few young flies (a few days old) had stem-cell tumors, while a very high percentage of old flies (6 weeks old) had at least one stem-cell tumor in the gut. Analysis of the genomic DNA of the stem-cell tumors revealed that many had deletions and rearrangements within and around the *N* locus. They further found that both homologous-recombination-based mechanisms and double-strand break-induced repair were involved in the *N* mutations. Their findings indicated that spontaneous somatic inactivation of the tumor-suppressor *N* drove neoplasias in 10% of the aged wild-type males (Siudeja et al., 2015).

2.3 Niche Appropriation Drives ISC Tumor Progression

Mutations that destabilize the asymmetric division or restrict the differentiation of stem cells are proposed to be a primary process by which cells are driven into malignant transformation (Gonzalez, 2007). However, it is not clear how the initial events (such as *N* mutation) drive tumor progression. A recent study in the *Drosophila* posterior midgut addressed this problem (Patel, Dutta, & Edgar, 2015).

2.3.1 Tumor Initiation

As described earlier, a single mutation in the *N* gene can drive stem-cell tumor formation in the adult *Drosophila* posterior midgut. Using this model, Patel et al. (2015) depleted *N* by RNA interference in stem cells and progenitors and found that some of the *N*-depleted flies did not contain stem-cell tumors, suggesting that merely restricting differentiation is not sufficient for tumor initiation. Previous studies showed that stress or enteric infection affect epithelial homeostasis and ISC tumor outgrowth (Apidianakis, Pitsouli, Perrimon, & Rahme, 2009; Jiang et al., 2009). Similarly, Patel et al. (2015) found that enteric infection with *Pseudomonas entomophila* could enhance the tumor initiation in *N*-depleted flies. In addition, as stress, damage, or enteric infection are known to activate JNK signaling (Jiang et al., 2009), Patel et al. (2015) found that the activation of JNK signaling in ECs also promoted tumor initiation in the *N*-depleted flies. They further demonstrated that stress, enteric infection, and JNK signaling affect ISC division by regulating the cell cycle. These findings showed that both the loss of differentiation and the presence of stress-induced stem-cell division are crucial for tumor initiation.

2.3.2 Tumor Progression

The above-described study in aged flies by Siudeja et al. (2015) showed that the N signaling-defective tumors in the *Drosophila* midgut grow over time. Patel et al. (2015) further examined the tumor-autonomous factors that advance tumor growth after initiation. The epidermal growth factor receptor (EGFR) signaling is known to regulate ISC proliferation (Biteau & Jasper, 2011; Buchon, Broderick, Poidevin, Pradervand, & Lemaitre,

2009; Jiang, Grenley, Bravo, Blumhagen, & Edgar, 2011). Therefore, Patel et al. (2015) examined the role of EGFR signaling in the expansion of N signaling-defective tumors and found that the EGFR ligand *spitz* (*spi*) was overexpressed in the tumors. They further found that the ISC tumors had elevated levels of mitogen-activated protein kinase (MAPK), and its target *argos* compared to normal ISCs. Their genetic experiments confirmed the involvement of EGFR signaling in stem-cell tumor expansion. Notably, *spi* is not required for normal ISC maintenance and possibly functions in ISC tumor clusters to stimulate ISC-like cell multiplication through a paracrine mechanism (Patel et al., 2015).

Patel et al. (2015) observed that during tumor growth, some of the ECs adjacent to and overlying the tumors became detached from the BM and then died. Further study revealed that the growing tumor induced proapoptotic genes in nontumor cells and that apoptosis supported tumor growth but was not needed for the EC detachment. Because epithelial stress is known to stimulate both JNK and Yorkie (Yki) signaling in the *Drosophila* midgut (Jiang et al., 2009; Karpowicz, Perez, & Perrimon, 2010; Shaw et al., 2010; Staley & Irvine, 2010), Patel et al. (2015) investigated the role of these signaling pathways in tumor development. They found that both JNK (*puc-lacZ*) and Yki (*ex-lacZ*) reporters were expressed near the detached ECs, apical to the ISC tumors; however, they were not expressed inside the ISC tumors. Furthermore, the knockdown of JNK or Yki signaling in the ECs suppressed tumor growth, while overexpression resulted in larger tumors. These findings indicated that growing tumors first push the adjacent ECs away, then JNK and Yki signaling activated in the detached ECs promote stem-cell tumor growth.

Next, Patel et al. (2015) sought to identify the signal(s) that directly stimulate stem-cell tumor growth. Previous studies showed that stresses on ECs could induce the expression of ISC mitogens, such as the EGFR ligands Vein (Vn) and Keren, and the cytokines unpaired 2 (Upd2) and unpaired 3 (Upd3) (Jiang et al., 2011, 2009). Therefore, Patel et al. (2015) investigated the expression of these mitogens in the N-defective tumors by mRNA-seq and quantitative PCR, and found increased expressions of *vn*, *upd2*, and *upd3* in the midgut tumors, specifically in the ECs and visceral muscle (VM). These data suggested that tumor growth induces mitogenic signals in the niche. Further, they found that Upd3 is induced by the detachment of ECs from the VM and plays a major role in the EC detachment-stimulated tumor growth.

These studies together suggest that stem-cell-based tumorigenesis in the adult *Drosophila* posterior midgut involves several steps. Spontaneous mutation of the tumor-suppressor *N* blocks stem-cell differentiation and may be the first event in tumor initiation. Environmental stress-activated JNK signaling and stem-cell division may be the second event, and these two events together coordinate tumor initiation. Next, the ISC-like cells in the small tumor clusters express *spi* and activate EGFR signaling to promote their autonomous expansion. Once the tumors attain a certain size, it pushes adjacent ECs away (possibly through integrin-mediated competition for attachment to the BM) to gain more space, the detached ECs further activate JNK and Yki signaling and *upd3* expression, and the Upd3 then activates Janus kinase–signal transducer and activator of transcription (JAK–STAT) signaling in the stem cells to promote further expansion of the tumor (Fig. 2).

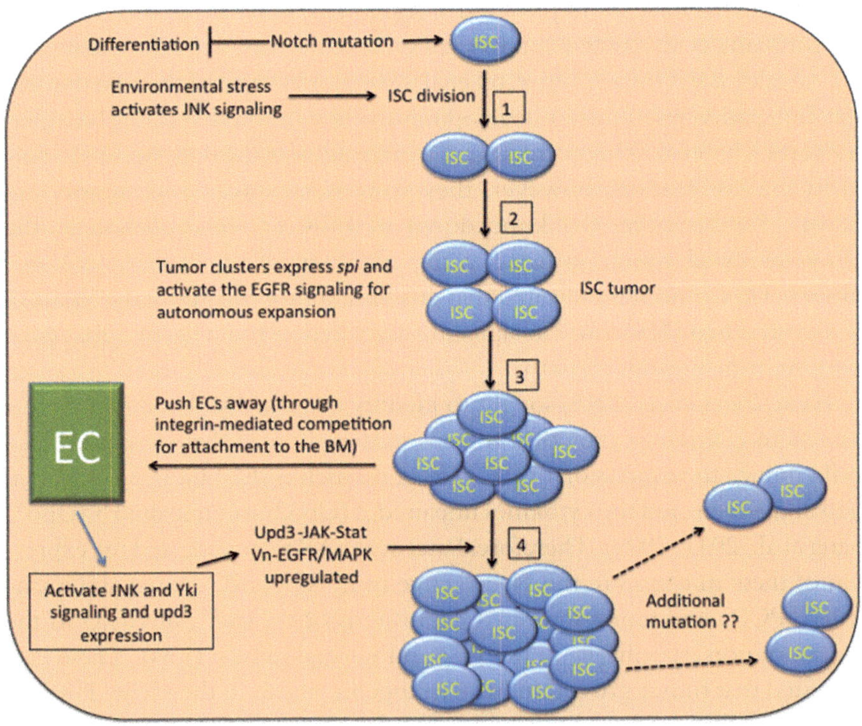

Fig. 2 Model of the sequence of events involved in N-dependent tumorigenesis in the *Drosophila* adult posterior midgut. *Modified after Patel, P. H., Dutta, D., & Edgar, B. A. (2015). Niche appropriation by Drosophila intestinal stem cell tumours.* Nature Cell Biology, *17, 1182–1192.*

3. TUMORIGENESIS THROUGH STEM-CELL COMPETITION IN TESTIS

3.1 Tumorigenesis Through Stem-Cell Competition in Mammals

Accumulating evidence suggests that all animal tissues and organs are maintained by stem cells. Stem cells are vital during organ development, tissue homeostasis, and regeneration. Recent studies suggest that competent stem cells are chosen for crucial functions through stem-cell competition (Bondar & Medzhitov, 2010; Issigonis et al., 2009; Jin et al., 2008; Martins et al., 2014; Rhiner et al., 2009).

Recently, it has been demonstrated that progenitor competition, in the mouse thymus, is a tumor-suppressor mechanism (Martins et al., 2014), where the "old" thymus-resident progenitors are continuously succeeded by "young" bone marrow-derived progenitors by natural cell competition. On the other hand, the competitive advantage of transformed stem cells (TSCs) or CSCs may be used to promote tumor formation, as reported in two other studies (Bondar & Medzhitov, 2010; Marusyk, Porter, Zaberezhnyy, & DeGregori, 2010).

3.2 Stem Cells in the Adult *Drosophila* Testis

Drosophila testis germline stem cells (GSCs) provide an excellent in vivo model system to study stem-cell behavior (Singh, Liu, Zhao, Zeng, & Hou, 2016). *Drosophila* testis tip contains the GSCs and somatic cyst stem cells (CySCs) (Singh et al., 2016; Fig. 3A). Each GSC is enclosed by two CySCs. Both the GSCs and CySCs adhere to niche (hub cells) by cell-adhesion components (Wang et al., 2006). The hub cells express the growth factor unpaired, a ligand that stimulates the JAK–STAT pathway in GSCs and CySCs to manage their self-renewal (Issigonis et al., 2009; Singh et al., 2010). For proper GSC division centrosome location and orientation of mitotic spindle are very important (Yamashita, Mahowald, Perlin, & Fuller, 2007).

In the CySCs, JAK–STAT signaling and its putative targets (such as zinc-finger homeodomain protein 1 (*zfh-1*) and *chinmo*) act both intrinsically to control CySC self-renewal and maintenance and nonautonomously to direct the self-renewal of neighboring GSCs (Flaherty et al., 2010; Leatherman & Dinardo, 2008). EGFR-Ras/Raf signaling also acts both intrinsically to regulate CySCs and nonautonomously to regulate adjacent

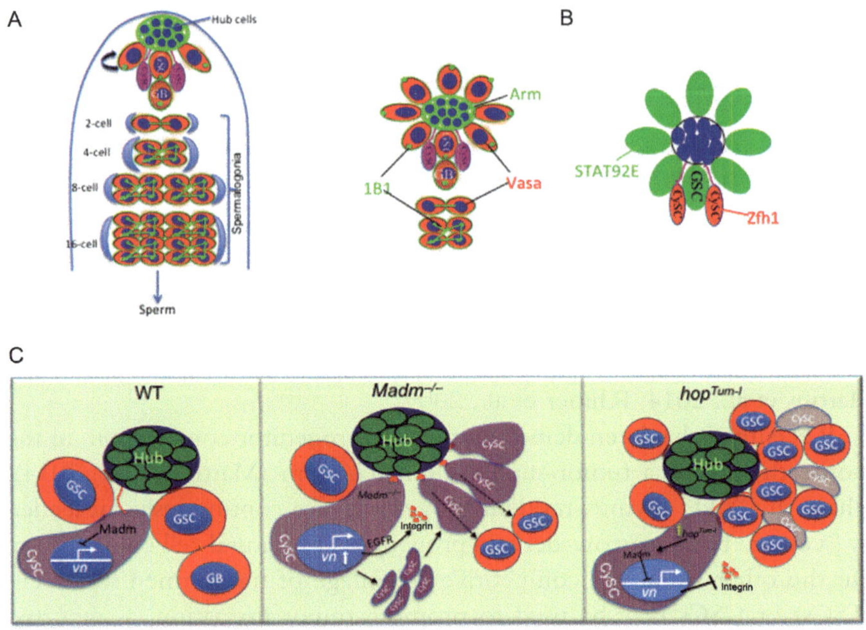

Fig. 3 (A) Diagram of the *Drosophila* adult testis. (B) Schematic diagrams of *Drosophila* testis cell types and the marker expression in germ cells including GSCs and CySCs. (C) Model for the Madm-guided GSC and CySC competition in the *Drosophila* adult testis. *Modified after Singh, S. R., Liu, Y., Zhao, J., Zeng, X., & Hou, S. X. (2016). The novel tumour suppressor Madm regulates stem cell competition in the Drosophila testis.* Nature Communications, *7, 10473.*

GSCs, but it has the opposite effects of JAK–STAT signaling (Amoyel, Simons, & Bach, 2014; Kiger, White-Cooper, & Fuller, 2000; Tran, Brenner, & DiNardo, 2000). However, it is unclear how these pathways are interconnected in CySCs to both control CySCs intrinsically and regulate GSCs nonautonomously.

In the wild-type testis, GSCs directly contact the niche, and CySCs, identified by their expression of Zfh-1, are located outside the GSCs (Fig. 3B). We recently reported that Madm (myeloid leukemia factor 1 (Mlf1)-adaptor molecule) is a novel tumor suppressor that regulates the competition of GSCs and CySCs for niche occupancy and may play a role in stem-cell tumor formation (Singh et al., 2016).

3.3 *Madm* Regulates Stem-Cell Competition in the Adult *Drosophila* Testis

Mlf1 is an oncogene (Yoneda-Kato et al., 1996). Madm and 14-3-3ζ were identified as binding partners of Mlf1 in a yeast two-hybrid screen (Lim et al.,

2002). The antagonistic effect of Madm and Mlf1 suggests that Madm might act as a tumor suppressor. Madm is also known as nuclear receptor-binding protein 1 (NRBP1) (Hooper, Baker, Ogbourne, Sutherland, & Antalis, 2000). The somatic loss of *NRBP1* causes tumorigenesis, with a prevalence of hematological and intestinal tumors, in mice (Wilson et al., 2012). The NRBP1 protein is downregulated in many human tumors, and its low expression is associated with a poor prognosis. These data together suggest that *NRBP1* is a conserved regulator of cell fate and a vital player in tumor suppression (Wilson et al., 2012).

In the *Drosophila* testis, GSCs and CySCs are associated with the same niche. Zfh-1 is expressed in CySCs and their immediate cyst cell daughters, and Madm is specifically expressed in CySCs. In the wild-type testis, GSCs are directly connected with hub cells, while the Zfh-1-positive CySCs are separated from the hub cells (Fig. 3B). Singh et al. (2016) found that in the Madm-deficient testis, Zfh-1-positive CySCs moved toward the hub cells and forced the removal of GSCs from the niche and the number of Zfh-1-positive cells dramatically increased over time (Fig. 3C). These findings suggest that Madm normally acts to restrict CySCs from outcompeting the GSCs for niche residency. They further demonstrated that the knockdown of *Madm* in CySCs resulted in overexpression of the EGFR ligand *vn*, which activated EGFR signaling noncell autonomously in the CySCs to elicit their overproliferation and integrin expression. Finally, they found that the high integrin level in CySCs enables them to outcompete the GSCs for niche occupancy.

3.4 JAK–STAT Signaling Regulates the Madm-Directed Stem-Cell Competition

Hopscotch$^{Tumorous\text{-}lethal}$ (*hop*$^{Tum\text{-}l}$) encodes a constitutively activated form of the *Drosophila* JAK kinase (Binari & Perrimon, 1994), and *zfh-1* and *chinmo* are putative targets of the JAK–STAT signaling pathway (Flaherty et al., 2010; Leatherman & Dinardo, 2008). The overexpression of *hop*$^{Tum\text{-}l}$, *zfh-1*, or *chinmo* in CySCs results in GSC tumors (Flaherty et al., 2010; Leatherman & Dinardo, 2008). The loss of EGFR–ERK signaling in CySCs also results in GSC tumors (Kiger et al., 2000; Tran et al., 2000). Recently, Singh et al. (2016) found that the GSC tumor formation under these conditions may occur through the suppression of integrin expression by the Madm-Vn pathway in CySCs; where Madm blocks the expression of the EGFR ligand, Vn. First, integrin overexpression in CySCs could significantly suppress the GSC tumor phenotypes of a dominant-negative form of EGFR (*EGFR*^{DN}), *Drosophila ERK* RNAi (*rl*^{RNAi}), *zfh-1*, or *hop*$^{Tum\text{-}l}$.

Second, the overexpression of *hop*$^{Tum-l}$ in CySCs resulted in the nuclear translocation of Madm and the suppression of *vn* expression. In addition, the EGFR-Ras pathway promotes integrin expression. Collectively, these data suggest that the expression *hop*$^{Tum-l}$ in CySCs promotes the nuclear translocation of Madm and decreases *vn* expression, which downregulate EGFR signaling and integrin expression in CySCs, allowing GSCs to outcompete the CySCs for niche occupancy and promoting GSC tumor formation (Fig. 3C). These results together suggest that the relative integrin level in CySCs determines whether GSCs or CySCs occupy the niche and that tumor suppressor and signal transduction pathways regulate stem-cell competition and stem-cell tumor formation by modulating the integrin expression.

3.4 *p53* and *Madm* Regulate Stem-Cell Competition

The HSCs and progenitor cells that were deficient in the tumor-suppressor *p53* were more competitive than their wild-type counterparts in irradiated mice, which resulted in long-term expansion of *p53*-deficient clones and increased lymphoma development (Bondar & Medzhitov, 2010; Marusyk et al., 2010). In this case, the competition appeared to be mediated in a noncell-autonomous manner and was probably based on *p53*-induced changes in the expression of genes that control the cell's interactions with its environment. These changes induced growth arrest and senescence-related gene expression in the outcompeted cells with higher p53 activity and simultaneously promoted the self-renewal or proliferation of the "winning" *p53*-deficient cells. A large group of genes was found to be differentially expressed in the competitive vs noncompetitive conditions; their products included proteins involved in HSCs adhesion and migration and in the modulation of receptor signaling (CD44). However, how *p53* controls stem-cell competition by regulating their expression is still unclear. The *p53*-deficient HSCs may secrete an undefined factor, which noncell autonomously promotes the proliferation and maintenance of *p53*-low cells and promotes growth arrest and senescence-related gene expression in *p53*-high cells (Bondar & Medzhitov, 2010; Marusyk et al., 2010). *p53* and *Madm* have some similarities in managing stem-cell competition through noncell-autonomous regulation mechanisms. In *Drosophila* testis, knocking down *Madm* in the CySCs promotes the expression of the EGFR ligand *vn*, which stimulates EGFR signaling and integrin expression noncell autonomously to promote the CySCs' proliferation and capacity to outcompete the GSCs for niche occupancy. These studies together suggest that stem-cell (or CSC) competition may be a basic mechanism of tumor formation. Further studies on tumor-suppressor-mediated stem-cell competition will help elucidate the molecular basis of tumor initiation and in designing novel strategies for treating cancer.

4. DIFFERENCES BETWEEN NORMAL AND CSCs

Most human cancers develop in somatic tissues. Accumulated evidence suggests that tumorigenesis often depends on CSCs. Understanding the difference between normal stem cells and CSCs is strategically important for cancer research. Accordingly, identifying the differences between normal stem cells and CSCs in various tissues and clarifying the molecular mechanisms by which normal stem cells are transformed into CSCs are essential for developing anti-CSC therapies. In *Drosophila*, Zeng, Singh, Hou, and Hou (2010) systematically examined the differences between normal and TSCs (CSC-like cells) in the adult malpighian tubules (MTs).

4.1 The Stem Cells in MTs

Drosophila MTs, the functional analog of the mammalian kidneys, provide a useful system for investigating adult stem-cell regulation and transformation in a genetic model organism. There are two pairs of MTs in adult *Drosophila* (Sözen, Armstrong, Yang, Kaiser, & Dow, 1997; Wessing & Eichelberg, 1978). Each tubule is divided into four compartments: initial, transitional, main (secretory), and proximal (reabsorptive). The proximal segment is further divided into the lower tubule and the ureter. The multipotent renal and nephric stem cells (RNSCs) were identified among the "tiny" cells in the lower tubule and ureter of the MT (Singh, Liu, & Hou, 2007; Fig. 4A). An RNSC generates a new RNSC and an immature renablast (RB) daughter. RBs in the lower tubule and ureter differentiate to become mature renalcytes, and RBs that move to the distal upper tubules can differentiate to become type I (principal) or II (stellate) cells in the transitional and initial segments (Fig. 4B). The RNSC fates are regulated by autocrine JAK–STAT signaling (Singh et al., 2007) and other signaling pathways (Li, Liu, & Cai, 2014, 2015; Zeng et al., 2010).

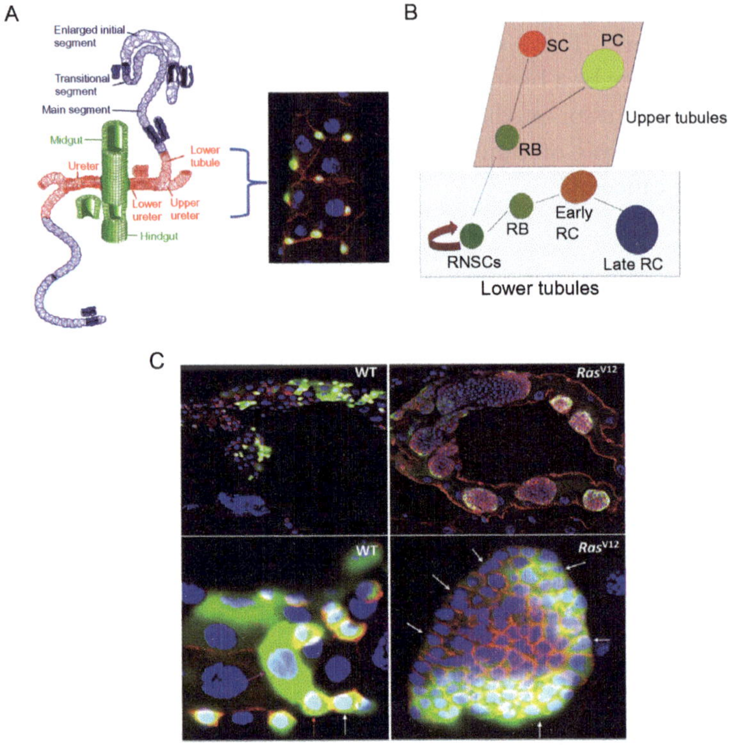

Fig. 4 (A) Drawing of the *Drosophila* MTs. STAT-GFP (a reporter of the JAK–STAT signaling, *green*) positive tiny cells are RNSCs. (B) Model of the RNSC lineage. (C) Activation of the oncogene Ras results in stem-cell tumor in adult *Drosophila* kidney. MTs with GFP-marked wild-type (*left panels*) and Ras^{V12} mutant (*right panels*) clones. *White arrows* point to RNSCs, *red arrow* points to RB, and *pink arrow* points to RC. Anti-GFP, *green*; anti-Arm, *red*; DAPI, *blue*. *Panel (A) Adapted from Wessing, A., & Eichelberg, D. (1978). Malpighian tubules, rectal papillae and excretion. In A. Ashburner & T. R. F. Wright (Ed.),* The genetics and biology of Drosophila *(Vol. 2c, pp. 1–42). London: Academic Press, panel (*B and C, left panels*) modified after Singh, S. R., Liu, W., & Hou, S. X. (2007). The adult Drosophila malpighian tubules are maintained by multipotent stem cells.* Cell Stem Cell, 1, *191–203, and panel (*C, right panels*) modified after Zeng, X., Singh, S. R., Hou, D., & Hou, S. X. (2010). Tumor suppressors Sav/Scrib and oncogene Ras regulate stem-cell transformation in adult Drosophila malpighian tubules.* Journal of Cellular Physiology, 224, *766–774.*

4.2 Ras-Transformed RNSCs

More than 30% of all types of human tumors are because of mutations that activate the oncogene Ras. In *Drosophila* MTs, Ras activation causes RNSCs to lose their differentiation and to overproliferate, resulting in neoplastic tumorous growth (Zeng et al., 2010; Fig. 4C).

4.3 Ras-TSCs Exhibit Hallmarks of Cancer

The advanced stages of human cancers exhibit several hallmarks (Hanahan & Weinberg, 2011). Among these hallmarks, the activation of a telomerase to allow unlimited replicative potential, and increased angiogenesis are not seen in *Drosophila* cancer models. Nevertheless, Zeng et al. (2010) showed that the Ras-transformed RNSCs exhibit all of the first four hallmarks of human cancers.

First, they found that the cell cycle and proliferation regulators Cyclin E (Cyc E), phosphorylated Cdc2, *Drosophila* Myc (dMyc), and phosphorylated ERK, were very weakly expressed in normal RNSCs but were highly expressed in the Ras^{V12}-transformed RNSCs. Second, the death-associated inhibitor of apoptosis 1 (DIAP1) and its transcriptional reporter *diap1-lacZ* were highly expressed in the Ras^{V12}-transformed RNSCs compared to wild-type RNSCs. Third, a few Ras^{V12}-transformed RNSCs migrated to the main segments of the MTs; in contrast, wild-type RNSCs remained in the lower tubule and ureter. The matrix metalloproteinase 1 (MMP1) was also dramatically expressed in the Ras^{V12}-transformed RNSCs compared to wild-type RNSCs. Moreover, an MMP1 inhibitor suppressed the Ras^{V12}-transformed RNSC phenotypes, suggesting that the TSCs were highly motile. However, the migration of the TSCs was limited to the main segments. Furthermore, the TSCs were highly proliferative but were not able to migrate to other organs when transplanted into the abdomen of wild-type adult hosts, suggesting that these tumors were not metastatic. However, careful experiments over long interval are needed to confirm their metastatic ability. Fourth, the cell-polarity markers Bazooka (Baz) and *Drosophila* atypical protein kinase C (DaPKC) were uniformly expressed throughout the cortex of Ras-transformed RNSCs, while they were only expressed in an apical crescent at metaphase in the wild-type RNSCs, suggesting that activated Ras may disrupt the polarity and asymmetric division of RNSCs (Zeng et al., 2010).

In summary, Ras^{V12}-transformed RNSCs exhibit accelerated proliferation, decreased apoptosis, disorganized cell polarity, a lack of differentiation, and elevated migration. Therefore, the Ras^{V12}-transformed RNSCs show most of the hallmarks of human cancer and may represent CSC-like cells (Zeng et al., 2010).

4.4 Signaling Downstream of Ras Regulates RNSC Transformation

Over the years, several Ras downstream effectors that control complex signaling networks have been identified. To determine which signaling pathways mediate the Ras activity in RNSC transformation, Zeng et al. (2010) screened a collection of UAS-transgenic lines available from public stock centers that express a dominant-negative form or RNAi of a known signaling component. They found that the expression of a dominant-negative form of Raf or RhoA or an RNAi of MEK (Dsor1) suppressed the Ras^{V12} phenotypes. In contrast, the expression of a dominant-negative form of Ral, PI3K, Bsk, or Rac1, or an RNAi of *cyc E* or *dMyc* did not alter the Ras^{V12} phenotypes. They also found that a *Stat92E* mutant did not block the Ras^{V12} phenotypes. Further, they also screened 22 commercially available protein kinase or proteinase inhibitors and found that inhibitors of protein kinase A (PKA), Tor, or MMP1 dramatically suppressed the Ras^{V12} phenotypes. The genetic loss of PKA or Tor also blocked the Ras^{V12}-induced RNSC tumor formation. Finally, by expressing constitutively active Raf, RhoA, or ERK (rl^{Sem}) or Cbl^{RNAi} (Cbl^{RNAi}, unpublished result) in RNSCs using the positively marked mosaic lineage-labeling technique, we found that only the constitutively active Raf and Cbl^{RNAi} induced stem-cell tumor formation.

In summary, the earlier results suggest that (1) Raf is an immediate target of Ras and that Cbl negatively regulates the Ras/Raf signaling. (2) RhoA or ERK signaling pathways cooperatively mediate Ras' function in RNSC transformation and that none of them alone is sufficient to transform the stem cells. (3) The *Drosophila* RNSCs are regulated by JAK–STAT signaling through an autocrine system, and mutations in *Stat92E* did not suppress the Ras-transformed RNSC tumor phenotype, indicating that the growing Ras-transformed RNSCs become independent of JAK–STAT signaling. Therefore, these TSCs become niche independent.

4.5 New Genes That Mediate Ras' Regulation of RNSC Transformation

To elucidate the molecular mechanism behind the Ras-mediated transformation of RNSCs, Zeng et al. (2010) performed transcriptome comparisons of wild-type RNSCs and Ras-transformed RNSCs using a microarray assay and identified 186 genes that showed significantly different expressions. They further examined the genetic interactions between Ras^{V12} and 147 genes using UAS-RNAi lines available from the Vienna Drosophila RNAi Center (VDRC). They identified 20 genes whose RNAi dramatically suppressed the Ras-transformed phenotypes, including genes encoding replication protein A2 (RPA2), β-tubulin at 60D (βTub60D), actin-binding protein Arpc3B, mitotic spindle protein Spc105R, transcription factor point (*Pnt*), and lipid phosphatase wunen-2 (Wun2).

The effect of the *Pnt* UAS-RNAi was not surprising, because *Pnt* is known to mediate Ras signaling in other systems (O'Neill, Rebay, Tjian, & Rubin, 1994). RPA2 is an independent prognostic indicator of colon cancer (Givalos et al., 2007). The suppression of Ras-transformed stem-cell phenotypes by RPA2-RNAi suggests that some change in DNA replication may be partially responsible for the CSC phenotypes. The other three proteins, βTub60D, Arpc3B, and Spc105R, are regulators of either the microtubule or the actin–myosin cytoskeletal network and may mediate cytoskeletal changes involved in the blockage of RNSC differentiation or in the facilitation of rapid stem-cell proliferation or increased stem-cell migration. The Wun2 regulates germ-cell repulsion and migration in the *Drosophila* female germline (Renault, Sigal, Morris, & Lehmann, 2004) and could play a role in RNSC migration. Thus, targeting the DNA replication mediated by RPA2, Arpc3B/Spc105R, or Wun2 may be an effective strategy for eliminating Ras-transformed CSCs.

In summary, Ras-TSCs display most of the hallmarks of human cancer. These changes are elicited through the upregulation of Cyc E, dMyc, DIAP1, MMP1, and several other genes. Downstream of Ras/Raf, and other signaling pathways, including MEK/MAPK, RhoA, PKA, and TOR, cooperatively mediates the Ras-induced stem-cell transformation (Fig. 5). This study also identified genes highly upregulated in the Ras-TSCs

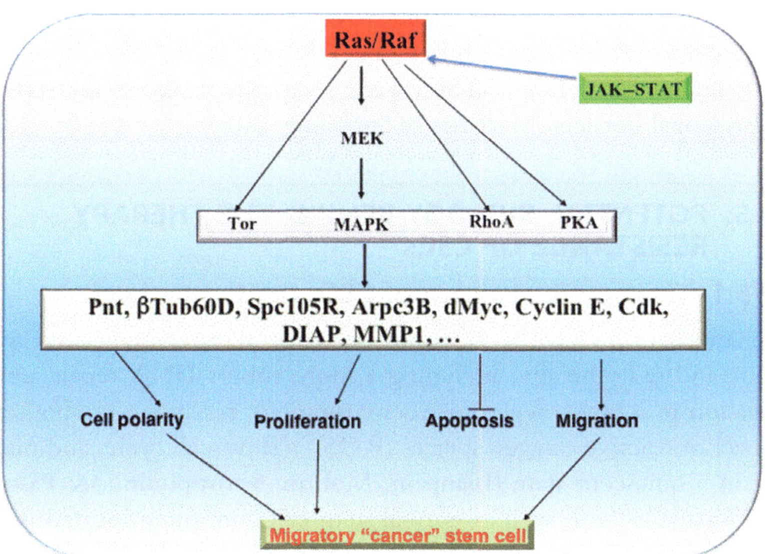

Fig. 5 Signaling pathways downstream of Ras that regulate RNSC transformation in MTs.

and may function in the downstream signaling to control behavior of TSCs. Further investigation of these genes would help in developing anti-tumorigenic stem-cell cancer therapies.

The tumors in *Drosophila* have been categorized into two groups: hyperplastic and neoplastic (Bilder, 2004). Hyperplastic tumors show expansive growth, but normal apico–basal polarity and differentiation ability. Expressing the constitutively activated form of *Ras*, *Raf*, or other signaling components in ISCs or activating JAK–STAT signaling in RNSCs generates hyperplastic tumors (Patel & Edgar, 2014). Neoplastic tumors are masses of overproliferating disorganized cells with lost their polarity and differentiation ability. The transplantation of a fraction of a neoplastic tumor into a wild-type fly is sufficient to seed new growth that in some cases metastatically invades other organs, killing the host (Bilder, 2004). The neoplastic tumors in *Drosophila* are more similar to high-grade malignant human tumors than are the hyperplastic tumors, and include the tumors of the Ras^{V12}; $scrib^{-/-}$ eye disc (Wu, Pastor-Pareja, & Xu, 2010), of larval neuroblasts with failed asymmetric division (Gonzalez, 2007), and of the Ras^{V12}; $csk^{-/-}$ eye disc fed on high dietary sucrose (Hirabayashi, Baranski, & Cagan, 2013). The transplantation of a small fraction of neoplastic tumors results in the invasion of other organs. The Ras^{V12}-transformed RNSCs appear to form neoplastic tumors. However, while the transplantation of a small fraction of the Ras^{V12} or Ras^{V12}; $scrib^{-/-}$ RNSC tumor into a wild-type fly was sufficient to seed new uncontrollable growth that filled the body cavity, the tumor masses never invaded other organs (Zeng et al., 2010), possibility because adult stem cells and TSCs were compartmentalized and restricted by subregional borders (Marianes & Spradling, 2013).

5. POTENTIAL BIOLOGY BEHIND THE THERAPY RESISTANCE OF CSCs

5.1 Therapy Resistance of CSCs in Mammals

Several mechanisms have been proposed to explain the resistance of CSCs to chemo- and radiotherapy, including a more robust DNA repair activity, localization to a low-oxygen microenvironment, resistance to apoptosis, a low level of reactive oxygen species (ROS), a slow-cell cycle, and maintenance in a quiescent state (Blanpain, Mohrin, Sotiropoulou, & Passegué, 2011). First, CSCs, as well as normal stem cells, are often located in low-oxygen (hypoxic) regions (Mohyeldin, Garzón-Muvdi, & Quiñones-Hinojosa, 2010). These hypoxic environments have the ability to upregulate ATP-binding cassette transporter genes that export small molecules (drugs) out of the cell (Trumpp & Wiestler, 2008). In addition, because oxygen is required to induce DNA damage by radiation, dormant CSCs located in hypoxic niches would be protected from radiation therapy (Brown & Wilson, 2004). Furthermore, hypoxia stabilizes hypoxia-inducible factor 1, which induces the transcription factor forkhead box O3 (Foxo3a) through the AKT signaling pathway, and Foxo3a further supports CSC survival by suppressing apoptosis (Mohyeldin et al., 2010; Naka et al., 2010). However, although attempts have been made to kill tumors using hypoxia-specific cytotoxins (Brown & Siim, 1996), they have not been clinically successful.

Second, CSCs rarely divide. In the mouse intestine, the slower cycling stem cells at position +4 above the crypt base are radioresistant cancer-initiating cells (Sangiorgi & Capecchi, 2008). In leukemia, growth arrest through the p53-mediated induction of *p21* was found to be critical for protecting CSCs from IR (Mohrin et al., 2010). This slow-cycling property would give CSCs more time to efficiently repair their damaged DNA to escape apoptosis (Bao et al., 2006; Viale et al., 2009). The rare division also makes CSCs more resistant to antiproliferative agents and signaling pathway inhibitors, which mostly target dividing and actively metabolizing cells.

Third, CSCs use quiescence as a protective mechanism against chemotherapy. Attempts to force quiescent stem cells to cycle by first exposing them to cytokines such as interferon-alpha and granulocyte colony-stimulating factor (G-CSF) and then killing them by different chemotherapeutic agents have yielded promising results in animal models and at the preclinical stage (Essers & Trumpp, 2010). However, some CSCs (extremely dormant fraction) still seem to be resistant to such combined therapies; for example, combined G-CSF and imatinib treatment failed to completely eliminate the CSCs in CML patients in a clinical pilot study (Drummond et al., 2009), and combined G-CSF and cytarabine treatment failed to cure xenograft mice of acute myeloid leukemia (Saito et al., 2010).

Fourth, CSCs are resistant to apoptosis. Glioma CSCs are highly radioresistant due to their overexpression of the antiapoptotic protein myeloid cell leukemia 1 (Tagscherer et al., 2008). Similarly, Ras^{V12}-transformed *Drosophila* RNSCs may become resistant to therapy through their overexpression of the apoptosis inhibitor DIAP1 (Zeng et al., 2010).

Fifth, CSCs have lower ROS levels than other cancer cells. Mouse mammary gland CSCs are more resistant to IR-induced DNA damage and apoptosis than other cancer cells, because of low ROS due to an increased glutathione metabolism and elevated free-radical scavenging systems (Diehn et al., 2009).

Finally, a more robust DNA repair activity may greatly contribute to the resistance of CSCs to chemo- and radiotherapy. The increased expression of many DNA repair-associated genes and more robust DNA repair activity has been observed in the CSCs of breast cancer (Zhang et al., 2008) and glioblastoma multiforme (Bao et al., 2006) compared to the other tumor cells.

5.2 Therapy Resistance of Normal and TSCs in *Drosophila*

5.2.1 *Female GSCs Use a "Dying Daughters Protect Their Mother" Strategy to Protect GSCs from IR-Induced Death in* Drosophila

Normal adult stem cells and CSCs share many properties, such as self-renewal and multipotency. Both types of stem cells are usually resistant to radiation (IR) and chemotherapy. In a recent study of normal GSCs in the *Drosophila* ovary, Xing, Su, and Ruohola-Baker (2015) found that female GSCs use a "dying daughters protect their mother" strategy to maintain GSCs under harmful conditions. During IR treatment, the dying differentiating daughter cells produce a putative Tie ligand, Pvf1 (human Angiopoietin), which activates the receptor tyrosine kinase Tie/Tie-2 signaling in GSCs. This activated signaling upregulates the microRNA bantam, which represses the FOXO-mediated transcription of the proapoptotic Smac/DIABLO ortholog Hid, to protect the GSCs from IR-induced apoptosis.

A similar mechanism may function in CSCs; the IR-induced death of the bulk of the tumor cells may send a signal to the CSCs to protect them from IR-induced apoptosis. If this is the case, blocking the protective signal may be a strategy for eliminating CSCs. Future experiments are needed to address whether such a mechanism functions in human cancer.

5.2.2 *ISCs Are Internally Resistant to Apoptosis but Sensitive to Lipolysis Disruption in* Drosophila

The induction of *reaper* (*rpr*, another proapoptotic gene in *Drosophila*) in differentiated ECs for 12 h causes widespread apoptosis but has little effect on ISCs/EBs, even after 7 days of induction in ISCs/EBs (Jiang et al., 2009), suggesting that ISCs/EBs are internally resistant to apoptosis.

Quiescent stem cells often reside in a secluded location surrounded by dense extracellular matrix and a dormant hypoxic storage niche (Trumpp & Wiestler, 2008), where they have less access to sugar and amino acid nutrition from the body's circulatory system. Like hibernating animals, the quiescent stem cells may mainly rely on lipid reserves for their energy supply, and blocking lipolysis may starve them to death. In support of this hypothesis, we recently found that knockdown of the vesicle-mediated coat protein complex I (COPI)/ADP-ribosylation factor 1 (Arf1)-mediated lipolysis pathway selectively kills ISCs, but not differentiated ECs (Zeng et al., 2015). If a similar mechanism functions in CSCs, targeting lipid metabolism could be an effective treatment for cancer.

Most of the current therapeutic strategies targeting CSCs involve inhibiting the self-renewal or survival pathways in these cells, including Notch, Hedgehog, Wnt, AKT, and transforming growth factor beta (Korkaya & Wicha, 2007; Naka et al., 2010). However, because these pathways are important in normal cells, inhibiting them could result in systemic toxicities, limiting the clinical usefulness of this approach. Normal cells mostly rely on sugar and amino acids for their energy supply, with lipolysis playing only a minor role in their survival. Thus, blocking lipolysis might selectively kill CSCs but not normal cells. Therefore, targeting the lipolysis pathway could be an effective approach for eliminating CSCs.

6. SUMMARY AND PERSPECTIVES

Even though the biological and clinical relevance of CSCs are still under debate, most solid tumors appear to follow the CSC model (Kreso & Dick, 2014). However, the origin of the CSCs/TICs varies in different types of tumors. The COCs can be either transformed normal stem cells or differentiated cells. It is still not clear how a few TICs evolve to become mature CSCs, responsible for tumor relapse and resistance to chemo/radiation therapies. Thus, to develop more effective cancer treatments, it is fundamentally important to identify the differences between normal stem cells and CSCs,and to understand the biology behind therapy resistance.

In the past few years, studies in *Drosophila* have made significant contributions to these new areas of cancer research. First, studies of stem-cell tumors in the *Drosophila* posterior midgut and testis revealed how stem-cell-tumors initiate and propagate through niche appropriation or stem-cell competition for niche occupancy. Second, studies on Ras^{V12}-transformed stem-cell tumors revealed differences between normal and TSCs. Third, studies on stem-cell resistance and sensitivity to death inducers using female GSCs and midgut ISCs shed new light on the biology behind therapy resistance. Most human tumors develop in adult somatic organs. Going forward, we believe that adult *Drosophila* stem-cell-based tumorigenesis, particularly in somatic tissues (such as the kidney, intestine/colon, or gastric organ), will provide useful models of advanced human tumors to help elucidate the properties of this disease.

The ultimate goal of cancer research is to find the most effective treatment for the disease with the least harmful side effects. Recent screens of chemical compounds in *Drosophila* have already contributed to FDA-approved drugs and drugs under clinical trials. A large-scale chemical screen in an adult *Drosophila* hyperplastic stem-cell tumor model was shown to be feasible (Markstein et al., 2014). However, most of the compounds identified targeting signaling events that specifically promote tumor growth. Further screens in better adult *Drosophila* neoplastic stem-cell tumor models are needed to identify broad antitumorigenic compounds that target the unique properties of all TSCs, to ultimately conquer this debilitating disease.

REFERENCES

Alcantara, L. S., Chen, J., Kwon, C. H., Jackson, E. L., Li, Y., Burns, D. K., et al. (2009). Malignant astrocytomas originate from neural stem/progenitor cells in a somatic tumor suppressor mouse model. *Cancer Cell, 15*, 45–56.

Al-Hajj, M., Wicha, M. S., Benito-Hernandez, A., Morrison, S. J., & Clarke, M. F. (2003). Prospective identification of tumorigenic breast cancer cells. *Proceedings of the National Academy of Sciences of the United States of America, 100*, 3983–3988.

Amoyel, M., Simons, B. D., & Bach, E. A. (2014). Neutral competition of stem cells is skewed by proliferative changes downstream of Hh and Hpo. *The EMBO Journal, 33*, 2295–2313.

Apidianakis, Y., Pitsouli, C., Perrimon, N., & Rahme, L. (2009). Synergy between bacterial infection and genetic predisposition in intestinal dysplasia. *Proceedings of the National Academy of Sciences of the United States of America, 106*, 20883–20888.

Asfaha, S., Hayakawa, Y., Muley, A., Stokes, S., Graham, T. A., Ericksen, R. E., et al. (2015). Krt19(+)/Lgr5(-) cells are radioresistant cancer-initiating stem cells in the colon and intestine. *Cell Stem Cell, 16*, 627–638.

Bao, S., Wu, Q., McLendon, R. E., Hao, Y., Shi, Q., Hjelmeland, A. B., et al. (2006). Glioma stem cells promote radioresistance by preferential activation of the DNA damage response. *Nature, 444*, 756–760.

Barker, N., Ridgway, R. A., van Es, J. H., van de Wetering, M., Begthel, H., van den Born, M., et al. (2009). Crypt stem cells as the cells-of-origin of intestinal cancer. *Nature, 457*, 608–611.

Bilder, D. (2004). Epithelial polarity and proliferation control: Links from the Drosophila neoplastic tumor suppressors. *Genes & Development, 18*, 1909–1925.

Binari, R., & Perrimon, N. (1994). Stripe-specific regulation of pair-rule genes by hopscotch, a putative Jak family tyrosine kinase in Drosophila. *Genes & Development, 8*, 300–312.

Biteau, B., & Jasper, H. (2011). EGF signaling regulates the proliferation of intestinal stem cells in Drosophila. *Development, 138*, 1045–1055.

Biteau, B., & Jasper, H. (2014). Slit/Robo signaling regulates cell fate decisions in the intestinal stem cell lineage of Drosophila. *Cell Reports, 7*, 1867–1875.

Blanpain, C., Mohrin, M., Sotiropoulou, P. A., & Passegué, E. (2011). DNA-damage response in tissue-specific and cancer stem cells. *Cell Stem Cell, 8*, 16–29.

Bondar, T., & Medzhitov, R. (2010). p53-mediated hematopoietic stem and progenitor cell competition. *Cell Stem Cell, 6*, 309–322.

Brown, J. M., & Siim, B. G. (1996). Hypoxia-specific cytotoxins in cancer therapy. *Seminars in Radiation Oncology, 6*, 22–36.

Brown, J. M., & Wilson, W. R. (2004). Exploiting tumour hypoxia in cancer treatment. *Nature Reviews. Cancer, 4*, 437–447.

Buchon, N., Broderick, N. A., Poidevin, M., Pradervand, S., & Lemaitre, B. (2009). Drosophila intestinal response to bacterial infection: Activation of host defense and stem cell proliferation. *Cell Host & Microbe, 5*, 200–211.

Calabrese, C., Poppleton, H., Kocak, M., Hogg, T. L., Fuller, C., Hamner, B., et al. (2007). A perivascular niche for brain tumor stem cells. *Cancer Cell, 11*, 69–82.

Creighton, C. J., Li, X., Landis, M., Dixon, J. M., Neumeister, V. M., Sjolund, A., et al. (2009). Residual breast cancers after conventional therapy display mesenchymal as well as tumor-initiating features. *Proceedings of the National Academy of Sciences of the United States of America, 106*, 13820–13825.

Diehn, M., Cho, R. W., Lobo, N. A., Kalisky, T., Dorie, M. J., Kulp, A. N., et al. (2009). Association of reactive oxygen species levels and radioresistance in cancer stem cells. *Nature, 458*, 780–783.

Drummond, M. W., Heaney, N., Kaeda, J., Nicolini, F. E., Clark, R. E., Wilson, G., et al. (2009). A pilot study of continuous imatinib vs pulsed imatinib with or without G-CSF in CML patients who have achieved a complete cytogenetic response. *Leukemia, 23*, 1199–1201.

Essers, M. A. G., & Trumpp, A. (2010). Targeting leukemic stem cells by breaking their dormancy. *Molecular Oncology, 4*, 443–450.

Fearon, E. R., & Vogelstein, B. (1990). A genetic model for colorectal tumorigenesis. *Cell, 61*, 759–767.

Flaherty, M. S., Salis, P., Evans, C. J., Ekas, L. A., Marouf, A., Zavadil, J., et al. (2010). *chinmo* is a functional effector of the JAK/STAT pathway that regulates eye development, tumor formation, and stem cell self-renewal in Drosophila. *Developmental Cell, 18*, 556–568.

Friedmann-Morvinski, D., Bushong, E. A., Ke, E., Soda, Y., Marumoto, T., Singer, O., et al. (2012). Dedifferentiation of neurons and astrocytes by oncogenes can induce gliomas in mice. *Science, 338*, 1080–1084.

Givalos, N., Gakiopoulou, H., Skliri, M., Bousboukea, K., Konstantinidou, A. E., Korkolopoulou, P., et al. (2007). Replication protein A is an independent prognostic indicator with potential therapeutic implications in colon cancer. *Modern Pathology, 20*, 159–166.

Gonzalez, C. (2007). Spindle orientation, asymmetric division and tumor suppression in Drosophila stem cells. *Nature Reviews. Genetics, 8*, 462–472.

Guo, Z., & Ohlstein, B. (2015). Stem cell regulation. Bidirectional Notch signaling regulates Drosophila intestinal stem cell multipotency. *Science, 350*, aab0988.

Hakim, R. S., Baldwin, K., & Smagghe, G. (2010). Regulation of midgut growth, development, and metamorphosis. *Annual Reviews of Entomology, 55*, 593–608.

Hanahan, D., & Weinberg, R. A. (2011). Hallmarks of cancer: The next generation. *Cell, 144*, 646–674.

Hirabayashi, S., Baranski, T. J., & Cagan, R. L. (2013). Transformed Drosophila cells evade diet-mediated insulin resistance through wingless signaling. *Cell, 154*, 664–675.

Hooper, J. D., Baker, E., Ogbourne, S. M., Sutherland, G. R., & Antalis, T. M. (2000). Cloning of the cDNA and localization of the gene encoding human NRBP, a ubiquitously expressed, multidomain putative adapter protein. *Genomics, 66*, 113–118.

Issigonis, M., Tulina, N., de Cuevas, M., Brawley, C., Sandler, L., & Matunis, E. (2009). JAK-STAT signal inhibition regulates competition in the Drosophila testis stem cell niche. *Science, 326*, 153–156.

Jamieson, C. H., Ailles, L. E., Dylla, S. J., Muijtjens, M., Jones, C., Zehnder, J. L., et al. (2004). Granulocyte-macrophage progenitors as candidate leukemic stem cells in blast-crisis CML. *The New England Journal of Medicine, 351*, 657–667.

Jiang, H., Grenley, M. O., Bravo, M. J., Blumhagen, R. Z., & Edgar, B. A. (2011). EGFR/Ras/MAPK signaling mediates adult midgut epithelial homeostasis and regeneration in Drosophila. *Cell Stem Cell*, *8*, 84–95.

Jiang, H., Patel, P. H., Kohlmaier, A., Grenley, M. O., McEwen, D. G., & Edgar, B. A. (2009). Cytokine/Jak/Stat signaling mediates regeneration and homeostasis in the Drosophila midgut. *Cell*, *137*, 1343–1355.

Jin, Z., Kirilly, D., Weng, C., Kawase, E., Song, X., Smith, S., et al. (2008). Differentiation-defective stem cells outcompete normal stem cells for niche occupancy in the Drosophila ovary. *Cell Stem Cell*, *2*, 39–49.

Karpowicz, P., Perez, J., & Perrimon, N. (2010). The Hippo tumor suppressor pathway regulates intestinal stem cell regeneration. *Development*, *137*, 4135–4145.

Kiger, A. A., White-Cooper, H., & Fuller, M. T. (2000). Somatic support cells restrict germline stem cell self-renewal and promote differentiation. *Nature*, *407*, 750–754.

Kohlmaier, A., Fassnacht, C., Jin, Y., Reuter, H., Begum, J., Dutta, D., & Edgar, B. A. (2015). Src kinase function controls progenitor cell pools during regeneration and tumor onset in the Drosophila intestine. *Oncogene*, *34*, 2371–2384.

Kolahgar, G., Suijkerbuijk, S. J., Kucinski, I., Poirier, E. Z., Mansour, S., Simons, B. D., & Piddini, E. (2015). Cell competition modifies adult stem cell and tissue population dynamics in a JAK-STAT-dependent manner. *Developmental Cell*, *34*, 297–309.

Korkaya, H., & Wicha, M. S. (2007). Selective targeting of cancer stem cells: A new concept in cancer therapeutics. *BioDrugs*, *21*, 299–310.

Kreso, A., & Dick, J. E. (2014). Evolution of the cancer stem cell model. *Cell Stem Cell*, *14*, 275–291.

Lapidot, T., Sirard, C., Vormoor, J., Murdoch, B., Hoang, T., Caceres-Cortes, J., et al. (1994). A cell initiating human acute myeloid leukaemia after transplantation into SCID mice. *Nature*, *367*, 645–648.

Leatherman, J. L., & Dinardo, S. (2008). Zfh-1 controls somatic stem cell self-renewal in the Drosophila testis and nonautonomously influences germline stem cell self-renewal. *Cell Stem Cell*, *3*, 44–54.

Leong, K. G., Wang, B. E., Johnson, L., & Gao, W. Q. (2008). Generation of a prostate from a single adult stem cell. *Nature*, *456*, 804–808.

Li, X., Lewis, M. T., Huang, J., Gutierrez, C., Osborne, C. K., Wu, M. F., et al. (2008). Intrinsic resistance of tumorigenic breast cancer cells to chemotherapy. *Journal of the National Cancer Institute*, *100*, 672–679.

Li, Z., Liu, S., & Cai, Y. (2014). Differential Notch activity is required for homeostasis of malpighian tubules in adult Drosophila. *Journal of Genetics and Genomics*, *41*, 649–652.

Li, Z., Liu, S., & Cai, Y. (2015). EGFR/MAPK signaling regulates the proliferation of Drosophila renal and nephric stem cells. *Journal of Genetics and Genomics*, *42*, 9–20.

Lim, R., Winteringham, L. N., Williams, J. H., McCulloch, R. K., Ingley, E., Tiao, J. Y., et al. (2002). MADM, a novel adaptor protein that mediates phosphorylation of the 14-3-3 binding site of myeloid leukemia factor 1. *The Journal of Biological Chemistry*, *277*, 40997–41008.

Malanchi, I., Santamaria-Martínez, A., Susanto, E., Peng, H., Lehr, H. A., Delaloye, J. F., et al. (2012). Interactions between cancer stem cells and their niche govern metastatic colonization. *Nature*, *481*, 85–89.

Marianes, A., & Spradling, A. C. (2013). Physiological and stem cell compartmentalization within the Drosophila midgut. *eLife*, *2*, e00886.

Markstein, M., Dettorre, S., Cho, J., Neumüller, R. A., Craig-Müller, S., & Perrimon, N. (2014). Systematic screen of chemotherapeutics in Drosophila stem cell tumors. *Proceedings of the National Academy of Sciences of the United States of America*, *111*, 4530–4535.

Marquardt, J. U., Andersen, J. B., & Thorgeirsson, S. S. (2015). Functional and genetic deconstruction of the cellular origin in liver cancer. *Nature Reviews. Cancer*, *15*, 653–667.

Martins, V. C., Busch, K., Juraeva, D., Blum, C., Ludwig, C., Rasche, V., et al. (2014). Cell competition is a tumour suppressor mechanism in the thymus. *Nature*, *509*, 465–470.
Marusyk, A., Porter, C. C., Zaberezhnyy, V., & DeGregori, J. (2010). Irradiation selects for p53-deficient hematopoietic progenitors. *PLoS Biology*, *8*, e1000324.
Micchelli, C. A., & Perrimon, N. (2006). Evidence that stem cells reside in the adult Drosophila midgut epithelium. *Nature*, *439*, 475–479.
Mohrin, M., Bourke, E., Alexander, D., Warr, M. R., Barry-Holson, K., Le Beau, M. M., et al. (2010). Hematopoietic stem cell quiescence promotes error-prone DNA repair and mutagenesis. *Cell Stem Cell*, 7, 174–185.
Mohyeldin, A., Garzón-Muvdi, T., & Quiñones-Hinojosa, A. (2010). Oxygen in stem cell biology: A critical component of the stem cell niche. *Cell Stem Cell*, 7, 150–161.
Mueller, M. T., Hermann, P. C., Witthauer, J., Rubio-Viqueira, B., Leicht, S. F., Huber, S., et al. (2009). Combined targeted treatment to eliminate tumorigenic cancer stem cells in human pancreatic cancer. *Gastroenterology*, *137*, 1102–1113.
Naka, K., Hoshii, T., Muraguchi, T., Tadokoro, Y., Ooshio, T., Kondo, Y., et al. (2010). TGF-beta-FOXO signalling maintains leukaemia-initiating cells in chronic myeloid leukaemia. *Nature*, *463*, 676–680.
Ohlstein, B., & Spradling, A. (2006). The adult Drosophila posterior midgut is maintained by pluripotent stem cells. *Nature*, *439*, 470–474.
Ohlstein, B., & Spradling, A. (2007). Multipotent Drosophila intestinal stem cells specify daughter cell fates by differential notch signaling. *Science*, *315*, 988–992.
O'Neill, E. M., Rebay, I., Tjian, R., & Rubin, G. M. (1994). The activities of two Ets-related transcription factors required for Drosophila eye development are modulated by the Ras/MAPK pathway. *Cell*, *78*, 137–147.
Pastor-Pareja, J. C., & Xu, T. (2013). Dissecting social cell biology and tumors using Drosophila genetics. *Annual Review of Genetics*, *47*, 51–74.
Patel, P. H., Dutta, D., & Edgar, B. A. (2015). Niche appropriation by Drosophila intestinal stem cell tumours. *Nature Cell Biology*, *17*, 1182–1192.
Patel, P. H., & Edgar, B. A. (2014). Tissue design: How Drosophila tumors remodel their neighborhood. *Seminars in Cell & Developmental Biology*, *28*, 86–95.
Phillips, T. M., McBride, W. H., & Pajonk, F. (2006). The response of CD24(-/low)/CD44 + breast cancer-initiating cells to radiation. *Journal of the National Cancer Institute*, *98*, 1777–1785.
Renault, A. D., Sigal, Y. J., Morris, A. J., & Lehmann, R. (2004). Soma-germ line competition for lipid phosphate uptake regulates germ cell migration and survival. *Science*, *305*, 1963–1966.
Rhiner, C., Díaz, B., Portela, M., Poyatos, J. F., Fernández-Ruiz, I., López-Gay, J. M., et al. (2009). Persistent competition among stem cells and their daughters in the Drosophila ovary germline niche. *Development*, *136*, 995–1006.
Ricci-Vitiani, L., Lombardi, D. G., Pilozzi, E., Biffoni, M., Todaro, M., Peschle, C., et al. (2007). Identification and expansion of human colon-cancer-initiating cells. *Nature*, *445*, 111–115.
Saito, Y., Uchida, N., Tanaka, S., Suzuki, N., Tomizawa-Murasawa, M., Sone, A., et al. (2010). Induction of cell cycle entry eliminates human leukemia stem cells in a mouse model of AML. *Nature Biotechnology*, *28*, 275–280.
Sangiorgi, E., & Capecchi, M. R. (2008). Bmi1 is expressed in vivo in intestinal stem cells. *Nature Genetics*, *40*, 915–920.
Shaw, R. L., Kohlmaier, A., Polesello, C., Veelken, C., Edgar, B. A., & Tapon, N. (2010). The Hippo pathway regulates intestinal stem cell proliferation during Drosophila adult midgut regeneration. *Development*, *137*, 4147–4258.
Singh, S. K., Hawkins, C., Clarke, I. D., Squire, J. A., Bayani, J., Hide, T., et al. (2004). Identification of human brain tumour initiating cells. *Nature*, *432*, 396–401.

Singh, S. R., Liu, W., & Hou, S. X. (2007). The adult Drosophila malpighian tubules are maintained by multipotent stem cells. *Cell Stem Cell, 1*, 191–203.
Singh, S. R., Liu, Y., Zhao, J., Zeng, X., & Hou, S. X. (2016). The novel tumour suppressor Madm regulates stem cell competition in the Drosophila testis. *Nature Communications, 7*, 10473.
Singh, S. R., Zheng, Z., Wang, H., Oh, S. W., Chen, X., & Hou, S. X. (2010). Competitiveness for the niche and mutual dependence of the germline and somatic stem cells in the Drosophila testis are regulated by the JAK/STAT signaling. *Journal of Cellular Physiology, 223*, 500–510.
Siudeja, K., Nassari, S., Gervais, L., Skorski, P., Lameiras, S., Stolfa, D., et al. (2015). Frequent somatic mutation in adult intestinal stem cells drives neoplasia and genetic mosaicism during aging. *Cell Stem Cell, 17*, 663–674.
Sonoshita, M., & Cagan, R. L. (2017). Modeling human cancers in *Drosophila*. *Current Topics in Developmental Biology, 121*, 287–309.
Sözen, M. A., Armstrong, J. D., Yang, M., Kaiser, K., & Dow, J. A. (1997). Functional domains are specified to single-cell resolution in a Drosophila epithelium. *Proceedings of the National Academy of Sciences of the United States of America, 94*, 5207–5212.
Staley, B. K., & Irvine, K. D. (2010). Warts and Yorkie mediate intestinal regeneration by influencing stem cell proliferation. *Current Biology, 20*, 1580–1587.
Tagscherer, K. E., Fassl, A., Campos, B., Farhadi, M., Kraemer, A., Böck, B. C., et al. (2008). Apoptosis-based treatment of glioblastomas with ABT-737, a novel small molecule inhibitor of Bcl-2 family proteins. *Oncogene, 27*, 6646–6656.
Tian, A., Shi, Q., Jiang, A., Li, S., Wang, B., & Jiang, J. (2015). Injury-stimulated Hedgehog signaling promotes regenerative proliferation of Drosophila intestinal stem cells. *The Journal of Cell Biology, 208*, 807–819.
Tran, J., Brenner, T. J., & DiNardo, S. (2000). Somatic control over the germline stem cell lineage during Drosophila spermatogenesis. *Nature, 407*, 754–757.
Trumpp, A., & Wiestler, O. D. (2008). Mechanisms of disease: Cancer stem cells—Targeting the evil twin. *Nature Clinical Practice Oncology, 5*, 337–347.
Viale, A., De Franco, F., Orleth, A., Cambiaghi, V., Giuliani, V., Bossi, D., et al. (2009). Cell-cycle restriction limits DNA damage and maintains self-renewal of leukaemia stem cells. *Nature, 457*, 51–56.
Wang, X., Kruithof-de Julio, M., Economides, K. D., Walker, D., Yu, H., Halili, M. V., et al. (2009). A luminal epithelial stem cell that is a cell of origin for prostate cancer. *Nature, 461*, 495–500.
Wang, H., Singh, S. R., Zheng, Z., Oh, S. W., Chen, X., Edwards, K., et al. (2006). Rap-GEF signaling controls stem cell anchoring to their niche through regulating DE-cadherin-mediated cell adhesion in the Drosophila testis. *Developmental Cell, 10*, 117–126.
Wang, B., Zhao, L., Fish, M., Logan, C. Y., & Nusse, R. (2015). Self-renewing diploid Axin2(+) cells fuel homeostatic renewal of the liver. *Nature, 524*, 180–185.
Wessing, A., & Eichelberg, D. (1978). Malpighian tubules, rectal papillae and excretion. In A. Ashburner & T. R. F. Wright (Eds.), *The genetics and biology of* Drosophila*: Vol. 2c.* (pp. 1–42). London: Academic Press.
Wilson, C. H., Crombie, C., van der Weyden, L., Poulogiannis, G., Rust, A. G., Pardo, M., et al. (2012). Nuclear receptor binding protein 1 regulates intestinal progenitor cell homeostasis and tumour formation. *The EMBO Journal, 31*, 2486–2497.
Wu, M., Pastor-Pareja, J. C., & Xu, T. (2010). Interaction between Ras(V12) and scribbled clones induces tumour growth and invasion. *Nature, 463*, 545–548.
Xing, Y., Su, T. T., & Ruohola-Baker, H. (2015). Tie-mediated signal from apoptotic cells protects stem cells in Drosophila melanogaster. *Nature Communications, 6*, 7058.

Yamashita, Y. M., Mahowald, A. P., Perlin, J. R., & Fuller, M. T. (2007). Asymmetric inheritance of mother versus daughter centrosome in stem cell division. *Science*, *315*, 518–521.

Yoneda-Kato, N., Look, A. T., Kirstein, M. N., Valentine, M. B., Raimondi, S. C., et al. (1996). The t(3;5)(q25.1;q34) of myelodysplastic syndrome and acute myeloid leukemia produces a novel fusion gene, NPM-MLF1. *Oncogene*, *12*, 265–275.

Zeng, X., Han, L., Singh, S. R., Liu, H., Neumüller, R. A., Yan, D., et al. (2015). Genome-wide RNAi screen identifies networks involved in intestinal stem cell regulation in Drosophila. *Cell Reports*, *10*, 1226–1238.

Zeng, X., & Hou, S. X. (2015). Enteroendocrine cells are generated from stem cells through a distinct progenitor in the adult Drosophila posterior midgut. *Development*, *142*, 644–653.

Zeng, X., Singh, S. R., Hou, D., & Hou, S. X. (2010). Tumor suppressors Sav/Scrib and oncogene Ras regulate stem-cell transformation in adult Drosophila malpighian tubules. *Journal of Cellular Physiology*, *224*, 766–774.

Zhang, M., Behbod, F., Atkinson, R. L., Landis, M. D., Kittrell, F., Edwards, D., et al. (2008). Identification of tumor-initiating cells in a p53-null mouse model of breast cancer. *Cancer Research*, *68*, 4674–4682.

Zhu, L., Gibson, P., Currle, D. S., Tong, Y., Richardson, R. J., Bayazitov, I. T., et al. (2009). Prominin 1 marks intestinal stem cells that are susceptible to neoplastic transformation. *Nature*, *457*, 603–607.

Zielke, N., Korzelius, J., van Straaten, M., Bender, K., Schuhknecht, G. F., Dutta, D., et al. (2014). Fly-FUCCI: A versatile tool for studying cell proliferation in complex tissues. *Cell Reports*, *7*, 588–598.

CHAPTER ELEVEN

The *Drosophila* Accessory Gland as a Model for Prostate Cancer and Other Pathologies

C. Wilson*[,1], A. Leiblich*[,†], D.C.I. Goberdhan*, F. Hamdy[†]
*University of Oxford, Oxford, United Kingdom
[†]University of Oxford, John Radcliffe Hospital, Oxford, United Kingdom
[1]Corresponding author: e-mail address: clive.wilson@dpag.ox.ac.uk

Contents

Current Topics in Developmental Biology, Volume 121
ISSN 0070-2153
http://dx.doi.org/10.1016/bs.ctdb.2016.06.001

Abstract

The human prostate is a gland of the male reproductive tract, which together with the seminal vesicles, is responsible for most seminal fluid production. It is a common site of cancer, and unlike other glands, it typically enlarges in aging men. In flies, the male accessory glands make many major seminal fluid components. Like their human equivalents, they secrete proteins from several conserved families, including proteases, lectins, and cysteine-rich secretory proteins, some of which interact with sperm and affect fertility. A key protein, sex peptide, is not conserved in vertebrates but plays a central role in mediating long-term effects on females after mating. Although postmitotic, one epithelial cell type in the accessory glands, the secondary cell, continues to grow in adults. It secretes microvesicles called exosomes from the endosomal multivesicular body, which, after mating, fuse with sperm. They also appear to affect female postmating behavior. Remarkably, the human prostate epithelium also secretes exosomes, which fuse to sperm in vitro to modulate their activity. Exosomes from prostate and other cancer cells are increasingly proposed to play fundamental roles in modulating the tumor microenvironment and in metastasis. Here we review a diverse accessory gland literature, which highlights functional analogies between the male reproductive glands of flies and humans, and a critical role for extracellular vesicles in allowing seminal fluid to promote male interests within the female. We postulate that secondary cells and prostate epithelial cells use common mechanisms to control growth, secretion, and signaling, which are relevant to prostate and other cancers, and can be genetically dissected in the uniquely tractable fly model.

1. INTRODUCTION

As discussed in the chapter "Modeling Human Cancers in *Drosophila*" by Sonoshita and Cagan in this volume, *Drosophila melanogaster* has provided a powerful system to model many different aspects of cancer biology. These include features described as hallmarks of cancer by Hanahan and Weinberg (2000, 2011), such as dysregulation of cell growth and proliferation, evasion from apoptosis, and invasive and metastatic properties (also reviewed in Rudrapatna, Cagan, & Das, 2012; Tipping & Perrimon, 2014). Genetic approaches in flies led to the identification and characterization of many of the intercellular signaling cascades that drive tumorigenesis (Perrimon, Pitsouli, & Shilo, 2012). In some cases, such as the control of epithelial polarity, they highlighted a new cell biological process that is part of the cancer regulatory network (Grifoni, Froldi, & Pession, 2013). Flies have also started to be employed in the study of tumor microenvironment and tumor–stroma interactions (Patel & Edgar, 2014). Furthermore, new candidate oncogenes and tumor suppressors, which are increasingly identified in genome sequence analysis of patient tumors (Martincorena & Campbell, 2015; Mitchell & Neal, 2015), can be tested in *Drosophila*.

Most cancer-related studies in flies have focused on developing tissues, particularly the larval imaginal discs and brain, because cells in these structures are still dividing (Tipping & Perrimon, 2014), unlike many adult cell types. However, proliferating follicular epithelial cells of the adult ovary have been particularly useful in analyzing the cancer-relevant mechanisms linking cell growth, signaling, polarity, and migration (Klusza & Deng, 2011; Rosales-Nieves & González-Reyes, 2014). Furthermore, in the last 10 years, numerous studies of stem cell regulation in the *Drosophila* midgut have not only revealed parallels with human intestinal stem cells (Nászai et al., 2015) but also suggested new control mechanisms that may be relevant to colorectal cancer (Panayidou & Apidianakis, 2013). These adult systems provide tissue- and cell type-specific insights into cancer mechanisms that complement data emerging from other fly tumor models (Bell & Thompson, 2014).

1.1 Modeling Adenocarcinoma

The majority of common human cancers are formed from glandular tissue, so-called adenocarcinomas, including most pancreatic (Ryan, Hong, & Bardeesy, 2014), lung (Ding et al., 2008), breast (Li & Daling, 2007), esophageal (Edgren, Adami, Weiderpass, & Nyrén, 2013), stomach (Shah et al., 2011), intestinal (Weitz et al., 2005), and prostate (Humphrey, 2012) tumors. Analysis of secretion from *Drosophila* glands and secretory epithelia has been undertaken in tissues like the Malpighian tubules (Cabrero et al., 2014) and larval salivary gland, where some detailed cell biological studies have been performed (Burgess et al., 2012, 2011; Torres, Rosa-Ferreira, & Munro, 2014). But generally, even analysis of genes that selectively promote tumorigenesis in glands, like the conserved receptor tyrosine kinase Ret, which is mutated in multiple endocrine neoplasia 2 (MEN2), has to date been undertaken in nonglandular fly tissue (Das & Cagan, 2013).

As discussed later, research into the human prostate and prostate cancer has lagged behind several other cancer types, particularly because of current limitations with mouse models. Recently, studies of the accessory gland in male flies have revealed surprising parallels with the human prostate epithelium (Corrigan et al., 2014; Ito et al., 2014; Leiblich et al., 2012; Xue & Noll, 2002), which indicate that this gland can be employed to model some prostate-specific cellular mechanisms, as well as processes potentially associated with other male reproductive glands like the seminal vesicles (see Sections 5–7). In addition, the remarkable subcellular structure of one epithelial cell type in this gland, known as the secondary cell, has opened up new opportunities to study secretory mechanisms, particularly those involving secretion of microvesicles (Corrigan et al., 2014). These mechanisms have not been extensively studied in any in vivo model to date but are likely to be relevant to all cancers and potentially other diseases affecting secretory functions.

2. THE PROSTATE

2.1 The Human Prostate

The prostate is a multilobular exocrine gland, which together with the other major male accessory glands, the seminal vesicles, produces most of the seminal fluid volume (Fig. 1A; Declan, Cahill, Chandra, & Davies, 2009). The human prostate epithelium appears pseudostratified and consists of two basic cell types, basal cells and more columnar luminal secretory cells (Long, Morrissey, Fitzpatrick, & Watson, 2005). The epithelium is arranged in branched tubules connected by ducts, which also have some secretory activity. The glandular tissue is surrounded by a fibromuscular stroma. The luminal content of the tubules is pumped into the urethra early during ejaculation and mixes with sperm. Prostate secretions play an important role in sperm capacitation, which is required to induce full sperm motility and fertilization capacity (Fraser, 1997; Miah, Salma, Hamano, & Schellander, 2015). A key role for the seminal vesicles, which produce about 70% of seminal fluid volume, is to provide fructose as an energy source to power sperm motility (Aumüller & Riva, 1992). However, the latter glands also appear to affect the signaling from the female reproductive tract to the preimplantation embryo (Bromfield et al., 2014), suggesting a complex interplay between male reproductive glands and the female reproductive system postmating.

Both normal prostate growth and secretion are dependent on androgens (Hayward & Cunha, 2000), linking prostate activity to the function of the testis, the major source of androgens in males. Prostate secretions include proteases, particularly those in the kallikrein class, most notably

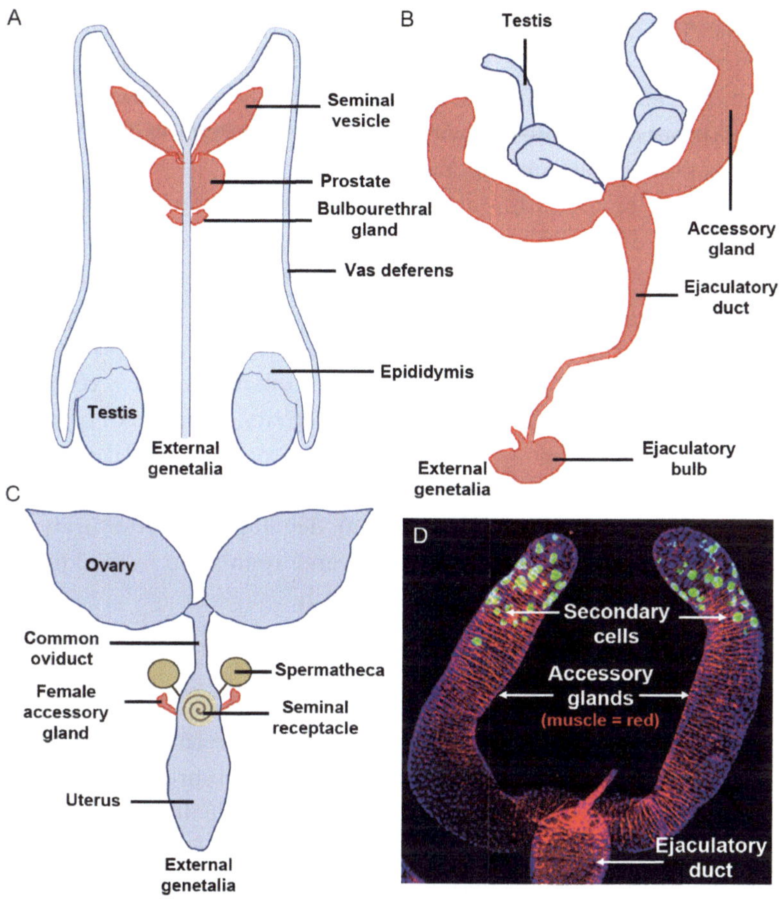

Fig. 1 Human and fly reproductive systems. (A and B) Schematics illustrate male reproductive systems of humans (A) and flies (B). Note *red*-shaded secretory organs in both organisms contribute much of the volume of seminal fluid, many key proteins, and nutrients, and are critical for normal fertility. (C) Fly female reproductive system includes two different sperm storage organs (*shaded green*), the paired spermathecae and the seminal receptacle. They gradually release sperm after mating under the control of SP, which is itself slowly released from its binding sites on sperm. Females also have two small secretory accessory glands. (D) Confocal image of paired accessory glands (*arrows*). In this whole mount, the 40 secondary cells in the epithelium express GFP (*green*), but the more abundant main cells do not. The preparation is stained with TRITC-phalloidin, which highlights the striated muscle layer (*red*) surrounding the gland that undergoes peristaltic contraction upon mating, and DAPI (*blue*), which marks the nuclei.

prostate-specific antigen (PSA), which has historically been employed as a biomarker for prostate cancer (Veveris-Lowe, Kruger, Walsh, Gardiner, & Clements, 2007). As discussed later, these enzymes are thought to cleave a number of substrates within the ejaculate, leading, for example, to the liquefaction of the seminal coagulum.

The prostate epithelium is also a source of secreted microvesicles called exosomes. These vesicles are formed inside multivesicular bodies (MVBs; Colombo, Raposo, & Théry, 2014), which are generally thought to be of late endosomal origin. Although one potential fate is for them to be degraded when the MVB fuses with the lysosome, they can also be released via MVB fusion to the plasma membrane. Human prostate exosomes, or prostasomes, are postulated to affect sperm activity and protect sperm from the female immune system (Aalberts, Stout, & Stoorvogel, 2013; Park et al., 2011; Ronquist, 2015). Since exosomes are increasingly implicated in cancer biology (Yu, Cao, Shen, & Feng, 2015), developing a better understanding of their normal physiological roles and regulation is becoming an important priority in modern cancer research.

2.2 Human Prostate Cancer

Although the prostate epithelium appears to have a relatively low turnover rate, the majority of men over 50 suffer from benign prostatic hyperplasia. The prostate becomes progressively enlarged, potentially leading to lower urinary tract symptoms, affecting urinary retention and frequency (Berry, Coffey, Walsh, & Ewing, 1984). A separate pathology that arises in aging men is adenocarcinoma of the prostate, which is one of the most common cancers affecting older men in the developed world (Siegel, Ma, Zou, & Jemal, 2014). Indeed, it is the second most common cause of male cancer deaths in the United Kingdom (Cancer Research UK, 2012). Many newly diagnosed cases are and will remain indolent, but when more aggressive disease progresses to its metastatic stages, drug treatments that block androgen signaling prove highly effective in most patients in the short term. However, androgen-independent tumor cells and castration-resistant prostate cancer inevitably emerge and ultimately lead to death (Schrijvers, 2007). Distinguishing the 2% of patients who are likely to advance to metastatic disease from those with indolent disease is a major clinical challenge (Attard et al., 2016; Jaiswal, Sarmad, Arora, Dasaraju, & Sarmad, 2015).

Surprisingly, androgen-independent cancer cells typically express the androgen receptor (AR) at high levels and require the receptor for growth, even though androgens themselves are not needed (Lamb, Massie, & Neal, 2014). How can the receptor continue to signal in the absence of its ligand? There are several possible explanations: *AR* gene amplification, which increases AR protein levels (Visakorpi et al., 1995); *AR* point mutations (which are much more frequent in metastatic tumors; Watson et al., 2010; Zhang et al., 2011) that can make the AR constitutively active or lead to altered steroid sensitivity (Duff & McEwan, 2005); androgen production by alternative mechanisms that circumvent drug inhibition (Titus, Schell, Lih, Tomer, & Mohler, 2005); and the influence of other signaling cascades on hormone-independent AR activity (Sharma et al., 2010; Vinall et al., 2011). Gene fusions between the AR-dependent gene *TMPRSS2* (*trans-membrane protease serine 2*) and *ETS* (*E26 transformation-specific*) genes, particularly *ERG* (*ETS-related gene*), are also implicated downstream of AR, but their role is controversial (Clark & Cooper, 2009). Key problems in current prostate cancer research are that prostate cancer cell lines have diverse androgen-dependent and -independent phenotypes, and normal AR-regulated prostate biology is not well understood, making it difficult to determine the initiating mechanisms by which androgen-independence emerges.

2.3 Mouse Models of Prostate Biology and Cancer

The mouse prostate is extremely small, and its lobular structure and epithelial morphology are quite different from the human prostate. It can therefore be difficult to relate tumor phenotypes in this organ to human cancers (Ittmann et al., 2013). Some prostate cancer models have been developed (Wu, Gong, Roy-Burman, Lee, & Culig, 2013), by expression of oncogenes like the SV40 large T antigen under androgen regulation in prostate epithelium (the transgenic adenocarcinoma of the mouse prostate model; Gingrich et al., 1997) or by prostate-specific loss of key tumor suppressor genes like *Pten* (Wang et al., 2003). However, the relevance of these models to human prostate cancer is still not fully established.

Orthotopic, intraosseous, and intracardiac inoculations of human cancer cells as xenografts are well established in mouse (van Weerden & Romijn, 2000), but these types of study have limitations. Not only are these experiments performed in an immunosuppressed host, but they often employ a single cancer cell line with a unique combination of genetic defects.

Cancer studies are frequently informed by the developmental biology of the cells and organs involved. Prostate epithelium-specific, Cre-induced systems for generating mutant cells at early developmental stages are available (Stanfel et al., 2006), and roles for transcription factors SOX9 and Pax2, and for FGF10, Notch, BMP, Wnt, and Hedgehog signaling (Leong & Gao, 2008; Omori et al., 2014; Peng & Joyner, 2015; Powers & Marker, 2013; Schwertfeger, 2009; Simons et al., 2012; Thomsen, Francis, & Swain, 2008; Xu, Hariharan, Rakshit, Dressler, & Wellik, 2012) have all been highlighted. However, an integrated picture of prostate development remains elusive. Even in adult epithelial tissue, the relationship between basal and luminal epithelial cells is not fully characterized (Long et al., 2005). Establishment of other in vivo systems that might be able to model at least some of the developmental, cellular, or even subcellular processes associated with the prostate epithelium could therefore provide valuable new insights into the human prostate in health and disease.

3. THE *DROSOPHILA* ACCESSORY GLAND—A KEY GLAND IN THE MALE REPRODUCTIVE SYSTEM

The male reproductive system in flies, like mammals, also contains a number of secretory structures that contribute to seminal fluid. Most prominent among these are the paired accessory glands that secrete most of the total seminal fluid volume. However, other epithelial cells in the ejaculatory duct and ejaculatory bulb also contribute (Fig. 1B). Each accessory gland is lined by a simple monolayer epithelium containing two cell types with distinct gene expression patterns (Bertram, Akerkar, Ard, Gonzalez, & Wolfner, 1992), about 1000 squamous main cells and roughly 40 more cuboidal secondary cells (Bairati, 1968; Fig. 1D). The epithelium is surrounded by a thin layer of striated muscle (Susic-Jung et al., 2012), which contracts during mating under neural control (Tayler, Pacheco, Hergarden, Murthy, & Anderson, 2012), ejecting the luminal contents of the gland through the ejaculatory duct and into the female's uterus after the transfer of sperm (Bertram, Neubaum, & Wolfner, 1996; Gilchrist & Partridge, 2000).

Male flies induce multiple behavioral changes in females to which they mate (Kubli, 2003; Sirot et al., 2009). Egg-laying rate is dramatically increased for many days, subsequent attempted matings by other males are rejected, the female's immune response, diet, and endocrine system are modulated, and ultimately lifespan is reduced. Mutations inhibiting accessory gland development or targeted ablation of main cells by expression of diphtheria toxin strongly suppress the effects on egg-laying and remating behavior (Kalb, DiBenedetto, & Wolfner, 1993; Xue & Noll, 2000), highlighting the critical function of the accessory gland in reproductive function.

As discussed later, it is only recently that the cell biology of the adult accessory gland epithelium has attracted increased attention. However, its functions in reproductive physiology have been studied for many years (Kubli, 2003; Sirot et al., 2009), revealing central roles in fertility and in providing the molecular weaponry required by males to enhance fecundity and drive sexual conflict with females. Some of the key findings are considered in the following sections, since they uncover a complex network of molecular interactions involving several families of proteins found in seminal fluid of all higher eukaryotes.

3.1 Sex Peptide—An Essential Component of the Accessory Gland Secretome

Extensive studies have shown that one molecule, sex peptide (SP or Acp70A [Accessory gland protein 70A]), a 36-amino acid peptide, which is secreted from main cells, is involved in many characterized aspects of female behavioral reprogramming (Chen et al., 1988; Kubli, 2008). *SP* mutant (or knockdown) males are unable to promote long-term increases in egg laying or reduce female receptivity to remating (Chapman et al., 2003; Chen et al., 1988). Furthermore, they fail to fully induce multiple other postmating changes, including those affecting the innate immune response (Peng, Zipperlen, & Kubli, 2005), hormone and pheromone production (Bontonou, Shaik, Denis, & Wicker-Thomas, 2015; Moshitzky et al., 1996), locomotor activity and sleep (Isaac, Li, Leedale, & Shirras, 2010), food intake (Carvalho, Kapahi, Anderson, & Benzer, 2006; Ribeiro & Dickson, 2010), and excretion (Apger-McGlaughon & Wolfner, 2013; Cognigni, Bailey, & Miguel-Aliaga, 2011).

The long-term postmating activities of SP require the transfer of sperm, the so-called sperm effect (Liu & Kubli, 2003; Manning, 1962). In *Drosophila*, sperm are stored after mating for at least 1 week in two organs, the paired spermathecae and the seminal receptacle (Fig. 1C). SP binds to sperm in females and is released gradually from the sperm storage organs by protease cleavage to mediate its long-term actions (Peng, Chen, et al., 2005). Wolfner's group has identified a set of genes expressed by the accessory gland, the long-term response (LTR) network, that is involved in trafficking SP into the sperm storage organs, presumably by permitting its stable binding to sperm (Fig. 2; Findlay et al., 2014; Ravi Ram & Wolfner, 2009).

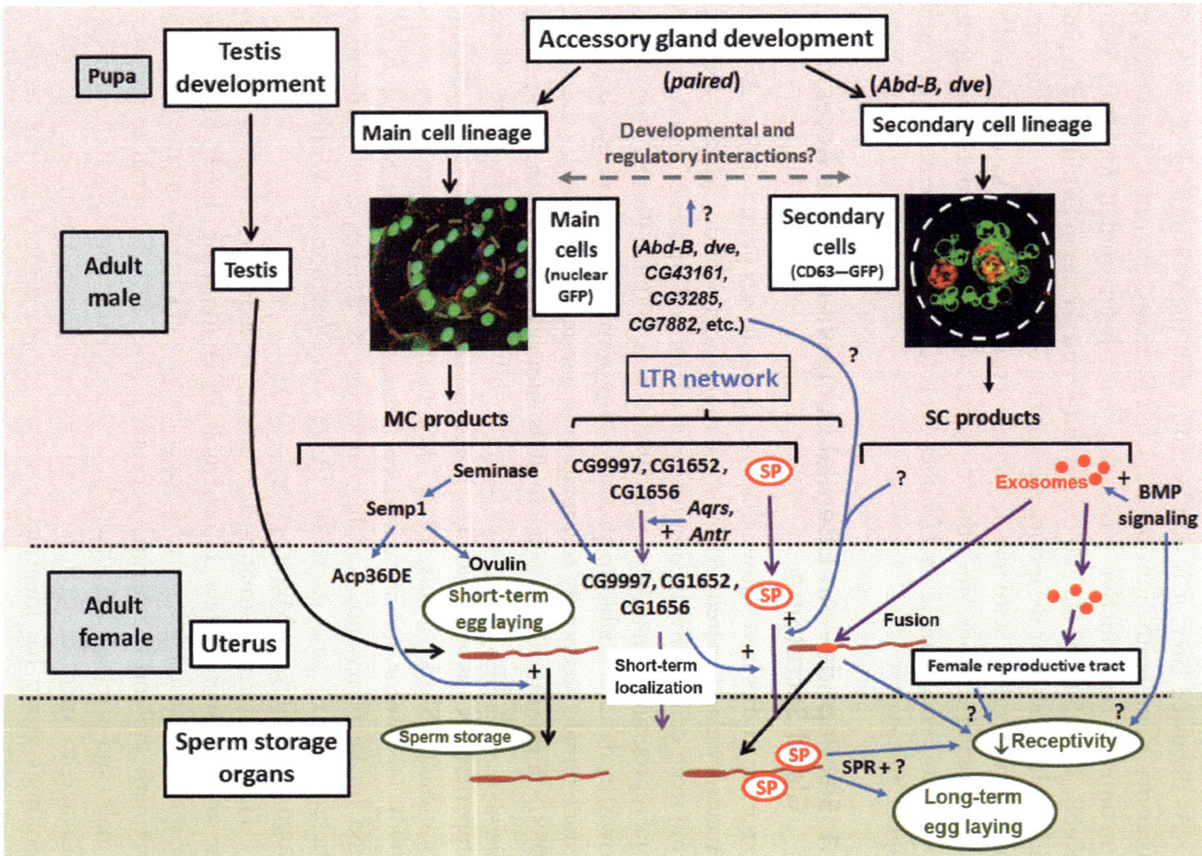

Fig. 2 See legend on opposite page.

Although some of these protein products enter the female storage organs, they do not appear to persist in females like SP. Interestingly, efficient depletion of sperm from storage organs to fertilize the continuous stream of new eggs produced by the ovaries is dependent on proteolytic release of SP from stored sperm (Avila, Ravi Ram, Bloch Qazi, & Wolfner, 2010). Careful analysis of the effects of SP indicates that it optimizes the male's short-term fitness benefits, so that on average, he sires more progeny, at least when mating with young females (Fricke, Green, Mills, & Chapman, 2013; Fricke, Wigby, Hobbs, & Chapman, 2009).

3.2 Sex Peptide Signaling in Females

Experiments in which mutant forms of SP are expressed in *SP* null males or where SP or SP mutant peptides are either injected or expressed ectopically in females have demonstrated that SP alone can induce most of the characterized female postmating responses, but that different parts of the molecule are required for different functions (e.g., Chen et al., 1988; Domanitskaya, Liu, Chen, & Kubli, 2007; Tsuda, Peyre, Asano, & Aigaki, 2015). One major breakthrough in our understanding was the identification of a sex peptide receptor (SPR) expressed in females that is required to mediate many of SP's effects (Yapici, Kim, Ribeiro, & Dickson, 2008). Subsequent work revealed specific SPR-expressing neurons in the female reproductive tract that are necessary and sufficient to induce SP-dependent postmating

Fig. 2 Regulation of female postmating responses by the fly accessory gland. Schematic outlines the complex genetic network associated with the accessory glands that regulates female postmating responses in flies. SP and secondary cell exosomes are marked in *red*, sperm are *brown*. Key molecular and exosome movements are highlighted with *purple arrows*, genetic interactions with *blue arrows*, structural and cellular elements of the reproductive system are in *black boxes*, and a selected group of postmating responses are also represented (*green ellipses*). Any developmental defect in secondary cell biology could have indirect actions on main cells that ultimately lead to effects on SP storage in females. Indeed, even adult-specific inhibition of BMP signaling or exosome secretion could affect main cells. However, the transfer of exosomes into females, their subsequent fusion with sperm and selective effect on only some SP-dependent postmating responses, suggests that these vesicles mediate at least some of their effects after mating. Note that the *iab-6* mutation in *Abd-B* leads to defects in glycosylation of main cell proteins such as ovulin. Acp36DE induces uterine contractions and is involved in anterior mating plug formation. Products of the ejaculatory bulb play a key role in forming the posterior mating plug. All these latter effects are not shown here for simplicity.

responses (Häsemeyer et al., 2009; Yang et al., 2009) and some of the circuitry regulated by these neurons (Rezával, Nojima, Neville, Lin, & Goodwin, 2014; Rezával et al., 2012; reviewed in Feng, Palfreyman, Häsemeyer, Talsma, & Dickson, 2014; Kubli & Bopp, 2012; Walker, Corrales-Carvajal, & Ribeiro, 2015). SP is also thought to bind to cells in other parts of the nervous system by entering the hemolymph (Ding, Haussmann, Ottiger, & Kubli, 2003).

Although the SP/SPR model provides a very neat explanation of the accessory gland's roles in reprogramming long-term female postmating responses, several observations suggest that it cannot represent the whole story. For example, myoinhibitory peptides are alternative ligands for the SPR and are expressed in females, so other aspects of signaling are affected in *SPR* mutant females (Kim et al., 2010). There is also evidence that SP can induce some postmating responses in females that lack *SPR* (Haussmann, Hemani, Wijesekera, Dauwalder, & Soller, 2013), suggesting other SP-dependent signaling mechanisms are involved.

3.3 Other Main Cell-Derived Peptides Have Effects on Female Behavior

Importantly, several other main cell-expressed genes have been implicated in inducing short-term female postmating responses (Fig. 2). For example, stimulation of egg laying in the first day postmating is dependent on the seminal fluid proteins ovulin (Acp26Aa; Herndon & Wolfner, 1995) and CG33943 (Ravi Ram & Wolfner, 2007). Another secreted main cell protein, Acp36DE, promotes sperm storage (Bloch Qazi & Wolfner, 2003) and is involved in inducing a series of conformational changes in the uterus both during and immediately after mating that may guide the movement of sperm (Avila & Wolfner, 2009). These functions may be linked to an additional role for Acp36DE in promoting formation of the anterior part of the mating plug, which contains some Acp36DE protein (Bertram et al., 1996). The posterior part of the mating plug is primarily formed from products of the ejaculatory bulb, including the proteins PMBII and PMBme (Avila, Cohen, et al., 2015; Bretman, Lawniczak, Boone, & Chapman, 2010). The mating plug is required to block immediate remating and retain sperm in the female reproductive tract.

Ovulin and Acp36DE must both be cleaved during mating to exert their functions in females. This cleavage is dependent on a protease cascade involving the proenzymes seminase (CG10586; LaFlamme, Ram, & Wolfner, 2012) and Semp1 (CG11864; Ravi Ram, Sirot, & Wolfner, 2006), which are both synthesized by main cells and processed during mating (Fig. 2). Full cleavage of substrates like ovulin only takes place in the female and may require female factors (LaFlamme, Avila, Michalski, & Wolfner, 2014). In addition, seminase regulates a second pathway independently of Semp1 that is required for SP storage on sperm in females, and therefore plays an important role in the long-term female postmating response (LaFlamme et al., 2012). It also appears to affect the transient localization of several other Acps (CG9997, CG1652, CG1656) in the LTR network to the seminal receptacle.

4. MOLECULAR AND FUNCTIONAL PARALLELS BETWEEN SEMINAL FLUID PROTEINS IN FLIES AND HUMANS

Drosophila genetics has been instrumental in developing a uniquely detailed genetic picture of interacting, main cell-derived molecules secreted into seminal fluid and their reproductive functions. But how relevant to the human prostate or other reproductive glands are the mechanisms involved? One clear similarity between *Drosophila* and human seminal fluids, which have both been subjected to extensive proteomics analysis, is the presence of specific classes of proteins that appear to have important roles in fertility. For example, members of the cysteine-rich secretory protein (Ernesto et al., 2015; Krätzschmar et al., 1996; Ravi Ram & Wolfner, 2009; Udby et al., 2005) and lectin (Garénaux et al., 2015; Ravi Ram & Wolfner, 2009) families are made by the fly accessory gland and prostate epithelium, and associate directly with sperm or interact with proteins that bind to sperm.

Another feature is the remarkable abundance of proteases and protease inhibitors in seminal fluid (Laflamme & Wolfner, 2013). In flies, at least 20% of identified seminal fluid proteins are proteolysis regulators (Findlay, MacCoss, & Swanson, 2009; Findlay, Yi, Maccoss, & Swanson, 2008). The proteases in humans and flies are characterized into several classes including trypsin- and chymotrypsin-like serine proteases (including kallikreins; some proteases secreted by the accessory gland, like CG4815, share greatest sequence similarity with human kallikrein family members), metalloproteases, cysteine proteases, and aspartic proteases.

4.1 Proteases and Their Inhibitors Have Multiple Functions in Seminal Fluid

In humans, some of the proteases secreted by the seminal vesicles promote coagulation of the ejaculate (Lilja, Oldbring, Rannevik, & Laurell, 1987). The seminal clot that forms is then broken down in a process called liquefaction by a cascade of kallikrein-like serine proteases, including PSA. These proteases are secreted by the prostate and activated when levels of Zn^{2+} ions fall as seminal fluid components are mixed (Pampalakis & Sotiropoulou, 2007). Liquefaction releases sperm, potentially increasing fertility.

Several other species, including chimpanzees and mice, use an equivalent coagulation mechanism to form a mating plug, which, like its fly equivalent, appears to be involved in reducing successful second matings and in retaining sperm within the female reproductive tract. Knockout studies of a protease inhibitor expressed primarily in the mouse seminal vesicles suggest that regulated proteolytic activity is required to form this plug properly (Murer et al., 2001). In this regard, the activities of the fly accessory gland potentially mirror those of both the seminal vesicles and the prostate in mammals.

Although the requirement for gland-derived protease activity in seminal fluid is well established in both mammals and flies, the precise functional parallels have yet to be fully assessed. Prostate kallikreins can release proteins from the seminal clot that potentially modulate immunity (Emami & Diamandis, 2010), promote inflammation (Sharkey et al., 2012), and inhibit bacterial growth (Edström et al., 2008). Proteins secreted from the *Drosophila* accessory gland, including SP, also affect female immunity and bacterial resistance (Peng, Zipperlen, et al., 2005; Short, Wolfner, & Lazzaro, 2012); the precise roles of proteases are not yet established, but the activities of multiple proteases are required for SP function. Protease activity is critical for binding of SP to sperm in females (Findlay et al., 2014; LaFlamme et al., 2012), for its subsequent release (Peng, Chen, et al., 2005) and as a result, for the gradual release of sperm from storage (Avila et al., 2010). Sperm storage also takes place to different extents in vertebrates, including humans, who appear to be able to maintain viable sperm for nearly a week. However, the mechanisms involved are poorly characterized (Holt & Fazeli, 2016).

4.2 Rapid Evolution of Seminal Fluid Proteins

One difficulty in comparing the components of seminal fluid in different species, including proteases, is that these molecules often evolve rapidly (Marques et al., 2012; Morrow & Innocenti, 2012; Sirot et al., 2014). This seems to be in part because male interests drive evolutionary changes to maximize offspring and enhance competition with other male mates. The male's interests may differ from those of the female, leading to conflict between the two sexes and potentially positive selection of specific variants, as is seen, for example, in ovulin orthologues found in different *Drosophila* species (Fay & Wu, 2000). Just like ovulin, SP orthologues are found in many *Drosophila* species, but there are no vertebrate equivalents. However, in some mammals, most notably llamas, alpacas, and bulls, seminal fluid contains an ovulation-inducing factor that appears to be equivalent to β-nerve growth factor (Kershaw-Young, Druart, Vaughan, & Maxwell, 2012; Ratto et al., 2012). It could, like SP, function by modulating the female's nervous system. In rabbits, this seminal protein is primarily expressed by the prostate (Maranesi et al., 2015), although its role in this animal as an ovulation-inducing factor is more controversial (Cervantes, Palomino, & Adams, 2015; Silva et al., 2011).

4.3 Males Strategically Allocate Seminal Fluid Proteins in Reproduction

Studies in *Drosophila* not only demonstrate that seminal fluid proteins like SP can enhance fitness benefits (Fricke et al., 2009), but also suggest that males can strategically allocate more SP and ovulin to specific females in the presence of a competitor male (Wigby et al., 2009). Even more remarkably, males can alter the composition of seminal fluid when mating with virgin or mated females. Relative to SP, males transfer less ovulin to previously mated females than they do to virgins (Sirot, Wolfner, & Wigby, 2011). SP will inhibit receptivity in both situations, whereas previously mated females will already be laying eggs, so the benefits of ovulin transfer are reduced. How such changes in seminal protein allocations are achieved, when both molecules are synthesized by the same cells primarily before mating begins, remains unclear. But these studies highlight the complex mechanisms by which seminal fluid content can be regulated and its powerful behavioral reprogramming effects that optimize male fecundity and may drive sexual conflict. Generating such a uniquely potent concoction is likely to require specialized adaptations of epithelial secretory mechanisms.

Recent analysis of epithelial cell biology in the fly accessory gland, particularly focusing on the secondary cells in each gland, has revealed some complex secretory and intercellular signaling mechanisms that share surprising parallels to the human prostate epithelium and are required to induce specific postmating responses (Corrigan et al., 2014; Ito et al., 2014; Leiblich et al., 2012).

5. DEVELOPMENT AND CELLULAR ORGANIZATION OF THE ACCESSORY GLAND

5.1 Early Development of the Accessory Gland

There are surprisingly few studies of accessory gland development in the literature. Ahmad and Baker (2002) reported that the gland is derived from mesodermal cells that in male larvae migrate into the genital imaginal disc. The migration requires signaling by the FGF homologue Branchless, which is expressed in the ectodermal imaginal disc; mesodermal cells, which synthesize the FGF receptor Breathless, move toward this signal. In this regard, the accessory gland epithelial layer might appear to parallel the mesoderm-derived seminal vesicles of mammals, rather than the endoderm-derived prostate epithelium (Thomson & Marker, 2006).

The transcription factor Paired (Prd), the founder member of the Pax family, is also essential for growth and proliferation in the accessory glands; when the embryonic lethality of *prd* mutants is selectively rescued, the resulting adult males are sterile and specifically lack, or have highly reduced, accessory glands (Xue & Noll, 2000, 2002). Interestingly, *prd* is still expressed in adult main cells and secondary cells, at roughly 100-fold higher levels than any other organ (Chintapalli, Wang, & Dow, 2007). This continued expression appears to be required for normal transcription of seminal protein genes, such as *SP* and *ovulin* (Xue & Noll, 2002). As mentioned earlier, specific *FGF* and *Pax* genes are involved in prostate development (Schwertfeger, 2009; Xu et al., 2012), but whether this reflects conserved mechanism or the independent deployment of two common developmental regulators in flies and mammals requires a more detailed analysis in both systems.

In fact, at first sight, the adult epithelium of the fly accessory gland has properties that indicate it would not make a good prostate cancer model. Its component cells are all postmitotic, having completed their last mitotic division about halfway through pupation (Taniguchi et al., 2014). About 5–10 h afterward, the cells go through one more mitosis without cytokinesis, so that adult main cells and secondary cells are all binucleate, a phenotype seen in some human tissues, most notably hepatocytes (Grizzi & Chiriva-Internati, 2007) and surface umbrella cells of the bladder transitional epithelium (White, Masters, & Woolf, 1997), but not the prostate. Very little is known about the mechanisms involved in binucleation. But Taniguchi et al. (2014) have shown that Mud, the fly homologue of the cytoskeletal regulator nuclear mitotic apparatus protein (NuMA), which is upregulated in some cancers (Hasholzner et al., 1999; Xavier de Carvalho et al., 2015), is involved in the accessory gland and can indeed induce binucleation in normally mononucleate *Drosophila* cells. Whether this mechanism is relevant to the binucleate cells observed in some cancers, particularly Hodgkin lymphoma (Farrell & Jarrett, 2011), remains to be tested.

5.2 Developmental Regulation of Secondary Cells

The morphology of adult secondary cells, which contain many large intracellular compartments involved in secretion (Fig. 3; Corrigan et al., 2014; Rylett, Walker, Howell, Shirras, & Isaac, 2007), is very different from main cells (Bairati, 1968). Recent studies have identified at least two transcriptional regulators involved in secondary cell formation, differentiation, and/or survival. The homeodomain transcriptional repressor Defective proventriculus (Dve) is expressed at high levels in secondary cells by mid-pupation, and only weakly in main cells (Minami et al., 2012). This strong secondary cell expression persists in adults. Loss of *dve* function reduces the number of secondary cells in the accessory gland, a phenotype that seems to involve some cell death (Minami et al., 2012). Those secondary cells that remain are small and lack large secretory compartments. *dve* also appears to play a role in main cell binucleation.

The *Hox* gene *Abd-B* is expressed specifically in secondary cells within the adult accessory gland, and mutations affecting the *iab-6* enhancer in this gene, which can independently drive secondary cell-specific reporter gene expression, lead to loss of this cell's characteristic secretory morphology (Gligorov, Sitnik, Maeda, Wolfner, & Karch, 2013). Importantly, *dve* and *Abd-B* are both required to induce long-term egg laying in mated females and to suppress receptivity to other males (Gligorov et al., 2013; Minami et al., 2012). Therefore, secondary cells appear to provide factors essential for long-term female postmating responses that function together with SP and other main cell products (Fig. 2). Females that mate with *iab-6* mutant males fail to store SP normally in their seminal receptacle (Gligorov et al., 2013). Furthermore, several main cell-derived Acps, including ovulin, are not normally glycosylated in the accessory gland. Whether this reflects an indirect effect of secondary cells on the glycosylation capacity of main cells, a secondary cell-specific modification of Acps in the accessory gland lumen, or an ability of secondary cells to take up main cell products and modify them remains unclear.

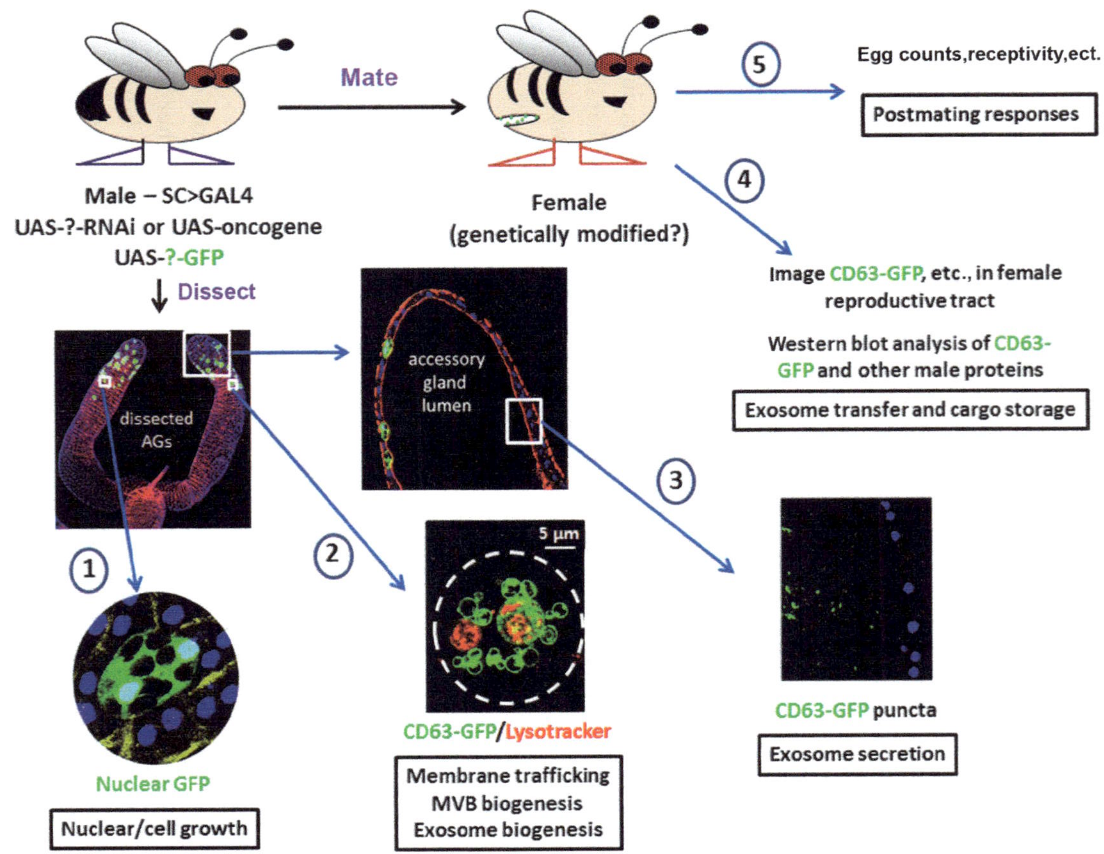

Fig. 3 See legend on opposite page.

5.3 *Abd-B*-Regulated Genes Control Secondary Cell Functions

A recent transcriptomics screen of genes expressed by the accessory glands of wild-type and *iab-6* mutant males identified 77 genes that are downregulated by at least fivefold and 115 genes that are at least fivefold upregulated (Sitnik, Gligorov, Maeda, Karch, & Wolfner, 2016). Interestingly, some genes like *CG11598*, which are expressed at high levels in both main cells and secondary cells, are strongly downregulated in mutant glands, while other genes encoding secreted proteins are highly upregulated. This is consistent with the idea that in development and/or in adults, there are major regulatory interactions between secondary cells and main cells (Fig. 2). Knockdown in secondary cells of some of the genes requiring *Abd-B* for their expression (*CG7882*, *CG9509*, and *CG14069*), using an *iab-6*-GAL4 driver, produces cellular phenotypes where large secretory compartments are absent or small, and also leads to a suppression of postmating responses (egg laying and receptivity changes) and reduced SP storage in mated females (Sitnik et al., 2016). However, knockdown of other genes that are downregulated in *Abd-B* mutant glands (e.g., *CG43161*, *CG3285*) affects these same postmating responses, but does not produce obvious cellular phenotypes.

All these studies highlight important functions for secondary cells in the induction of postmating responses and SP storage. However, because they employ knockdown throughout accessory gland development, they could be explained by defects that are generated during pupation, potentially altering the development of secondary cells, and indirectly affecting the development of main cells or other cells. Experiments focused only on the adult have revealed more specific roles for these cells in postmating responses and some prostate-like cell biological properties.

Fig. 3 The secondary cell as a versatile in vivo model to study endolysosomal trafficking, exosome biogenesis, and functions. Diagram illustrates the different approaches that can be employed to study processes associated with exosome biogenesis and functions using the fly accessory glands, ranging from analysis of growth (1) and membrane trafficking/vesicle formation (2) in SCs, to exosome secretion into the gland lumen (3), to analysis of exosome transfer (4), and effects on postmating responses (5) in females. The living secondary cell shown in (2) is marked with CD63-GFP and stained with Lysotracker Red to reveal large acidic compartments, which, as in this case, are typically greater than 5 μm in diameter in older males. Note that the male and female can both be genetically modified in different ways to affect exosome-secreting and target cells, respectively. Exosomes and the compartments that produce them can also be imaged by more standard approaches, such as transmission electron microscopy (Corrigan et al., 2014).

6. FUNCTIONS OF THE ADULT SECONDARY CELL

6.1 Aging Adult Secondary Cells Continue to Grow

Morphological analysis of the aging adult accessory gland reveals that unlike main cells, which become more squamous as the glandular lumen expands, secondary cells enlarge in nuclear and total cell volume (Leiblich et al., 2012). Although nuclei in some cell types enlarge by endoduplication (Zielke, Edgar, & DePamphilis, 2013), secondary cell nuclei do not appear to replicate their DNA as they grow, mirroring several other cell types in flies that alter their nuclear size in response to growth signals in the absence of DNA synthesis (Gao & Pan, 2001). Interestingly, secondary cell growth is affected by sexual activity, accelerating in multiply-mated males.

6.2 BMP Signaling Controls Adult Secondary Cell Growth and Migration

Cell type-specific expression is most commonly achieved in *Drosophila* using the GAL4/UAS modular misexpression system (Brand & Perrimon, 1993). By employing a ubiquitously expressed, temperature-sensitive form of the GAL4 antagonist, GAL80, GAL4-induced expression can be activated only in adults by shifting the culture temperature to 29°C at eclosion (McGuire, Mao, & Davis, 2004). Using this approach, Leiblich et al. (2012) demonstrated that adult secondary cell growth requires BMP signaling. Blocking BMP signaling in these cells has no effect on female egg laying, but does suppress female receptivity to remating, indicating that these two processes, which both normally require stored SP, are differentially controlled by adult male-specific, secondary cell-dependent factors (Fig. 2). One alternative explanation is that other seminal proteins like ovulin act independently of SP to promote long-term effects on ovulation. However, this seems relatively unlikely, since such a mechanism could only work in this specific scenario, because these factors cannot induce long-term ovulation in the absence of SP or when SCs fail to develop normally (Gligorov et al., 2013; Minami et al., 2012).

Surprisingly, if aging males mate, a small subset of secondary cells also sporadically delaminates from the apical surface of the epithelium in a BMP-dependent fashion (Leiblich et al., 2012). These cells migrate along the epithelium to the proximal end of the gland and are transferred to females upon mating. This transfer is not essential for normal fertility, because it does not occur in young males or following every mating in older males. One possibility is that in those females that receive secondary cells, the cells continue to secrete into the uterus prior to egg laying and increase the potency of the aging ejaculate. Alternatively, these cells may be a by-product of the BMP-mediated events required to suppress female receptivity, which occur in secondary cells that remain embedded in the glandular epithelium.

6.3 BMP Signaling in the Human Prostate

BMP signaling is involved in prostate growth and development (Omori et al., 2014). It has growth-inhibitory (Ding et al., 2011) and -stimulatory (Lee et al., 2011) effects in prostate cancer and has been implicated in promoting hormone resistance in bone metastases (Lee et al., 2014). Prostate epithelial cells are also found in seminal fluid (Barren et al., 1998), though it has been assumed they slough off and are not the product of an active delamination process. Ito et al. (2014) have used the secondary cell system to screen for genes that regulate migration and shown that identified candidate genes, which are evolutionarily conserved, also affect invasive properties of prostate cancer cells, although not in the way that the screen might have predicted. To better understand the roles of BMP signaling in secondary cells, Corrigan et al. (2014) undertook a more detailed cell biological analysis to characterize cell biological defects associated with altered signaling.

6.4 Secretion by Adult Secondary Cells

There are about 15 large ($\geq$3 μm in diameter) intracellular compartments in 3-day-old adult secondary cells, most of which contain dense-core granules (DCGs; Rylett et al., 2007). Such granules are present in many higher eukaryotic cells involved in regulated secretion, like pancreatic beta-cells, and store bioactive small molecules, peptides, and proteins, as well as proteases that can cleave proteins into active forms (Kim, Gondré-Lewis, Arnaoutova, & Loh, 2006). The bioactive molecules in secondary cell DCGs are not known. However, the fly homologue of angiotensin-I-converting enzyme, which is also synthesized by prostate epithelial cells and appears to be linked to prostate cancer susceptibility (Nassis et al., 2001; Xie, You, & Chen, 2014), is present in these granules (Rylett et al., 2007).

6.5 Secondary Cells Secrete Exosomes That Inhibit Female Receptivity

Using live confocal imaging with different YFP-tagged Rab GTPases, which are conserved markers of specific membrane-bound compartments in all eukaryotic cells (Stenmark, 2009), Corrigan et al. (2014) demonstrated that the DCG-containing compartments are marked by the recycling endosome marker Rab11. However, about three other large compartments have acidic lumens and are marked by Rab7, identifying them as late endosomes or lysosomes (Fig. 3). When a GFP-tagged form of the human transmembrane exosome marker CD63 is expressed in these cells, almost all of the large compartments are marked, and GFP puncta are observed inside the large acidic compartments and in the lumen of the gland. Transmission electron microscopy confirms the presence of vesicles in these compartments and the lumen, indicating that the GFP-labeled structures are exosomes. Luminal puncta can be counted to measure levels of exosome secretion (Fig. 3).

The CD63-GFP-positive exosomes are transferred to females upon mating, where they fuse with sperm and appear to interact with the lining of the female reproductive tract (Corrigan et al., 2014). Multiple genetic manipulations that knock down conserved genes involved in human exosome biogenesis, including members of the ESCRT (endosomal sorting complexes required for transport) family, which control formation of intraluminal vesicles in MVBs, and Rabs, which regulate endosomal compartment trafficking to the cell surface (Colombo et al., 2014), suppress exosome secretion from adult secondary cells. Remarkably, these manipulations also prevent males from fully inhibiting female receptivity after mating, but do not affect the induction of egg laying, potentially highlighting a selective role in modifying specific female postmating behaviors.

Further analysis revealed that BMP signaling is absolutely required for exosome secretion and controls membrane trafficking through the endolysosomal system, suggesting that its effects on female receptivity may be mediated by exosomes (Corrigan et al., 2014). Since the long-term effects of mating on egg laying are retained in the absence of exosomes, SP is likely to still bind to stored sperm. Hence, these findings indicate that BMP-dependent exosomes are either involved in regulating a specific aspect of SP function or that they act in a parallel pathway that selectively influences female receptivity (Fig. 2). Interestingly, by expressing SP in specific neurons of wild-type and *SPR* mutant females, Haussmann et al. (2013) have also provided evidence that egg laying and receptivity are not regulated by identical neural circuits or signaling cascades, supporting the hypothesis that SP mediates its actions via more than one downstream pathway.

6.6 Secondary Cell Exosomes and Prostasomes in Fertility and Sexual Conflict

Prostasomes from human prostate epithelial cells also target sperm, at least in vitro, and have been reported to deliver several molecules that promote motility and capacitation (Park et al., 2011), although this remains controversial (Pons-Rejraji et al., 2011). They may also coat the sperm surface and modulate the female immune response to sperm (Ronquist, 2015). In flies, the absence of secondary cell exosomes does not significantly affect the number of offspring produced by sperm, but does suppress a postmating behavioral effect that involves interaction between sperm and female tissues, namely, female receptivity. Whether these exosomes fuse with and directly reprogramme female cells in addition to sperm remains unclear. However, such a mechanism would provide a novel weapon to drive sexual conflict by potentially delivering intracellular signaling components, miRNAs, and membrane-bound receptors to overcome cellular processes that favor female interests. Several studies (reviewed in Sirot et al., 2009) reveal a complex regulation of female responses after mating, not just affecting behaviors and metabolism, but also modulating the female reproductive tract directly. The latter changes involve multiple male signals and include release of neurotransmitters from female nerve termini (e.g., Heifetz, Lindner, Garini, & Wolfner, 2014) and activation of signaling in sperm storage organs (Avila, Mattei, & Wolfner, 2015). It will be interesting to test which, if any, of these responses are exosome dependent.

Progesterone receptors and Ca^{2+} signaling machinery have been shown to be transferred to human sperm by prostasomes (Park et al., 2011), providing a means of rapidly upregulating specific signaling cascades in sperm after mating. However, the active molecules in secondary cell exosomes are yet to be identified. Males of another *Drosophila* species, *Drosophila mojavensis*, transfer mRNAs from the accessory gland to females (Bono, Matzkin, Kelleher, & Markow, 2011), though their cellular origin is unknown. The levels of RNA or proteins transferred by exosomes are almost certainly low, and probably only the most abundant will be detectable by standard transcriptomics or proteomics techniques unless large quantities of exosomes can be isolated. A major debate in the exosome field is how exosomes can deliver sufficiently high levels of bioactive molecules to alter cell behavior under physiological conditions (e.g., Chevillet et al., 2014). The genetic tractability of the fly system should allow this conundrum to be resolved, and provide the opportunity to independently manipulate exosome-secreting and potential target cells (in males and females, respectively) to dissect out the mechanisms involved (Fig. 3).

7. MODELING PROSTATE CANCER BIOLOGY IN SECONDARY CELLS

7.1 Studying Exosome Regulation and Functions in Flies

A flurry of recent reports has implicated exosomes and other secreted extracellular vesicles in multiple aspects of cancer biology, including modulation of the tumor microenvironment, transfer of drug resistance and malignant cell properties, immunosuppression, and priming and establishment of premetastatic sites (e.g., Costa-Silva et al., 2015; Hoshino et al., 2015; Ji et al., 2015; Peinado et al., 2012; Zhang et al., 2015; Zhou et al., 2014; Zomer et al., 2015). If exosomes from human male reproductive glands are involved in modulating female physiology and cell function, as they are in flies, it is perhaps not surprising that when they are released at high levels into the male's circulation by prostate cancer cells, which have lost their polarity, they might mediate multiple cancer-promoting effects on some target cells.

As discussed earlier, seminal fluid proteins with roles in reproduction are not highly evolutionarily conserved, but the cellular mechanisms controlling exosome biogenesis and secretion are (Corrigan et al., 2014). Perhaps the greatest challenge in current exosome research is to understand the detailed cell biology of exosome biogenesis. Both ESCRT-dependent (Baietti et al., 2012) and -independent (Stuffers, Sem Wegner, Stenmark, & Brech, 2009; Trajkovic et al., 2008) mechanisms have been proposed, and a number of different Rab GTPases appear to control the process of secretion either coordinately or independently in different cell types (Colombo et al., 2014). Intraluminal vesicles and the compartments in which they are formed are typically at the limit of fluorescence microscopy resolution, so most studies rely on electron microscopy to visualize these structures. Alternatively the compartments are artificially enlarged by genetic manipulation with a constitutively activated form of the early endosomal regulator Rab5 (Baietti et al., 2012; Trajkovic et al., 2008), a treatment that inevitably alters the identity of MVBs.

The substructure of multivesicular endosomes and lysosomes, and the generation and dynamics of intraluminal vesicles can be visualized in real time by confocal microscopy in secondary cells (Corrigan et al., 2014), overcoming this hurdle. This system should allow some of the key questions in exosome biology to be addressed: For example, are intraluminal vesicles loaded with different cargos in the same or different compartments? Do these vesicles traffic differently out of the cell? What distinguishes vesicles targeted for degradation in the lysosome from those destined to be secreted? Answers to all these questions will not only be relevant to prostate cancer but also to the biology of many other cancers, where exosome biogenesis mechanisms are poorly characterized.

7.2 Signaling and Exosome Biogenesis

The discovery that BMP signaling plays such a critical role in exosome secretion from *Drosophila* SCs already highlights a strong connection between intracellular signaling and its control of intercellular exosome-mediated communication. With defective intracellular signaling playing such a key role in cancer biology, dissecting out links of this kind is likely to inform our understanding of how tumors modulate their microenvironment. It may also indicate new ways of extending the use of exosomes for cancer diagnostics (e.g., Melo et al., 2015), employing emerging high-throughput techniques (He, Crow, Roth, Zeng, & Godwin, 2014) to detect specific signaling signatures or responses to drugs. Indeed, the complex roles of BMP signaling in prostate cancer (Ding et al., 2011; Lee et al., 2011, 2014) need to be reevaluated in the context of its possible effects on exosome regulation.

7.3 Growth and Exosome Biogenesis

The enlarged late endosomes and lysosomes of secondary cells also illustrate the importance of endocytic trafficking in cells adapted to high level exosome secretion (Corrigan et al., 2014). These same compartments are required for amino acid-dependent activation of the growth and metabolic regulator, mechanistic target of rapamycin complex 1 (mTORC1; Bar-Peled & Sabatini, 2014; Goberdhan, 2010; Goberdhan, Wilson, & Harris, 2016), which plays a central role in cancer. This may provide an explanation for the observation that cancer cells, in which mTORC1 signaling is frequently upregulated, often secrete increased numbers of exosomes. And since mTORC1 signaling itself controls endolysosomal trafficking (Kim et al., 2015; Peña-Llopis & Brugarolas, 2011), it reveals a complex link between growth signaling and exosome secretion that requires further analysis.

7.4 Steroid Signaling in the Male Reproductive System

One critical aspect of prostate biology that cannot be fully modeled in fly accessory glands is androgen signaling. However, the major steroid hormone in *Drosophila*, ecdysone, which has primarily been characterized as a regulator of developmental transitions (reviewed in Yamanaka, Rewitz, & O'Connor, 2013), has more recently been shown to be regulated by the socio-sexual environment of adult males (Ganter et al., 2011; Ishimoto, Sakai, & Kitamoto, 2009; Schwedes & Carney, 2012). It also affects somatic and germline stem cell maintenance in the testis (Li, Ma, Cherry, & Matunis, 2014; Qian et al., 2014) and accessory gland secretion (Wolfner et al., 1997). Since male flies can alter the levels of different seminal fluid proteins in response to the flies around them (Fricke et al., 2009; Wigby et al., 2009), it is appealing to speculate that ecdysone signaling might provide a link between a male's environment and accessory gland secretion. In fact, Hentze et al. (2013) have suggested that secondary cells might act as a source of ecdysone. Studying both ecdysone synthesis and the possible roles of the ecdysone receptor, which shares broad structural similarity to the AR, in the accessory gland are important goals in the immediate future.

7.5 The Secondary Cell as a General Model for Exosome Biology

Analysis of exosome biogenesis in the fly system is likely to also inform our understanding of exosome control in other normal and cancer cell types. Even the idea that exosomes are secreted to reprogramme another individual is not unique to male reproductive glands. Exosomes are abundant in breast milk and are proposed to play roles in modulating immunity in newborn babies (Melnik, John, & Schmitz, 2014). Furthermore, visualizing exosome biogenesis in living cells and observing the specific changes that take place when these cells are genetically manipulated will also reveal whether these vesicles play other roles inside cells and how these coordinate with their extracellular signaling functions. Therefore, despite the diminutive size of the fly and its accessory gland, the 40 secondary cells and their enlarged vesicle-containing compartments offer a new way to address some of the most challenging questions in cancer biology and potentially a direct link to prostate subcellular biology.

ACKNOWLEDGMENTS

We apologize to those authors whose articles we were not able to cite because of space limitations. We thank Sumeth Perera for helpful comments on the manuscript. The authors gratefully acknowledge the support of the BBSRC (BB/K017462/1, BB/L007096/1), Cancer Research UK (C191591/A6181, C19591/A9093, C7713/A6174, C19591/A19076), C38302/A12278 grants through the Cancer Research UK Oxford Centre Development Fund, and the John Fell Fund, Oxford, as well as studentships and scholarships from the Wellcome Trust, CRUK, Urology Foundation, and the MRC in developing their research in this area.

REFERENCES

Aalberts, M., Stout, T. A., & Stoorvogel, W. (2013). Prostasomes: Extracellular vesicles from the prostate. *Reproduction*, *147*, R1–R14.

Ahmad, S. M., & Baker, B. S. (2002). Sex-specific deployment of FGF signaling in *Drosophila* recruits mesodermal cells into the male genital imaginal disc. *Cell*, *109*, 651–661.

Apger-McGlaughon, J., & Wolfner, M. F. (2013). Post-mating change in excretion by mated *Drosophila melanogaster* females is a long-term response that depends on sex peptide and sperm. *Journal of Insect Physiology*, *59*, 1024–1030.

Attard, G., Parker, C., Eeles, R. A., Schröder, F., Tomlins, S. A., Tannock, I., et al. (2016). Prostate cancer. *Lancet*, *387*, 70–82. pii, S0140-6736(14)61947-4.

Aumüller, G., & Riva, A. (1992). Morphology and functions of the human seminal vesicle. *Andrologia*, *24*, 183–196.

Avila, F. W., Cohen, A. B., Ameerudeen, F. S., Duneau, D., Suresh, S., Mattei, A. L., et al. (2015a). Retention of ejaculate by *Drosophila melanogaster* females requires the male-derived mating plug protein PEBme. *Genetics*, *200*, 1171–1179.

Avila, F. W., Mattei, A. L., & Wolfner, M. F. (2015b). Sex peptide receptor is required for the release of stored sperm by mated *Drosophila melanogaster* females. *Journal of Insect Physiology*, *76*, 1–6.

Avila, F. W., Ravi Ram, K., Bloch Qazi, M. C., & Wolfner, M. F. (2010). Sex peptide is required for the efficient release of stored sperm in mated *Drosophila* females. *Genetics*, *186*, 595–600.

Avila, F. W., & Wolfner, M. F. (2009). Acp36DE is required for uterine conformational changes in mated *Drosophila* females. *Proceedings of the National Academy of Sciences of the United States of America*, *106*, 15796–15800.

Baietti, M. F., Zhang, Z., Mortier, E., Melchior, A., Degeest, G., Geeraerts, A., et al. (2012). Syndecan-syntenin-ALIX regulates the biogenesis of exosomes. *Nature Cell Biology*, *14*, 677–685.

Bairati, A. (1968). Structure and ultrastructure of the male reproductive system in *Drosophila melanogaster*: The genital duct and accessory glands. *Monitore Zoologico Italiano*, *2*, 105–182.

Bar-Peled, L., & Sabatini, D. M. (2014). Regulation of mTORC1 by amino acids. *Trends in Cell Biology*, *24*, 400–406.

Barren, R. J., 3rd., Holmes, E. H., Boynton, A. L., Gregorakis, A., Elgamal, A. A., Cobb, O. E., et al. (1998). Method for identifying prostate cells in semen using flow cytometry. *Prostate*, *36*, 181–188.

Bell, G. P., & Thompson, B. J. (2014). Colorectal cancer progression: Lessons from *Drosophila*? *Seminars in Cell & Developmental Biology*, *28*, 70–77.

Berry, S. J., Coffey, D. S., Walsh, P. C., & Ewing, L. L. (1984). The development of human benign prostatic hyperplasia with age. *The Journal of Urology*, *132*, 474–479.

Bertram, M. J., Akerkar, G. A., Ard, R. L., Gonzalez, C., & Wolfner, M. F. (1992). Cell type-specific gene expression in the *Drosophila melanogaster* male accessory gland. *Mechanisms of Development*, *38*, 33–40.

Bertram, M. J., Neubaum, D. M., & Wolfner, M. F. (1996). Localization of the *Drosophila* male accessory gland protein Acp36DE in the mated female suggests a role in sperm storage. *Insect Biochemistry and Molecular Biology*, *26*, 971–980.

Bloch Qazi, M. C., & Wolfner, M. F. (2003). An early role for the *Drosophila melanogaster* male seminal protein Acp36DE in female sperm storage. *The Journal of Experimental Biology*, *206*, 3521–3528.

Bono, J. M., Matzkin, L. M., Kelleher, E. S., & Markow, T. A. (2011). Postmating transcriptional changes in reproductive tracts of con- and heterospecifically mated *Drosophila mojavensis* females. *Proceedings of the National Academy of Sciences of the United States of America*, *108*, 7878–7883.
Bontonou, G., Shaik, H. A., Denis, B., & Wicker-Thomas, C. (2015). Acp70A regulates *Drosophila* pheromones through juvenile hormone induction. *Insect Biochemistry and Molecular Biology*, *56*, 36–49.
Brand, A. H., & Perrimon, N. (1993). Targeted gene expression as a means of altering cell fates and generating dominant phenotypes. *Development*, *118*, 401–415.
Bretman, A., Lawniczak, M. K., Boone, J., & Chapman, T. (2010). A mating plug protein reduces early female remating in *Drosophila melanogaster*. *Journal of Insect Physiology*, *56*, 107–113.
Bromfield, J. J., Schjenken, J. E., Chin, P. Y., Care, A. S., Jasper, M. J., & Robertson, S. A. (2014). Maternal tract factors contribute to paternal seminal fluid impact on metabolic phenotype in offspring. *Proceedings of the National Academy of Sciences of the United States of America*, *111*, 2200–2205.
Burgess, J., Del Bel, L. M., Ma, C. I., Barylko, B., Polevoy, G., Rollins, J., et al. (2012). Type II phosphatidylinositol 4-kinase regulates trafficking of secretory granule proteins in *Drosophila*. *Development*, *139*, 3040–3050.
Burgess, J., Jauregui, M., Tan, J., Rollins, J., Lallet, S., Leventis, P. A., et al. (2011). AP-1 and clathrin are essential for secretory granule biogenesis in *Drosophila*. *Molecular Biology of the Cell*, *22*, 2094–2105.
Cabrero, P., Terhzaz, S., Romero, M. F., Davies, S. A., Blumenthal, E. M., & Dow, J. A. (2014). Chloride channels in stellate cells are essential for uniquely high secretion rates in neuropeptide-stimulated *Drosophila* diuresis. *Proceedings of the National Academy of Sciences of the United States of America*, *111*, 14301–14306.
Cancer Research UK. (2012). *Cancer statistics*. http://www.cancerresearchuk.org/content/cancer-statistics-for-the-uk.
Carvalho, G. B., Kapahi, P., Anderson, D. J., & Benzer, S. (2006). Allocrine modulation of feeding behavior by the sex peptide of *Drosophila*. *Current Biology*, *16*, 692–696.
Cervantes, M. P., Palomino, J. M., & Adams, G. P. (2015). *In vivo* imaging in the rabbit as a model for the study of ovulation-inducing factors. *Laboratory Animals*, *49*, 1–9.
Chapman, T., Bangham, J., Vinti, G., Seifried, B., Lung, O., Wolfner, M. F., et al. (2003). The sex peptide of *Drosophila melanogaster*: Female post-mating responses analyzed by using RNA interference. *Proceedings of the National Academy of Sciences of the United States of America*, *100*, 9923–9928.
Chen, P. S., Stumm-Zollinger, E., Aigaki, T., Balmer, J., Bienz, M., & Böhlen, P. (1988). A male accessory gland peptide that regulates reproductive behavior of female *D. melanogaster*. *Cell*, *54*, 291–298.
Chevillet, J. R., Kang, Q., Ruf, I. K., Briggs, H. A., Vojtech, L. N., Hughes, S. M., et al. (2014). Quantitative and stoichiometric analysis of the microRNA content of exosomes. *Proceedings of the National Academy of Sciences of the United States of America*, *111*, 14888–14893.
Chintapalli, V. R., Wang, J., & Dow, J. A. (2007). Using FlyAtlas to identify better *Drosophila melanogaster* models of human disease. *Nature Genetics*, *39*, 715–720.
Clark, J. P., & Cooper, C. S. (2009). ETS gene fusions in prostate cancer. *Nature Reviews. Urology*, *6*, 429–439.
Cognigni, P., Bailey, A. P., & Miguel-Aliaga, I. (2011). Enteric neurons and systemic signals couple nutritional and reproductive status with intestinal homeostasis. *Cell Metabolism*, *13*, 92–104.
Colombo, M., Raposo, G., & Théry, C. (2014). Biogenesis, secretion, and intercellular interactions of exosomes and other extracellular vesicles. *Annual Review of Cell and Developmental Biology*, *30*, 255–289.

Corrigan, L., Redhai, S., Leiblich, A., Fan, S.-J., Perera, S. M. W., Patel, R., et al. (2014). BMP-regulated exosomes from *Drosophila* male reproductive glands reprogramme female behaviour. *The Journal of Cell Biology*, *206*, 671–688.

Costa-Silva, B., Aiello, N. M., Ocean, A. J., Singh, S., Zhang, H., Thakur, B. K., et al. (2015). Pancreatic cancer exosomes initiate pre-metastatic niche formation in the liver. *Nature Cell Biology*, *17*, 816–826.

Das, T. K., & Cagan, R. L. (2013). A *Drosophila* approach to thyroid cancer therapeutics. *Drug Discovery Today. Technologies*, *10*, e65–e71.

Declan, J. P., Cahill, D. J. P., Chandra, A., & Davies, C. (2009). Bladder, prostate and urethra. In S. Standring & J. Healy (Eds.), *Gray's anatomy* (pp. 2280–2311). Amsterdam: Elsevier.

Ding, L., Getz, G., Wheeler, D. A., Mardis, E. R., McLellan, M. D., Cibulskis, K., et al. (2008). Somatic mutations affect key pathways in lung adenocarcinoma. *Nature*, *455*, 1069–1075.

Ding, Z., Haussmann, I., Ottiger, M., & Kubli, E. (2003). Sex-peptides bind to two molecularly different targets in *Drosophila melanogaster* females. *Journal of Neurobiology*, *55*, 372–384.

Ding, Z., Wu, C. J., Chu, G. C., Xiao, Y., Ho, D., Zhang, J., et al. (2011). SMAD4-dependent barrier constrains prostate cancer growth and metastatic progression. *Nature*, *470*, 269–273.

Domanitskaya, E. V., Liu, H., Chen, S., & Kubli, E. (2007). The hydroxyproline motif of male sex peptide elicits the innate immune response in *Drosophila* females. *The FEBS Journal*, *274*, 5659–5668.

Duff, J., & McEwan, I. J. (2005). Mutation of histidine 874 in the androgen receptor ligand-binding domain leads to promiscuous ligand activation and altered p160 coactivator interactions. *Molecular Endocrinology*, *19*, 2943–2954.

Edgren, G., Adami, H. O., Weiderpass, E., & Nyrén, O. (2013). A global assessment of the oesophageal adenocarcinoma epidemic. *Gut*, *62*, 1406–1414.

Edström, A. M., Malm, J., Frohm, B., Martellini, J. A., Giwercman, A., Mörgelin, M., et al. (2008). The major bactericidal activity of human seminal plasma is zinc-dependent and derived from fragmentation of the semenogelins. *Journal of Immunology*, *181*, 3413–3421.

Emami, N., & Diamandis, E. P. (2010). Potential role of multiple members of the kallikrein-related peptidase family of serine proteases in activating latent TGF beta 1 in semen. *Biological Chemistry*, *391*, 85–95.

Ernesto, J. I., Weigel Muñoz, M., Battistone, M. A., Vasen, G., Martínez-López, P., Orta, G., et al. (2015). CRISP1 as a novel CatSper regulator that modulates sperm motility and orientation during fertilization. *The Journal of Cell Biology*, *210*, 1213–1224.

Farrell, K., & Jarrett, R. F. (2011). The molecular pathogenesis of Hodgkin lymphoma. *Histopathology*, *58*, 15–25.

Fay, J. C., & Wu, C. I. (2000). Hitchhiking under positive Darwinian selection. *Genetics*, *155*, 1405–1413.

Feng, K., Palfreyman, M. T., Häsemeyer, M., Talsma, A., & Dickson, B. J. (2014). Ascending SAG neurons control sexual receptivity of *Drosophila* females. *Neuron*, *83*, 135–148.

Findlay, G. D., MacCoss, M. J., & Swanson, W. J. (2009). Proteomic discovery of previously unannotated, rapidly evolving seminal fluid genes in *Drosophila*. *Genome Research*, *19*, 886–896.

Findlay, G. D., Sitnik, J. L., Wang, W., Aquadro, C. F., Clark, N. L., & Wolfner, M. F. (2014). Evolutionary rate covariation identifies new members of a protein network required for *Drosophila melanogaster* female post-mating responses. *PLoS Genetics*, *10*, e1004108.

Findlay, G. D., Yi, X., Maccoss, M. J., & Swanson, W. J. (2008). Proteomics reveals novel *Drosophila* seminal fluid proteins transferred at mating. *PLoS Biology*, *6*, e178.

Fraser, L. R. (1997). Fertilisation promoting peptide: A key player in male fertility/subfertility? *Expert Opinion on Investigational Drugs*, *6*, 1797–1801.
Fricke, C., Green, D., Mills, W. E., & Chapman, T. (2013). Age-dependent female responses to a male ejaculate signal alter demographic opportunities for selection. *Proceedings of the Biological Sciences*, *280*, 20130428.
Fricke, C., Wigby, S., Hobbs, R., & Chapman, T. (2009). The benefits of male ejaculate sex peptide transfer in *Drosophila melanogaster*. *Journal of Evolutionary Biology*, *22*, 275–286.
Ganter, G. K., Panaitiu, A. E., Desilets, J. B., Davis-Heim, J. A., Fisher, E. A., Tan, L. C., et al. (2011). *Drosophila* male courtship behavior is modulated by ecdysteroids. *Journal of Insect Physiology*, *57*, 1179–1184.
Gao, X., & Pan, D. (2001). TSC1 and TSC2 tumor suppressors antagonize insulin signaling in cell growth. *Genes & Development*, *15*, 1383–1392.
Garénaux, E., Kanagawa, M., Tsuchiyama, T., Hori, K., Kanazawa, T., Goshima, A., et al. (2015). Discovery, primary, and crystal structures and capacitation-related properties of a prostate-derived heparin-binding protein WGA16 from boar sperm. *The Journal of Biological Chemistry*, *290*, 5484–5501.
Gilchrist, A. S., & Partridge, L. (2000). Why it is difficult to model sperm displacement in *Drosophila melanogaster*: The relation between sperm transfer and copulation duration. *Evolution*, *54*, 534–542.
Gingrich, J. R., Barrios, R. J., Kattan, M. W., Nahm, H. S., Finegold, M. J., & Greenberg, N. M. (1997). Androgen-independent prostate cancer progression in the TRAMP model. *Cancer Research*, *57*, 4687–4691.
Gligorov, D., Sitnik, J. L., Maeda, R. K., Wolfner, M. F., & Karch, F. (2013). A novel function for the *Hox* gene *Abd-B* in the male accessory gland regulates the long-term female post-mating response in *Drosophila*. *PLoS Genetics*, *9*, e1003395.
Goberdhan, D. C. (2010). Intracellular amino acid sensing and mTORC1-regulated growth: New ways to block an old target? *Current Opinion in Investigational Drugs*, *11*, 1360–1367.
Goberdhan, D. C., Wilson, C., & Harris, A. L. (2016). Amino acid sensing by mTORC1: Intracellular transporters mark the spot. *Cell Metabolism*, *23*, 580–589.
Grifoni, D., Froldi, F., & Pession, A. (2013). Connecting epithelial polarity, proliferation and cancer in *Drosophila*: The many faces of *lgl* loss of function. *The International Journal of Developmental Biology*, *57*, 677–687.
Grizzi, F., & Chiriva-Internati, M. (2007). Human binucleate hepatocytes: Are they a defence during chronic liver diseases? *Medical Hypotheses*, *69*, 258–261.
Hanahan, D., & Weinberg, R. A. (2000). The hallmarks of cancer. *Cell*, *100*, 57–70.
Hanahan, D., & Weinberg, R. A. (2011). Hallmarks of cancer: The next generation. *Cell*, *144*, 646–674.
Häsemeyer, M., Yapici, N., Heberlein, U., & Dickson, B. J. (2009). Sensory neurons in the *Drosophila* genital tract regulate female reproductive behavior. *Neuron*, *61*, 511–518.
Hasholzner, U., Stieber, P., Zimmermann, A., Burges, A., Hofmann, K., Schmitt, U. M., et al. (1999). Nuclear mitotic apparatus protein (NuMA) in benign and malignant diseases. *Anticancer Research*, *19*, 2415–2420.
Haussmann, I. U., Hemani, Y., Wijesekera, T., Dauwalder, B., & Soller, M. (2013). Multiple pathways mediate the sex-peptide-regulated switch in female *Drosophila* reproductive behaviours. *Proceedings of the Biological Sciences*, *280*, 20131938.
Hayward, S. W., & Cunha, G. R. (2000). The prostate: Development and physiology. *Radiologic Clinics of North America*, *38*, 1–14.
He, M., Crow, J., Roth, M., Zeng, Y., & Godwin, A. K. (2014). Integrated immunoisolation and protein analysis of circulating exosomes using microfluidic technology. *Lab on a Chip*, *14*, 3773–3780.

Heifetz, Y., Lindner, M., Garini, Y., & Wolfner, M. F. (2014). Mating regulates neuromodulator ensembles at nerve termini innervating the *Drosophila* reproductive tract. *Current Biology*, *24*, 731–737.

Hentze, J. L., Moeller, M. E., Jørgensen, A. F., Bengtsson, M. S., Bordoy, A. M., Warren, J. T., et al. (2013). Accessory gland as a site for prothoracicotropic hormone controlled ecdysone synthesis in adult male insects. *PLoS One*, *8*, e55131.

Herndon, L. A., & Wolfner, M. F. (1995). A *Drosophila* seminal fluid protein, Acp26Aa, stimulates egg laying in females for 1 day after mating. *Proceedings of the National Academy of Sciences of the United States of America*, *92*, 10114–10118.

Holt, W. V., & Fazeli, A. (2016). Sperm storage in the female reproductive tract. *Annual Review of Animal Biosciences*, *4*, 291–310.

Hoshino, A., Costa-Silva, B., Shen, T. L., Rodrigues, G., Hashimoto, A., Tesic Mark, M., et al. (2015). Tumour exosome integrins determine organotropic metastasis. *Nature*, *527*, 329–335.

Humphrey, P. A. (2012). Histological variants of prostatic carcinoma and their significance. *Histopathology*, *60*, 59–74.

Isaac, R. E., Li, C., Leedale, A. E., & Shirras, A. D. (2010). *Drosophila* male sex peptide inhibits siesta sleep and promotes locomotor activity in the post-mated female. *Proceedings of the Biological Sciences*, *277*, 65–70.

Ishimoto, H., Sakai, T., & Kitamoto, T. (2009). Ecdysone signaling regulates the formation of long-term courtship memory in adult *Drosophila melanogaster*. *Proceedings of the National Academy of Sciences of the United States of America*, *106*, 6381–6386.

Ito, S., Ueda, T., Ueno, A., Nakagawa, H., Taniguchi, H., Kayukawa, N., et al. (2014). A genetic screen in *Drosophila* for regulators of human prostate cancer progression. *Biochemical and Biophysical Research Communications*, *451*, 548–555.

Ittmann, M., Huang, J., Radaelli, E., Martin, P., Signoretti, S., Sullivan, R., et al. (2013). Animal models of human prostate cancer: The consensus report of the New York meeting of the Mouse Models of Human Cancers Consortium Prostate Pathology Committee. *Cancer Research*, *73*, 2718–2736.

Jaiswal, S., Sarmad, R., Arora, S., Dasaraju, R., & Sarmad, K. (2015). Prostate cancer for the internist. *North American Journal of Medical Sciences*, 7, 429–435.

Ji, R., Zhang, B., Zhang, X., Xue, J., Yuan, X., Yan, Y., et al. (2015). Exosomes derived from human mesenchymal stem cells confer drug resistance in gastric cancer. *Cell Cycle*, *14*, 2473–2483.

Kalb, J. M., DiBenedetto, A. J., & Wolfner, M. F. (1993). Probing the function of *Drosophila melanogaster* accessory glands by directed cell ablation. *Proceedings of the National Academy of Sciences of the United States of America*, *90*, 8093–8097.

Kershaw-Young, C. M., Druart, X., Vaughan, J., & Maxwell, W. M. (2012). β-Nerve growth factor is a major component of alpaca seminal plasma and induces ovulation in female alpacas. *Reproduction, Fertility, and Development*, *24*, 1093–1097.

Kim, Y. J., Bartalska, K., Audsley, N., Yamanaka, N., Yapici, N., Lee, J. Y., et al. (2010). MIPs are ancestral ligands for the sex peptide receptor. *Proceedings of the National Academy of Sciences of the United States of America*, *107*, 6520–6525.

Kim, T., Gondré-Lewis, M. C., Arnaoutova, I., & Loh, Y. P. (2006). Dense-core secretory granule biogenesis. *Physiology (Bethesda)*, *21*, 124–133.

Kim, Y. M., Jung, C. H., Seo, M., Kim, E. K., Park, J. M., Bae, S. S., et al. (2015). mTORC1 phosphorylates UVRAG to negatively regulate autophagosome and endosome maturation. *Molecular Cell*, *57*, 207–218.

Klusza, S., & Deng, W. M. (2011). At the crossroads of differentiation and proliferation: Precise control of cell-cycle changes by multiple signaling pathways in *Drosophila* follicle cells. *Bioessays*, *33*, 124–134.

Krätzschmar, J., Haendler, B., Eberspaecher, U., Roosterman, D., Donner, P., & Schleuning, W. D. (1996). The human cysteine-rich secretory protein (CRISP) family.

Primary structure and tissue distribution of CRISP-1, CRISP-2 and CRISP-3. *European Journal of Biochemistry*, *236*, 827–836.
Kubli, E. (2003). Sex-peptides: Seminal peptides of the *Drosophila* male. *Cellular and Molecular Life Sciences*, *60*, 1689–1704.
Kubli, E. (2008). Sexual behaviour: A receptor for sex control in *Drosophila* females. *Current Biology*, *18*, R210–R212.
Kubli, E., & Bopp, D. (2012). Sexual behavior: How sex peptide flips the postmating switch of female flies. *Current Biology*, *22*, R520–R522.
LaFlamme, B. A., Avila, F. W., Michalski, K., & Wolfner, M. F. (2014). A *Drosophila* protease cascade member, seminal metalloprotease-1, is activated stepwise by male factors and requires female factors for full activity. *Genetics*, *196*, 1117–1129.
LaFlamme, B. A., Ram, K. R., & Wolfner, M. F. (2012). The *Drosophila melanogaster* seminal fluid protease "seminase" regulates proteolytic and post-mating reproductive processes. *PLoS Genetics*, *8*, e1002435.
Laflamme, B. A., & Wolfner, M. F. (2013). Identification and function of proteolysis regulators in seminal fluid. *Molecular Reproduction and Development*, *80*, 80–101.
Lamb, A. D., Massie, C. E., & Neal, D. E. (2014). The transcriptional programme of the androgen receptor (AR) in prostate cancer. *BJU International*, *113*, 358–366.
Lee, Y. C., Cheng, C. J., Bilen, M. A., Lu, J. F., Satcher, R. L., Yu-Lee, L. Y., et al. (2011). BMP4 promotes prostate tumor growth in bone through osteogenesis. *Cancer Research*, *71*, 5194–5203.
Lee, G. T., Kang, D. I., Ha, Y. S., Jung, Y. S., Chung, J., Min, K., et al. (2014). Prostate cancer bone metastases acquire resistance to androgen deprivation via WNT5A-mediated BMP-6 induction. *British Journal of Cancer*, *110*, 1634–1644.
Leiblich, A., Marsden, L., Gandy, C., Corrigan, L., Jenkins, R., Hamdy, F., et al. (2012). Bone morphogenetic protein- and mating-dependent secretory cell growth and migration in the *Drosophila* accessory gland. *Proceedings of the National Academy of Sciences of the United States of America*, *109*, 19292–19297.
Leong, K. G., & Gao, W. Q. (2008). The Notch pathway in prostate development and cancer. *Differentiation*, *76*, 699–716.
Li, C. I., & Daling, J. R. (2007). Changes in breast cancer incidence rates in the United States by histologic subtype and race/ethnicity, 1995 to 2004. *Cancer Epidemiology, Biomarkers & Prevention*, *16*, 2773–2780.
Li, Y., Ma, Q., Cherry, C. M., & Matunis, E. L. (2014). Steroid signaling promotes stem cell maintenance in the *Drosophila* testis. *Developmental Biology*, *394*, 129–141.
Lilja, H., Oldbring, J., Rannevik, G., & Laurell, C. B. (1987). Seminal vesicle-secreted proteins and their reactions during gelation and liquefaction of human semen. *The Journal of Clinical Investigation*, *80*, 281–285.
Liu, H., & Kubli, E. (2003). Sex-peptide is the molecular basis of the sperm effect in *Drosophila melanogaster*. *Proceedings of the National Academy of Sciences of the United States of America*, *100*, 9929–9933.
Long, R. M., Morrissey, C., Fitzpatrick, J. M., & Watson, R. W. (2005). Prostate epithelial cell differentiation and its relevance to the understanding of prostate cancer therapies. *Clinical Science (London, England)*, *108*, 1–11.
Manning, A. (1962). A sperm factor affecting the receptivity of *Drosophila melanogaster* females. *Nature*, *194*, 252–253.
Maranesi, M., Zerani, M., Leonardi, L., Pistilli, A., Arruda-Alencar, J., Stabile, A. M., et al. (2015). Gene expression and localization of NGF and its cognate receptors NTRK1 and NGFR in the sex organs of male rabbits. *Reproduction in Domestic Animals*, *50*, 918–925.
Marques, P. I., Bernardino, R., Fernandes, T., NISC Comparative Sequencing Program, Green, E. D., Hurle, B., et al. (2012). Birth-and-death of KLK3 and KLK2 in primates: Evolution driven by reproductive biology. *Genome Biology and Evolution*, *4*, 1331–1338.

Martincorena, I., & Campbell, P. J. (2015). Somatic mutation in cancer and normal cells. *Science*, *349*, 1483–1489.
McGuire, S. E., Mao, Z., & Davis, R. L. (2004). Spatiotemporal gene expression targeting with the TARGET and gene-switch systems in *Drosophila*. *Science's STKE*, *2004*(220), pl6.
Melnik, B. C., John, S. M., & Schmitz, G. (2014). Milk: An exosomal microRNA transmitter promoting thymic regulatory T cell maturation preventing the development of atopy? *Journal of Translational Medicine*, *12*, 43.
Melo, S. A., Luecke, L. B., Kahlert, C., Fernandez, A. F., Gammon, S. T., Kaye, J., et al. (2015). Glypican-1 identifies cancer exosomes and detects early pancreatic cancer. *Nature*, *523*, 177–182.
Miah, A. G., Salma, U., Hamano, K., & Schellander, K. (2015). Physiological roles of relaxin in prefertilizing activities of spermatozoa. *Animal Reproduction Science*, *161*, 1–15.
Minami, R., Wakabayashi, M., Sugimori, S., Taniguchi, K., Kokuryo, A., Imano, T., et al. (2012). The homeodomain protein defective proventriculus is essential for male accessory gland development to enhance fecundity in *Drosophila*. *PLoS One*, *7*, e32302.
Mitchell, T., & Neal, D. E. (2015). The genomic evolution of human prostate cancer. *British Journal of Cancer*, *113*, 193–198.
Morrow, E. H., & Innocenti, P. (2012). Female postmating immune responses, immune system evolution and immunogenic males. *Biological Reviews of the Cambridge Philosophical Society*, *87*, 631–638.
Moshitzky, P., Fleischmann, I., Chaimov, N., Saudan, P., Klauser, S., Kubli, E., et al. (1996). Sex-peptide activates juvenile hormone biosynthesis in the *Drosophila melanogaster* corpus allatum. *Archives of Insect Biochemistry and Physiology*, *32*, 363–374.
Murer, V., Spetz, J. F., Hengst, U., Altrogge, L. M., de Agostini, A., & Monard, D. (2001). Male fertility defects in mice lacking the serine protease inhibitor protease nexin-1. *Proceedings of the National Academy of Sciences of the United States of America*, *98*, 3029–3033.
Nassis, L., Frauman, A. G., Ohishi, M., Zhuo, J., Casley, D. J., Johnston, C. I., et al. (2001). Localization of angiotensin-converting enzyme in the human prostate: Pathological expression in benign prostatic hyperplasia. *The Journal of Pathology*, *195*, 571–579.
Nászai, M., Carroll, L. R., & Cordero, J. B. (2015). Intestinal stem cell proliferation and epithelial homeostasis in the adult *Drosophila* midgut. *Insect Biochemistry and Molecular Biology*, *67*, 9–14.
Omori, A., Miyagawa, S., Ogino, Y., Harada, M., Ishii, K., Sugimura, Y., et al. (2014). Essential roles of epithelial bone morphogenetic protein signaling during prostatic development. *Endocrinology*, *155*, 2534–2544.
Pampalakis, G., & Sotiropoulou, G. (2007). Tissue kallikrein proteolytic cascade pathways in normal physiology and cancer. *Biochimica et Biophysica Acta*, *1776*, 22–31.
Panayidou, S., & Apidianakis, Y. (2013). Regenerative inflammation: Lessons from *Drosophila* intestinal epithelium in health and disease. *Pathogens*, *2*, 209–231.
Park, K. H., Kim, B. J., Kang, J., Nam, T. S., Lim, J. M., Kim, H. T., et al. (2011). Ca^{2+} signaling tools acquired from prostasomes are required for progesterone-induced sperm motility. *Science Signaling*, *4*, ra31.
Patel, P. H., & Edgar, B. A. (2014). Tissue design: How *Drosophila* tumors remodel their neighborhood. *Seminars in Cell & Developmental Biology*, *28*, 86–95.
Peinado, H., Alečković, M., Lavotshkin, S., Matei, I., Costa-Silva, B., Moreno-Bueno, G., et al. (2012). Melanoma exosomes educate bone marrow progenitor cells toward a pro-metastatic phenotype through MET. *Nature Medicine*, *18*, 883–891.
Peña-Llopis, S., & Brugarolas, J. (2011). TFEB, a novel mTORC1 effector implicated in lysosome biogenesis, endocytosis and autophagy. *Cell Cycle*, *10*, 3987–3988.
Peng, J., Chen, S., Büsser, S., Liu, H., Honegger, T., & Kubli, E. (2005a). Gradual release of sperm bound sex-peptide controls female postmating behavior in *Drosophila*. *Current Biology*, *15*, 207–213.

Peng, Y. C., & Joyner, A. L. (2015). Hedgehog signaling in prostate epithelial-mesenchymal growth regulation. *Developmental Biology*, *400*, 94–104.

Peng, J., Zipperlen, P., & Kubli, E. (2005b). *Drosophila* sex-peptide stimulates female innate immune system after mating via the Toll and Imd pathways. *Current Biology*, *215*, 1690–1694.

Perrimon, N., Pitsouli, C., & Shilo, B. Z. (2012). Signaling mechanisms controlling cell fate and embryonic patterning. *Cold Spring Harbor Perspectives in Biology*, *4*, a005975.

Pons-Rejraji, H., Artonne, C., Sion, B., Brugnon, F., Canis, M., Janny, L., et al. (2011). Prostasomes: Inhibitors of capacitation and modulators of cellular signalling in human sperm. *International Journal of Andrology*, *34*, 568–580.

Powers, G. L., & Marker, P. C. (2013). Recent advances in prostate development and links to prostatic diseases. *Wiley Interdisciplinary Reviews. Systems Biology and Medicine*, *5*, 243–256.

Qian, Y., Dominado, N., Zoller, R., Ng, C., Kudyba, K., Siddall, N. A., et al. (2014). Ecdysone signaling opposes epidermal growth factor signaling in regulating cyst differentiation in the male gonad of *Drosophila melanogaster*. *Developmental Biology*, *394*, 217–227.

Ratto, M. H., Leduc, Y. A., Valderrama, X. P., van Straaten, K. E., Delbaere, L. T., Pierson, R. A., et al. (2012). The nerve of ovulation-inducing factor in semen. *Proceedings of the National Academy of Sciences of the United States of America*, *109*, 15042–15047.

Ravi Ram, K., Sirot, L. K., & Wolfner, M. F. (2006). Predicted seminal astacin-like protease is required for processing of reproductive proteins in *Drosophila melanogaster*. *Proceedings of the National Academy of Sciences of the United States of America*, *103*, 18674–18679.

Ravi Ram, K., & Wolfner, M. F. (2007). Sustained post-mating response in *Drosophila melanogaster* requires multiple seminal fluid proteins. *PLoS Genetics*, *3*, e238.

Ravi Ram, K., & Wolfner, M. F. (2009). A network of interactions among seminal proteins underlies the long-term postmating response in *Drosophila*. *Proceedings of the National Academy of Sciences of the United States of America*, *106*, 15384–15389.

Rezával, C., Nojima, T., Neville, M. C., Lin, A. C., & Goodwin, S. F. (2014). Sexually dimorphic octopaminergic neurons modulate female postmating behaviors in *Drosophila*. *Current Biology*, *24*, 725–730.

Rezával, C., Pavlou, H. J., Dornan, A. J., Chan, Y. B., Kravitz, E. A., & Goodwin, S. F. (2012). Neural circuitry underlying *Drosophila* female postmating behavioral responses. *Current Biology*, *22*, 1155–1165.

Ribeiro, C., & Dickson, B. J. (2010). Sex peptide receptor and neuronal TOR/S6K signaling modulate nutrient balancing in *Drosophila*. *Current Biology*, *20*, 1000–1005.

Ronquist, G. (2015). Prostasomes: Their characterisation: Implications for human reproduction: Prostasomes and human reproduction. *Advances in Experimental Medicine and Biology*, *868*, 191–209.

Rosales-Nieves, A. E., & González-Reyes, A. (2014). Genetics and mechanisms of ovarian cancer: Parallels between *Drosophila* and humans. *Seminars in Cell and Developmental Biology*, *28*, 104–109.

Rudrapatna, V. A., Cagan, R. L., & Das, T. K. (2012). *Drosophila* cancer models. *Developmental Dynamics*, *241*, 107–118.

Ryan, D. P., Hong, T. S., & Bardeesy, N. (2014). Pancreatic adenocarcinoma. *The New England Journal of Medicine*, *371*, 2140–2141.

Rylett, C. M., Walker, M. J., Howell, G. J., Shirras, A. D., & Isaac, R. E. (2007). Male accessory glands of *Drosophila melanogaster* make a secreted angiotensin I-converting enzyme (ANCE), suggesting a role for the peptide-processing enzyme in seminal fluid. *The Journal of Experimental Biology*, *210*, 3601–3606.

Schrijvers, D. (2007). Androgen-independent prostate cancer. *Recent Results in Cancer Research*, *175*, 239–249.

Schwedes, C. C., & Carney, G. E. (2012). Ecdysone signaling in adult *Drosophila melanogaster*. *Journal of Insect Physiology*, *58*, 293–302.

Schwertfeger, K. L. (2009). Fibroblast growth factors in development and cancer: Insights from the mammary and prostate glands. *Current Drug Targets, 10*, 632–644.

Shah, M. A., Khanin, R., Tang, L., Janjigian, Y. Y., Klimstra, D. S., Gerdes, H., et al. (2011). Molecular classification of gastric cancer: A new paradigm. *Clinical Cancer Research, 17*, 2693–2701.

Sharkey, D. J., Macpherson, A. M., Tremellen, K. P., Mottershead, D. G., Gilchrist, R. B., & Robertson, S. A. (2012). TGF-β mediates proinflammatory seminal fluid signaling in human cervical epithelial cells. *Journal of Immunology, 189*, 1024–1035.

Sharma, A., Yeow, W. S., Ertel, A., Coleman, I., Clegg, N., Thangavel, C., et al. (2010). The retinoblastoma tumor suppressor controls androgen signaling and human prostate cancer progression. *The Journal of Clinical Investigation, 120*, 4478–4492.

Short, S. M., Wolfner, M. F., & Lazzaro, B. P. (2012). Female *Drosophila melanogaster* suffer reduced defense against infection due to seminal fluid components. *Journal of Insect Physiology, 58*, 1192–1201.

Siegel, R., Ma, J., Zou, Z., & Jemal, A. (2014). Cancer statistics, 2014. *CA: A Cancer Journal for Clinicians, 64*, 9–29.

Silva, M., Niño, A., Guerra, M., Letelier, C., Valderrama, X. P., Adams, G. P., et al. (2011). Is an ovulation-inducing factor (OIF) present in the seminal plasma of rabbits? *Animal Reproduction Science, 127*, 213–221.

Simons, B. W., Hurley, P. J., Huang, Z., Ross, A. E., Miller, R., Marchionni, L., et al. (2012). Wnt signaling though beta-catenin is required for prostate lineage specification. *Developmental Biology, 371*, 246–255.

Sirot, L. K., Findlay, G. D., Sitnik, J. L., Frasheri, D., Avila, F. W., & Wolfner, M. F. (2014). Molecular characterization and evolution of a gene family encoding both female- and male-specific reproductive proteins in *Drosophila. Molecular Biology and Evolution, 31*, 1554–1567.

Sirot, L. K., LaFlamme, B. A., Sitnik, J. L., Rubinstein, C. D., Avila, F. W., Chow, C. Y., et al. (2009). Molecular social interactions: *Drosophila melanogaster* seminal fluid proteins as a case study. *Advances in Genetics, 68*, 23–56.

Sirot, L. K., Wolfner, M. F., & Wigby, S. (2011). Protein-specific manipulation of ejaculate composition in response to female mating status in *Drosophila melanogaster. Proceedings of the National Academy of Sciences of the United States of America, 108*, 9922–9926.

Sitnik, J. L., Gligorov, D., Maeda, R. K., Karch, F., & Wolfner, M. F. (2016). Secondary cell expressed genes in the male accessory gland are needed for the female post-mating response in *Drosophila melanogaster. Genetics, 202*, 1029–1041. pii, genetics. 115.181644.

Stanfel, M. N., Moses, K. A., Carson, J. A., Zimmer, D. B., DeMayo, F., Schwartz, R. J., et al. (2006). Expression of an *Nkx3.1-CRE* gene using ROSA26 reporter mice. *Genesis, 44*, 550–555.

Stenmark, H. (2009). Rab GTPases as coordinators of vesicle traffic. *Nature Reviews. Molecular Cell Biology, 10*, 513–525.

Stuffers, S., Sem Wegner, C., Stenmark, H., & Brech, A. (2009). Multivesicular endosome biogenesis in the absence of ESCRTs. *Traffic, 10*, 925–937.

Susic-Jung, L., Hornbruch-Freitag, C., Kuckwa, J., Rexer, K. H., Lammel, U., & Renkawitz-Pohl, R. (2012). Multinucleated smooth muscles and mononucleated as well as multinucleated striated muscles develop during establishment of the male reproductive organs of *Drosophila melanogaster. Developmental Biology, 2370*, 86–97.

Taniguchi, K., Kokuryo, A., Imano, T., Minami, R., Nakagoshi, H., & Adachi-Yamada, T. (2014). Isoform-specific functions of Mud/NuMA mediate binucleation of *Drosophila* male accessory gland cells. *BMC Developmental Biology, 14*, 46.

Tayler, T. D., Pacheco, D. A., Hergarden, A. C., Murthy, M., & Anderson, D. J. (2012). A neuropeptide circuit that coordinates sperm transfer and copulation duration in

Drosophila. Proceedings of the National Academy of Sciences of the United States of America, 109, 20697–20702.
Thomsen, M. K., Francis, J. C., & Swain, A. (2008). The role of *Sox9* in prostate development. *Differentiation, 76,* 728–735.
Thomson, A. A., & Marker, P. C. (2006). Branching morphogenesis in the prostate gland and seminal vesicles. *Differentiation, 74,* 382–392.
Tipping, M., & Perrimon, N. (2014). *Drosophila* as a model for context-dependent tumorigenesis. *Journal of Cellular Physiology, 229,* 27–33.
Titus, M. A., Schell, M. J., Lih, F. B., Tomer, K. B., & Mohler, J. L. (2005). Testosterone and dihydrotestosterone tissue levels in recurrent prostate cancer. *Clinical Cancer Research, 11,* 4653–4657.
Torres, I. L., Rosa-Ferreira, C., & Munro, S. (2014). The Arf family G protein Arl1 is required for secretory granule biogenesis in *Drosophila. Journal of Cell Science, 127,* 2151–2160.
Trajkovic, K., Hsu, C., Chiantia, S., Rajendran, L., Wenzel, D., Wieland, F., et al. (2008). Ceramide triggers budding of exosome vesicles into multivesicular endosomes. *Science, 319,* 1244–1247.
Tsuda, M., Peyre, J. B., Asano, T., & Aigaki, T. (2015). Visualizing molecular functions and cross-species activity of sex-peptide in *Drosophila. Genetics, 200,* 1161–1169.
Udby, L., Lundwall, A., Johnsen, A. H., Fernlund, P., Valtonen-André, C., Blom, A. M., et al. (2005). beta-Microseminoprotein binds CRISP-3 in human seminal plasma. *Biochemical and Biophysical Research Communications, 333,* 555–561.
van Weerden, W. M., & Romijn, J. C. (2000). Use of nude mouse xenograft models in prostate cancer research. *Prostate, 43,* 263–271.
Veveris-Lowe, T. L., Kruger, S. J., Walsh, T., Gardiner, R. A., & Clements, J. A. (2007). Seminal fluid characterization for male fertility and prostate cancer: Kallikrein-related serine proteases and whole proteome approaches. *Seminars in Thrombosis and Hemostasis, 33,* 87–99.
Vinall, R. L., Mahaffey, C. M., Davis, R. R., Luo, Z., Gandour-Edwards, R., Ghosh, P. M., et al. (2011). Dual blockade of PKA and NF-kB inhibits H2 relaxin-mediated castrate-resistant growth of prostate cancer sublines and induces apoptosis. *Hormones & Cancer, 2,* 224–238.
Visakorpi, T., Hyytinen, E., Koivisto, P., Tanner, M., Keinänen, R., Palmberg, C., et al. (1995). *In vivo* amplification of the androgen receptor gene and progression of human prostate cancer. *Nature Genetics, 9,* 401–406.
Walker, S. J., Corrales-Carvajal, V. M., & Ribeiro, C. (2015). Postmating circuitry modulates salt taste processing to increase reproductive output in *Drosophila. Current Biology, 25,* 2621–2630.
Wang, S., Gao, J., Lei, Q., Rozengurt, N., Pritchard, C., Jiao, J., et al. (2003). Prostate-specific deletion of the murine *Pten* tumor suppressor gene leads to metastatic prostate cancer. *Cancer Cell, 4,* 209–221.
Watson, P. A., Chen, Y. F., Balbas, M. D., Wongvipat, J., Socci, N. D., Viale, A., et al. (2010). Constitutively active androgen receptor splice variants expressed in castration-resistant prostate cancer require full-length androgen receptor. *Proceedings of the National Academy of Sciences of the United States of America, 107,* 16759–16765.
Weitz, J., Koch, M., Debus, J., Höhler, T., Galle, P. R., & Büchler, M. W. (2005). Colorectal cancer. *Lancet, 365,* 153–165.
White, S. J., Masters, A. R. W., & Woolf, A. S. (1997). Renal and bladder epithelial cells. In A. Harris (Ed.), *Epithelial cell culture* (pp. 58–78). Cambridge: Cambridge University Press.

Wigby, S., Sirot, L. K., Linklater, J. R., Buehner, N., Calboli, F. C., Bretman, A., et al. (2009). Seminal fluid protein allocation and male reproductive success. *Current Biology, 19*, 751–757.

Wolfner, M. F., Partridge, L., Lewin, S., Kalb, J. M., Chapman, T., & Herndon, L. A. (1997). Mating and hormonal triggers regulate accessory gland gene expression in male *Drosophila. Journal of Insect Physiology, 43*, 1117–1123.

Wu, X., Gong, S., Roy-Burman, P., Lee, P., & Culig, Z. (2013). Current mouse and cell models in prostate cancer research. *Endocrine-Related Cancer, 20*, R155–R170.

Xavier de Carvalho, A., Maiato, H., Maia, A. F., Ribeiro, S. A., Pontes, P., Bickmore, W., et al. (2015). Reed-Sternberg cells form by abscission failure in the presence of functional Aurora B kinase. *PLoS One, 10*, e0124629.

Xie, Y., You, C., & Chen, J. (2014). An updated meta-analysis on association between angiotensin I-converting enzyme gene insertion/deletion polymorphism and cancer risk. *Tumour Biology, 35*, 6567–6579.

Xu, B., Hariharan, A., Rakshit, S., Dressler, G. R., & Wellik, D. M. (2012). The role of *Pax2* in mouse prostate development. *Prostate, 72*, 217–224.

Xue, L., & Noll, M. (2000). *Drosophila* female sexual behavior induced by sterile males showing copulation complementation. *Proceedings of the National Academy of Sciences of the United States of America, 97*, 3272–3275.

Xue, L., & Noll, M. (2002). Dual role of the *Pax* gene paired in accessory gland development of *Drosophila. Development, 129*, 339–346.

Yamanaka, N., Rewitz, K. F., & O'Connor, M. B. (2013). Ecdysone control of developmental transitions: Lessons from *Drosophila* research. *Annual Review of Entomology, 58*, 497–516.

Yang, C. H., Rumpf, S., Xiang, Y., Gordon, M. D., Song, W., Jan, L. Y., et al. (2009). Control of the postmating behavioral switch in *Drosophila* females by internal sensory neurons. *Neuron, 61*, 519–526.

Yapici, N., Kim, Y. J., Ribeiro, C., & Dickson, B. J. (2008). A receptor that mediates the post-mating switch in *Drosophila* reproductive behaviour. *Nature, 451*, 33–37.

Yu, S., Cao, H., Shen, B., & Feng, J. (2015). Tumor-derived exosomes in cancer progression and treatment failure. *Oncotarget, 6*, 37151–37168.

Zhang, X., Morrissey, C., Sun, S., Ketchandji, M., Nelson, P. S., True, L. D., et al. (2011). Androgen receptor variants occur frequently in castration-resistant prostate cancer metastases. *PLoS One, 6*, e27970.

Zhang, L., Zhang, S., Yao, J., Lowery, F. J., Zhang, Q., Huang, W. C., et al. (2015). Microenvironment-induced PTEN loss by exosomal microRNA primes brain metastasis outgrowth. *Nature, 527*, 100–104.

Zhou, M., Chen, J., Zhou, L., Chen, W., Ding, G., & Cao, L. (2014). Pancreatic cancer derived exosomes regulate the expression of TLR4 in dendritic cells via miR-203. *Cellular Immunology, 292*, 65–69.

Zielke, N., Edgar, B. A., & DePamphilis, M. L. (2013). Endoreplication. *Cold Spring Harbor Perspectives in Biology, 5*, a012948.

Zomer, A., Maynard, C., Verweij, F. J., Kamermans, A., Schäfer, R., Beerling, E., et al. (2015). *In vivo* imaging reveals extracellular vesicle-mediated phenocopying of metastatic behavior. *Cell, 161*, 1046–1057.

CHAPTER TWELVE

Drosophila melanogaster Models of Galactosemia

J.M.I. Daenzer, J.L. Fridovich-Keil[1]
Emory University School of Medicine, Atlanta, GA, United States
[1]Corresponding author: e-mail address: jfridov@emory.edu

Contents

Abstract

The galactosemias are a family of autosomal recessive genetic disorders resulting from impaired function of the Leloir pathway of galactose metabolism. Type I, or classic galactosemia, results from profound deficiency of galactose-1-phosphate uridylyltransferase, the second enzyme in the Leloir pathway. Type II galactosemia results from profound deficiency of galactokinase, the first enzyme in the Leloir pathway. Type III galactosemia results from partial deficiency of UDP galactose 4′-epimerase, the third enzyme in the Leloir pathway. Although at least classic galactosemia has been recognized clinically for more than 100 years, and detectable by newborn screening for more than 50 years, all three galactosemias remain poorly understood. Early detection and dietary restriction of

Current Topics in Developmental Biology, Volume 121
ISSN 0070-2153
http://dx.doi.org/10.1016/bs.ctdb.2016.07.009

galactose prevent neonatal lethality, but many affected infants grow to experience a broad range of developmental and other disabilities. To date, there is no intervention known that prevents or reverses these long-term complications. *Drosophila melanogaster* provides a genetically and biochemically facile model for these conditions, enabling studies that address mechanism and open the door for novel approaches to intervention.

1. GALACTOSE METABOLISM VIA THE LELOIR PATHWAY

Galactose is a naturally occurring monosaccharide that is found at high levels in milk and dairy products, and at low levels in many other foods. Galactose is also synthesized endogenously. In species ranging from *Escherichia coli* to *Drosophila* and humans, galactose is metabolized predominantly via the Leloir pathway (Fig. 1; Holden, Rayment, & Thoden, 2003), which is comprised of three enzymes.

The first Leloir pathway enzyme, galactokinase (GALK, EC 2.7.1.6), catalyzes the phosphorylation of galactose to galactose-1-phosphate (Gal-1P) at the expense of ATP. The second enzyme, galactose-1-phosphate uridylyltransferase (GALT, EC 2.7.7.12), catalyzes a two-step reaction. In the first step, GALT cleaves UDPglucose (UDPGlc), releasing

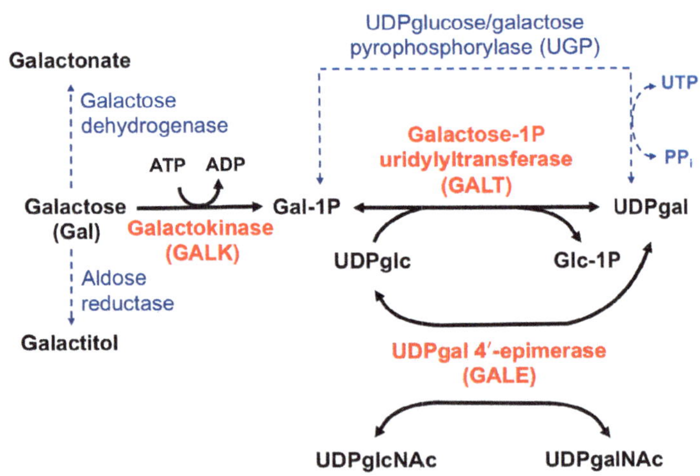

Fig. 1 The Leloir pathway of galactose metabolism (*red*) and recognized bypass routes (*blue*). The highly conserved Leloir pathway is comprised of three enzymes: galactokinase (GALK), galactose-1-phosphate uridylyltransferase (GALT), and UDP galactose 4′-epimerase (GALE). Mutations resulting in deficiency of any of these three enzymes lead to a form of galactosemia.

glucose-1-phosphate (Glc-1P) and retaining UMP covalently bound to a histidine residue in the active site. In the second step, Gal-1P enters the active site and receives the UMP group, restoring the active site histidine and creating UDPgalactose (UDPGal), which is released. The third Leloir enzyme, UDPgalactose 4′-epimerase (GALE, EC 5.1.3.2), interconverts UDPgal and UDPglc. Acting together, the three enzymes of the Leloir pathway therefore convert galactose into Glc-1P at the expense of ATP (Fig. 1).

In many metazoans, including both humans and *Drosophila*, GALE also interconverts a larger substrate pair: UDP-*N*-acetylgalactosamine (UDPGalNAc) and UDP-*N*-acetylglucosamine (UDPGlcNAc). GALE therefore not only functions in the Leloir pathway but also maintains the ratios of UDPGal to UDPGlc, and UDPGalNAc to UDPGlcNAc, and allows for the endogenous synthesis of galactose when exogenous sources are lacking. All four of the UDP sugars recognized by GALE serve as key substrates for the biosynthesis of glycoproteins and glycolipids in humans, *Drosophila*, and many other species.

1.1 Alternate Pathways of Galactose Metabolism

While the Leloir pathway is the predominant route of galactose metabolism, minor "bypass" pathways also exist (Fig. 1). These include oxidation of galactose to galactonate by galactose dehydrogenase, reduction of galactose to galactitol by aldose reductase, and conversion of galactose-1P to UDPGal by UDPglucose pyrophosphorylase (UGP) (reviewed in Walter & Fridovich-Keil, 2014). These bypass pathways are generally considered insignificant when the Leloir pathway is functioning, but when the Leloir pathway is blocked by deficiency of GALK, GALT, or GALE, these alternate pathways may become extremely important.

1.2 Three Types of Galactosemias

Inherited mutations leading to deficiency of any of the three enzymes of the Leloir pathway (GALK, GALT, or GALE) result in a form of galactosemia. The clinical presentation and severity of the disorder may vary depending on which enzyme is affected, the degree of impairment, and other genetic and environmental factors—most of which remain poorly understood. Of note, while sometimes mistakenly confused with lactose intolerance, a common and generally mild adult-onset condition resulting from reduced ability to digest lactose into its constituent monosaccharides (glucose and galactose), galactosemia results from the inability to fully metabolize galactose itself, presents clinically in infancy, and can be lethal.

1.2.1 Type I, or Classic Galactosemia

The most clinically severe form of galactosemia in most populations is called Type I, or classic galactosemia (CG; reviewed in Berry, 2014; Walter & Fridovich-Keil, 2014). CG results from profound deficiency of GALT, the middle enzyme of the Leloir pathway (Fig. 1). Population-based newborn screening demonstrates that the prevalence of CG is about 1/50,000 in the United States (Pyhtila, Shaw, Neumann, & Fridovich-Keil, 2015); however, prevalence can vary markedly among different ancestral groups.

Infants with CG may appear normal at birth but develop a rapid progression of potentially lethal sequelae within days of beginning to consume milk—either breast milk or a milk-based formula. Symptoms can escalate in days from vomiting and diarrhea to jaundice, liver disease, failure to thrive, and *E. coli* sepsis, which is a common cause of neonatal death.

If diagnosed early, an affected infant may be spared the trauma of acute disease by rapid dietary restriction of galactose, generally achieved by switching the baby from milk to a low-galactose soy or elemental formula (Berry, 2012). This simple intervention prevents or reverses the potentially lethal symptoms of CG, but fails to prevent most affected infants from experiencing a constellation of complications as they grow. Common complications of CG despite early and continued dietary restriction of galactose include growth delays in both boys and girls, cognitive and/or behavioral disabilities, speech difficulties, and motor disturbance. At least 80% or more of affected girls and young women also experience primary or premature ovarian insufficiency. At present, residual GALT activity is the only documented modifier of outcome severity in CG (Ryan et al., 2013; Spencer et al., 2013). Patients with even trace residual GALT activity detectable in a model system (yeast; Riehman, Crews, & Fridovich-Keil, 2001) tend to show milder outcomes, and those with 10% or more GALT activity may be very mildly affected (reviewed in Berry, 2014; Walter & Fridovich-Keil, 2014). Other modifiers of outcome, beyond galactose exposure in infancy and residual GALT activity, likely exist but remain undefined.

1.2.2 Type II, or GALK-Deficiency Galactosemia

Type II galactosemia results from profound deficiency of GALK (reviewed in Walter & Fridovich-Keil, 2014). Because many newborn screening programs are not designed to detect GALK deficiency, the population prevalence and natural history of this condition remain poorly understood.

Based on the limited information available, profound GALK deficiency was long considered extremely rare with common clinical sequelae restricted to cataracts that could be prevented by early and continued dietary restriction of galactose (Bosch et al., 2002). This conclusion was challenged in 2011, however, when newborn screening and follow-up results from Germany reported a population prevalence of GALK deficiency close to that of GALT deficiency (Hennermann et al., 2011). Of note, the prevalence of GALK deficiency appears to vary considerably by ancestral group, and the increased prevalence detected in Germany was attributed to changing demographics of the newborn population rather than altered screening procedures. Of concern, follow-up studies of the infants diagnosed with GALK deficiency in Germany demonstrated that close to 30% exhibited cognitive disabilities by childhood, and this long-term complication was associated with continued galactose exposure, but not consanguinity (Hennermann et al., 2011). Further studies will be required to confirm the generalizability of these results.

1.2.3 Type III, or Epimerase (GALE) Deficiency Galactosemia

Type III galactosemia results from partial deficiency of GALE, the third Leloir pathway enzyme (Fridovich-Keil, Bean, He, & Schroer, 2011). Like GALK deficiency, GALE deficiency goes undetected by many newborn screening programs so what is known about the prevalence and natural history of the disorder is limited.

Like CG, clinical severity in GALE deficiency reflects many factors, including the level of residual GALE activity present. Unlike CG, however, no live births have been reported with complete loss of GALE, and the most severely affected patients described (Walter et al., 1999) nonetheless demonstrate detectable GALE, at least in some tissues (Openo et al., 2006; Wohlers, Christacos, Harreman, & Fridovich-Keil, 1999; Wohlers & Fridovich-Keil, 2000). GALE deficiency is now considered a continuum disorder, ranging from the ostensibly benign peripheral epimerase deficiency (Gitzelmann, 1972), characterized by biochemical impairment apparently restricted to the red blood cells (RBCs) and circulating white blood cells (Gitzelmann & Steinmann, 1973; Mitchell, Haigis, Steinmann, & Gitzelmann, 1975), through intermediate epimerase deficiency (Openo et al., 2006), in which biochemical impairment is profound in RBCs, but partial in other cell types, to generalized epimerase deficiency (Holton, Gillett, Macfaul, & Young,

1981), which presents with symptoms reminiscent of CG following neonatal milk exposure, and GALE deficiency that is profound, but not complete, in all tissues examined.

Patients with peripheral or intermediate GALE deficiency generally remain clinically well despite milk exposure in infancy, and patients with generalized GALE deficiency respond quickly to dietary restriction of milk. The long-term outcomes associated with different levels of GALE deficiency remain unclear, complicated by consanguinity among many of the patients reported with generalized epimerase deficiency (Walter et al., 1999), and the small numbers and limited follow-up information available for patients with the intermediate disorder (Fridovich-Keil et al., 2011).

1.3 Galactose Metabolism in *Drosophila melanogaster*

Like humans, *D. melanogaster* metabolize galactose predominantly via the Leloir pathway. The *D. melanogaster* genes that encode GALK, GALT, and GALE are *CG5288* on chromosome 3, referred to here as *dGALK* and in FlyBase (http://flybase.org/) as *Galk*; *CG9232* on chromosome 2, referred to here as *dGALT* and in FlyBase (http://flybase.org/) as *Galt*; and *CG12030* on chromosome 3, referred to here as *dGALE* and in FlyBase (http://flybase.org/) as *Gale*, respectively (Chien, Reiter, Bier, & Gribskov, 2002; Kushner et al., 2010). The predicted protein product of *dGALK* shows 27% amino acid sequence identity and 44% amino acid sequence similarity with human GALK. The predicted protein product of *dGALT* shows 57% amino acid sequence identity and 72% amino acid sequence similarity with human GALT. Finally, the predicted protein product of *dGALE* shows 60% amino acid sequence identity and 76% amino acid sequence similarity with human GALE. While the three Leloir genes in *D. melanogaster* were originally identified on the basis of sequence homology, they have all now been confirmed based on loss of the expected enzyme activity following disruption or deletion of the corresponding gene (explained below; Daenzer et al., in preparation). *D. melanogaster* is therefore an excellent model system in which to study the galactosemias.

2. A *D. MELANOGASTER* MODEL OF CLASSIC GALACTOSEMIA

2.1 Creating GALT-Null *Drosophila*

In 2010, Kushner et al. (2010) reported the creation of a *dGALT*-null allele ($dGALT^{\Delta AP2}$) by imprecise excision of an existing P-element insertion,

KG00049, in the 5′ untranslated region (UTR) of the *CG9232* gene. The excision removed 1647 bp of sequence, including almost the entire *dGALT* gene (Kushner et al., 2010). The mutated allele was designated *dGALT*$^{\Delta AP2}$. Because *dGALT* is encoded within an intron of *cup*, a fly gene important for female fertility (Keyes & Spradling, 1997), Kushner and colleagues further tested whether a chromosome carrying the *dGALT*$^{\Delta AP2}$ allele could complement a strong allele of *cup* (*cup*^{01355}); it did (Kushner et al., 2010). Biochemical studies of lysates prepared from *dGALT*$^{\Delta AP2}$ homozygotes confirmed complete loss of GALT activity and normal, or near-normal, levels of GALK and GALE activity (Kushner et al., 2010). Further studies, described later, also demonstrated that a human *GALT* transgene expressed in these *dGALT*-null flies complemented both the loss of GALT activity and the mutant phenotypes observed (Kushner et al., 2010).

2.2 Characterizing the Galactose Sensitivity of GALT-Null Larvae

Patients with classic galactosemia typically die as infants if galactose, generally in the form of milk, is not removed from their diet (Berry, 2014). Once switched to a low-galactose diet such as soy formula, these infants survive and, at least initially, appear to thrive.

To determine whether a similar sensitivity to dietary galactose would be evident in GALT-null *Drosophila*, Kushner and colleagues tested the ability of larvae homozygous for the *dGALT*$^{\Delta AP2}$ allele to survive to adulthood in the presence vs absence of galactose (Kushner et al., 2010). First, crosses of heterozygotes on molasses food were tested; these yielded viable homozygotes in almost the expected proportion, confirming that loss of GALT in flies, as in humans, is not incompatible with life.

Next, Kushner and colleagues tested the ability of GALT-null larvae to survive on food containing glucose spiked with different levels of galactose. Their results demonstrated a clear dose-response sensitivity of the GALT-null, but not control, larvae. Specifically, on food containing 555 mM glucose the majority of GALT-null larvae, like controls, survived to adulthood. On food containing 555 mM glucose plus 222 mM galactose, however, while the control animals continued to thrive, none, or almost none, of the GALT-null larvae survived to adulthood. When an intermediate level of galactose was used, the surviving fraction was intermediate. When a comparable level of mannose was spiked into the food in place of galactose there was no detectable impact on survival of either GALT-null or control animals, demonstrating the specificity of the result for galactose.

To confirm that the galactose sensitivity of $dGALT^{\Delta AP2}$ homozygotes resulted from loss of GALT and not some other off-target effect of the deletion, Kushner and colleagues created a human *GALT* transgene and crossed this allele, with an appropriate driver, into flies lacking endogenous GALT. As expected, expression of human GALT rescued survival of $dGALT^{\Delta AP2}$ homozygotes on food containing otherwise lethal levels of galactose.

To test whether GALT-null *Drosophila* larvae, like neonatal patients, could be rescued by switching from high (222 mM) to low (0 mM) galactose food, Kushner and colleagues conducted a time course experiment in which larvae were physically transferred from one food source to another—and then followed to see if they survived. Specifically, cohorts of L1 larvae were collected from egg-laying plates manually using a small spatula and transferred to the desired food (e.g., glucose food) for the desired length of time, then transferred again into individual wells of a 96-well plate containing the second food type (e.g., glucose plus 222 mM galactose food) where they could be monitored daily for viability and development. The results clearly demonstrated that GALT-null larvae could be rescued by transferring to 0 mM galactose food after 2 days on high-galactose food (L1 to L2 stage). After 4 days on high-galactose food only a fraction could be rescued (L2 to L3 stage). After 6 days on high-galactose food (L3 stage) almost none could be rescued. Conversely, a fraction of GALT-null larvae transferred to high-galactose food after up to 4 days on 0 mM galactose food (L2 to L3) nonetheless survived, and larvae transferred to high-galactose food after 6 days on 0 mM galactose food (L3s) almost uniformly survived. These data suggest that the window of greatest galactose sensitivity is in the first days of life, and that the "damage" done by early galactose exposure is largely reversible if caught sufficiently early. This result is fully consistent with studies documenting that outcomes of patients diagnosed with classic galactosemia as infants after beginning to drink milk are comparable to outcomes of affected infants who never drank milk (Hughes et al., 2009).

2.3 A Motor Defect in Adult GALT-Null *Drosophila*

A significant fraction of patients with classic galactosemia demonstrate motor complications, especially as adults (Waisbren et al., 2012; Walter & Fridovich-Keil, 2014), despite early detection and continued dietary restriction of galactose. To test whether GALT-null flies also exhibit some form of motor disturbance, Kushner and colleagues subjected cohorts of mutant and control animals raised on low-galactose food to a repetitive climbing task using a "counter current" device first introduced by Benzer in the 1960s (Benzer, 1967). Consistent with the results of Benzer, Kushner and colleagues found that >60% of control flies climbed well, ending up in the final tube of the device. In contrast, cohorts of GALT-null flies ended up distributed across the series of tubes, with only about 20% making it to the final tube. As earlier, expression of a human *GALT* transgene rescued the climbing defect otherwise seen in $dGALT^{\Delta AP2}$ homozygotes.

2.4 Mediators of Long-Term Outcome Severity in GALT-Null *Drosophila*

One of the striking features of classic galactosemia is the variability of long-term outcomes among patients. Ryan, DuBoff, Feany, and Fridovich-Keil (2012) applied the GALT-null *Drosophila* model in a set of experiments testing the impact of two candidate modifiers of long-term outcome severity: trace residual GALT activity and low-level dietary exposure to galactose. The outcome followed was the ability of adult flies to traverse a countercurrent device to reach the final tube.

Trace GALT activity was achieved by using a leaky human *GALT* transgene, $UAS\text{-}hGALT^{10A11}$, in the absence of a *GAL4* driver. When present in single copy this transgene conferred 2.3% wild-type GALT activity; when present in double copy this transgene conferred 6.5% wild-type GALT activity (Ryan et al., 2012). In both cases, adult flies expressing these trace levels of GALT activity demonstrated significantly improved climbing ability as measured using the countercurrent device. Consistent with this result, subsequent studies in patients with classic galactosemia demonstrated that *GALT* genotypes linked with at least 0.4% predicted residual GALT activity were associated with significantly milder scholastic (Ryan et al., 2013) and ovarian (Spencer et al., 2013) outcomes.

Impact of low-level galactose exposure on long-term outcome was tested by comparing the climbing phenotypes of flies raised on food containing either glucose as the sole sugar, or glucose plus a sublethal dose of galactose (50 mM). Of note, while insufficient to cause mortality in GALT-null larvae, 50 mM galactose added to the fly food did cause a >20-fold increase in the level of Gal-1P detected in lysates (Ryan et al., 2012). Nonetheless, being reared in the presence of 50 mM galactose did not cause any significant impact on the climbing phenotype of control or GALT-null flies (Ryan et al., 2012). This result was consistent with anecdotal data from patients, suggesting that more lenient diets that include slightly increased galactose are not deleterious to long-term outcomes (Jumbo-Lucioni et al., 2012; van Calcar et al., 2014). These data are also consistent with reports demonstrating a lack of association between patient Gal-1P levels and long-term outcome severity (reviewed in Walter & Fridovich-Keil, 2014).

2.5 Oxidative Stress and GALT Deficiency

Jumbo-Lucioni et al. (2013) applied the GALT-null *Drosophila* model to test the hypothesis that galactose exposure leads to increased oxidative stress, and that GALT deficiency makes animals more vulnerable to that stress. A number of different experimental approaches were applied. First, these authors tested the impact of oxidants (paraquat and DMSO) and antioxidants (vitamin C and α-mangostin) on survival of control and GALT-null *Drosophila* larvae exposed to food containing 555 mM glucose with or without 200 mM galactose. Both oxidants compromised the survival of GALT-null but not control larvae, and both antioxidants improved survival of GALT-null but not control larvae. Of note, these changes in survival were not accompanied by significant changes in the levels of Gal-1P accumulated, suggesting that the mechanism of oxidant or antioxidant impact was either independent, or downstream, of Gal-1P.

Finally, Jumbo-Lucioni et al. (2013) tested the impact of galactose, paraquat, and vitamin C on the levels and ratios of oxidized and reduced glutathione and cysteine, and on the expression of two glutathione-*S*-transferase genes known to be responsive to oxidative stress. The results demonstrated that galactose exposure alone caused modest changes in markers of oxidative stress, and that the combination of galactose plus paraquat appeared almost synergistic. While changes were seen in both controls and GALT-null *Drosophila*, the levels of these changes differed between cases and controls, suggesting that loss of GALT may impact the pathways of either sensing or response to oxidative stress. These results also raised the possibility that patients with classic galactosemia might be unusually vulnerable to environmental oxidants or other exposures.

In a follow-up study Jumbo-Lucioni, Ryan, et al. (2014) extended from their earlier findings by testing the impact of each of two manganese porphyrin mimics of superoxide dismutase (SOD), MnTnBuOE-2-PyP5+ and MnTE-2-PyP5+ (Batinic-Haberle et al., 2011), on both acute and long-term outcomes in GALT-null *Drosophila*. Both SOD mimics significantly improved the survival rates of GALT-null but not control larvae exposed to galactose. As seen earlier, this impact occurred despite the continued accumulation of high Gal-1P in the galactose-exposed GALT-null animals. One mimic, MnTE-2-PyP5+, also notably improved the climbing phenotype of GALT-null adult flies, suggesting that oxidative stress may contribute to long-term as well as acute outcomes in classic galactosemia.

2.6 Potential Impact of GALT Deficiency on Synaptic Architecture and Synaptomatrix Glycosylation at the Neuromuscular Junction in *Drosophila*

Recently, Jumbo-Lucioni, Parkinson, and Broadie (2014) reported further studies of GALT-null *Drosophila*, raising the possibility of altered synaptic architecture in the mutant animals. Specifically, these authors found altered bouton number and also altered lectin binding at the neuromuscular junction in GALT-null larvae. These authors also reported an abnormal coordination/movement phenotype in GALT-null larvae. Surprisingly, many of the phenotypes reported were both relieved by deletion of GALK and also phenocopied in wild-type animals exposed to galactose. Given that infants who do not have galactosemia routinely drink large quantities of milk, which contains high levels of galactose, without experiencing neuromuscular problems, it is difficult to know how to relate these results in *Drosophila* to human outcomes. Further studies will be required to connect these findings with potential mechanisms of outcome among galactosemia patients.

3. A *D. MELANOGASTER* MODEL OF EPIMERASE DEFICIENCY GALACTOSEMIA

3.1 Creation of GALE-Deficient *Drosophila*

Sanders, Sefton, Moberg, and Fridovich-Keil (2010) developed the first whole-animal model of GALE deficiency using *D. melanogaster*. These authors used two severe loss-of-function alleles and one hypomorphic allele of *dGALE* to characterize the effects of GALE loss. One loss-of-function allele, $dGALE^{f00624}$, was the result of a P-element insertion in the second intron of *dGALE*. This allele ($PBac\{WH\}CG12030^{f00624}$) was found in the Exelixis collection (Thibault et al., 2004). Flies homozygous for $dGALE^{f00624}$ displayed 100% lethality, and lysates from flies heterozygous for this allele exhibited a 50% reduction in GALE activity compared to lysates from wild-type animals.

Two additional alleles were created by imprecise excision of a P-element, $P\{EPgy2\}CG12030^{EY22205}$, located in the 5′ UTR of the *dGALE* gene. One of these excisions, $dGALE^{\Delta y}$, resulted in a 1339-bp deletion in *dGALE* that included the first coding exon. This allele displayed a homozygous lethal phenotype that was rescued by expression of a human *GALE* transgene (described later; Sanders et al., 2010), and lysates from flies heterozygous for $dGALE^{\Delta y}$ demonstrated approximately 50% GALE enzymatic activity compared with lysates from wild-type animals. Another allele, $dGALE^{h}$, resulted from partial excision of the P-element such that 1500 bp of P-element sequence was left behind in the 5′ UTR; the rest of the *dGALE* gene remained unchanged. This disruption of the 5′ UTR yielded a hypomorphic allele; flies homozygous for $dGALE^{h}$ demonstrated approximately 8% of wild-type GALE activity.

3.2 Characterizing the Phenotype of GALE Deficiency in *Drosophila*

The work described in this section was published by Sanders et al. (2010) unless otherwise noted.

3.2.1 dGALE *Is Required Throughout Development*

Both profound loss-of-function *dGALE* alleles, $dGALE^{f00624}$ and $dGALE^{\Delta y}$, demonstrated a homozygous lethal phenotype, and *trans*-heterozygotes were also not viable. These mutants displayed a variable period of death such that some animals died during embryonic development prior to hatching, while others survived to the second instar larval stage (L2). Because maternally loaded *dGALE* mRNA or protein could be present during early stages of development, Sanders et al. created $dGALE^{f00624}$ germline clone mutants which were unable to contribute maternal loading of GALE to their offspring. The *dGALE*-null offspring from these mothers died uniformly in late embryogenesis. Notably, a wild-type paternal *dGALE* allele rescued the germline clone mutants. Together, these data demonstrated that *dGALE* is essential during late embryogenesis, but is not required before the onset of zygotic transcription.

To look at the requirement for GALE in developmental stages beyond embryogenesis, Sanders et al. used RNAi knockdown of *dGALE* in combination with a temperature sensitive allele of *GAL80* to generate a series of *Drosophila* that experienced knockdown of *dGALE* on successive days of development. Knockdown during larval or pupal stages of development resulted in less than 8% residual GALE activity compared to control animals. GALE loss was lethal at all stages of embryonic and larval development. During pupal stages of development, *dGALE* knockdown was not lethal;

however, the adult flies displayed problems including a reduced life span and infertility. These adult phenotypes were dependent on the stage of pupal development at which GALE was lost, with the phenotype being more severe the earlier *dGALE* knockdown was initiated. The effects of *dGALE* knockdown initiated after eclosion of adult flies could not be evaluated, as the knockdown was inefficient when initiated at this stage.

3.2.2 *GALE Activity Is Required in Specific Tissues in* Drosophila

To assess tissue-specific requirements for GALE, Sanders et al. (2010) used two approaches: first, knocking down endogenous *dGALE* expression in specific tissues via RNAi; and second, expressing a human *GALE* transgene in various tissues of otherwise *dGALE*-null animals. Knockdown of *dGALE* was tested using *GAL4* drivers specific to five tissues or tissue combinations. Knockdown of *dGALE* in the salivary gland, neurons, larval brain and fat body, or eye did not result in any clear phenotypes. In contrast, knockdown in the embryonic proventriculus, anterior midgut, posterior midgut, Malpighian tubules, and small intestine by the *drm-GAL4* driver resulted in lethality, indicating that GALE is required in some or all of these tissues. The same drivers were also used to express hGALE in a tissue-specific manor in *dGALE*-null animals to test whether human GALE (hGALE) expression in any of these tissues was sufficient for rescue. hGALE expression driven by *drm-GAL4* in the embryonic proventriculus, anterior midgut, posterior midgut, Malpighian tubules, and small intestine resulted in a partial rescue; however, the proportion of animals rescued was approximately one-tenth the expected number. None of the other drivers tested resulted in any degree of rescue. The authors speculated that the requirement for *dGALE* in the gut and tubules might be related to the expression of glycans, since GALE produces UDPgalNAc, the obligate first sugar donor in mucin-type O-linked glycosylation.

3.2.3 *Galactose Sensitivity of* dGALE *Hypomorphs*

Infants with generalized epimerase deficiency galactosemia demonstrate acute sensitivity to dietary galactose (Henderson & Holton, 1983; Holton et al., 1981; Sardharwalla, Wraith, Bridge, Fowler, & Roberts, 1988). Sanders et al. (2010) tested whether *Drosophila* with low-level GALE activity would also display sensitivity to galactose. Specifically, flies homozygous for the hypomorphic allele *dGALE*^{h}, which retains approximately 8% GALE activity, were challenged in development with a diet containing 555 mM glucose plus 111 mM galactose. In progeny of crosses where the expected ratio of $dGALE^h$ homozygous offspring was 50%, only about 25% of the surviving animals were homozygotes. However, when the food contained only glucose, or glucose plus mannose, rather than galactose, the expected proportion of homozygous offspring was observed. These results confirmed that galactose exposure caused lethality in development of *Drosophila* lacking normal GALE activity.

3.2.4 GALE-*Deficient* Drosophila *Accumulate a High Level of Gal-1P*

dGALE knockdown larvae developing on food containing 555 mM glucose plus 111 mM galactose accumulated approximately 20-fold more Gal-1P than corresponding animals developing on food containing only 555 mM glucose, with no galactose. This increase is much larger than was seen in control larvae expressing wild-type GALE, which accumulated only about twice as much Gal-1P on galactose-containing food compared with "glucose-only" food. The elevated levels of Gal-1P in *GALE*-deficient flies were consistent with data from *GALE*-impaired yeast (Douglas & Hawthorne, 1964; Ross, Davis, & Fridovich-Keil, 2004), mammalian tissue culture cells (Schulz, Ross, Malmstrom, Krieger, & Fridovich-Keil, 2005), and patients (Openo et al., 2006), all of which accumulate Gal-1P upon exposure to high levels of environmental galactose.

3.3 Using *Drosophila* to Dissect the Differential Roles of GALE in Development

The work described in this section is from Daenzer, Sanders, Hang, and Fridovich-Keil (2012) unless otherwise noted.

3.3.1 Uncoupling the Two Activities of GALE *in* Drosophila

GALE from both humans and *Drosophila* interconverts two sets of substrates: UDPgal/UDPglc and UDPgalNAc/UDPglcNAc (Fig. 1). To uncouple and test the developmental roles of these activities individually, Daenzer and colleagues created flies with activity toward only one or the other substrate set. This substrate specificity was achieved by effectively replacing *dGALE* expression with expression of either of two prokaryotic epimerase genes, each encoding a product capable of interconverting only one substrate pair: *eGALE*, which interconverts only UDPgal/UDPglc, and *wbgU*, which interconverts only UDPgalNAc/UDPglcNAc. Expressing these microbial transgenes in *dGALE*-deficient *Drosophila* resulted in animals with only one or the other GALE activity.

3.3.2 Both GALE Activities Play Essential Roles During Development

As explained earlier, Sanders et al. (2010) showed that GALE is essential for *Drosophila* development, and that dGALE-deficient animals die in embryogenesis. By expressing *eGALE* only, *wbgU* only, or both *eGALE* and *wbgU* together in dGALE-deficient embryos, Daenzer and colleagues confirmed that GALE activity toward both substrate pairs is essential for survival of *Drosophila*. Specifically, neither microbial transgene alone enabled survival, but together they did, and further, hGALE, which like dGALE recognizes both substrate pairs, was also sufficient.

Using a modified version of the conditional knockdown technique described in Sanders et al. (2010), Daenzer et al. (2012) created a time course series of *Drosophila* that experienced *dGALE* knockdown with concurrent induction of either *hGALE*, *eGALE*, *wbgU*, or both *eGALE and wbgU* at successive stages of development. As expected, RNAi knockdown of *dGALE* during both embryonic and larval stages of development was lethal, but was rescued by expression of either *hGALE* or *eGALE* + *wbgU*. Surprisingly, expression of either *eGALE* or *wbgU* alone was also sufficient for survival when *dGALE* knockdown was initiated during the early stages of development, though to a lesser extent. This result suggested that residual dGALE activity remaining after knockdown, or existing substrate and product pools already accumulated in the animals, may have reduced the requirement for transgene activity. However, animals experiencing *dGALE* knockdown with replacement by either *eGALE* only or *wbgU* only, at any stage prior to late pupation, exhibited marked fecundity defects. Interestingly, the fecundity defects differed between males and females, and were dependent on which GALE activity was missing. Specifically, *dGALE* knockdown during early to mid-pupation resulted in fecundity defects in both male and female flies. GALE activity toward UDPgal/UDPglc alone, but not UDPgalNAc/UDPglcNAc alone, was sufficient to rescue the male defect. However, both GALE activities were required to rescue female fecundity.

3.3.3 GALE Activity Toward UDPgal/UDPglc Is Required for Normal Life Span of Adult Drosophila *Exposed to Dietary Galactose*

Sanders et al. (2010) demonstrated that *dGALE* hypomorphs developing in the presence of environmental galactose displayed a reduction in viability. Daenzer et al. (2012) extended from that finding by measuring the life spans of *dGALE*-knockdown animals expressing neither, only one, or both GALE activities. Specifically, flies experiencing *dGALE* knockdown during early to mid-pupal stages were allowed to develop on a standard molasses food diet and, upon eclosion, were moved to food containing either 555 mM glucose or 555 mM glucose plus 175 mM galactose. The life span of *dGALE*-knockdown animals eating glucose-only food did not differ significantly from that of control animals. However, when exposed to galactose as adults, the *dGALE*-knockdown animals displayed a dramatic reduction in life span such that in the absence of knockdown close to 50% of animals remained viable at 35 days, but in the presence of knockdown close to half the animals had died by 10 days and almost no animals remained alive at 35 days. Expression of GALE activity toward UDPgalNAc/UDPglcNAc alone (*wbgU* transgene) was not sufficient to overcome the galactose-dependent reduction in life span. In contrast, *dGALE*-knockdown animals in which GALE activity toward UDPgal/UDPglc was restored (*eGALE* transgene) survived equally well in the presence and absence of dietary galactose. These data confirmed that loss of activity toward UDPgal/UDPglc was responsible for the galactose-induced reduction of life span in *dGALE*-knockdown animals.

3.3.4 The Two GALE *Activities Impact Galactose Metabolite Levels Differently*

While it is clear that *dGALE*-impaired *Drosophila* experience a variety of acute and long-term outcomes, the pathophysiology of these outcomes remains unclear. Daenzer and colleagues gained some insights into pathophysiology from studying the galactose metabolite levels in their differentially impaired animals. For example, *Drosophila* in which *dGALE* knockdown occurred early in development accumulated very high levels of Gal-1P when exposed to galactose as larvae. These larvae also accumulated very high levels of UDPgal. Similarly, larvae deficient only in activity toward UDPgal/UDPglc accumulated very high levels of both Gal-1P and UDPgal when they developed in the presence, but not in the absence, of high dietary galactose. In contrast, animals deficient only in GALE activity toward UDPgalNAc/UDPglcNAc did not accumulate abnormal levels of either Gal-1P or UDPgal. Nonetheless, these animals were not viable. Clearly, elevated Gal-1P and UDPgal cannot be the only cause of pathophysiology in *GALE* deficiency.

4. A *D. MELANOGASTER* MODEL OF KINASE DEFICIENCY GALACTOSEMIA

A deletion allele of d*GALK* was created by Garza and colleagues in the Fridovich-Keil lab by imprecise excision of a nearby P-element insertion;

this allele was used by Jumbo-Lucioni, Parkinson, et al. (2014) and is described in a forthcoming publication from the Fridovich-Keil lab (Daenzer et al., in preparation). A number of different d*GALK* alleles or GALK knockdown experiments were also described recently by Lee, Yu, Wolf, and Rockman (2014), who reported that loss of GALK modified the cardiac outcomes otherwise seen in response to expression of constitutively active calcineurin in *Drosophila*. The mechanism of this interaction remains unclear, as does its relevance to galactosemia.

5. CONCLUSIONS

In the past 6 years, *D. melanogaster* models of GALT and GALE deficiency have offered powerful insights into Type I and Type III galactosemia, and studies of GALK loss in *Drosophila* have raised the possibility that this gene and encoded enzyme may play important roles in biological processes beyond those typically associated with metabolic disease. Future studies leveraging the genetic and biochemical facility of these and additional models will be aimed at identifying genetic and environmental modifiers of outcomes beyond the Leloir pathway and galactose exposure, understanding mechanism, and seeking to identify pharmacological agents to prevent or reverse the complications of galactosemia in patients.

ACKNOWLEDGMENTS

The authors gratefully acknowledge funding from the National Institutes of Health (1R01DK107900 to J.L.F.-K.).

REFERENCES

Batinic-Haberle, I., Rajic, Z., Tovmasyan, A., Reboucas, J. S., Ye, X. D., Leong, K. W., et al. (2011). Diverse functions of cationic Mn(III) N-substituted pyridylporphyrins, recognized as SOD mimics. *Free Radical Biology and Medicine*, *51*(5), 1035–1053.

Benzer, S. (1967). Behavioral mutants of Drosophila isolated by countercurrent distribution. *Proceedings of the National Academy of Sciences of the United States of America*, *58*(3), 1112–1119.

Berry, G. (2012). Galactosemia: When is it a newborn screening emergency? *Molecular Genetics and Metabolism*, *106*(1), 7–11.

Berry, G. (2014). Classic galactosemia and clinical variant galactosemia. In R. Pagon, et al. (Eds.), GeneReviews®. Seattle, WA: University of Washington, Seattle. http://www.ncbi.nlm.nih.gov/books/NBK1518/.

Bosch, A. M., Bakker, H. D., van Gennip, A. H., van Kempen, J. V., Wanders, R. J., & Wijburg, F. A. (2002). Clinical features of galactokinase deficiency: A review of the literature. *Journal of Inherited Metabolic Disease*, *25*(8), 629–634.

Chien, S., Reiter, L. T., Bier, E., & Gribskov, M. (2002). Homophila: Human disease gene cognates in Drosophila. *Nucleic Acids Research*, *30*(1), 149–151.

Daenzer, J. M., Sanders, R. D., Hang, D., & Fridovich-Keil, J. L. (2012). UDP-galactose 4′-epimerase activities toward UDP-Gal and UDP-GalNAc play different roles in the development of Drosophila melanogaster. *PLoS Genetics*, *8*(5), e1002721.

Douglas, H. C., & Hawthorne, D. C. (1964). Enzymatic expression and genetic linkage of genes controlling galactose utilization in Saccharomyces. *Genetics*, *49*, 837–844.

Fridovich-Keil, J., Bean, L., He, M., & Schroer, R. (2011). Epimerase deficiency galactosemia. In R. Pagon, et al. (Eds.), GeneReviews. Seattle, WA: University of Washington, Seattle. http://www.ncbi.nlm.nih.gov/books/NBK51671/.

Gitzelmann, R. (1972). Deficiency of uridine diphosphate galactose 4-epimerase in blood cells of an apparently healthy infant. Preliminary communication. *Helvetica Paediatrica Acta*, *27*(2), 125–130.

Gitzelmann, R., & Steinmann, B. (1973). Uridine diphosphate galactose 4-epimerase deficiency. II. Clinical follow-up, biochemical studies and family investigation. *Helvetica Paediatrica Acta*, *28*(6), 497–510.

Henderson, M. J., & Holton, J. B. (1983). Further observations in a case of uridine diphosphate galactose-4-epimerase deficiency with a severe clinical presentation. *Journal of Inherited Metabolic Disease*, *6*, 17–20.

Hennermann, J. B., Schadewaldt, P., Vetter, B., Shin, Y. S., Monch, E., & Klein, J. (2011). Features and outcome of galactokinase deficiency in children diagnosed by newborn screening. *Journal of Inherited Metabolic Disease*, *34*(2), 399–407.

Holden, H. M., Rayment, I., & Thoden, J. B. (2003). Structure and function of enzymes of the Leloir pathway for galactose metabolism. *The Journal of Biological Chemistry*, *278*(45), 43885–43888.

Holton, J. B., Gillett, M. G., Macfaul, R., & Young, R. (1981). Galactosemia: A new severe variant due to uridine-diphosphate galactose-4-epimerase deficiency. *Archives of Disease in Childhood*, *56*(11), 885–887.

Hughes, J., Ryan, S., Lambert, D., Geoghegan, O., Clark, A., Rogers, Y., et al. (2009). Outcomes of siblings with classical galactosemia. *The Journal of Pediatrics*, *154*(5), 721–726.

Jumbo-Lucioni, P. P., Garber, K., Kiel, J., Baric, I., Berry, G. T., Bosch, A., et al. (2012). Diversity of approaches to classic galactosemia around the world: A comparison of diagnosis, intervention, and outcomes. *Journal of Inherited Metabolic Disease*, *35*(6), 1037–1049.

Jumbo-Lucioni, P. P., Hopson, M. L., Hang, D., Liang, Y., Jones, D. P., & Fridovich-Keil, J. L. (2013). Oxidative stress contributes to outcome severity in a Drosophila melanogaster model of classic galactosemia. *Disease Models & Mechanisms*, *6*(1), 84–94.

Jumbo-Lucioni, P., Parkinson, W., & Broadie, K. (2014). Overelaborated synaptic architecture and reduced synaptomatrix glycosylation in a Drosophila classic galactosemia disease model. *Disease Models & Mechanisms*, 7(12), 1365–1378.

Jumbo-Lucioni, P. P., Ryan, E. L., Hopson, M. L., Bishop, H. M., Weitner, T., Tovmasyan, A., et al. (2014). Manganese-based superoxide dismutase mimics modify both acute and long-term outcome severity in a Drosophila melanogaster model of classic galactosemia. *Antioxidants & Redox Signaling*, *20*(15), 2361–2371.

Keyes, L. N., & Spradling, A. C. (1997). The Drosophila gene fs(2)cup interacts with otu to define a cytoplasmic pathway required for the structure and function of germ-line chromosomes. *Development*, *124*(7), 1419–1431.

Kushner, R. F., Ryan, E. L., Sefton, J. M., Sanders, R. D., Lucioni, P. J., Moberg, K. H., et al. (2010). A Drosophila melanogaster model of classic galactosemia. *Disease Models & Mechanisms*, *3*(9–10), 618–627.

Lee, T. E., Yu, L., Wolf, M. J., & Rockman, H. A. (2014). Galactokinase is a novel modifier of calcineurin-induced cardiomyopathy in Drosophila. *Genetics*, *198*, 591–603.

Mitchell, B., Haigis, E., Steinmann, B., & Gitzelmann, R. (1975). Reversal of UDP-galactose 4-epimerase deficiency of human leukocytes in culture. *Proceedings of the National Academy of Sciences of the United States of America*, 72(12), 5026–5030.

Openo, K., Schulz, J., Vargas, C., Orton, C., Epstein, M., Schnur, R., et al. (2006). Epimerase-deficiency galactosemia is not a binary condition. *American Journal of Human Genetics*, *78*(1), 89–102.

Pyhtila, B. M., Shaw, K. A., Neumann, S. E., & Fridovich-Keil, J. L. (2015). Newborn screening for galactosemia in the United States: Looking back, looking around, and looking ahead. *JIMD Reports*, *15*, 79–93.

Riehman, K., Crews, C., & Fridovich-Keil, J. L. (2001). Relationship between genotype, activity, and galactose sensitivity in yeast expressing patient alleles of human galactose-1-phosphate uridylyltransferase. *Journal of Biological Chemistry*, *276*(14), 10634–10640.

Ross, K. L., Davis, C. N., & Fridovich-Keil, J. L. (2004). Differential roles of the Leloir pathway enzymes and metabolites in defining galactose sensitivity in yeast. *Molecular Genetics and Metabolism*, *83*(1–2), 103–116.

Ryan, E. L., DuBoff, B., Feany, M. B., & Fridovich-Keil, J. L. (2012). Mediators of a long-term movement abnormality in a Drosophila melanogaster model of classic galactosemia. *Disease Models & Mechanisms*, *5*(6), 796–803.

Ryan, E. L., Lynch, M. E., Taddeo, E., Gleason, T. J., Epstein, M. P., & Fridovich-Keil, J. L. (2013). Cryptic residual GALT activity is a potential modifier of scholastic outcome in school age children with classic galactosemia. *Journal of Inherited Metabolic Disease*, *36*(6), 1049–1061.

Sanders, R. D., Sefton, J. M., Moberg, K. H., & Fridovich-Keil, J. L. (2010). UDP-galactose 4′ epimerase (GALE) is essential for development of Drosophila melanogaster. *Disease Models & Mechanisms*, *3*(9–10), 628–638.

Sardharwalla, I. B., Wraith, J. E., Bridge, C., Fowler, B., & Roberts, S. A. (1988). A patient with severe type of epimerase deficiency galactosemia. *Journal of Inherited Metabolic Disease*, *11*(Suppl. 2), 249–251.

Schulz, J. M., Ross, K. L., Malmstrom, K., Krieger, M., & Fridovich-Keil, J. L. (2005). Mediators of galactose sensitivity in UDP-galactose 4′-epimerase-impaired mammalian cells. *The Journal of Biological Chemistry*, *280*(14), 13493–13502.

Spencer, J. B., Badik, J. R., Ryan, E. L., Gleason, T. J., Broadaway, K. A., Epstein, M. P., et al. (2013). Modifiers of ovarian function in girls and women with classic galactosemia. *The Journal of Clinical Endocrinology and Metabolism*, *98*(7), E1257–E1265.

Thibault, S. T., Singer, M. A., Miyazaki, W. Y., Milash, B., Dompe, N. A., Singh, C. M., et al. (2004). A complementary transposon tool kit for Drosophila melanogaster using P and piggyBac. *Nature Genetics*, *36*(3), 283–287.

van Calcar, S. C., Bernstein, L. E., Rohr, F. J., Scaman, C. H., Yannicelli, S., & Berry, G. T. (2014). A re-evaluation of life-long severe galactose restriction for the nutrition management of classic galactosemia. *Molecular Genetics and Metabolism*, *112*(3), 191–197.

Waisbren, S. E., Potter, N. L., Gordon, C. M., Green, R. C., Greenstein, P., Gubbels, C. S., et al. (2012). The adult galactosemic phenotype. *Journal of Inherited Metabolic Disease*, *35*(2), 279–286.

Walter, J. H., & Fridovich-Keil, J. L. (2014). Galactosemia. In D. Valle, A. L. Beaudet, B. Vogelstein, K. W. Kinzler, S. E. Antonarakis, A. Ballabio, K. Gibson, & G. Mitchell (Eds.), New York, NY: McGraw-Hill. http://ommbid.mhmedical.com/content.aspx?bookid=971&Sectionid=62672411. Accessed July 25, 2016.

Walter, J. H., Roberts, R. E., Besley, G. T., Wraith, J. E., Cleary, M. A., Holton, J. B., et al. (1999). Generalised uridine diphosphate galactose-4-epimerase deficiency. *Archives of Disease in Childhood*, *80*(4), 374–376.

Wohlers, T. M., Christacos, N. C., Harreman, M. T., & Fridovich-Keil, J. L. (1999). Identification and characterization of a mutation, in the human UDP-galactose-4-epimerase gene, associated with generalized epimerase-deficiency galactosemia. *American Journal of Human Genetics*, *64*(2), 462–470.

Wohlers, T., & Fridovich-Keil, J. L. (2000). Studies of the V94M-substituted human UDP-galactose-4-epimerase enzyme associated with generalized epimerase-deficiency galactosemia. *Journal of Inherited Metabolic Disease*, *23*, 713–729.

CHAPTER THIRTEEN

Drosophila as a Model for Diabetes and Diseases of Insulin Resistance

P. Graham, L. Pick[1]
University of Maryland, College Park, MD, United States
[1]Corresponding author: e-mail address: lpick@umd.edu

Contents

Abstract

Despite the importance of insulin signaling pathways in human disease, initial concerns that insect physiology and sugar metabolism differ enough from humans that flies would not model human disease hampered research in this area. However, during the past 10–15 years, evidence has accumulated that flies can indeed model various aspects of diabetes and related human disorders. This cluster of diseases impact insulin and insulin signaling pathways, fields which have been discussed in many excellent review articles in recent years. In this chapter, we restrict our focus to specific examples of diabetes-related disease models in *Drosophila*, discussing the advantages and limitations of these models in light of physiological similarities and differences between insects and mammals. We discuss features of metabolism and sugar regulation that are shared between flies and mammals, and specific *Drosophila* models for Type 1 and Type 2 diabetes, Metabolic syndrome, and related abnormalities including insulin resistance and heart disease. We conclude that fly models for diabetes and related disorders enhance our ability to identify genes and discern functional interactions that can be exploited for disease intervention.

Current Topics in Developmental Biology, Volume 121
ISSN 0070-2153
http://dx.doi.org/10.1016/bs.ctdb.2016.07.011

1. INTRODUCTION

Diabetes and related metabolic disorders are growing health problems worldwide. In the United States alone, incidence of diabetes has increased approximately fourfold between 2008 and 2014 to more than 20 million affected individuals (Centers for Disease Control and Prevention, 2016a). Incidence has been increasing for both adults and children, with differential impact on different ethnic groups. Most severely affected are American Indians and Alaskan Natives, followed by Non-Hispanic Blacks, Hispanics, and Asian Americans. Non-Hispanic Whites show the lowest incidence. Perhaps most alarming in the United States, the CDC SEARCH study noted a 30% rise in diabetic children in recent years (Centers for Disease Control and Prevention, 2016b). Diabetes comprises a set of metabolic disorders defined by increased levels of circulating blood sugar (hyperglycemia) caused by abnormal insulin secretion and/or signaling (American Diabetes Association, 2014). The most common form of childhood diabetes, Type I diabetes (T1D), accounts for 5–10% of all diabetes cases. T1D is an autoimmune disorder that causes destruction of insulin-producing β-cells of the pancreas, resulting in decreased or complete loss of insulin. Among children, the incidence of T1D has increased in recent years (a 20% increase between 2001 and 2009) with the largest rise in Non-Hispanic white children.

Type 2 diabetes (T2D), originally named adult-onset diabetes, accounts for >90% of overall diabetes cases. T2D now affects both children and adults and has a more complex etiology than T1D. It is a disease of insulin resistance, such that hyperglycemia persists despite the presence of high levels of circulating insulin. The incidence of T2D has increased 30% between 2001 and 2009 with American Indians and Alaskan Natives and Non-Hispanic Blacks being disproportionately affected. Correlated, and likely causal in the increasing numbers of individuals with T2D, is the increased incidence of obesity and sedentary lifestyle. It is estimated that approximately 79 million adults (35%) and approximately 13 million children (17%) in the United States are obese—this despite recent widely reported decreases in obesity rates among children this past year. Long-term complications of diabetes include neuropathy, retinopathy, heart disease, and stroke. The cost is measured not only in pain and suffering, but there is also an estimated $245 billion in medical costs and lost wages per year. The differential distribution of disease among ethnic groups points to both environmental and genetic factors influencing the incidence of disease.

Prediabetes, a precursor to T2D, is characterized by insulin resistance and represents a syndrome in its own right. This condition is thought to be frequently undiagnosed, with an estimated 86 million adults (one-third of the US population) affected. Prediabetes can often be controlled by changes in diet and exercise but roughly one quarter of individuals with prediabetes progress to T2D over a 5-year period. A number of other metabolic disorders fall under the general umbrella of diabetes-related disorders. These include gestational diabetes, maturity onset diabetes of the young (MODY) and Metabolic syndrome (MetS) (also known as Syndrome X). Because of their wide prevalence, diabetes and related disorders have been studied and reviewed extensively, as have the roles of insulin signaling pathways in both mammals and *Drosophila* (for examples, Baker & Thummel, 2007; Das & Dobens, 2015; Garofalo, 2002; Goberdhan & Wilson, 2003; Kahn et al., 2005; Lasko, 2002; Owusu-Ansah & Perrimon, 2014; Padmanabha & Baker, 2014; Tatar, 2004; Wu & Brown, 2006 #141). In this chapter, we have focused on recent studies in which *Drosophila* models have been developed for specific disease features and types.

2. SUGAR METABOLISM IN *DROSOPHILA* AND HUMANS

The degree to which any animal can serve as a model for human disease depends upon the degree of similarity among pathways and physiological responses. In some cases, *Drosophila* can serve as an "in vivo test tube" to test gene function, despite the fact that the physiology differs. For example, the *Drosophila* eye is an ideal organ to test cellular functions of genes involved in cell growth and viability, even if those genes do not normally function in the fly eye (Bonini & Fortini, 2002; He et al., 2014). Here, we are examining the use of *Drosophila* as a whole animal model for disease and thus need to ask how conserved are both gene function and physiology. Where do similarities reflect homology and common ancestry, and where are similarities merely superficial? Do signaling pathways end in similar outputs or have signals been co-opted to different downstream processes in different species?

The mechanisms that maintain the balance between stored and circulating forms of energy appear to be largely shared among animals (reviewed in Baker & Thummel, 2007). In mammals, when energy is abundant (after eating), excess glucose is stored as glycogen, primarily in skeletal muscle and the liver. Free fatty acids (FFAs) are stored as triglycerides (TGs) predominantly in adipose tissue but also in the liver (Fig. 1). When energy is scarce (after fasting

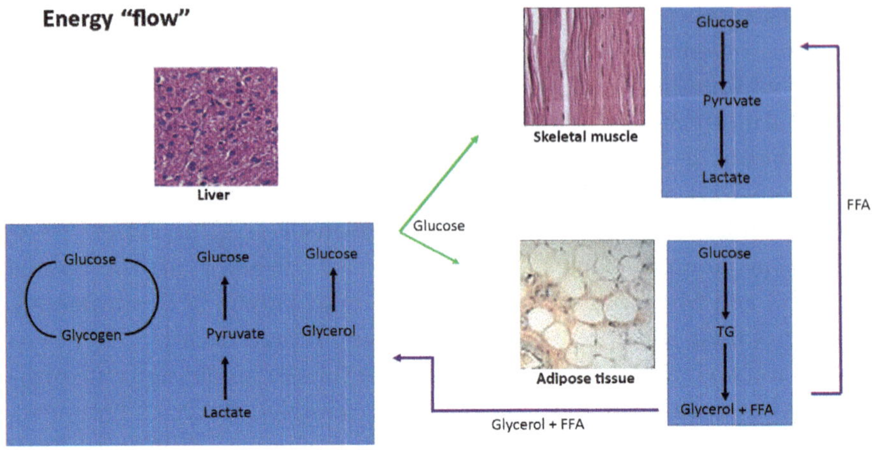

Fig. 1 Flow of energy in mammals. Energy is stored as glycogen, primarily in skeletal muscle and the liver. Free fatty acids (FFAs) are stored as triglycerides (TGs) predominantly in adipose tissue. When energy is scarce, these reserves are mobilized and released for use. The principle organ to mobilize glucose is the liver. The liver can produce glucose in two ways—by breaking down glycogen or by gluconeogenesis. Lactate is produced by skeletal muscles when they use glucose, and glycerol is produced by adipocytes when they mobilize the TGs. The principle cells that normally mobilize FFAs are the adipocytes. *Images from www.slideshare.net/roger961/adipose-cells-kibbe-sandin-112105 and www.slideshare.net/aravindhravi88/histology-of-liver-by-aravindh-dpi/16.*

or exercise), these reserves are mobilized and released for use. Glucose is mobilized first, then FFA. Although skeletal muscle has large reserves of glycogen, the principle organ to mobilize glucose is the liver. The liver can produce glucose in two ways—by breaking down glycogen or by gluconeogenesis from lactate and/or glycerol. Lactate is produced by skeletal muscles when they use glucose, and glycerol is produced by adipocytes when they mobilize the TGs. The principle cells that normally mobilize FFAs are the adipocytes.

Mobilization of both glucose and FFA is regulated by the hormones insulin and glucagon, which act in opposition to one another to keep circulating glucose levels stable. When circulating sugar levels rise, insulin is released from the β-cells in the pancreas, triggering glucose uptake and fuel storage (synthesis of glycogen, TGs, and proteins), and repressing secretion of glucagon. When circulating sugar levels drop, glucagon is released from the α-cells of the pancreas triggering the opposite effects on glycogen, TGs, and proteins and repressing insulin secretion.

Circulating glucose levels are sensed by a family of sugar transporters, GLUT proteins, encoded by the *SLC2* family of genes (reviewed in

Mueckler & Thorens, 2013). While glucose can be taken up into the brain and other tissues by facilitated diffusion via broadly expressed transporters, insulin and insulin-dependent glucose uptake play the major role in homeostatic regulation of circulating sugar levels. Insulin release from the β-cells is exquisitely sensitive to circulating glucose levels, sensed by the glucose transporters GLUT1 and/or GLUT2 at the β-cell surface (reviewed in Joost & Thorens, 2001; Rutter, Pullen, Hodson, & Martinez-Sanchez, 2015). Circulating insulin then promotes the uptake of glucose into peripheral tissues via activation of the insulin-dependent sugar transporter, GLUT4 (Huang & Czech, 2007; Kahn & Cushman, 1986; Lizunov, Matsumoto, Zimmerberg, Cushman, & Frolov, 2005; Lizunov et al., 2012).

Similar processes appear to occur in *Drosophila* in that insulin-like proteins (Ilps) are released in response to high levels of circulating sugar and a glucagon-like molecule, adipokinetic hormone (AKH) is released in response to low levels of circulating sugar (Colombani et al., 2003; Haselton et al., 2010; Kim & Rulifson, 2004; Lee & Park, 2004). These similarities are discussed in more detail later but we first point out some important differences between the fly and mammalian systems. First, in insects, simple sugars from food are taken up passively from the digestive tract directly into the fat body where they are converted to trehalose, a non-reducing sugar. Trehalose can be stored and/or released into the hemolymph as the primary circulating sugar in insects. Only very low levels of glucose are present insect hemolymph, on the order of one one-hundredth the levels of trehalose (Nation, 2002; Ugrankar et al., 2015; Wyatt & Kale, 1957). Because of this, complications of diabetes that result from the nonspecific glycation of proteins due to the reactivity of high concentrations of glucose, will not be modeled in the fly unless levels of reactive sugars are artificially elevated in this system.

Second, some researchers have questioned the extent to which circulating sugar levels are homeostatically regulated at all in insects (Thompson, Borchardt, & Wang, 2003). Several points suggested that circulating sugars levels might not be homeostatically regulated: In several studies using different insects, levels of circulating sugar were shown to rise with increasing dietary sugar and these sugar levels were shown to reflect metabolic activity of the insect, and to some extent, food source (Abou-Seif et al., 1993; Blatt & Roces, 2001; Maurizio, 1965; Thompson, 1999; Thompson & Redak, 2000; Thompson et al., 2003). Given the fact that high levels of circulating trehalose are likely chemically neutral, it could be argued that there would be little pressure to hormonally regulate them. In addition, insects require the ability to rapidly mobilize energy to fuel flight and much of insect metabolism is thought to be centered around this requirement (Candy, Becker, & Wegener, 1997). This requirement for the immediate availability of large stores of energy—that may be needed to power very long flights—appears to be unique to insects. Finally, the fat body acts as a functional homolog of both the liver and adipocytes in insects suggesting a consolidation of physiological function (Arrese & Soulages, 2010). However, as discussed in the remainder of this chapter, and in contrast to earlier findings and expectations, considerable evidence suggests that flies share with mammals evolutionary ancient mechanisms to regulate sugar homeostasis through conserved and functionally similar pathways.

3. *DROSOPHILA* MODELS OF TYPE 1 DIABETES

Insulin-like peptides were discovered in *Drosophila* and other invertebrates in the late 1970s and 1980s (Duve, Thorpe, & Lazarus, 1979; LeRoith, Lesniak, & Roth, 1981; O'Connor & Baxter, 1985; Tager, Markese, Kramer, Speirs, & Childs, 1976), although their roles were not appreciated at the time. Some years later, an insulin-receptor-like protein was purified and the corresponding gene cloned from *Drosophila* (Nishida, Hata, Nishizuka, Rutter, & Ebina, 1986; Petruzzelli, Herrera, Arenas-Garcia et al., 1986; Petruzzelli, Herrera, Garcia-Arenas, & Rosen, 1986). This receptor (DInR or InR) was shown to function similarly to the mammalian insulin receptor in that it has tyrosine kinase activity and autophosphorylates in response to human insulin, but not other peptide hormones (Fernandez-Almonacid & Rosen, 1987). These studies and many others demonstrated that insulin and insulin-like signaling (IIS) pathways are shared between flies and humans (reviewed in Das & Dobens, 2015; Kannan & Fridell, 2013; Oldham, 2011; Taguchi & White, 2008; Teleman, 2010). Interestingly, insects express a large number of Ilps, from eight in *Drosophila* to many more in other invertebrates, but only one insulin-like receptor (reviewed in Wu & Brown, 2006). The different Ilps likely play numerous roles in the animal, although to some extent at least, functions in metabolism appear to be redundant (Gronke, Clarke, Broughton, Andrews, & Partridge, 2010). It thus appears that insects have numerous ligands for one receptor, while mammals have receptors with partially redundant functions, but a restricted number of ligands. The cause of this divergence between insect and mammalian lineages remains an unsolved question in the field.

The first indication that *Drosophila* could serve as a useful model for studying diabetes came from studies of Rulifson, Kim, and Nusse (2002) who showed that ablation of a cluster of brain cells known as insulin-producing cells (IPCs) resulted in animals that display features of T1D (Rulifson et al., 2002). Specifically, these animals displayed an increased level of circulating sugar compared to wild-type controls—the hallmark of diabetes—and this important finding was observed independently by other groups (Broughton et al., 2005; Ikeya, Galic, Belawat, Nairz, & Hafen, 2002). The increase in sugar levels after IPC-ablation was rescued by expression of a *Drosophila* insulin-like peptide (DILP). The authors went on to propose that IPCs are equivalent to the β-pancreatic islet cells that produce insulin in mammals. In addition to the features of T1D, animals in which IPCs were ablated were small and developmentally delayed, reflecting roles of IIS in both growth and metabolism in insects (Rulifson et al., 2002).

A more recent study examined effects of IPC ablation in adult flies. Similar to results in larvae, circulating sugar levels are higher in these flies than in control animals (Haselton et al., 2010). However, in an elegant set of experiments, Haselton et al. (2010) made use of the fact that adult feeding can be manipulated more readily than larval feeding to perform an Oral Glucose Tolerance Test (OGTT), a test used to diagnose human diabetes. Specifically, animals were fasted, then fed on a glucose solution, and circulating sugar levels were measured over time. Wild-type *Drosophila* displayed a mammalian-like response with low-circulating sugar levels following the fast, followed by an initial increase after glucose feeding, followed by clearing and return to baseline levels. Ablation of IPCs resulted in higher circulating sugar levels and slower clearance, a response that was abrogated by injection of bovine insulin. This study arguably provides more convincing evidence for flies accurately modeling T1D than those done in larvae, as it made use of a test similar to OGTT, a diagnostic for human diabetes.

Does loss of DILP function explain phenotypes seen for IPC ablation? While IPCs clearly produce DILPs, these cells may have other functions as well that contribute to the "diabetic" phenotypes seen in IPC-ablated larvae and adults. To test this, our lab generated a genomic deletion that simultaneously removed five DILPs (*Df[dilp1-5]*). This produced animals with "diabetic" symptoms, as well as growth defects and developmental delay similar to those seen for IPC ablation (Zhang et al., 2009). These animals were small (Fig. 2A) and displayed a feature of T1D known as "starvation in the midst of plenty" in that they activated starvation responses even while actively feeding (Fig. 2B–E). Further, *Df[dilp1-5]* animals had reduced levels

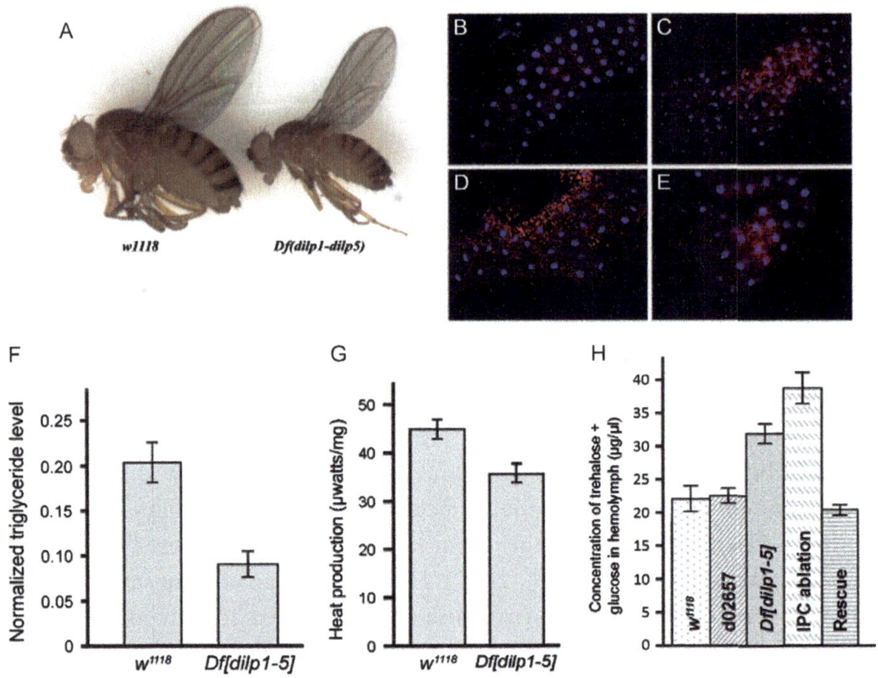

Fig. 2 Deletion of *dilps 1-5* generates flies with features of T1D. (A) A genomic deletion that removes genes encoding DILPs 1-5 (*Df[dilp1-5]*) is homozygous viable but flies are developmentally delayed and small. (B–E) *Df[dilp1-5]* homozygotes induce fat body autophagy while actively eating. Lysotracker staining (*red*) of dissected fat bodies from third-instar larvae. Hoechst 33342 (*blue*) reveals nuclei. For B, C, and E, larvae were actively eating, burrowed in the food, and had full guts. Genotypes: (B) Control parental line d02657; (C) *Df[dilp1-5]/Df[dilp1-5]*; (D) Control parental line d02657; third-instar larva removed from food and starved for 3 h in the presence of water (E) *hsGAL4>UASdilp2; Df [dilp1-5]/Df[dilp1-5]*. Feeding *Df[dilp1-5]* homozygotes (C) resembled starved control animals (D). (F) Total body triglyceride levels are lower in *Df[dilp1-5]* homozygotes. Total body triglyceride level and total body protein from w^{1118} or *Df[dilp1-5]* adult male flies were measured; data shows the triglyceride levels normalized to the total protein level. Error bars indicate standard error. (G) Overall metabolic activity was reduced in *Df[dilp1-5]* homozygotes. (H) Circulating sugar levels are elevated in *Df[dilp1-5]* homozygotes. Sugar levels (trehalose + glucose) were determined in hemolymph extracted from early third-instar larvae. Levels are indicated for negative controls: w^{1118} and parental line d02657; for experimental samples: *Df[dilp1-5]/Df[dilp1-5]*, IPC-ablated animals, *dilp2-GAL4>UAS-rpr,* and rescued *hsGAL4>UASdilp2; Df[dilp1-5]/Df[dilp1-5] animals*. Larval hemolymph was collected from 10 to 15 animals and pooled for each genotype. The circulating sugar levels of *Df[dilp1-5]* homozygotes were higher than controls in early and late third-instar larvae (not shown). Levels were lowered by ubiquitous expression of DILP2. Sugar levels were highest in animals in which IPCs had been ablated (*dilp2GAL4>UASrpr*). Error bars indicate standard error. *From Zhang, H., Liu, J., Li, C. R., Momen, B., Kohanski, R. A., & Pick, L. (2009). Deletion of Drosophila insulin-like peptides causes growth defects and metabolic abnormalities.* Proceedings of the National Academy of Sciences of the United States of America, 106*(46), 19617–19622.*

of TGs (Fig. 2F), displayed reduced overall metabolic activity (Fig. 2G), and increased levels of circulating sugar (Fig. 2H). Interestingly, circulating sugar levels were slightly but consistently higher in IPC-ablated animals than *Df [dilp1-5]* animals, suggesting the possibility that additional functions are indeed affected by IPC-ablation. Further study of different combinations of *dilp* mutants suggested that *dilps 2,3* and *5* (Broughton et al., 2005; Ikeya et al., 2002), are primarily responsible for regulation of circulating sugar levels (Gronke et al., 2010), with considerable redundancy among DILPs.

These studies support the notion that T1D can be effectively modeled in the fly because IIS functions in a similar fashion in flies and mammals to regulate circulating sugar levels. However, it is important to note that *Df[dilp1-5]* homozygous animals, although developmentally delayed and poorly fertile, were viable. In fact they can be maintained in long-term culture under standard lab conditions. This contrasts with lethality seen for mice lacking insulin, which survive to birth but rapidly develop diabetes and die neonatally (Duvillie et al., 1997). This difference between flies and mammals reveals both advantages and disadvantages of the fly system. On one hand, this discrepancy reveals a limitation of this particular fly model, in that lethality seen in mammals is not mimicked. On the other hand, since *Df[dilp1-5]* flies can be maintained in culture, they provide a model for experimentation, including screens for gene interactions, environmental impacts, and therapeutics.

4. MODELING INSULIN-DEPENDENT SUGAR UPTAKE AND INSULIN RELEASE IN *DROSOPHILA*

In mammals, insulin mediates sugar uptake by peripheral tissues primarily through activation of the insulin-dependent sugar transporter GLUT4 (Cushman et al., 1998; Dawson, Aviles-Hernandez, Cushman, & Malide, 2001). The "diabetic" fly models discussed earlier suggested that similar mechanisms operate in insects. As a first step to test whether flies have the machinery to mount an insulin-dependent sugar uptake response, our lab generated transgenic flies in which doubly tagged human GLUT4 was expressed in fat cells (Crivat et al., 2013). An HA tag was inserted in the first exofacial loop of GLUT4 to monitor expression at the cell surface (Dawson et al., 2001) and a C-terminal GFP tag was added to monitor transgene expression. We found that fat cells from these transgenic animals responded to mammalian insulin, mobilizing hGLUT4 trafficking, and translocation to the membrane (Fig. 3). This indicates that flies have all the necessary molecules and signals needed to direct mammalian-like vesicular trafficking in response to insulin. Animals that had been sugar-restricted showed an enhanced GLUT4 trafficking response to insulin, suggesting that flies reared on standard food in the laboratory are insulin-resistant (see also below). Thus, *Drosophila* appear capable of regulating the uptake of circulating sugar in response to hormonal signals in a manner analogous to that used by mammals. Which endogenous sugar transporter(s) function in IIS in the fly and how glucose and trehalose levels are regulated, independently and/or relative to each other, remain to be determined.

Release of insulin from the β-cells of the pancreas in mammals depends upon glucose sensing by GLUT1 and/or GLUT2 (Joost & Thorens, 2001; Rutter et al., 2015), which triggers glycolysis and the subsequent release of ATP from the mitochondria. The increase in ATP regulates K_{ATP} channels in the cell membrane causing the cell to depolarize. Depolarization activates Ca^{2+} channels and leads to exocytosis of insulin from the β-cells (reviewed in Nassel & Broeck, 2015).

Similar mechanisms appear to trigger the release of insulin from *Drosophila* IPCs. Studies by Fridell found that IPCs respond to glucose with an influx of Ca^{2+} and action potentials similar to those seen in mammalian β-cells (Fridell et al., 2009; Kreneisz, Chen, Fridell, & Mulkey, 2010). Furthermore, this response was triggered by a specific K_{ATP} channel activator, glibenclamide, implicating K_{ATP} channels in the response. Further, GLUT1 appears to play similar roles in flies and mammals. Park et al. (2014) generated a tagged version of DILP2 (ilp2HF) to monitor its secretion (Park et al., 2014). They found a marked increase in ilp2HF-circulating levels upon refeeding after a 24-h fast that is likely a result of glucose sensing by GLUT1 in the IPCs, as IPC-specific knockdown of *Glut1* decreased circulating ilp2HF. This investigation also provided additional evidence that ion channels are important for response to glucose in that expression of *Kir2.1*, a potassium channel that can silence electrical activity of *Drosophila* neurons and neuroendocrine cells (Kim & Rulifson, 2004) in adult IPCs led to significantly reduced levels of circulating Ilp2HF. Together, these studies suggest that at a mechanistic level, the "diabetic" increase in circulating sugar levels seen in *Drosophila* models do in fact reflect evolutionarily conserved, hormonally mediated sugar uptake responses.

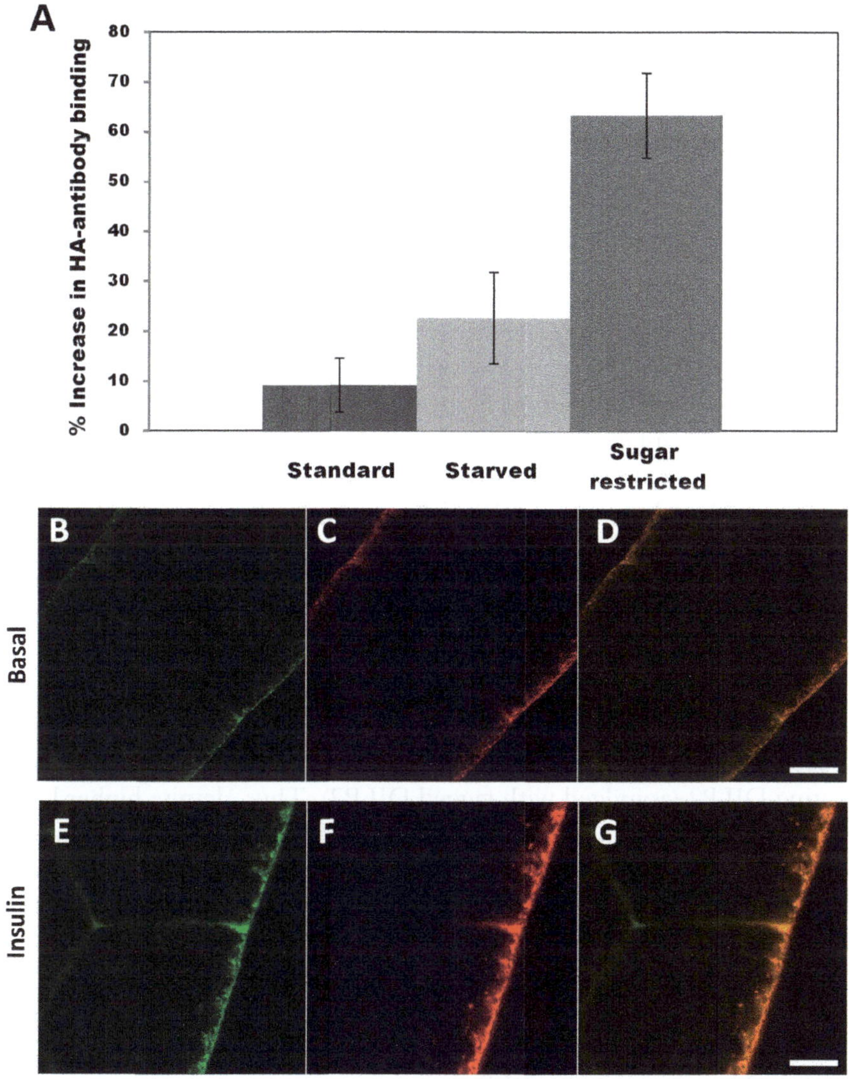

Fig. 3 *Drosophila* harbor the machinery to mediate an insulin-dependent sugar uptake response. (A) Surface exposure of HA-GLUT4 upon insulin stimulation. Fat bodies were collected from larvae reared on different dietary regimes, as indicated. After immunostaining of nonpermeabilized fat body cells, values were determined by measuring Alexa 647 conjugated anti-HA antibody fluorescence and averaging calculations of corrected integrated density. Values are expressed as percent of HA-GLUT4 fluorescence of insulin stimulation over basal conditions. Fat bodies were collected from three animals for each dietary regime. (B–G) Confocal microscopy of HA-GLUT4-GFP expression in fat body cells from animals reared on a sugar-restricted diet in the absence or

(Continued)

5. *DROSOPHILA* AS A MODEL FOR INSULIN RESISTANCE AND TYPE 2 DIABETES

To our knowledge, the first paper to explicitly model T2D in *Drosophila* did so by testing the role of a high-sugar diet (HSD) on acquisition of insulin resistance (Musselman et al., 2011). In this study, the authors showed that larvae reared from hatching on HSDs have increased levels of circulating sugar, both glucose and trehalose. These results are consistent with findings in honeybees and other insects that hemolymph sugar levels increase in response to HSDs (see above). However, even in nondiabetic humans, circulating sugar levels spike after a high-sugar meal. Typically, a medical diagnosis of insulin resistance involves a glucose clearing or glucose tolerance test in which the ability to clear glucose from the blood after a period of fasting is assessed (see Haselton et al., 2010). This type of test is difficult to perform with *Drosophila* at larval stages, since larvae eat constantly until the wandering stage when they stop feeding prior to pupation. Thus, the finding that animals continuously feeding on HSDs were hyperglycemic is not sufficient to conclude that they were in a diabetic-like state.

However, the authors provided additional evidence that these animals displayed features of insulin resistance (Musselman et al., 2011). First, rearing on HSD resulted in higher levels of expression of *dilp 2,3,5* RNA and circulating DILP2 (as judged with tagged DILP2). Thus, despite higher levels of DILP(s) in circulation, sugar levels remained high—a feature resembling mammalian insulin resistance. Second, they observed decreased levels of phospho-Akt in response to administration of exogenous insulin in flies reared on HSD, suggesting a weakened ability to respond to insulin signaling after chronic levels of high sugar in the diet. Finally, flies reared on HSD had somewhat higher levels of TGs and FFAs as compared to controls fed an isocaloric diet, and the morphology of lipid stores changed, with fewer, larger droplets (Musselman et al., 2013).

Fig. 3—cont'd presence of insulin. (B, C, and D) Basal conditions; (E, F, and G) after addition of 0.1 U/mL insulin. GLUT4 was visualized in nonpermeabilized cells by GFP fluorescence (*green* B and E) or with anti-HA antibody (*red* C and F) to monitor membrane translocation. Scale bar is 5 μm. Note increase in GLUT4 at the cell surface (*red* F). *From Crivat, G., Lizunov, V. A., Li, C. R., Stenkula, K. G., Zimmerberg, J., Cushman, S. W., & Pick, L. (2013). Insulin stimulates translocation of human GLUT4 to the membrane in fat bodies of transgenic Drosophila melanogaster.* PloS One, *8(11), e77953. doi:10.1371/journal.pone.0077953.*

Together, these studies showed that flies reared on chronic HSD are hyperglycemic, insulin-resistant and obese, all features of T2D and the precursors thereof. However, attention to detail is necessary when carrying out and interpreting these types of experiments. Musselman et al. (2013) point out that the sugar levels used in the high-sugar feeding experiments are similar to that of a banana, a natural food for these flies. This raises the suggestion that flies are "naturally" diabetic in the wild. Similarly, standard lab conditions may lead to "T2D flies," as Crivat et al. found that mobilization of a human GLUT4 transgene in response to insulin was enhanced upon sugar-restriction (Crivat et al., 2013). Furthermore, levels of circulating glucose were shown to be influenced by larval density (Ugrankar et al., 2015), an important factor since density may vary in lab conditions in a genotype-dependent fashion, confounding effects of genetic differences. Finally, larvae undergo a natural starvation stage at the end of the third instar—a physiologically "normal" fasting that may result in metabolic changes different from forced fasting at other developmental stages or in mammals. Thus, overall, while it is clear that modeling T2D in flies is possible, such modeling requires extreme caution in controlling for genetic differences, rearing conditions and extrapolating to the human condition. To learn more about modeling T2D in *Drosophila*, we suggest a recent review on this topic, published while this chapter was in press (Alfa & Kim, 2016).

6. ARE GLUCOSE AND TREHALOSE METABOLISM REGULATED INDEPENDENTLY IN *DROSOPHILA*?

In most studies that examined levels of circulating sugar described above, total sugars were reported. The standard assay (Rulifson et al., 2002) was to extract hemolymph and incubate it with trehalase to generate glucose, which is then measured spectrophotometrically. The justification for this approach was twofold: First, trehalose levels are up to 100-fold higher in circulation than glucose in insects, and second, trehalose may be broken down to glucose during the extraction process, thus artificially raising estimates of circulating glucose levels. More recent studies have attempted to separately measure each of these circulating sugars.

Pasco and Leopold observed that larvae fed a HSD increased circulating glucose but not trehalose within minutes of high-sugar feeding (Pasco & Leopold, 2012). Larvae reared for longer term on moderately HSD showed increases in both glucose and trehalose by the end of this life stage. Consistent with the findings of Musselman et al. (2013) discussed earlier, larvae reared on HSD displayed features of insulin resistance, including increased levels of circulating DILPs, as well InR expression. Further, isolated fat bodies from these animals exhibited a decreased response to exogenous insulin, suggestive of insulin resistance resulting from long-term feeding on a HSD.

A starker difference in responses seen for glucose and trehalose was observed by Ugrankar et al. (2015) who found an increase in glucose but not trehalose in larvae fed HSD. Circulating glucose levels were also increased in animals carrying mutations in *Glut1*, *chico*, and *PI3K*, while trehalose levels were not affected. These results are consistent with earlier findings in silk moth that trehalose synthesis is inhibited by free trehalose, which also promotes glycogen synthesis. This feedback regulation would result in relatively constant trehalose levels over a range of feeding conditions (Murphy & Wyatt, 1965). However, they appear to differ from previous studies that documented stable changes in trehalose levels in response to loss of DILP function, IPC-ablation, and changes in diet.

What is suggested by Ugrankar et al. (2015) is a more flexible response for glucose than trehalose in *Drosophila*. Speculating, one could imagine that pathways regulating glucose levels in flies may be those similar to mammalian regulators, while trehalose may be under a regulatory program that is more insect specific. Additionally, circulating glucose may be more tightly regulated than circulating trehalose in insects because it is a reducing sugar that can have negative impact, even though the overall levels of glucose are quite low.

7. *DROSOPHILA* AS A MODEL FOR OBESITY-RELATED HEART DISEASE

In addition to negative impacts of HSDs, high-fat diets (HFD) are strongly correlated with obesity, heart disease, and T2D in humans (Szendroedi & Roden, 2009; van Herpen & Schrauwen-Hinderling, 2008). HFD caused flies to respond initially (2 days) with increased DILP2 and decreased total glucose (Birse et al., 2010) but, when the HFD was continued for 3 or more days, DILP2 levels dropped, glucose levels rose, and pAKT levels decreased, suggesting that *Drosophila* respond to HFD with a physiology similar to mammals. Flies also accumulated lipids (specifically TGs) in both the fat body and the midgut in a dose-dependent manner, mimicking major risk factors for T2D and related conditions.

Drosophila are the only invertebrate model organism with a beating heart for which the tools and assays needed to study heart function are available (reviewed in Diop & Bodmer, 2015; Wessells, Fitzgerald, Cypser, Tatar, & Bodmer, 2004). As for IIS, early studies identified genes involved in specification of the *Drosophila* heart that are shared with vertebrates (Bodmer & Venkatesh, 1998). Furthermore, although *Drosophila* have a simple tubular heart and an open circulatory system, there seems to be a common physiology with the vertebrate heart. Early studies demonstrated that mutations in ion channels associated with human cardiac arrhythmias also influence heart rate and/or rhythm in *Drosophila* (see Ganetzky, 2000). More recently, it was shown that chronic high-sugar feeding also led to heart defects in flies (Na et al., 2013) and a HFD increased TG levels and altered contraction patterns, similar to the effects seen in mammals (Birse et al., 2010). In addition, genetically altering metabolic regulators common to both vertebrates and *Drosophila* led to obesity and related heart defects. The defects caused by a HFD appear to result from deposition of TG in the heart itself, as they were abrogated by genetically altering lipid metabolism in the entire fly, the FB or in the heart (Birse et al., 2010). Together, these findings lead to the somewhat surprising conclusion that *Drosophila* serve as an ideal model to dissect mechanisms underlying diet-induced heart disease.

8. *DROSOPHILA* AS A MODEL FOR METABOLIC SYNDROME

Metabolic syndrome (MetS) has been defined by a cluster of clinical presentations that include at least three of five risk factors: Fasting glucose at levels indicative of insulin resistance or even T2D, high blood pressure, elevated TGs, low HDL-cholesterol, and obesity (Alberti, Zimmet, Shaw, & IDF Epidemiology Task Force Consensus Group, 2005). These metabolic abnormalities, also referred to as insulin resistance syndrome, present a high risk of cardiovascular disease. Reed and coworkers have taken a novel approach to identifying the genetic, environmental, and gene × environment (specifically diet, Gene × Diet [G × D]) contributions to features of MetS contributions that are difficult to disentangle in human populations. Using a series of inbred *Drosophila* lines, they compared effects of diet on traits including weight, TG levels, circulating sugar, and survival. In a comparison of 146 inbred lines obtained from two independent wild populations, they observed that gene and G × D interactions outweighed diet alone in contributing to observed variance, with G × D interactions accounting for ~15% of the variance for the traits most closely resembling those seen in MetS (Reed et al., 2010). They also found that metabolic trait variance increased when flies were reared on a HFD. Interestingly, they found that across genotypes, flies reared on the lowest sugar diets weighed the most, which may reflect a suppression of feeding in high-sugar conditions, as seen in other studies (Lebreton, Witzgall, Olsson, & Becher, 2014; Rovenko et al., 2015).

Reed and coworkers went on to analyze global changes in transcriptome and metabolic profiles in response to varying diets, assessing variations within and across populations at a systems level. Among other findings, these authors identified small groups of candidate genes and metabolites that had not been previously associated with metabolic disorders for further examination. Finally, several studies from these researchers suggested that when populations were exposed to new environments, latent, or cryptic genetic variation was revealed that cannot be observed in long-term stable environments (Gibson & Reed, 2008; Reed et al., 2014, 2010). This may be particularly significant for diseases like MetS, insulin resistance, and T2D, for which dietary changes appear to have had large impacts on disease rates in human populations over a relatively short-time scale (Reed et al., 2014; Williams et al., 2015). Overall, these studies provide a wealth of information on the complexities of metabolic variation in flies that remains to be mined in the future and which will likely have high relevance for human populations.

9. FROM CORRELATION TO CAUSATION: *DROSOPHILA* AS A MODEL TO STUDY GENE FUNCTION IN METABOLISM

Given the complexity of diabetes and related disorders, genome-wide association studies (GWAS) have been used to identify genetic variants associated with susceptibility to disease, using data from human populations. This is a powerful technique that establishes a correlation between genomic markers and disease risk. However, GWAS alone points to a large number of candidate genes that then need to be evaluated for function. The power of *Drosophila* genetics makes it an ideal system to screen through numerous candidates to identify causal genes.

Pendse et al. (2013) were among the first to demonstrate that flies can be used to test the function of candidate genes identified in GWAS for diabetes and related disorders, serving as a proof of principle that *Drosophila* can be used for large-scale screening to narrow down genes associated with SNPs linked to metabolic disease. Starting with 83 orthologs of 71 T2D candidate human genes, they used RNAi knockdown to screen for roles in sugar metabolism based upon larval sucrose tolerance. This screen identified a number of expected candidate genes and further analyzed the role of one of these, dHHEX, which was shown to regulate circulating glucose levels in *Drosophila*.

Similarly, Park et al. (2014) analyzed loss-of-function mutations in fly orthologs of genes identified as candidates for T2D in human GWAS studies (Park et al., 2014). IPC-specific knockdown of one of these, the fly ortholog of *Glis3* (fly *Imd*) resulted in a decrease in circulating DILP2, similar to the situation in mice where *Glis3* regulates insulin expression (Yang, Chang, & Chan, 2013). IPC-specific knockdown of a second gene that had not been characterized in mice, BCL11A, increased circulating DILP2 but not mRNA levels, suggesting a role for this gene in insulin secretion. Although more work remains to be done to determine the mechanisms of action of these genes, they clearly support the notion that flies are a facile model system to study the function of candidate genes. Flies will, of course, be an ideal system to study mechanism of action of these genes as well.

In addition to screening GWAS candidates, flies can also be used as the initial organism to screen for genes potentially involved in metabolic disorders. For example, Pasco and Leopold (2012) revealed a new role for lipocalins in *Drosophila* metabolism, a class of molecules previously implicated in insulin resistance in mammals. Ugrankar et al. (2015) performed large scale RNAi screens of ~1000 genes and identified >150 that resulted in hyperglycemia. This pool included 76 novel candidates, one of which (*CSNK1a1*) was tested in mice. Interestingly, loss-of-function mutations of *CSNK1a* in the adipose tissue of mice caused hyperglycemia, a clear demonstration that genetic screens in *Drosophila* can be used to identify genes important in mammalian sugar metabolism.

10. CONCLUDING REMARKS

Drosophila have been used to model a large variety of diseases, including both T1D and T2D, as well as related metabolic syndromes. These models reveal that many aspects of sugar metabolism are shared between insects and humans, despite obvious physiological differences. *Drosophila* bring the advantage of powerful genetic tools that enable gene identification and mechanistic studies. All of the tools have been applied to model diseases of insulin resistance and have identified several new candidate genes for human disease. Future studies with these model systems will build upon these fundamental approaches to identify additional genes and pathways impacting insulin resistance. The use of high-throughput screens and studies of pharmacologic interventions in these fly models have the potential to elucidate mechanisms of drug action and to identify viable therapeutic interventions for human patients.

ACKNOWLEDGMENTS

Work on insulin signaling in the Pick lab was supported by the March of Dimes Birth Defects Foundation (1-FY-10-368) and the National Institutes of Health (ROI EY14290).

REFERENCES

Abou-Seif, M. A., Maier, V., Fuchs, J., Mezger, M., Pfeiffer, E. F., & Bounias, M. (1993). Fluctuations of carbohydrates in haemolymph of honeybee (*Apis mellifica*) after fasting, feeding and stress. *Hormone and Metabolic Research*, *25*(1), 4–8.

Alberti, K. G., Zimmet, P., & Shaw, J.IDF Epidemiology Task Force Consensus Group. (2005). The metabolic syndrome—A new worldwide definition. *Lancet*, *366*(9491), 1059–1062. http://dx.doi.org/10.1016/S0140-6736(05)67402-8.

Alfa, R. W., & Kim, S. K. (2016). Using Drosophila to discover mechanisms underlying type 2 diabetes. *Disease Models & Mechanisms*, *9*(4), 365–376. http://dx.doi.org/10.1242/dmm.023887.

American Diabetes Association. (2014). Diagnosis and classification of diabetes mellitus. *Diabetes Care*, *37*(Suppl. 1), 581–590.

Arrese, E. L., & Soulages, J. L. (2010). Insect fat body: Energy, metabolism, and regulation. *Annual Review of Entomology*, *55*, 207–225. http://dx.doi.org/10.1146/annurev-ento-112408-085356.

Baker, K. D., & Thummel, C. S. (2007). Diabetic larvae and obese flies-emerging studies of metabolism in Drosophila. *Cell Metabolism*, *6*(4), 257–266. http://dx.doi.org/10.1016/j.cmet.2007.09.002.

Birse, R. T., Choi, J., Reardon, K., Rodriguez, J., Graham, S., Diop, S., … Oldham, S. (2010). High-fat-diet-induced obesity and heart dysfunction are regulated by the TOR pathway in Drosophila. *Cell Metabolism*, *12*(5), 533–544. http://dx.doi.org/10.1016/j.cmet.2010.09.014.

Blatt, J., & Roces, F. (2001). Haemolymph sugar levels in foraging honeybees (Apis mellifera carnica): Dependence on metabolic rate and in vivo measurement of maximal rates of trehalose synthesis. *The Journal of Experimental Biology*, *204*(Pt 15), 2709–2716.

Bodmer, R., & Venkatesh, T. V. (1998). Heart development in Drosophila and vertebrates: Conservation of molecular mechanisms. *Developmental Genetics*, *22*(3), 181–186. http://dx.doi.org/10.1002/(SICI)1520-6408(1998)22:3<181::AID-DVG1->3.0.CO;2-2.

Bonini, N. M., & Fortini, M. E. (2002). Applications of the Drosophila retina to human disease modeling. *Results and Problems in Cell Differentiation*, *37*, 257–275.

Broughton, S. J., Piper, M. D., Ikeya, T., Bass, T. M., Jacobson, J., Driege, Y., … Partridge, L. (2005). Longer lifespan, altered metabolism, and stress resistance in Drosophila from ablation of cells making insulin-like ligands. *Proceedings of the National Academy of Sciences of the United States of America*, *102*(8), 3105–3110.

Candy, D. J., Becker, A., & Wegener, G. (1997). Coordination and integration of metabolism and insect flight. *Comparative Biochemistry and Physiology*, *117B*, 497–512.

Centers for Disease Control and Prevention (2016a). http://www.cdc.gov/diabetes.
Centers for Disease Control and Prevention (2016b). https://www.searchfordiabetes.org/dspHome.cfm. *SEARCH*.
Colombani, J., Raisin, S., Pantalacci, S., Radimerski, T., Montagne, J., & Leopold, P. (2003). A nutrient sensor mechanism controls Drosophila growth. *Cell*, *114*(6), 739–749.
Crivat, G., Lizunov, V. A., Li, C. R., Stenkula, K. G., Zimmerberg, J., Cushman, S. W., & Pick, L. (2013). Insulin stimulates translocation of human GLUT4 to the membrane in fat bodies of transgenic Drosophila melanogaster. *PloS One*, *8*(11). e77953. http://dx.doi.org/10.1371/journal.pone.0077953.
Cushman, S. W., Goodyear, L. J., Pilch, P. F., Ralston, E., Galbo, H., Ploug, T., … Klip, A. (1998). Molecular mechanisms involved in GLUT4 translocation in muscle during insulin and contraction stimulation. *Advances in Experimental Medicine and Biology*, *441*, 63–71.
Das, R., & Dobens, L. L. (2015). Conservation of gene and tissue networks regulating insulin signalling in flies and vertebrates. *Biochemical Society Transactions*, *43*(5), 1057–1062. http://dx.doi.org/10.1042/BST20150078.
Dawson, K., Aviles-Hernandez, A., Cushman, S. W., & Malide, D. (2001). Insulin-regulated trafficking of dual-labeled glucose transporter 4 in primary rat adipose cells. *Biochemical and Biophysical Research Communications*, *287*(2), 445–454. http://dx.doi.org/10.1006/bbrc.2001.5620.
Diop, S. B., & Bodmer, R. (2015). Gaining insights into diabetic cardiomyopathy from Drosophila. *Trends in Endocrinology and Metabolism*, *26*(11), 618–627. http://dx.doi.org/10.1016/j.tem.2015.09.009.
Duve, H., Thorpe, A., & Lazarus, N. R. (1979). Isolation of material displaying insulin-like immunological biological activity from the brain of the blowfly Calliphora vomitoria. *The Biochemical Journal*, *184*(2), 221–227.
Duvillie, B., Cordonnier, N., Deltour, L., Dandoy-Dron, F., Itier, J. M., Monthioux, E., … Bucchini, D. (1997). Phenotypic alterations in insulin-deficient mutant mice. *Proceedings of the National Academy of Sciences of the United States of America*, *94*(10), 5137–5140.
Fernandez-Almonacid, R., & Rosen, O. M. (1987). Structure and ligand specificity of the *Drosophila melanogaster* insulin receptor. *Molecular and Cellular Biology*, *7*, 2718–2727.
Fridell, Y. W., Hoh, M., Kreneisz, O., Hosier, S., Chang, C., Scantling, D., … Helfand, S. L. (2009). Increased uncoupling protein (UCP) activity in Drosophila insulin-producing neurons attenuates insulin signaling and extends lifespan. *Aging (Albany NY)*, *1*(8), 699–713.
Ganetzky, B. (2000). Genetic analysis of ion channel dysfunction in Drosophila. *Kidney International*, *57*(3), 766–771. http://dx.doi.org/10.1046/j.1523-1755.2000.00913.x.
Garofalo, R. S. (2002). Genetic analysis of insulin signaling in Drosophila. *Trends in Endocrinology and Metabolism*, *13*(4), 156–162.
Gibson, G., & Reed, L. K. (2008). Cryptic genetic variation. *Current Biology*, *18*(21), R989–R990. http://dx.doi.org/10.1016/j.cub.2008.08.011.
Goberdhan, D. C., & Wilson, C. (2003). The functions of insulin signaling: Size isn't everything, even in Drosophila. *Differentiation*, *71*(7), 375–397.
Gronke, S., Clarke, D. F., Broughton, S., Andrews, T. D., & Partridge, L. (2010). Molecular evolution and functional characterization of Drosophila insulin-like peptides. *PLoS Genetics*, *6*(2). e1000857. http://dx.doi.org/10.1371/journal.pgen.1000857.
Haselton, A., Sharmin, E., Schrader, J., Sah, M., Poon, P., & Fridell, Y. W. (2010). Partial ablation of adult Drosophila insulin-producing neurons modulates glucose homeostasis and extends life span without insulin resistance. *Cell Cycle*, *9*(15), 3063–3071. http://dx.doi.org/10.4161/cc.9.15.12458.

He, B. Z., Ludwig, M. Z., Dickerson, D. A., Barse, L., Arun, B., Vilhjalmsson, B. J., … Kreitman, M. (2014). Effect of genetic variation in a Drosophila model of diabetes-associated misfolded human proinsulin. *Genetics*, *196*(2), 557–567. http://dx.doi.org/10.1534/genetics.113.157800.

Huang, S., & Czech, M. P. (2007). The GLUT4 glucose transporter. *Cell Metabolism*, *5*(4), 237–252. http://dx.doi.org/10.1016/j.cmet.2007.03.006.

Ikeya, T., Galic, M., Belawat, P., Nairz, K., & Hafen, E. (2002). Nutrient-dependent expression of insulin-like peptides from neuroendocrine cells in the CNS contributes to growth regulation in Drosophila. *Current Biology*, *12*(15), 1293–1300.

Joost, H. G., & Thorens, B. (2001). The extended GLUT-family of sugar/polyol transport facilitators: Nomenclature, sequence characteristics, and potential function of its novel members (review). *Molecular Membrane Biology*, *18*(4), 247–256.

Kahn, B. B., & Cushman, S. W. (1986). Cell biology of insulin's stimulatory action on glucose transport and its perturbation in altered metabolic states. *The Annals of the New York Academy of Sciences*, *488*, 356–369.

Kahn, R. C., King, G. L., Moses, A. C., Weir, G. C., Jacobson, A. M., & Smith, R. J. (2005). *Joslin's diabetes mellitus* (14th ed.). . Boston, MA: Lippincott Williams & Wilkins.

Kannan, K., & Fridell, Y. W. (2013). Functional implications of Drosophila insulin-like peptides in metabolism, aging, and dietary restriction. *Frontiers in Physiology*, *4*, 288. http://dx.doi.org/10.3389/fphys.2013.00288.

Kim, S. K., & Rulifson, E. J. (2004). Conserved mechanisms of glucose sensing and regulation by Drosophila corpora cardiaca cells. *Nature*, *431*(7006), 316–320.

Kreneisz, O., Chen, X., Fridell, Y. W., & Mulkey, D. K. (2010). Glucose increases activity and Ca^{2+} in insulin-producing cells of adult Drosophila. *Neuroreport*, *21*(17), 1116–1120. http://dx.doi.org/10.1097/WNR.0b013e3283409200.

Lasko, P. (2002). Diabetic flies? Using Drosophila melanogaster to understand the causes of monogenic and genetically complex diseases. *Clinical Genetics*, *62*(5), 358–367.

Lebreton, S., Witzgall, P., Olsson, M., & Becher, P. G. (2014). Dietary glucose regulates yeast consumption in adult Drosophila males. *Frontiers in Physiology*, *5*, 504. http://dx.doi.org/10.3389/fphys.2014.00504.

Lee, G., & Park, J. H. (2004). Hemolymph sugar homeostasis and starvation-induced hyperactivity affected by genetic manipulations of the adipokinetic hormone-encoding gene in Drosophila melanogaster. *Genetics*, *167*(1), 311–323.

LeRoith, D., Lesniak, M. A., & Roth, J. (1981). Insulin in insects and annelids. *Diabetes*, *30*(1), 70–76.

Lizunov, V. A., Matsumoto, H., Zimmerberg, J., Cushman, S. W., & Frolov, V. A. (2005). Insulin stimulates the halting, tethering, and fusion of mobile GLUT4 vesicles in rat adipose cells. *The Journal of Cell Biology*, *169*(3), 481–489.

Lizunov, V. A., Stenkula, K. G., Lisinski, I., Gavrilova, O., Yver, D. R., Chadt, A., Cushman, S. W. … (2012). Insulin stimulates fusion, but not tethering, of GLUT4 vesicles in skeletal muscle of HA-GLUT4-GFP transgenic mice. *American Journal of Physiology Endocrinology and Metabolism*, *302*(8), E950–E960. http://dx.doi.org/10.1152/ajpendo.00466.2011.

Maurizio, A. (1965). Studies on the sugar spectrum of the hemolymph of the honeybee (*Apis mellifica* L.). I. The blood sugar spectrum of adult bees. *Journal of Insect Physiology*, *11*(6), 745–763.

Mueckler, M., & Thorens, B. (2013). The SLC2 (GLUT) family of membrane transporters. *Molecular Aspects of Medicine*, *34*(2-3), 121–138. http://dx.doi.org/10.1016/j.mam.2012.07.001.

Murphy, T. A., & Wyatt, G. R. (1965). The enzymes of glycogen and trehalose synthesis in silk moth fat body. *The Journal of Biological Chemistry*, *240*, 1500–1508.

Musselman, L. P., Fink, J. L., Narzinski, K., Ramachandran, P. V., Hathiramani, S. S., Cagan, R. L., & Baranski, T. J. (2011). A high-sugar diet produces obesity and insulin

resistance in wild-type Drosophila. *Disease Models & Mechanisms*, *4*(6), 842–849. http://dx.doi.org/10.1242/dmm.007948.
Musselman, L. P., Fink, J. L., Ramachandran, P. V., Patterson, B. W., Okunade, A. L., Maier, E., … Baranski, T. J. (2013). Role of fat body lipogenesis in protection against the effects of caloric overload in Drosophila. *The Journal of Biological Chemistry*, *288*(12), 8028–8042. http://dx.doi.org/10.1074/jbc.M112.371047.
Na, J., Musselman, L. P., Pendse, J., Baranski, T. J., Bodmer, R., Ocorr, K., & Cagan, R. (2013). A Drosophila model of high sugar diet-induced cardiomyopathy. *PLoS Genetics*, *9*(1), e1003175. http://dx.doi.org/10.1371/journal.pgen.1003175.
Nassel, D. R., & Broeck, J. V. (2015). Insulin/IGF signaling in Drosophila and other insects: Factors that regulate production, release and post-release action of the insulin-like peptides. *Cellular and Molecular Life Sciences*, *73*(2), 271–290. http://dx.doi.org/10.1007/s00018-015-2063-3.
Nation, J. L. (2002). *Insect physiology and biochemistry*. Boca Raton: CRC Press LLC.
Nishida, Y., Hata, M., Nishizuka, Y., Rutter, W. J., & Ebina, Y. (1986). Cloning of a Drosophila cDNA encoding a polypeptide similar to the human insulin receptor precursor. *Biochemical and Biophysical Research Communications*, *141*, 474–481.
O'Connor, K. J., & Baxter, D. (1985). The demonstration of insulin-like material in the honey bee, *Apis mellifera*. *Comparative Biochemistry and Physiology*, *81B*, 755–760.
Oldham, S. (2011). Obesity and nutrient sensing TOR pathway in flies and vertebrates: Functional conservation of genetic mechanisms. *Trends in Endocrinology and Metabolism*, *22*(2), 45–52. http://dx.doi.org/10.1016/j.tem.2010.11.002.
Owusu-Ansah, E., & Perrimon, N. (2014). Modeling metabolic homeostasis and nutrient sensing in Drosophila: Implications for aging and metabolic diseases. *Disease Models & Mechanisms*, 7(3), 343–350. http://dx.doi.org/10.1242/dmm.012989.
Padmanabha, D., & Baker, K. D. (2014). Drosophila gains traction as a repurposed tool to investigate metabolism. *Trends in Endocrinology and Metabolism*, *25*(10), 518–527. http://dx.doi.org/10.1016/j.tem.2014.03.011.
Park, S., Alfa, R. W., Topper, S. M., Kim, G. E., Kockel, L., & Kim, S. K. (2014). A genetic strategy to measure circulating Drosophila insulin reveals genes regulating insulin production and secretion. *PLoS Genetics*, *10*(8), e1004555. http://dx.doi.org/10.1371/journal.pgen.1004555.
Pasco, M. Y., & Leopold, P. (2012). High sugar-induced insulin resistance in Drosophila relies on the lipocalin Neural Lazarillo. *PloS One*, 7(5), e36583. http://dx.doi.org/10.1371/journal.pone.0036583.
Pendse, J., Ramachandran, P. V., Na, J., Narisu, N., Fink, J. L., Cagan, R. L., … Baranski, T. J. (2013). A Drosophila functional evaluation of candidates from human genome-wide association studies of type 2 diabetes and related metabolic traits identifies tissue-specific roles for dHHEX. *BMC Genomics*, *14*, 136. http://dx.doi.org/10.1186/1471-2164-14-136.
Petruzzelli, L., Herrera, R., Arenas-Garcia, R., Fernandez, R., Birnbaum, M. J., & Rosen, O. M. (1986a). Isolation of a *Drosophila* genomic sequence homologous to the kinase domain of the human insulin receptor and detection of the phosphorylated *Drosophila* receptor with an anti-peptide antibody. *Proceedings of the National Academy of Sciences of the United States of America*, *83*, 4710–4714.
Petruzzelli, L., Herrera, R., Garcia-Arenas, R., & Rosen, O. M. (1986b). Acquisition of insulin-dependent protein tyrosine kinase activity during *Drosophila* embryogenesis. *The Journal of Biological Chemistry*, *260*, 16072–16075.
Reed, L. K., Lee, K., Zhang, Z., Rashid, L., Poe, A., Hsieh, B., … Gibson, G. (2014). Systems genomics of metabolic phenotypes in wild-type Drosophila melanogaster. *Genetics*, *197*(2), 781–793. http://dx.doi.org/10.1534/genetics.114.163857.
Reed, L. K., Williams, S., Springston, M., Brown, J., Freeman, K., DesRoches, C. E., … Gibson, G. (2010). Genotype-by-diet interactions drive metabolic phenotype variation

in Drosophila melanogaster. *Genetics*, *185*(3), 1009–1019. http://dx.doi.org/10.1534/genetics.109.113571.
Rovenko, B. M., Perkhulyn, N. V., Gospodaryov, D. V., Sanz, A., Lushchak, O. V., & Lushchak, V. I. (2015). High consumption of fructose rather than glucose promotes a diet-induced obese phenotype in Drosophila melanogaster. *Comparative Biochemistry and Physiology. Part A, Molecular & Integrative Physiology*, *180*, 75–85. http://dx.doi.org/10.1016/j.cbpa.2014.11.008.
Rulifson, E. J., Kim, S. K., & Nusse, R. (2002). Ablation of insulin-producing neurons in flies: Growth and diabetic phenotypes. *Science*, *296*(5570), 1118–1120.
Rutter, G. A., Pullen, T. J., Hodson, D. J., & Martinez-Sanchez, A. (2015). Pancreatic beta-cell identity, glucose sensing and the control of insulin secretion. *The Biochemical Journal*, *466*(2), 203–218. http://dx.doi.org/10.1042/BJ20141384.
Szendroedi, J., & Roden, M. (2009). Ectopic lipids and organ function. *Current Opinion in Lipidology*, *20*(1), 50–56.
Tager, H. S., Markese, J., Kramer, K. J., Speirs, R. D., & Childs, C. N. (1976). Glucagon-like and insulin-like hormones of the insect neurosecretory system. *The Biochemical Journal*, *156*(3), 515–520.
Taguchi, A., & White, M. F. (2008). Insulin-like signaling, nutrient homeostasis, and life span. *Annual Review of Physiology*, *70*, 191–212.
Tatar, M. (2004). The neuroendocrine regulation of Drosophila aging. *Experimental Gerontology*, *39*(11-12), 1745–1750.
Teleman, A. A. (2010). Molecular mechanisms of metabolic regulation by insulin in Drosophila. *The Biochemical Journal*, *425*(1), 13–26. http://dx.doi.org/10.1042/BJ20091181.
Thompson, S. N. (1999). Blood sugar formation from dietary carbohydrate is facilitated by the pentose phosphate pathway in an insect Manduca sexta Linnaeus. *Biochimica et Biophysica Acta*, *1472*(3), 565–575.
Thompson, S. N., Borchardt, D. B., & Wang, L. W. (2003). Dietary nutrient levels regulate protein and carbohydrate intake, gluconeogenic/glycolytic flux and blood trehalose level in the insect *Manduca sexta* L. *Journal of Comparative Physiology. B*, *173*(2), 149–163.
Thompson, S. N., & Redak, R. A. (2000). Interactions of dietary protein and carbohydrate determine blood sugar level and regulate nutrient selection in the insect *Manduca sexta* L. *Biochimica et Biophysica Acta*, *1523*(1), 91–102.
Ugrankar, R., Berglund, E., Akdemir, F., Tran, C., Kim, M. S., Noh, J., … Graff, J. M. (2015). Drosophila glucome screening identifies Ck1alpha as a regulator of mammalian glucose metabolism. *Nature Communications*, *6*, 7102. http://dx.doi.org/10.1038/ncomms8102.
van Herpen, N. A., & Schrauwen-Hinderling, V. B. (2008). Lipid accumulation in non-adipose tissue and lipotoxicity. *Physiology and Behavior*, *94*(2), 231–241. http://dx.doi.org/10.1016/j.physbeh.2007.11.049.
Wessells, R. J., Fitzgerald, E., Cypser, J. R., Tatar, M., & Bodmer, R. (2004). Insulin regulation of heart function in aging fruit flies. *Nature Genetics*, *36*(12), 1275–1281.
Williams, S., Dew-Budd, K., Davis, K., Anderson, J., Bishop, R., Freeman, K., … Reed, L. K. (2015). Metabolomic and gene expression profiles exhibit modular genetic and dietary structure linking metabolic syndrome phenotypes in Drosophila. *G3 (Bethesda, Md.)*, *5*(12), 2817–2829. http://dx.doi.org/10.1534/g3.115.023564.
Wu, Q., & Brown, M. R. (2006). Signaling and function of insulin-like peptides in insects. *Annual Review of Entomology*, *51*, 1–24.
Wyatt, G. R., & Kale, G. F. (1957). The chemistry of insect hemolymph. II. Trehalose and other carbohydrates. *The Journal of General Physiology*, *40*(6), 833–847.

Yang, Y., Chang, B. H., & Chan, L. (2013). Sustained expression of the transcription factor GLIS3 is required for normal beta cell function in adults. *EMBO Molecular Medicine*, *5*(1), 92–104. http://dx.doi.org/10.1002/emmm.201201398.
Zhang, H., Liu, J., Li, C. R., Momen, B., Kohanski, R. A., & Pick, L. (2009). Deletion of Drosophila insulin-like peptides causes growth defects and metabolic abnormalities. *Proceedings of the National Academy of Sciences of the United States of America*, *106*(46), 19617–19622.

INDEX

Note: Page numbers followed by "*f*" indicate figures, and "*t*" indicate tables.

D

Edwards Brothers Malloy
Ann Arbor MI. USA
January 6, 2017